Ecological Studies, Vol. 156

Analysis and Synthesis

Ecological Studies

Volumes published since 1995 are listed at the end of this book.

Springer

New York
Berlin
Heidelberg
Barcelona
Hong Kong
London
Milan
Paris
Singapore
Tokyo

Mark E. Fenn L.I. de Bauer
Tomás Hernández-Tejeda
Editors

Urban Air Pollution and Forests

Resources at Risk in the Mexico City Air Basin

Foreword by Mario J. Molina

With 97 Illustrations, 10 in Full Color

Springer

Mark E. Fenn
USDA Forest Service
Pacific Southwest Research Station
Riverside, CA 92507, USA
mfenn@deltanet.com
mfenn@fs.fed.us

L.I. de Bauer
Especialidad de Agrometeorología
Instituto de Recursos Naturales
Colegio de Postgraduados
CP 56230 Montecillo, Mexico
libauer@colpos.mx

Tomás Hernández-Tejeda
SAGARPA
Instituto Nacional de Investigaciones
Forestales, Agrícolas y Pecuarias (INIFAP)
Centro Nacional de Investigación
Disciplinaria en Conservación y
Mejoramiento de Ecosistemas Forestales
(CENID-COMEF)
CP 04110 Mexico City, Mexico
textom99@yahoo.com
thtejeda@colpos.mx

Cover illustration: Modified from Figure 6.2, showing, in the upper portion, the location of air-quality monitoring stations in the different sectors of the Mexico City Metropolitan Area (RAMA, Red Automática de Monitoreo Atmosférico or Automated Network for Atmospheric Monitoring), and the main forested areas in the Basin. Courtesy of Alfonso Salas-Cruz and Ricardo Torres-Jardón, both of the Centro de Ciencias de la Atmósfera, Universidad Nacional Autónoma de México.

Library of Congress Cataloging-in-Publication Data
Urban air pollution and forests: resources at risk in the Mexico City Air Basin / editors, Mark E. Fenn, L.I. de Bauer, Tomás Hernández-Tejeda.
p. cm.—(Ecological studies; v. 156)
Includes bibliographical references.
ISBN 0-387-95337-X (alk. paper)
1. Trees—Effect of air pollution on—Mexico—Mexico City Metropolitan Area. 2. Forest ecology—Mexico—Mexico City Metropolitan Area. 3. Air—Pollution—Environmental aspects—Mexico—Mexico City Metropolitan Area. I. Fenn, Mark E. II. Bauer, Ma. de Lourdes de la I. de. III. Hernández-Tejeda, Tomás. IV. Series.
SB745 .U73 2002
634.9′619′097253—dc21 2001042989

Printed on acid-free paper.

Production coordinated by Chernow Editorial Services, Inc., and managed by Francine McNeill; manufacturing supervised by Erica Bresler.
Typeset by SNP Best-set Typesetter Ltd., Hong Kong.
Printed and bound by Sheridan Books, Inc., Ann Arbor, MI.
Printed in the United States of America.

9 8 7 6 5 4 3 2 1

ISSN 0070-8356
ISBN 0-387-95337-X SPIN 10845410

Springer-Verlag New York Berlin Heidelberg
A member of BertelsmannSpringer Science+Business Media GmbH

Foreword

At present, roughly half of the world's population lives in urban centers. There are now more than 20 cities with a population of over 10 million inhabitants, compared to less than 5 about 50 years ago. This tendency toward urbanization is expected to continue, particularly in the developing world. A consequence of this growing trend is that millions of people are being exposed to harmful levels of urban air pollutants caused mainly by emissions from motor vehicles and from industrial and domestic activities involving the combustion of fossil fuels.

The driving force for the design and implementation of emission control strategies aimed at improving air quality has been the protection of the health of the population in urban centers. There are, however, other consequences of the presence of air pollutants besides the direct effect on human health. Reduced visibility, damage to monuments and buildings, and many other such consequences indirectly affect our quality of life. Another set of consequences involves damage to ecological systems. In fact, the nature of "photochemical smog" was first uncovered in the 1950s in connection with observations of its harmful effects on crops and plants in the vicinity of Los Angeles.

The potential damage to ecological systems caused by air pollution is an issue that is particularly worrisome if one considers the current growing population and urbanization trends. The degradation of air quality is no longer just a local, urban problem; it is beginning to acquire regional and even global proportions, and to affect large portions of the Northern Hemisphere.

The focus of this book is the impact of air pollution on forests in the Mexico City Air Basin and its vicinity. The chapters are written by leading experts in the field, and the book begins with a few introductory chapters on environmental and sociological aspects of the air pollution problem in the Mexico City Air Basin. Next are several chapters on the air pollution trends and control strategies and the ecological effects. The book ends with two chapters summarizing the impacts of multiple air pollutants on forests and an overall perspective on the problem.

This book provides an invaluable service, not only by calling attention to the harmful effects of air pollution on ecological systems around Mexico City but also by documenting the existing knowledge on the topic. The information presented in the book should contribute to the overall assessment of the air-quality problem and to the design of improved strategic control options to protect these important natural resources.

It is a pleasure for me to write the Foreword, and I congratulate the editors and the authors of the book for their important contributions.

Mario J. Molina

Preface

Mexico City has been the cultural, social, economic, religious, and political center of Mexico since pre-Hispanic times. However, the rapid increase in population in this megacity during the latter half of the twentieth century has led to unusually severe pressure on the limited natural resources of the Basin of Mexico. The current population within the Basin of Mexico of more than 18 million puts a severe strain on the capacity of the government and public utilities to provide services to the expanding population. Not surprisingly, this has caused severe environmental and ecological problems for the Basin and adjacent areas, including the chronic occurrence of air pollution concentrations that are unhealthy for plants, animals, and humans. For example, recent studies in forested watersheds near Los Angeles, California, and downwind of Mexico City demonstrate that the quality of drinking water extracted from these watersheds is affected by air pollution. Chronic deposition of atmospheric nitrogen, the main source of which is the use of fossil fuels, results in elevated nitrate levels from springs and streams. Furthermore, preliminary evidence suggests that the accumulation of lead in watersheds to the south and southwest of Mexico City can result in lead export in drainage waters. The severity and geographic extent of this problem is largely unknown at this time. These and other studies illustrate that urban air pollution impacts natural resources and the services provided by ecosystems within the geographic zone of influence.

We wish to insert a brief explanation here of the terminology commonly used to describe this unique region. The most common terms are the *Basin of Mexico* or the *Valley of Mexico*. Although the Basin of Mexico is the most correct term when referring to the larger area encompassed by the mountain ranges, these terms have often been used interchangeably. Some have even referred to this geographic region as the *Valley of the Basin of Mexico* or the *Mexico City Basin*. In some chapters of this volume, the term *Mexico City Air Basin* is used in reference to the ground-level atmosphere of the Basin.

Severe air pollution is a world-famous problem in the Mexico City Air Basin as a result of a combination of factors: over 3 million vehicles, including 1.4 million pre-1991 models; numerous domestic, industrial, commercial, and natural (e.g., fugitive dust) sources of air pollutants; abundant solar radiation that provides the energy for smog-producing photochemical reactions; thermal atmospheric inversions; a basin surrounded by volcanic mountain ranges; and frequent stagnant air masses. Atmospheric inversions and the mountain ranges circumventing the Basin prevent the transport of atmospheric pollutants from the Basin. The drainage of several lakes that formerly covered much of the Valley of Mexico has resulted in severe dust pollution during certain times of the year, in spite of continuous efforts by the government over the last 40 years to alleviate this problem.

Several commonalities between Mexico City and Los Angeles, California, link these North American urban regions. Both have a strong Spanish-American/Mexican heritage, both are exposed to severe photochemical smog as mentioned above, both experience atmospheric inversions, and both are surrounded by mountains inhabited by a dominant ozone-sensitive pine species exhibiting classic symptoms of ozone injury. Also, as mentioned above, in both areas, watersheds exposed to urban smog are exporting high levels of nitrate, although levels rarely exceed the drinking water standard. Atmospheric particulate concentrations commonly reach unhealthy levels in both areas, although extreme exposures are more common in the Mexico City Air Basin.

Increased tree mortality has also been documented in the most polluted forest stands downwind of these two metropolitan regions, following the period of major urban growth. Considerable evidence demonstrates a prominent role for air pollution in both cases. However, tree mortality was much more rapid and dramatic in the case of the decline of sacred fir in the Desierto de los Leones National Park near Mexico City, as entire stands were decimated. These dead zones became known as "cemeteries." Needless to say, the sudden death of thousands of fir trees in the highly valued Desierto de los Leones National Park in the 1980s was alarming to many and greatly increased public awareness of the impacts of air pollution on forests within the Basin. The causal mechanisms leading to the rapid onset of sacred fir decline in the 1980s and the continuing decline of the species are not definitively known, but the more dramatic appearance of the syndrome in the Mexican forests is exacerbated by the fact that fir decline generally occurs in pure fir stands. In contrast, increased mortality of ozone-stressed ponderosa pine trees in the San Bernardino Mountains east of Los Angeles occurred in mixed conifer stands, in which ponderosa pine was one species within a mix of four to five major overstory species.

Differences between the Los Angeles and Mexico City regions include the fact that ocean breezes carry pollutant loads to mountains downwind of Los Angeles, but there is no nearby ocean in the Mexico City Air Basin. The predominant wind pattern in the Mexico City Air Basin is to the southwest, which results in severe pollution exposure to the highly valued forests located in the south southwestern portion of the Basin. Rainfall patterns also differ between the two urban areas, the prolonged summer drought in the Los Angeles Basin contrasting with the summer rainy season in the Mexico City Air Basin. In Mexico City, sulfur dioxide emissions have decreased in recent years as a result of reductions in the sulfur content of fuels, yet sulfur deposition to forests in the southwest of the Basin remains high. This is in contrast with the Los Angeles Basin where sulfur pollutants are minimal compared to nitrogen oxides and ozone. In both regions, lead emissions have decreased dramatically with the introduction of unleaded fuels, although this is a much more recent trend in the Mexico City Air Basin.

Ozone concentrations and nitrogen oxide emissions have steadily decreased in recent years in the Los Angeles Basin, although greater reductions are needed. Occurrences of emergency levels of ozone have recently decreased in Mexico City, but ozone levels still remain dangerously high throughout much of the year. However, we are encouraged by an increasing recognition by the public and government officials of Mexico City of the importance of improving air quality. They have reacted by enacting and enforcing stricter regulatory controls. Only in this way can the concentrations of all criteria air pollutants begin to approach levels that are within the established air-quality standards on a consistent basis. The trend for future impacts of air pollutants on forests and other natural resources in the Basin of Mexico largely depends on the effectiveness of future emissions control programs and the level of public support for these measures.

Studies of the impacts of air pollutants in forests and crops in the Mexico City Air Basin have been carried out with the allocation of minimal financial resources to this area of study. The first published studies took place in the early 1970s mainly in the form of biomonitoring, that is, using established bioindicator plant species and field surveys to document the occurrence of phytotoxic levels of air pollutants in the urban and montane regions of the Basin. Largely, through the cooperation of researchers from various institutions in Mexico, the United States, Canada, and Europe, and Mexican students studying abroad, the scientific investigation of the effects of air pollution on forests and vegetation of the Mexico City Air Basin has slowly advanced over the past 30 years. Prominent international scientists who have contributed to air pollution studies on forests and crops in the Mexico City Air Basin, whose roles in advancing the science we wish to acknowledge, include Sagar Krupa, William J. Manning, Paul Miller, John Skelly, and others.

The accumulated body of work regarding the effects of air pollution on forests and vegetation in the Mexico City Air Basin has been independently published in a myriad of disparate sources, many of which are available to only a very select audience (e.g., relatively unknown graduate theses, government reports, scientific proceedings papers, articles, and papers spread throughout various Mexican and international journals). It is the aim of this volume to summarize and synthesize this work, to bring this information together into one readily available source, and to guide future researchers in filling many of the important gaps in our understanding of the impacts of air pollution on the natural resources of the Basin of Mexico. It is also our hope that this book will further highlight and raise awareness of the environmental and ecological problems occurring in the region affected by one of the world's largest metropolitan centers.

Mark E. Fenn
L.I. de Bauer
Tomás Hernández-Tejeda

Acknowledgments

We wish to acknowledge the ongoing support of the Atmospheric Change and Forests Working Group of the North American Forest Commission (within the FAO) in furthering the study of the impacts of air pollution on forests within the Basin of Mexico. The support of Roger Cox of the Canadian Forest Service in sponsoring graduate researchers and in cooperation with other air pollution studies in Mexico is also appreciated. We gratefully acknowledge the support of personnel at the Parque Cultural y Recreativo Desierto de los Leones in providing a secure location and power supply to maintain a long-term meteorological and ozone-monitoring station (the only one of its kind in a montane site within the Mexico City Air Basin) since 1990. This monitoring station was established during the administration of the former director Jorge P. Galvan. We also thank Bertha Gutierrez and colleagues at the park for field support in streamwater sampling for the past several years.

The long-term support of Elvia Niebla for cooperative air pollution and global change research in Mexico and the financial support of the Vegetation Management and Protection Research staff of the U.S. Forest Service toward the publication of color figures in this volume are also greatly appreciated. We give thanks to the Pacific Southwest Research Station of the United States Department of Agriculture (USDA) Forest Service and the National Council for Science and Technology (CONACYT, MEXICO) for financial support throughout the years. We express our appreciation to INIFAP (the National Research Institute for Forestry and Agriculture; MEXICO) for their long-term institutional support of air pollution studies in the Mexico City Air Basin and to the Pacific Southwest Research Station, USDA Forest Service (again), for long-standing support of international collaborative air pollution studies. We also wish to thank Susan Schilling for her invaluable technical assistance and numerous hours spent producing and improving many of the map figures found in various chapters.

Lastly, we express our sincere appreciation to the following individuals for their time and efforts in serving as chapter reviewers: Michael Arbaugh, William Carter, Andrew Chang, Pablo Cicero-Fernández, Linda Geiser, Ken Hubbert, Ernesto Jáuregui, José Lugo, Hector Maldonado, Laura Merrill, Luisa T. Molina, Robert Musselman, Marcia Narog, James Pick, Irene Pisanty, Mark Poth, John Skelly, Robert Slott, Gustavo Sosa, Larry St. Clair, Brent Takemoto, Patrick Temple, Ray Thompson, David Weise, Angel Zambrano, and Karl Zeller.

Mark E. Fenn
L.I. de Bauer
Tomás Hernández-Tejeda

Contents

Contributors

Aguilar, Adrián Guillermo	Instituto de Geografía, Universidad Nacional Autónoma de México, Circuito Exterior, Ciudad Universitaria, CP 04510 Mexico City, Mexico. e-mail: adrian@servidor.unam.mx
Alvarado-Rosales, Dionicio	Especialidad de Fitopatología, Instituto de Fitosanidad, Colegio de Postgraduados, CP 56230 Montecillo, Mexico. e-mail: dionicio@colpos.mx
de Bauer, L.I.	Especialidad de Agrometeorología, Instituto de Recursos Naturales, Colegio de Postgraduados, CP 56230 Montecillo, Mexico. e-mail: libauer@colpos.mx
Bernal-Salazar, Sergio	Especialidad de Botánica, Instituto de Recursos Naturales, Colegio de Postgraduados, CP 56230 Montecillo, Mexico. e-mail: bernals@colpos.mx

Bravo-Alvarez, Humberto — Centro de Ciencias de la Atmósfera, Universidad Nacional Autónoma de México, Circuito Exterior, Ciudad Universitaria, CP 04510 Mexico City, Mexico. e-mail: hbravo@servidor.unam.mx

Castro-Servín, Juana María — SAGARPA, Instituto Nacional de Investigaciones Forestales, Agrícolas y Pecuarias (INIFAP), Centro Nacional de Investigación Disciplinaria en Conservación y Mejoramiento de Ecosistemas Forestales (CENID-COMEF), CP 04110 Mexico City, Mexico. e-mail: castrojm@hotmail.com

Escolero-Fuentes, Oscar — Instituto de Geofísica, Universidad Nacional Autónoma de México, Circuito Exterior, Ciudad Universitaria, CP 04510 Mexico City, Mexico. e-mail: oescolero@sgt.cna.gob.mx

Ezcurra, Exequiel — Instituto Nacional de Ecología, SEMARNAT Colonia Tlacopac, CP 01040 Mexico City, Mexico. e-mail: eezcurra@ine.gob.mx

Fenn, Mark E. — USDA Forest Service, Pacific Southwest Research Station, Riverside, CA 92507, USA. e-mail: mfenn@deltanet.com or mfenn@fs.fed.us

García-Calderón, José Luis — Departamento de Hidrobiología, Universidad Autónoma Metropolitana, Unidad Iztapalapa, CP 09340 Mexico City, Mexico. e-mail: jlgc@xanum.uam.mx

Gasca, Jorge — Instituto Mexicano del Petróleo, CP 07730 Mexico City, Mexico. e-mail: jgazca@imp.mx

Goodson, Chris — Department of Environmental Sciences, University of California, Riverside, CA 92521, USA.

Hernández-Tejeda, Tomás	SAGARPA, Instituto Nacional de Investigaciones Forestales, Agrícolas y Pecuarias (INIFAP), Centro Nacional de Investigación Disciplinaria en Conservación y Mejoramiento de Ecosistemas Forestales (CENID–COMEF), CP 04110 Mexico City, Mexico. e-mail: textom99@yahoo.com or thtejeda@colpos.mx
Herrera-Campos, M.A.	Instituto de Biología, Universidad Nacional Autónoma de México, Circuito Exterior, Ciudad Universitaria, CP 04510 Mexico City, Mexico. e-mail: mahc@ibiologia.unam.mx
Jáuregui, Ernesto	Centro de Ciencias de la Atmósfera, Departamento de Meteorología General, Universidad Nacional Autónoma de México, Circuito Exterior, Ciudad Universitaria, CP 04510 Mexico City, Mexico. e-mail: ejos@ccaunam.atmosfcu.unam.mx
Krage, Natalie	Department of Environmental Sciences, University of California, Riverside, CA 92521, USA.
Laguette-Rey, Héctor D.	Massachusetts Department of Environmental Protection, Wilmington, MA 01887-2941, USA. e-mail: Hector.Laguette-Rey@state.ma.us
de la Lanza Espino, Guadalupe	Instituto de Biología, Universidad Nacional Autónoma de México, Ciudad Universitaria, CP 04510 Mexico City, Mexico. e-mail: gdlle@servidor.unam.mx
Marín, Luis E.	Instituto de Geofísica, Universidad Nacional Autónoma de México, Circuito Exterior, Ciudad Universitaria, CP 04510 Mexico City, Mexico. e-mail: lmarin@mail.com
Mazari-Hiriart, Marisa	Instituto de Ecología, Universidad Nacional Autónoma de México, Ciudad Universitaria, CP 04510 Mexico City, Mexico. e-mail: mazari@servidor.unam.mx

Meixner, Thomas	Department of Environmental Sciences, University of California, Riverside, CA 92521, USA. e-mail: tmeixner@mail.ucr.edu
Miller, Paul R.	USDA Forest Service (Retired), Pacific Southwest Research Station, Riverside, CA 92507, USA. e-mail: millerpr@aol.com
Nash III, T.H.	Department of Plant Biology, Arizona State University, Tempe, AZ 85281-1605, USA. e-mail: tom.nash@asu.edu
Ortíz-García, Carlos Fredy	Periférico Carlos A. Molina, Colegio de Postgraduados, Campus Tabasco, CP 86570 Cárdenas, Tabasco, Mexico. e-mail: cfortiz@colpos.mx
Pisanty, Irene	Laboratorio Especializado de Ecología, Facultad de Ciencias, Universidad Nacional Autónoma de México, Ciudad Universitaria, CP 04510 Mexico City, Mexico. e-mail: ipisanty@servidor.unam.mx
Quevedo, Abel	Programa de Edafología, Instituto de Recursos Naturales, Colegio de Postgraduados, CP 56230 Montecillo, Mexico. e-mail: anolasco@colpos.mx
Rodríguez, Claudio	Especialidad de Agrometeorología, Instituto de Recursos Naturales, Colegio de Postgraduados, CP 56230 Montecillo, Mexico. e-mail: anolasco@colpos.mx
Rodríguez-Franco, Carlos	SAGARPA, Instituto Nacional de Investigaciones Forestales, Agrícolas y Pecuarias (INIFAP), Centro Nacional de Investigación Disciplinaria en Conservación y Mejoramiento de Ecosistemas Forestales (CENID-COMEF), Col. Viveros de Coyoacán, CP 04110 Mexico City, Mexico. e-mail: crfranco@aya.yale.edu
Ruiz, M. Esther	Instituto Mexicano del Petróleo, CP 07730 Mexico City, Mexico. e-mail: mruiz@www.imp.mx

Terrazas, Teresa	Especialidad de Botánica, Instituto de Recursos Naturales, Colegio de Postgraduados, CP 56230 Montecillo, Mexico. e-mail: winchi@colpos.mx
Torres-Jardón, Ricardo	Centro de Ciencias de la Atmósfera, Universidad Nacional Autónoma de México, Circuito Exterior, Ciudad Universitaria, CP 04510 Mexico City, Mexico. e-mail: ricardo_tjardon@yahoo.com
Trinidad-Santos, Antonio	Especialidad de Edafología, Instituto de Recursos Naturales, Colegio de Postgraduados, CP 56230 Montecillo, Mexico. e-mail: trinidad@colpos.mx
Zambrano, Angel	Instituto Mexicano del Petróleo, CP 07730 Mexico City, Mexico. e-mail: azambran@www.imp.mx
Zeller, Karl	USDA Forest Service, Rocky Mountain Research Station, Ft. Collins, CO 80526, USA. e-mail: kzeller@lamar.colostate.edu

1. Past and Present Environmental and Sociological Factors Affecting Forests in the Mexico City Air Basin

1. Historical Summary of the Geology, Climate, Hydrology, Culture, and Natural Resource Utilization in the Basin of Mexico

Guadalupe de la Lanza Espino and José Luis García-Calderón

The Valley of Mexico is a closed basin with an artificial drainage outlet situated at 2240 masl and is bordered by mountains that reach a maximum elevation of 5540 m. The basin was partially covered by a lake, which is now practically extinct but, for most of the past 100,000 years, covered 2000 km^2. Near the end of the nineteenth century, lakewater still covered an area half as large as that covered by water at the beginning of the sixteenth century (Table 1.1). The Valley of Mexico continues as the main thoroughfare of human activity in the region, and for the last 700 years, has been the economic, political, and social center of Mesoamerica.

The basin, 9600 km^2 in size, includes the capital of the Mexican republic (which encompasses nearly two million km^2) and one fifth of the population and produces one fourth of the gross domestic product. To arrive at this point, the basin has experienced many changes. In this chapter, some of the most relevant stages of this transformation will be highlighted. Management of the aquatic resources, which are the principal focus for management of the basin, and the changes in forest cover, an indicator of the state of health of the environment, will be emphasized.

Geologic History of the Basin of Mexico

Mexico is located where the North American continent narrows resembling a cone. At 32° latitude, near the headwaters of the Rio Bravo, more than 3000 km separate the Pacific Ocean from the Gulf of Mexico, whereas at the Tehuantepec

Table 1.1. Lakes of the Basin of Mexico at the end of the nineteenth and beginning of the twentieth centuries

Lake	Surface area (ha)	Height above Lake Texcoco (m)	Volume ($m^3 \times 10^6$)	Quality of water
Zumpango	2,632	6.06	14.5	Brackish
Xaltocan, Sn. Cristóbal	12,144	3.54	47.4	Brackish
Texcoco	23,854	0.0	429.4	Saltwater
Chalco	11,408	3.08	233.9	Freshwater
Xochimilco	6,336	3.14	133	Freshwater
Total	56,374		858.2	

Source: Orozco-Berra (1862), Iglesias (1902).

Isthmus, at 16° to 18° latitude, this separation is barely 200 km. The rugged terrain of Mexico encompasses five (Challenger 1998) to six ecological regions (Toledo 1997; Toledo and Ordóñez 1993; Toledo et al. 1987) with 17 phytogeographic provinces (Rzedowski 1978) and 32 vegetation types corresponding to the various climates (Miranda and Hernández-Xolocotzi 1963). The topography, resulting from tectonism, has also been affected, mostly in the last million years, by volcanic processes that have left a rugged surface delimited by two large mountain ranges (the Eastern Sierra Madre and the Western Sierra Madre), which compress toward the south as the cone narrows. Intermountain regions are occupied by alluvial and lacustrine sediments, frequently of the endoreic type found in the Basin of Mexico. Here, the Cretaceous sediments are covered by magmatic emanations, ashes, breccia, and other products from seven volcanic periods that occurred successively during the last 25 million years (Mosser 1963, 1975), resulting in elevations greater than 5000 m. These successive events covered the earlier ones so that little evidence of the first events remain except in some parts of the Neovolcanic Axis (the only mountain range of the continent running from the Pacific to the Atlantic). In the southern portion of the Basin of Mexico, in a 4600 km^2 area, up to 300 volcanic structures can be counted (Lugo 1984).

The Basin of Mexico, with an area of 9600 km^2, occupies the southern end of the endoreic environments of the Northern Hemisphere. The basin was formed by the materials produced by the formation of a volcanic apparatus (the Chichinautzin Volcano) and the volcano's emanations, which blocked off two runnoff currents draining to the south into the Balsas River, which in turn drains its waters into the Pacific Ocean. These recent events, occurring not much more than 100,000 years ago, have hidden the evidence of Cretaceous events (110–65 million years BP), which occurred when the region now known as the Basin of Mexico was covered by a shallow sea, in which a series of high mountains appeared due to a sequence of earth foldings at the beginning of the Eocene (65–45 million years BP). Towards the end of the Eocene, intense volcanic activity was generated accompanied by land subsidence that modified the region and this was followed by stages of vulcanism and tectonism. From the seven recorded stages that occurred in the last 25 million years, the last three stages stand out. During the fifth stage, the Sierra de Las Cruces, Monte Alto y Bajo, as well as the basement of the Sierra Nevada, were formed; during the sixth, the Sierra de Calpulalpan and the Ajusco, Gordo, and El Pino mountains originated (Pliopleistocene, 10 to 1 million years BP); and during the seventh (Upper Pleistocene, less than 1 million years BP), most of the Ajusco range (Chinchinautzin), the Santa Catarina Sierra, and the Popocatépetl and Iztaccíhuatl volcanoes were formed. During the Pleistocene, the Ajusco closed the fluvial drainage to form the endoreic basin, which initiated sediment accumulation that eventually covered more than half of the valley (Rzedowski and Rzedowski 1979).

Vegetation and Forest Resources

According to Rzedowski and Rzedowski (1979), the Valley of Mexico, because of its southerly location in the country, is a region rich in flora with up to 1500 phanerogamic species (Reiche 1965), comparable to the United States and European countries. Based on diverse records, it is estimated that in the past, climates were more humid, and genera of holarctic origin, *Abies*, *Alnus*, *Pinus*, and *Quercus* among others, can still be found. In addition, species of a southern affinity have long coexisted with these above the arboreal limits, where the largest concentration of species common to the Andes can be found. This evidence supports the idea that prior colder conditions favored a larger extension of alpine regions. Notwithstanding the fact that 10 main plant communities, located at different altitudes (Table 1.2), have been recorded in the basin, the region is not an important center of plant evolution, at least not when the region is considered in isolation. An example can be found in the origin of the Mexican xerophilous flora. These seem to date back to ancient times and, to a great extent, are only found locally, although evidence exists to suggest that at some period there was an influx of an abundance of xerophytes originating from South America.

Table 1.2. Main plant communities in the Basin of Mexico

Community	Elevation (m)	Rainfall (mm)	Temperature (°C)
Abies forest	2700–3500	1000–1400	7.5–13.5
Cloud forests or mesophyllous forests	2500–2800	>1000	12–14
Pinus forests	2350–4000	700–1200	10–14
Quercus forests	2350–3100	700–1200	
Juniperus forests	2450–2800	600–800	11–14
Quercus shrubland	2350–3100	700–900	9–13
Grasslands	2250–4300		
Xerophilous scrub	2250–2700	400–700	12–16
Halophytic vegetation	<2250		
Aquatic and subaquatic vegetation	2240		

Source: Rzedowski and Rzedowski (1979).

Among the most representative forests of the Valley of Mexico, *Abies*-dominated communities can be found at altitudes between 2700 and 3500 m and higher in well-drained deep soils rich in organic matter and humid all year long with rainfall of more than 1000 mm and mean temperatures between 7.5° C and 13.5° C. *Abies* is found in the mountain ranges of the southern half of the Valley. Pine forests are found at altitudes between 2350 and 4000 m, mainly associated with other species. They prosper in sites with annual rainfall between 700 and 1200 mm in deep, shallow, or even rocky soils. At 2350 to 2600 m elevation, only *Pinus leiophylla* is found. At 2500 to 3100 m, *P. montezumae* and P. *rudis* are found, and *P. teocote* is found in more humid climates. According to Rzedowski and Rzedowski (1979), oak forests were more extensive in the past than now. They prosper at altitudes between 2350 and 3100 m in habitats similar to those inhabited by *Pinus*. The dominant species below 2500 m are *Q. laeta*, *Q. deserticola*, and *Q. crassipes*. *Juniperus* forests are open and low-canopy communities which can barely be considered forests. They are found on hillsides in more-or-less flat terrain of the north, northeast, and eastern sections of the Valley between 2450 and 2800 m. The mean temperature in *Juniperus* forests ranges from 11° C to 14° C with annual rainfall of 600 to 800 mm. The most common species is *J. depeana*.

Present-Day Climate

Due to its latitudinal location (19°–21°), the Valley of Mexico has a highland tropical climate with trade winds carrying humid air that produces summer rainfalls, and a dry season in which western winds predominate. Strong solar radiation with clear skies causes increased evapotranspiration which reaches a yearly mean of 1400 to 2200 mm. Because of its complex and rugged orography, the valley has high climatic diversity within a relatively small area. In general terms, precipitation increases with elevation, although at elevations between 2700 and 3500 m, this relationship is often reversed because the upper montane areas are frequently above the cloud level (Rzedowski and Rzedowski 1979). According to Koepen's classification, the valley includes the following three climates:

BSkwg, a dry climate with temperatures above 18° C in the hottest month, rains in the summer, and the highest mean monthly temperature just before the summer solstice, covers almost the entire basin, and with a pronounced dry season in the Pachuca Sierra, where rainfall is no more than 400 mm due to the orographic rain shadow, a mean annual rainfall of less than 600 mm.

Cwbg, is a temperate rainy climate with summer rainfall, a temperature below 22° C in the hottest month, the highest mean monthly temperature just before the summer solstice, and an annual mean rainfall above 600 mm. This climate occurs mainly on plains or flatlands.

ET, a tundra climate with a mean monthly temperature in the hottest month above 0° C, covers the highest elevations of the Nevada, Ajusco, and Las Cruces Sierras. The Las Cruces Sierras have an average annual rainfall of up to 1400 mm.

Historical Aspects of Climate and Hydrology in the Basin

Based on hydrological characteristics, the Basin can be subdivided into various zones. To the south is the highly permeable Chichinautzin Sierra. The Basin is bordered by the Churubusco River to the east, the City of Mexico in the southwest, the Tepotzotlán-Cuauhtitlán Rivers to the northwest, and the Avenida de Pachuca River to the north together with the Teotihuacán River in the same region (Fig. 1.1). Furthermore, due to its geomorphologic characteristics, derived from highly permeable basalts, infiltration in the high mountain elevations is favored, producing abundant groundwater springs.

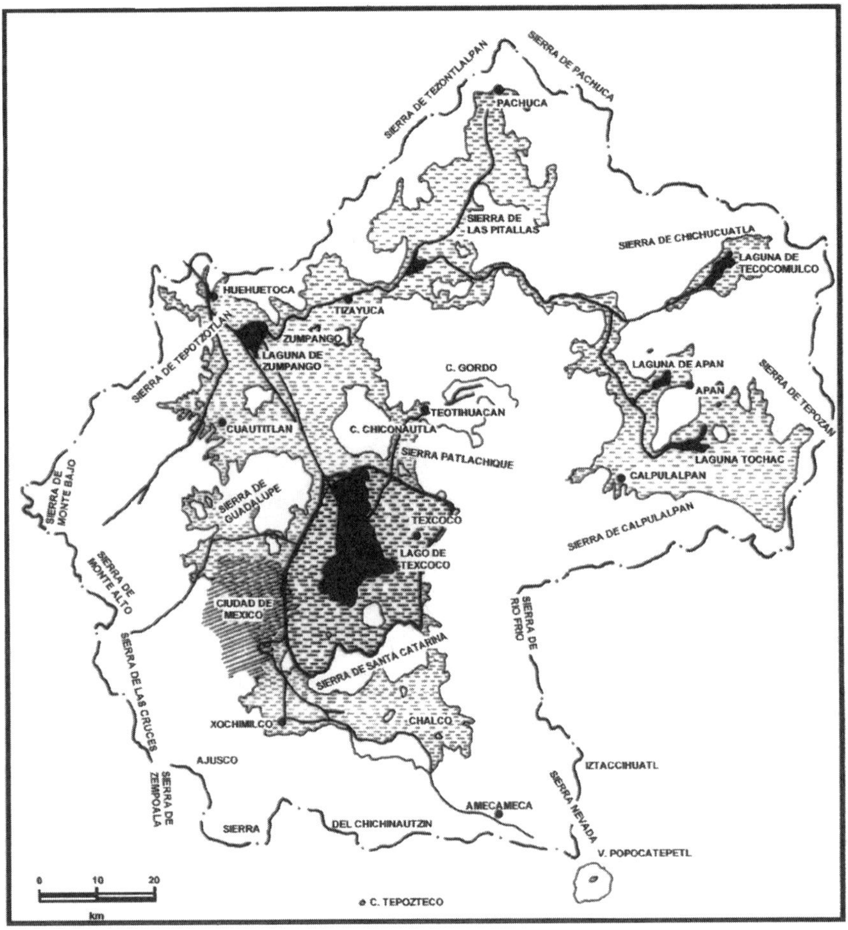

Figure 1.1. The Basin of the Valley of Mexico. ___.___.___. Basin area; ________ Boundary of the ancient lake. (From Mooser 1975.)

In the sixteenth century, the great lake constituted a mirror of water encompassing 1200 km^2 which formed three lacustrine subbasins: Zumpango–Xaltocan–San Cristóbal in the north, Texcoco in the center, and Chalco–Xochimilco in the south. The Chalco–Xochimilco basin contained fresh water, water in the basin of Texcoco was salty and bitter, and water in the northern basin was brackish.

The most ancient limnologic changes recorded by stratigraphic studies (Watts and Bradbury 1982) date back 100,000 years (Bradbury 1971) and consist of a succession of saline and fresh-water environments (the hydrologically closed saline environment did not drain, lost volume through evaporation, and accumulated dissolved matter). Wide variations in water volume occurred accompanied by intervals of varying productivity conditioned by various climatic factors. In addition, it is estimated that around 6000 years BP, environmental changes similar to those induced by natural agents such as climate or hydrological dynamics were introduced by human activities in the riparian zones associated with springs (Bradbury 1971; Buckler et al. 1998).

Between 3000 and 1500 years BP, the climate was humid and cold, and the basin was dominated by *Pinus–Quercus–Abies* forests. Between 1100 and 800 years BP, the predominating climate was hotter and subhumid with deeper lacustrine environments and *Pinus* and *Quercus* forests. This was followed by an increase in temperature to a point where the temperature stabilized 300 years ago. Current climatic characteristics date back 200 years. The gradual temperature increase has displaced forests towards higher regions more favorable for *Pinus* and *Quercus*, or for *Abies*.

History of Human Utilization of Natural Resources in a Lacustrine Environment

This environment was gradually occupied by man beginning more than 10,000 years ago. More precise archaeological records, commencing with the Preclassic period (3500 years BP), provide evidence of an expanding population that reached one million inhabitants at the beginning of the sixteenth century. This population was part of the 20 million inhabitants of Mesoamerica (Hassig 1985), a population which was not surpassed until the 1940s. The stress exerted by the population on the environment, because of the need to provide resources for the population, led to the development of environmental-management strategies that are still manifested in various ecological systems, such as those occurring at some very remote sites. For example, in the Purepecha community (Toledo et al. 1980), which developed in the basin of Lake Pátzcuaro, a multiple-use strategy for managing the ecosystem was employed along an altitudinal dimension extending from the lake to the mountain, providing multiple resources. This pattern was repeated in different Mesoamerican environments (Boege 1988; Challenger 1998; Leff 1990; Rojas 1990).

This multiple-use strategy for managing the ecosystem allowed efficient utilization of a large variety of resources. A lake could provide fish, mollusks, amphibians, and insects, to which vegetative resources such as *Spirulina*, known as blue seaweed, should be added. Some 50 bird species were also associated with the lake, some of which were used as food. In the shallow flooded areas as well as in the low-lying areas near the lake, original methods, such as the *chinampas* (sometimes known as floating gardens) (Armillas 1971; Challenger 1998; Parsons 1976), were developed for highly efficient agricultural production from the limited space used for agriculture. Nonetheless, it is believed that at the time of contact with the Spaniards, the carrying capacity of the land for sustaining the population had been surpassed (Williams 1989). The areas above 2500 m, covered by forests, represented a reserve from which energy and construction materials were obtained. This variety of resources was used to achieve environmental management from the lacustrine zone on up to the highest elevations of the basin. These environmental resource types were distributed in concentric bands along an elevational gradient: the urban area was integrated with the *chinampa* agriculture; croplands were located in the adjacent plains and were bordered upslope by the forest belt.

For the lakes, a management system was developed which considered human labor to be in cooperation with the environment and not in opposition to it. For several centuries, during which diverse succeeding human groups inhabited the region, an extensive culture developed around the lacustrine environment. In one century of development, the Mexica group increased their influence throughout Mesoamerica by founding a city on a small island that was extended by human labor until it reached 11,500 ha in size (González-Rul 1998) and a population of 300 to 350 thousand inhabitants (Gurria 1974). This was a city established within the Mesoamerican tradition, synthesizing in an urban environment the ritualistic, administrative, and commercial attributes of Mesoamerica (Sanders and Webster 1988). Tenochtitlan is recognized as the inheritor of a thousand-year-old tradition dating back to Teotihuacan, the first great urban center of the High Plateau. Artifacts from Teotihuacan are observed in the offerings found in excavations of the Great Temple in the center of current-day Mexico City. This city state, socially diverse and stratified, controlled a territory of more than a million km^2. It was also the commercial center, where on a daily basis, thousands of shipments arrived with all kinds of articles and food supplies from diverse points in the Basin and from the territories that rendered tribute. A large part of the food supplies came from the Chalco–Xochimilco region in the south, where a large part of the food consumed in the metropolis was produced from more than 10,000 ha of the swamp (*chinampas*) agricultural area.

Commercial expansion was made possible by the lacustrine environment which provided an efficient mode of transport in the absence of animal-powered transport. Manpower could be used to transport 20 kg (a little over 50 lbs) of goods over land for a distance of 10 to 15 km in one day. On a boat, on the other hand, nearly a ton of goods could be shipped over the same distance in 10 to 12 hours. This fact converted the agricultural lands into satellites of the metropolis, notwithstanding the 20 km distance between the urban center and the agricultural areas.

The agricultural system called *chinampas* parallels the concepts used for a system developed in Italy in the Venetian lagoon. This system holds that in order to obtain sustainable development "the lagoon has to be re-created in the lagoon". An urban lacustrine environment was also achieved in the Valley of Mexico, which encompassed a land with around one million inhabitants (Corona 1977; Sanders et al. 1979). Furthermore, the city of México–Tenochtitlan was a manufacturing center of sufficient importance to be considered as a preindustrial mercantile city.

To harmonize the colonization of this extended territory, an infrastructure adequate to support such a daring adventure was needed. Thus, the city became involved in alliances and political activities for control of the basin while at the same time building a large infrastructure, including aqueducts, dikes, and causeways of more than 50 km in length, on which the European conquerors walked (Cortés 1994; Diaz del Castillo 1970; Palerm 1973). The large hydraulic system regulated water flows arriving at the western end of the lake system, partially separating it from the eastern saline section with a clockwise circulation pattern (Fig. 1.2).

The Netzahualcoyotl dike helped control the water flow, preventing its entrance into the city of Mexico (Fig. 1.3), but it did not obstruct water coming in from the Chalco–Xochimilco Lakes, which since it lay a little more than 3 m above the level of the Lake Texcoco, drained into Lake Texcoco (Fig. 1.2). In contrast, salt water entering from the north flowed counterclockwise to the south and was discharged by the siphon effect exerted by the water coming from the Chalco–Xochimilco system (Fig. 1.3). These natural water mass movements, although imprecise and varying, were harnessed by hydraulic works to further the coexistence of neighboring sites containing saline water and *chinampas* supplied with mostly fresh water. As an example of this, large aqueducts were used not only to supply water for the city's needs but also to fill the canals, thus providing a fresh water buffer that protected the *chinampas* along the periphery from the salt front of Lake Texcoco (García-Calderón and De la Lanza 1995).

Disintegration of the pre-Hispanic hydraulic management system represented a major ecological disaster, especially for a culture in which the aquatic environment was the main thoroughfare for its productive activities. Furthermore, the culture had not yet availed itself of the use of the wheel. The endoreicity of the Basin is known to have been maintained until the end of the nineteenth century. Fragmentation of the large lake into five units (Table 1.1), with conditions varying from fresh to saline waters, was finally accomplished by the construction of the Large Drainage Canal and, more recently, with the Deep Sewage Drainage System built during the period of increasing urbanization in the 1970s. The Deep Sewage Drainage System drains into the Tula River, a tributary of the Pánuco River. These engineering feats caused the disappearance of water systems or reduced them to mere relics. Nevertheless, species of biological importance and economic potential in these aquatic systems have been conserved (Lot-Helgueras and Novelo-Retana 1978; Novelo-Retana and Gallegos 1988).

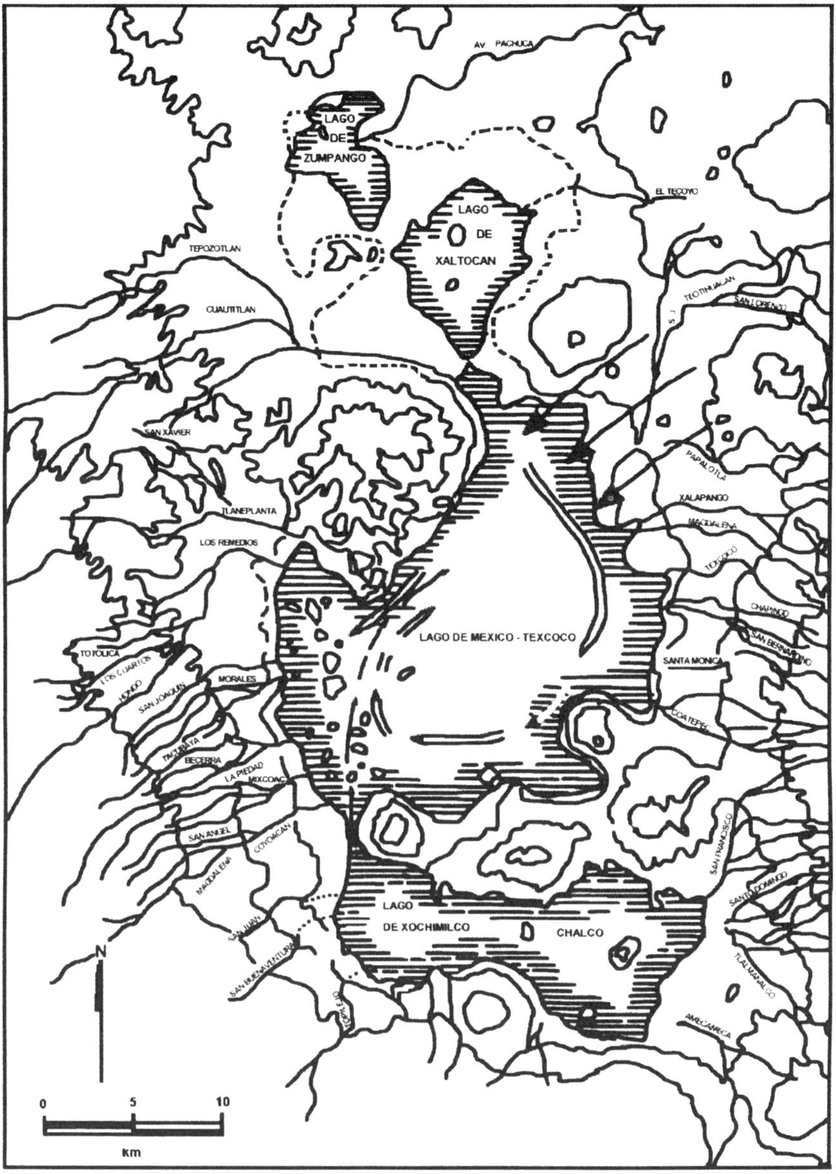

Figure 1.2. Water circulation pattern in the ancient lake in the Valley of Mexico Basin. (Reproduced by Permission of Instituto Nacional de Antropología e Historia, Mexico.)

As a result of the conquest, not only was the the hydraulic management system destroyed, but in addition, enormous destruction occurred in all of the Basin. Forests were burned to prepare the land for cultivation and cattle grazing, water diversion was increased to fill greater numbers of holding ponds. Canals were now considered as obstacles instead of means of transportation and communica-

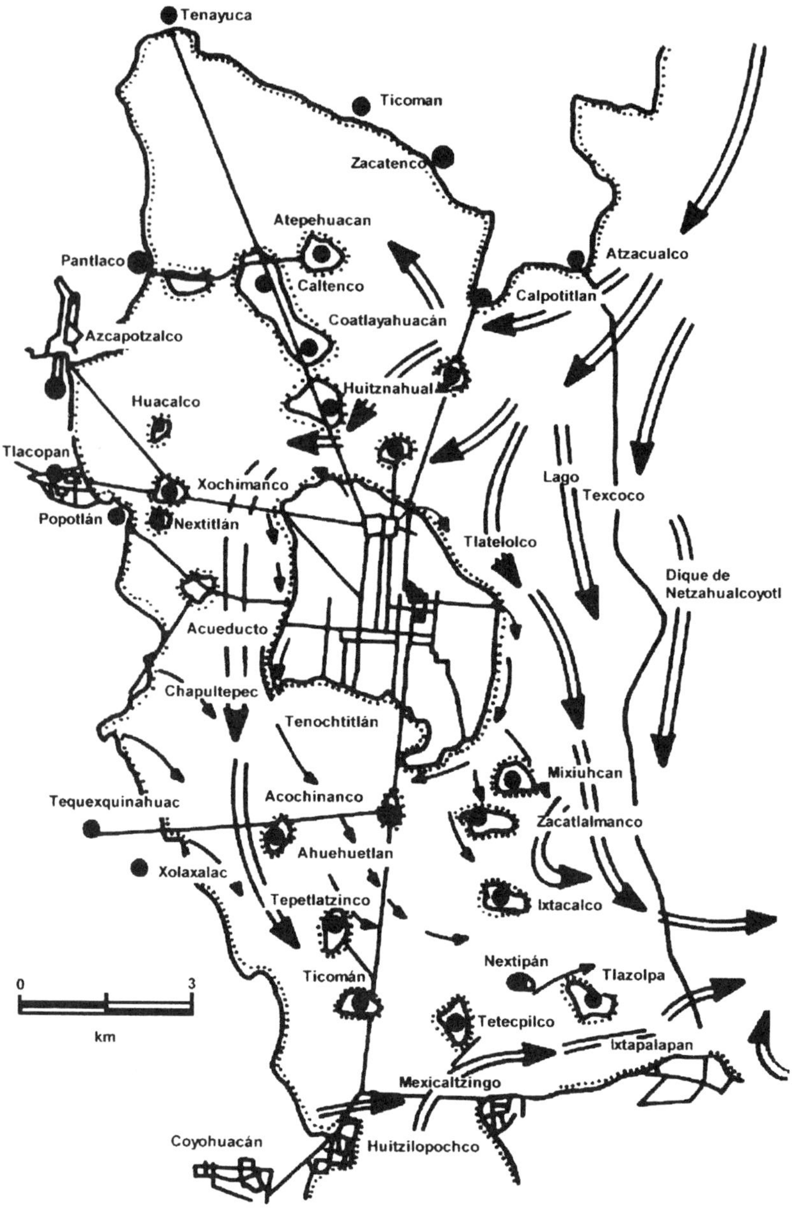

Figure 1.3. Management of water circulation on the western side of the ancient lake of Texcoco. (Reproduced by Permission of Centre Français d'Études Mexicaines et Centraméricaines, Mexico.)

tion. The soft subsoil provided poor support for stone buildings, and the available technology was inadequate for such construction on the lacustrine foundations. Because of the different levels of the lakes, those lakes situated at higher altitude overflowed towards Mexico City, and in 1555 there was a great flood. Because the waters from the lakes were considered by the Spaniards to be of low quality because they were hot and fetid in summer and sufficiently cold in winter to cause sickness, water from the mountains, for example, from Santa Fe in the highlands of the west, was preferred. However the neglect or mismanagement of the lakes, aside from efforts directed toward finding a way to drain the lakes, resulted in a series of floods beginning in 1553 and occurring again in 1580, 1604, 1607 and 1629 (Garza 1987). Flooding problems continued during the 300 years of European domination, and greatly increased the desire to find a way to drain the water from the basin.

Pre-Hispanic and Colonial Utilization of Forest Resources

Reduction of forest cover in the Mexico City Basin has been caused by fires, diseases, and uncontrolled cutting and clearing for agriculture and pasture. These have caused most of the 30 or so species which have historically been preferred for their usefulness as energy sources and building materials to be in serious decline if not already absent as a natural resource. However, environmental destruction is not only a recent occurrence. Environmental transformation since pre-Hispanic times is widely recorded, both in written accounts and in the findings of archaeological investigations (Niederberger 1987) carried out in urban areas or on agricultural lands. These findings attest to intentional burning, intentional desiccation of lakes and springs, as well as extensive modification of the plant cover along with consequent soil erosion and high rates of material export to the lakes. Niederberger (1987) concludes that climatic changes of the last 10,000 years, beginning at the time of the supposed arrival of man to the basin, brought about the first changes in plant distribution. However, the limited knowledge of the previous era, free of human disturbances and with a climate similar to the present, can only be considered as theoretical conjecture.

Early Horizon (1550 BC–AD 100)

The scarce population of the basin was centered in coastal environments where incipient agriculture was complimented with hunting and fishing activities to such a degree that the major effects on forested areas were natural impacts, among which volcanic eruptions are prominent. Volcanic activity in central Mexico has been very high. The lava of the Xitle Volcano eruption (300 BC) not only covered thousands of hectares that included forests but also caused fires over a much greater area. Among these natural phenomena, fires are a consequence of severe droughts, which affect not only forest environments but water bodies as well (Armillas 1969; Bernal 1984; Florescano and Swan 1995; Metcalfe et al. 1994;

Street-Perrott et al. 1989). Droughts occurred periodically, mainly as a result of increased atmospheric temperatures and infrequent rainfall (conditions which cause high evaporation rates from lakes and the forest), causing a severe lowering of the water level in the lakes of the highlands.

Middle Horizon (AD 300–750)

During the peak of the Teotihuacan culture, charcoal ovens must have been common. According to Vazquez-Yanes (1982), they were used to make stucco by those who could light fires. Moncayo (1975) reports that large trees were girdled, leaving them as standing dead trees thus easier to bring down.

Water surrounding the cities was considered part of the city and was given the same name as the city. In the native *nahuatl* the binomial *atl tepetl*, "the water, the mountain", was used in naming the cities, as was the case for the island of Mexico. Water was an unpredictable force that could threaten human settlements at any time because it frequently escaped human control. Nonetheless, its significance was ambivalent, because water also signified the power of creation and the source of fecundity.

Late Horizon (AD 1350–1519)

It is important to recognize that for the native people, in addition to being a source of life, the lakes also represented major problems, such as the floods that occurred in 1382 and 1499. These catastrophes alternated with other disasters caused by droughts, which although they are not well documented, were a recurring phenomenon affecting the environment, sometimes as events with only minor impacts, sometimes as catastrophic events. Droughts exerted significant effects on forests and on production activities such as agriculture and as a result, caused important social impacts. Although it is difficult to know the exact dates of droughts because of the lack of detailed written records from that time, it has been possible to propose an outline of drought occurrence (Table 1.3).

In 1430, The Triple Alliance merged the three most important kingdoms in the basin (the two mainland kingdoms, Texcoco to the east of the lake and Tlacopan to the west, and Mexico on the island of Tenochtitlan). There is evidence that after the alliance, when territorial expansion took place, increased amounts of wood were needed to construct the temples and palaces in the urban zone, especially in the City of Mexico (González-Rul 1998). These temples and palaces were ornamented with cedar and cypress wood (Torquemada 1975). Notwithstanding the intensive utilization of timber, there was an effort to normalize the exploitation of forest resources, as evidenced by the regulations established by Netzahualcoyotl, king of Texcoco, or by his kinsman Nopaltzin, who even established the death penalty for burning grasslands or hillsides. Similar rules existed in other ethnic groups. For example, the Maya agricultural system miminized effects on natural vegetation by leaving the soil fallow for several years. This

Table 1.3. Droughts in pre-Hispanic Mexico

Probable dates	Affected regions
1500 BC–AD 300	Valley of Mexico. Teotihuacan
1052	Valley of Mexico
1064	Valley of Mexico, Coatepec
11th Century	Valley of Mexico
1287	Valley of Mexico, Otlaltepec
1328	Valley of Mexico, El Peñolcillo
1332	Coyoacan, Chalco
1347	Valley of Mexico, Chalco
1448	Valley of Mexico, Chalco
1450	Valley of Mexico and many other places
1454	Valley of Mexico
1460	Tenochtitlan
1502	Valley of Mexico, Cocollan and Atenchicalcan
1514	Tenochtitlan

Source: Sánchez (1995).

practice enhanced vegetative regeneration via stump resprouting and prevented clearcutting of useful trees.

This first stage of forest use ended at the beginning of the sixteenth century, when the Basin of Mexico was still composed of large forested areas with cypress, cedar, pine, oak, and *oyamel* (sacred fir) forests that were used for construction and, in the case of the *oyamel*, for its resins (Valero de García-Lascuraín 1990). In fact, every 80 days, Tenochtitlán received from the three regions of the Alliance 1200 wooden beams, 1200 thick planks, and 1200 thin planks, used in walls, ceilings, and foundations. This illustrates the large amount of wood used in the city. Forest clearing was very extensive in the southern portion of the Valley. However, Martín del Campo (1954) states that these uses, including fire wood, were not excessive, since it is known that, at least in the time of king Netzahualcoyotl, devastating forest clearing was prohibited in Texcoco.

Forest exploitation was in the hands of the lower social classes, and in some cases, they combined agriculture with their services of delivering wood to their lords. Forested areas were located mainly in the south of the Basin (De Rojas 1986; Sahagún 1975), and the most important markets were Coyoacán, Xochimilco, Acolman, and Ecatepec (Gibson 1984).

The stress on resources must have been very large before the Conquest, because according to Hassig (1985), the population reached up to 25 million people in the center of the country. This caused soil erosion affecting-agricultural yields in several regions (Denevan 1980). It is still not known, whether cultivation in terraced fields was designed to prevent erosion or was instituted in response to it. There are records indicating that the soil of central Mexico was already partially eroded when the Spaniards arrived and that the carrying capacity of the Valley of Mexico, under the multiple-use system in place at the time, was already near its limits. In spite of these factors, Hassig (1985) and Denevan (1980) indicate

Table 1.4. Major hemorrhagic fever epidemics in Mexico

Year	Number of deaths in the country (Gerhard 1986)	Number of deaths in the Basin of Mexico (Acuña-Soto et al. 2000)[a]
1545	12,000,000–15,000,000	800,000
1576	2,000,000	35,000
1736	200,000	40,000
1813	100,000	8,271

[a] The estimated number of deaths in the Basin of Mexico includes the areas of Xochimilco, México, Texcoco, Chalco, Cuautitlan, Mexicaltzingo and Coyoacán (according to Acuña-Soto et al. 2000).

that the pre-Columbian systems were based on ecological management of the natural resources using methods developed through centuries of trial and error, methods which allowed them to sustain a high population density without seriously damaging the environment.

After the conquest, agricultural activity intensified in previously-forested lands. The need for charcoal and firewood as fuel sources and the demand for other resources and cattle grazing greatly reduced the forest cover (Gutierrez 1989). As Villaseñor (1980) pointed out, the Spaniards buried poles along the sides of paved roads as a result of the constant flooding, because the ground did not provide a secure foundation. This practice required great quantities of wood. However, diverse laws were enacted for the protection of forests. One 1536 law required tree planting. In any case, the Colonial phase marked a stage in which there was a sudden increase in stress on the forested zones followed by a marked decrease due to the accelerated decline in the population of the Basin as a result of slavery, environmental imbalances, and successive disease outbreaks. At least 11 outbreaks of hemorrhagic fevers (known by the Nahuatl Indians as *huey cocoliztli* or the "great pestilence") between 1545 and 1813 have been reported, four of which (1545, 1576, 1736, and 1813) devastated the country (Table 1.4). These epidemics have been associated with periods of extreme drought and flooding, which to date have been little studied (Acuña-Soto et al. 2000). Conventional wisdom has long held that imported diseases such as smallpox and measles were responsible for tens of millions of deaths of native Mexicans, but recent studies support the hypothesis that many of the deaths were from the hemorrhagic fever outbreaks, which may have been caused by an indigenous agent such as a virus spread by rodents (Showstack 2001).

Regulations of the City Council of Mexico City helped to control forest harvesting by prohibiting tree cutting in a 20 km zone surrounding the city (Payno 1870). However, exploitation of the forests by the indigenous people continued, and in a 340-year period between 1524 and 1864, timber demand for construction, mines, and as fuel required 174 million trees. In 1870, Mexico City, with 170,000 inhabitants, consumed 4600 ton of wood and 13,800 ton of charcoal annually (Payno 1870). The dimensions of the problem can be approximated if we consider that, by the end of the eighteenth century, 25,000 trees per year were used just for stakes. In fact, more than 60 stakes were produced per 3 m^2 (Alzate 1792).

In the nineteenth century, as indicated by Rzedowski et al. (1964), the mining boom was another devastating factor for the forest regions of Pachuca in the north of the basin. This area had been spared during Colonial times because of the regulation that prohibited the native people, but not other Mexicans or foreigners, from cutting wood in areas neighboring the mines (Payno 1870). Industrialization, in the last quarter of the century, fostered the development of industrial establishments in the southeastern section of the city, which accelerated the deterioration of the forest resources.

Human Impacts and Management of Forests in the Twentieth Century

Aproximately 25 million inhabitants, one fourth of the nation's population, reside in the greater metropolitan zone in the center of the country. More than seventy percent of these people (18 million) live within the Basin of Mexico itself. This urban agglomeration has a profound effect on the surrounding rural and forested areas. Urbanization has lead to the disappearance of some types of plants, as a result of drastic anthropogenic modification of the environment, and caused displacement of forests to the mountain slopes. Within just the last 35 years, 17% of the forests, 40% of the mattoral, and 42% of the *chinampas* agricultural lands have been lost, and every year 7.4% of the corn fields are lost (Aguilar et al. 1995; Ezcurra et al. 1999). This loss is related to the urbanization process and the expansion of the industrial zone. Greater Mexico City now produces one fifth of the gross domestic product of the country, and this productivity also causes severe environmental degradation and contamination of soil, air, and water. Unless this trend is reversed, by the year 2010, there will only be 1.5 m^2 of green area per inhabitant (Government of the City of Mexico, 1997), although currently more than a fourth of the basin consists of forest cover (Table 1.5).

This reduction in green areas, especially severe in the last few decades, marks the third stage in the environmental deterioration occurring over the last 150 years. The first stage began in the latter half of the nineteenth century when industrialization began in the southern portion of the basin. Of the 17 textile and thread factories in Mexico City, five were in the south (Garza 1987), and these required the use of energy sources and materials from the local forests and streams.

Table 1.5. Land use in the Basin of Mexico

Land use	Surface area (ha)	Proportion (%)
Trees	276,798	27.4
Shrubs and herbaceous vegetation	88,531	8.8
Agriculture	541,623	53.6
Water bodies	5,956	0.6
Human settlements	96,712	9.6
Total	1,009,620	100

Source: Melo-Gallegos and Oropeza (1986).

Environmental deterioration accelerated as a result of exploitation of forests (previously in the hands of indigenous communities) by industrial activities that replaced agricultural production in many areas. Along with the introduction of the railroad, new industries led to accelerated urbanization, a process which still continues (Camarena 1996; Trujillo 1996). In this way, deterioration of the basin increased. Before the start of the twentieth century, signs of the upcoming catastrophe were already present.

At the beginning of the twentieth century, the second phase of environmental deterioration occurred with the desiccation of 10,000 hectares of Chalco Lake for the purpose of converting the land to agriculture. This project involved the construction of 203 km of canals, including the Great Drainage Canal Constructed in 1900 (Tortolero 2000), for drainage and for irrigation (Tortolero 1996).

Deforestation, loss of hydraulic control, and desiccation of the lakes gradually introduced new environmental deterioration beginning in the sixteenth century. Dust storms caused by winds blowing the sediments from the dry lake bed of Lake Texcoco have been observed since the eighteenth century to be a source of diseases. Another source is the deficient sewage system of the city, which deposited its effluents in the riparian areas to the east of the city causing fetid odors along a 2-km stretch (Lobato 1876).

When the construction of the Large Sewage Drainage Canal finally became a reality, the endeavor of getting the lake water out of the basin, so often desired, was finally achieved. But this accentuated the desertification process to such an extent that in the early 1970s, new plans and programs were designed to regulate urban expansion, to attenuate impacts on the environment, and to rescue what remained of the lakes. These efforts are still being carried out, although with variable and even debatable success, but they have served as the basis for new plans and programs intended to achieve a more holistic solution (González de León et al. 1998). In the late 1970s and early 1980s, forest decline, reportedly a consequence of the diverse impacts of air pollution, was observed in forests of the southwestern end of the basin (Alvarado-Rosales and Hernández-Tejeda 2002; Rodríguez 1996).

Currently, there are 88,500 ha of ecological conservation areas, most in the southern portion of the basin, which contribute to aquifer recharge and climatic regulation. Notwithstanding the forest set-aside areas, the forests in the basin are highly vulnerable, as seen by the more than 2000 forest fires that occur each year. The loss of vegetation cover is also observed as soil erosion to various degrees, with 10% of the land surface severely affected with losses greater than 200 tons/ha/yr (INEGI 1999).

Summary

The origin of the Basin of the Valley of Mexico is volcanic–tectonic, associated with the development of the Neovolcanic Axis in which it occupies the central area. This central area, located at 98°15′ to 99°30′ W and 19°00′ to 20°15′ N, has an area of 9600 km^2 at an altitude of 2240 m surrounded by mountains. It is an endoreic system of lakes at different elevations. The limnological changes in the Basin started 100,000 years BP with a succession of salty and freshwater environments accompanied by wide variations in the volume and the range of productivity, all conditioned by a variety of factors associated with the climate. Between 3000 and 1500 years BP, the climate was cold and damp, and floodings on the plains enhanced the growth of *Pinus–Quercus–Abies* forests. Later, between 1100 and 800 years BP, the weather was warmer and semihumid, the level of the lakes was high, and the forests were dominated by *Pinus* and *Quercus*. The gradual increase in temperature has caused the relocation of forests uphill where conditions have become favorable for *Pinus*, *Quercus*, and *Abies*. The origin of the flora in the Basin is related to that of the flora of the mountains and plateau of Mexico in general, and for that reason, no genus restricted solely to the basin is known.

Extensive agricultural cultivation was initiated between 1600 and 1200 BP, although human occupation of the basin occurred several thousand years BP. By the beginning of the sixteenth century, the pre-Hispanic civilization in the Basin effectively managed environmental resource types which were distributed in concentric bands around the lake covering more than 100,000 ha. The concentric bands were located along an elevational gradient: the urban area was integrated with the *chinampas* (floating garden) agriculture; croplands were located in the adjacent plains and were bordered upslope by the forest belt. Forests were located above 2500 m and provided the pre-Hispanic population of a million inhabitants with materials and energy. Demographic pressures and environmental changes (climatic and volcanic), did not prevent the development and maintenance of an advanced culture, though they imposed important restrictions on the expansion of the population.

The disappearance of native plants, which illustrates the degree of ecological disturbance in the Basin, is the result of drastic anthropogenic influences such as urbanization, industrialization, contamination of the environment, and over-use of the forests. As a result, forests have survived only in the higher montane sites. It is difficult to evaluate the degree of long-term change in vegetation in the Valley of Mexico, particularly in the earlier periods. It is also difficult to evaluate long-term changes in vegetation regulated by climatic change, by changes in the level of the lakes, and by alterations in physiography. Changes in vegetation due to these natural factors have been greatly accelerated by man. Thus, any proposed re-creations of past changes become approximations at best. However, it is believed that the *Quercus* forests once covered an area as great as that of *Pinus* and *Abies*. Urban growth and industrial growth in the last 100 years have broken the connection between forests and forest aquifers and much of the basin.

Acknowledgments. The authors gratefully acknowlege the assistance of Salvador Hernández Pulido in producing the figures and transcribing the manuscript into an electronic format. Likewise we wish to thank Mark Fenn for his help with the English language and for his patience.

References

Acuña-Soto, R., Calderón-Romero, L., and Maguire, J.H. 2000. Large epidemics of hemorrhagic fevers in Mexico 1545–1815. *Am. J. Trop. Med. Hyg.* 62:733–739.

Aguilar, A.G., Ezcurra, E., García, T., Mazari-Hiriart, M., and Pisanty, I. 1995. The Basin of Mexico. In *Comparisons of Threatened Environments*, eds. J.X. Kasperson, R.E. Kasperson, and B.L. Turner II, pp. 304–366. Tokyo: United Nations University Press.

Alvarado-Rosales, D., and Hernández-Tejeda, T. 2002. Decline of sacred fir in the Desierto de los Leones National Park. In *Urban Air Pollution and Forests: Resources at Risk in the Mexico City Air Basin*. Ecol. Stud., eds. M.E. Fenn, L.I. de Bauer, and T. Hernández-Tejeda, pp. 243–260. New York: Springer-Verlag.

Alzate R., J.A. 1792. Estanques para pescado. *La Naturaleza* 6:198–199.

Armillas, P. 1969. The arid frontier of Mexican civilization. *Trans. NY Acad. Sci.* 31:697–704.

Armillas, P. 1971. Gardens on swamps. *Science* 174:653–661.

Bernal, I. 1984. *Tenochtitlan en una Isla*. México: Fondo de Cultura Económica.

Boege, E. 1988. *Los Mazatecos ante la Nación. Contradicciones de la Identidad Étnica en el México Actual*. México: Siglo Veintiuno Editores.

Bradbury, J.P. 1971. Paleolimnology of Lake Texcoco. México. Evidence from diatoms. *Limnol. Oceanogr.* 16:180–200.

Buckler, E.S., Pearsall, D.M., and Holstford, T.P. 1998. Climate, plant ecology, and Central Mexican archaic subsistence. *Curr. Anthr.* 39:152–164.

Camarena, M. 1996. Fábricas, naturaleza y sociedad en San Ángel (1850–1910). In *Tierra Agua y Bosques: Historia y Medio Ambiente en el México Central*, ed. A. Tortolero, pp. 317–341. México: Instituto de Investigaciones José María Luis Mora.

Challenger, A. 1998. *Utilización y Conservación de los Ecosistemas Terrestres de México. Pasado, Presente y Futuro*. México: Comisión Nacional para el Uso y Aprovechamiento de la Biodiversidad and Universidad Nacional Autónoma de México (CONABIO UNAM).

Cook, S.F., and Borah, W. 1980. *Ensayos Sobre la Historia de la Población, México y California*. México: Siglo Veintiuno Editores.

Corona, E.C. 1977. Los sistemas de chinampas y las formaciones de estado en la Cuenca de México. *Biología* 7:27–33.

Cortés, H. 1994. *Cartas de Relación*. México: Porrúa.

Denevan, W.M. 1980. Chapter 9. Latin America. In *World Systems of Traditional Resource Management*, ed. G.A. Klee, pp. 217–244. New York: John Wiley and Sons.

De Rojas, J.L. 1986. *México Tenochtitlan. Economía y Sociedad en el Siglo XVI*. México: Fondo de Cultura Económica.

Díaz del Castillo, B. 1970. *Historia Verdadera de la Conquista de la Nueva España*. México: Porrúa.

Ezcurra, E., Mazari-Hiriart, M., Pisanty, I., and Aguilar, A.G. 1999. *The Basin of Mexico: Critical Environmental Issues and Sustainability*. New York: United Nations University Press.

Florescano, E., and Swan, S. 1995. *Breve Historia de la Sequía en México*. Xalapa: Universidad Veracruzana.

Foreman, F. 1955. Palynology in Southern North America. *Bull. Geol. Soc. Am.* 66:475–509.

García-Calderón, J.L., and De la Lanza, G. 1995. La cuenca de México. In *Lagos y Presas de México*, ed. G. De la Lanza, and J.L. García-Calderón, pp. 27–50. México: Centro de Ecología y Desarrollo.

Garza, G. 1987. La ciudad de México como centro comercial e industrial en el siglo XIX. In *Atlas de la Ciudad de México*, ed. G. Garza, pp. 85–88. México: Departamento del Distrito Federal.

Gerhard, P. 1986. *Geografía histórica de la Nueva España 1514–1821*, pp. 4–36. Mexico, D.F.: Instituto Nacional de Investigaciones Históricas, Universidad Nacional Autónoma de México.
Gibson, C. 1984. *Los Aztecas Bajo el Dominio Español 1519–1810*. México: Siglo Veintiuno Editores.
González A., L. 1988. *Plano Reconstructivo de la Región de Tenochtitlan*. Mexico: Instituto Nacional de Antropología e Historia.
González de León, T., Rosas, A., Kalach, A., and Quadri, G. 1998. *La Ciudad y Sus Lagos*. México: Clío.
González-Rul, F. 1998. *Urbanismo y Arquitectura en Tlatelolco*. México: Instituto Nacional de Antropología e Historia (INAH).
Government of the City of Mexico. 1997. *Programa de Ordenación de la Zona Metropolitana del Valle de México*. México: Gobierno de la Ciudad de México.
Gurría L., J. 1974. La conquista de México. In *Historia de México 4*, ed. M. Leon Portilla, pp. 17–40. México: Salvat Editores de México.
Gutierrez, P.A. 1989. *Conservacionismo y Desarrollo del Recurso Forestal: Texto Guía Forestal*. México: Trillas.
Hassig, R. 1985. *Trade, Tribute, and Transportation: The Sixtenth Century Political Economy of the Valley of Mexico*. New York: University of Oklahoma Press.
Hicks, F. 1982. Tetzcoco in the early sixteenth century: The state, the city, and the calpolli. *Am. Ethn.* 9:230–249.
Iglesias, M. 1902. *Memoria Histórica, Técnica y Administrativa de las Obras de Desagüe del Valle de México, 1449–1900*. México: Oficina Impresora de Estampillas.
Instituto Nacional de Estadística, Geografía e Informática (INEGI). 1999. *Estadísticas del Medio Ambiente del Distrito Federal y Zona Metropolitana*. Mexico: INEGI.
Leff, E. (ed.) 1990. *Medio Ambiente y Desarrollo en México*. México: Universidad Nacional Autónoma de México.
Lobato, J.C. 1876. Comparación de la clasificación higiénica de la capital de México. *Bol. Soc. Geog. Est.* 3:11–130.
Lot-Helgueras, A., Novelo-Retana, A. 1978. *Laguna de Tecocomulco Hidalgo, Méx.* Mexico: Sociedad Botánica de México.
Lugo H., J. 1984. *Geomorfología del sur de la Cuenca de México*. México: Instituto de Geografía. Universidad Nacional Autónoma de México.
Martín del Campo, R. 1954. Productos biológicos del Valle de México. *Rev. Mex. Est. Antrop.* 14:53–77.
Melo-Gallegos, C., and Oropeza, O. 1986. Vegetación, In *Atlas de la Ciudad de Mexico*, ed. G. Garza, pp. 33–36. Mexico: Departamento del Distrito Federal.
Metcalfe, S.E., Street-Perrott, F.A., O'Hara, S.L., Hales, P.E., and Perrott, R.A. 1994. The paleolimnological record of environmental change: Examples from the arid frontier of Mesoamerica. In *Environmental Change in Drylands: Biogeographical and Geomorphological Perspectives*, ed. A.C. Millington, and K. Pye, pp. 131–145. New York: John Wiley and Sons.
Miranda, F., and Hernández-Xolocotzi, E. 1963. Los tipos de vegetación de México y su clasificación. *Bol. Soc. Bot. México* 28:29–179.
Moncayo R., F. 1975. Las culturas indígenas mexicanas y los bosques. *México y sus Bosques* 14:13–22.
Mosser, F. 1963. La Cuenca Lacustre del Valle de México. In *Mesas Redondas Sobre Problemas del Valle de México*, ed. J.L. Tamayo, pp. 1–48. México: Instituto Mexicano de Recursos Naturales Renovables.
Mosser, F. 1975. Historia geológica de la Cuenca de México. In *Memoria de las Obras del Sistema de Drenaje Profundo del Distrito Federal*, ed. R.R. Elizondo, pp. 7–38. México: Departamento del Distrito Federal.
Niederberger, B.C. 1987. *Paleopaysages et Archeologie Pre-urbaine du Bassin de Mexico (Mexique)*. México: Centre d'etudes Mexicaines et Centroamericaines.

Novelo-Retana, A., and Gallegos, M. 1988. Estudio de la flora y la vegetación acuática relacionada con el sistema de chinampas en el Valle de México. Biótica 13:121–139.

Orozco-Berra, M. 1862. Memoria para la carta hidrográfica del Valle de México. *Bol. Soc. Mex. Geog. Est.* 9:337–480.

Palerm, A. 1973. *Obras Hidraúlicas Prehispánicas en el Sistema Lacustre del Valle de México*. México: Secretaría de Educación Pública–Istituto Nacional de Antropología e Historia.

Parsons, J.R. 1976. The role of chinampa agriculture in the food supply of aztec Tenochtitlan. In *Cultural Change and Continuity: Essays in Honor of James Bennett Griffin*, ed. C.E. Cleland, pp. 233–262. New York: Academic Press.

Payno, M. 1870. Bosques y arbolados. *Bol. Soc. Mex. Geog. Est.* 2:77–91.

Reiche, C. 1965. *Flora Excursoria del Valle Central de México*. México: Instituto Politécnico Nacional.

Rodríguez, D.A. 1996. *Incendios Forestales*. México: Mundi Prensa.

Rojas, T. 1990. *Agricultura Indígena: Pasado y Presente*. México: Ediciones de la Casa Chata. Centro de Investigaciones y Estudios Superiores en Antropología Social (CIESAS).

Rzedowski, J. 1978. *Vegetación de México*. México: Editorial Limusa.

Rzedowski, J., and Rzedowski, G.C. 1979. *Flora Fanerogámica del Valle de México I*. México: Compañía Editorial Continental (CECSA).

Rzedowski, J., Guzmán, G., Hernández-Corzo, A., and Muñíz, R. 1964. Cartografía de los principales tipos de vegetación de la mitad septentrional del Valle de México. *An. Esc. Nac. Cienc. Biol.* 13:31–57.

Sahagun, B. 1975. *Historia General de las Casas de Nueva España*. México: Porrúa.

Sánchez, M.E. 1995. Sequías registradas en la época prehispánica. In *Breve Historia de la Sequía en México*, ed. E. Flores-Cano, and S. Swan, pp. 165–169. Xalapa, Mexico: Universidad Veracruzana.

Sanders, W.T., and Webster, D. 1988. The Mesoamerican urban tradition. *Am. Anthr.* 90:521–546.

Sanders, W.T., Parsons, J.R., and Stanley, R.S. 1979. *The Basin of Mexico. Ecological Processes in the Evolution of a Civilization*. New York: Academic Press.

Secretaría de Recursos Hidraúlicos. (SARH). 1992. *Inventario Nacional Forestal de Gran Visión. Reporte Principal*. México: SARH.

Showstack, R. 2001. Natural disasters linked to collapse of some indigenous populations. EOS, Transactions of the American Geophysical Union 82:25, 27.

Street-Perrott, F.A., Perrott, R.A., and Harkness, D.D. 1989. Anthropogenic soil erosion around Lake Pátzcuaro, Michoacán, Mexico, during the preclassic and late postclassic–hispanic periods. *Am. Anti.* 54:759–765.

Toledo, V.M. 1997. La diversidad ecológica de México. In *El Patrimonio Nacional de México*, Vol. I, ed. E. Florescano, pp. 111–138. México: Consejo Nacional para la Cultura y las Artes. Fondo de Cultura Económica.

Toledo, V.M., and Ordóñez, M.A. 1993. The biodiversity scenario of Mexico. In *Biological Diversity of Mexico: Origins and Distribution*, eds. T.P. Ramamoorthy, R. Bye, A. Lot-Helgueras, and J. Fa, pp. 757–777. New York: Oxford University Press.

Toledo, V.M., Caballero, J., Mapes, C., Barrera, N., Argueta, A., and Núñez, M.A. 1980. Los purépechas de Pátzcuaro: Una aproximación ecológica. *Am. Indígena* 40:17–54.

Toledo, V.M., Carabias, J., Mapes, C., and Toledo, C. 1987. *Ecología y Autosuficiencia Alimentaria*. México: Siglo Veintiuno Editores.

Torquemada, J. 1975. *Monarquía Indiana*. México: Universidad Nacional Autónoma de México.

Tortolero, A. 1996. Los usos del agua en la región de Chalco 1893–1913: del Antiguo Régimen a la gran hidraúlica. In *Tierra, Agua y Bosques: Historia y Medio Ambiente en el México Central*, ed. A. Tortolero, pp. 219–251. México: Instituto de Investigaciones José María Luis Mora.

Tortolero, A. 2000. *El Agua y su Historia. México y sus Desafíos Hacia el Siglo XXI.* México: Siglo Veintiuno Editores.
Trujillo, M. 1996. Producción fabril y medio ambiente en las inmediaciones del Valle de México 1850–1880. In *Tierra, Agua y Bosques: Historia y Medio Ambiente en el México Central*, ed. A. Tortolero, pp. 343–360. México: Instituto de Investigaciones José María Luis Mora.
Valero de García-Lascuraín, A.K. 1990. Los indios de Tenochtitlan, la ciudad imperial mexica. *An. Est. Am.* 47:29–63.
Vazquez-Yanes, C. 1982. *Deterioro Ambiental.* México: Compañía Editorial Continental.
Villaseñor, R. 1980. Desarrollo histórico del subsector forestal. In *Memoria Aspectos Económicos y Sociales de la Actividad Forestal*, Publ. Esp. 18, pp. 35–45. México: Instituto Nacional de Investigaciones Forestales.
Watts, W.A., and Bradbury, J.P. 1982. Paleoecological studies at lake Patzcuaro on the West-Central Mexican Plateau and at Chalco in the Basin of Mexico. *Quat. Res.* 17: 56–70.
Williams, B.J. 1989. Contact period rural overpopulation in the Basin of Mexico: Carrying-capacity models tested with documentary data. *Am. Anti.* 54:715–732.

2. Socioeconomic Change and Its Impact on Forest Resources in the Basin of Mexico

Exequiel Ezcurra, Marisa Mazari-Hiriart, Irene Pisanty, and Adrián Guillermo Aguilar

Surrounded by the mountains of Mexico's Central Volcanic Axis, the Basin of Mexico remains, on all counts, the center of the nation. It is also the habitat in which one of the largest urban complexes in the world has developed (Pick and Butler 1997). Following up and summarizing previous studies (Aguilar et al. 1995; Ezcurra and Mazari-Hiriart 1996; Ezcurra et al. 1999), in this paper we shall discuss some of the socioeconomic issues related to the changes in forest ecosystems in the Basin of Mexico. The depletion of natural resources in the Basin of Mexico is not just a recent phenomenon. Rooted deep in the historic past, it has triggered massive emigration and various sociocultural collapses in the past (Ezcurra 1990a, 1990b, 1992; Whitmore and Turner 1986; Whitmore et al. 1990). Yet the scale, pace, and period of change currently under way are unprecedented, and the impact of urban growth on what remains of the surrounding forests is of primary importance.

Mexico City is a massive urban area. Originally contained within the boundaries of the Federal District (DF), during the 1950s, it spilled beyond these borders into the adjacent municipalities of the State of Mexico. The urbanized area contains 16 administrative subunits, called *delegaciones*, in the DF, as well as 29 municipalities (*municipios*) in the neighboring State of Mexico. Thus, the limit of the urban physical unit does not coincide with political administrative divisions. Larger, of course, is the Basin of Mexico, a hydrologically defined unit that encompasses 84 municipalities in the DF and the states of Mexico, Hidalgo, and Puebla. Mexico City, in which 93% of the total population is concentrated, represents the most important socioeconomic unit within the Basin. Thus, the Basin of Mexico and Mexico City are almost synonyms in demographic terms, but the latter is a subset of the former in geographic terms (Valverde and Aguilar 1987).

Historic Forest Use in the Basin of Mexico

Teotihuacan

In approximately the year 100 AD, Teotihuacan, situated in the northeast of the Basin, had some 30,000 inhabitants. Five centuries later, in 650 AD, its population of 150,000 made it one of the largest cities in the world (Millon 1970; Parsons 1976). A century later, the population of Teotihuacan had collapsed to fewer than 10,000. The cause of this decline is not well understood. Some researchers attribute it to the rebellion of subjugated groups; others, to the exhaustion of the natural resources exploited by the Teotihuacans. Ecologically, both hypotheses point to a similar underlying problem: The need for war tribute was the result of the inability of the Teotihuacans to survive on their own natural resources and of the need to import natural resources from other areas to support their economy. In either case, the exhaustion of local resources and the conflicts for the appropriation of foreign commodities appear as the driving forces of the collapse. According to Sanders (1976a, see also Sanders; Parsons and Santley 1979), the overexploitation of the forests and shrub-lands in the semiarid environment surrounding Teotihuacan and the lack of a sufficiently developed technology to exploit the fertile but flood-prone terrains of the Basin lake beds were decisive determinants in the sudden collapse of this civilization.

Good evidence shows that the Teotihuacan area was strongly disturbed at the time of the decline and remains so even today. The area is strongly desertified and completely lacks native woody vegetation. It is possible to calculate, for example, the amount of wood needed to calcinate the mortar and stucco used to construct the gigantic pyramids of the city as on the order of tens of thousands of tons (Cook 1947). Adding to this amount the amount of wood necessary to satisfy the domestic needs of 150,000 persons (a conservative estimate is 30,000 tons per year), it becomes clear that the development of Teotihuacan brought a large-scale process of deforestation in the north of the Basin, with erosion and loss of good agricultural soils as consequences. The total amount of wood used by Teotihuacan during its last century implies the cutting of 30,000 to 60,000 ha of dry pine and oak forests (Ezcurra 1992). Everything seems to indicate that resource exhaustion in general, and deforestation in particular, were key determinants of the decline of Teotihuacan.

Forests and Aztec Agriculture

Climate was restrictive for human settlements in the temperate zones of Mexico, where the dry season, starting in October and ending in May or June, allowed only one harvest for most crops. Many of the early inhabitants regarded the muddy bottom of the lacustrine system of the Basin as unsuitable for settlement. When the Aztec immigrants arrived in the Basin, they were displaced into the lowlands by established dominant groups already in the Basin. The lakes, however, proved to be the solution for coping with the seasonal droughts. The Aztecs developed a unique agricultural system, known as *chinampas*, consisting of artificial islands built with the rich silt of the swampy lakes. There was no soil limitation for the *chinampa* system, for as soon as soil fertility became impoverished by the crops, the silt was replaced by a new layer of rich lake-bottom sediment. Water shortages were also nonexistent throughout the year, since the shallow soil of the artificial islands absorbed water directly from the water table of the lakes. Nor were the *chinampas* washed out by water because trees with large root systems, mainly a willow (*Salix bomplandiana*), were planted along the borders to protect the plots from the eroding effects of water. Finally, the rich aquatic fauna of the canals also provided a valuable and rich source of additional food for the *chinampa* farmers.

Thus, by harvesting aquatic products and by digging the muddy sediments from the canals into the fields, the *chinampa* system was able to recycle in a very efficient manner the nutrients leached from the fertile soils of the agricultural fields. Although this system required a large amount of maintenance work (building the *chinampa*, cleaning the canals, protecting seedbeds from freezing during the cold winter mornings), it turned out to be one of the most productive Mesoamerican agricultural systems (Armillas 1971; Coe 1964; Sanders 1976a; Whitmore and Turner 1992).

The Colonial Period

After the Aztec conquest in 1521, the Mexican colony became a synthesis of both Aztec and Spanish traditions. Some persistent cultural differences, however, led to transformations of the landscape in general, and of the Basin of Mexico's forests. From early colonial times, it became clear that the Spanish city plan was not compatible with the lacustrine landscape of the Basin (Sala Catalá 1986). The conquest of Mexico by the Spaniards brought, among many other things, an introduction of alien plant and animal species as well as unfamiliar techniques for the use of natural resources. With the introduction of cattle came the first use of animals for labor and transportation. The filling of the Aztec canals and the *chinampa* fields to build elevated roads obstructed the surface drainage of the city and created large expanses of stagnant water. At the same time, the grazing and intensive logging of the slopes surrounding the Basin resulted in increased surface runoff and silting during the rainy season (de Humboldt 1811). The colonial city began to be troubled by severe floods. The first one occurred in 1553; floods recurred in 1580, 1604, 1629, and thereafter at shorter intervals. Growing deforestation was, to a large degree, responsible for these floods. Although many scientists warned the colonial authorities of the consequences of forest loss, the governmental response was to drain the Basin through the construction of an ever-growing, complex, and immensely expensive system of canals and tunnels. The works to drain the basin began in the early colony and have never stopped.

The Twentieth Century

After the end of the Mexican Revolution, in 1924, Mexico City began to grow rapidly. The pace of environmental deterioration increased rapidly with the development of the modern city. As in many other parts of Latin America, industrialization during this century accelerated migration towards the city, where job opportunities were higher. In its rapid growth, Mexico City began to engulf the satellite towns that surrounded the old Aztec capital, transforming itself into the immense megalopolis of today. Even the forests in the mountain ranges south and west of the city, little affected by the process of urbanization until the 1970s, began to suffer the consequences of explosive urban growth.

Realizing the environmental importance of the surrounding forests, President Lázaro Cárdenas (1934–1940) created various national parks in the mountains enclosing the basin. Two national parks, Desierto de los Leones and Cumbres del Ajusco, were created to the west and south as a way of restricting the deforestation of the basin slopes. Unfortunately, during the presidency of Miguel Alemán (1946–1952) a good part of the Ajusco park was given to a paper mill, which embarked on an ambitious timber–logging program (DDF 1986). The consequent elimination of the park and the ensuing deforestation of neighboring areas opened the way for urban expansion. The measure was again reversed in December 1988 when President Miguel de la Madrid decreed the ecological corridor Ajusco–Chichinautzin a protected area, justifying the measure on the need to maintain the recharge of the basin's aquifers, among other reasons.

Although substantial forested areas were logged during the colonial period, it was during the late twentieth century that accelerated deforestation began to have noticeable and dramatic impacts. More than 9000 hectares of trees disappeared in only 15 years between 1985 and 2000, and some insect pests that thrive on the trees debilitated by air pollution are attacking the protected forests. The widespread use of introduced and often invasive species such as *Eucalytus* to reforest the Basin has resulted in a loss of the native plant communities (Segura-Burciaga and Martínez-Ramos 1994) and of the animal species that relied on the native plants for food or shelter.

In its rapid growth, the city occupied land unsuitable for urbanization, such as the desiccated lake bed of Texcoco to the east and the forested mountain slopes in the south and southwest of the Basin (Figs. 2.1 and 2.2). To some extent the physical barriers imposed by the mountain ranges in the south and west have restricted urban growth. Extension of the urban expansion to the mountain slopes, however, has resulted in forest disturbance and deterioration of the important environmental services they provide.

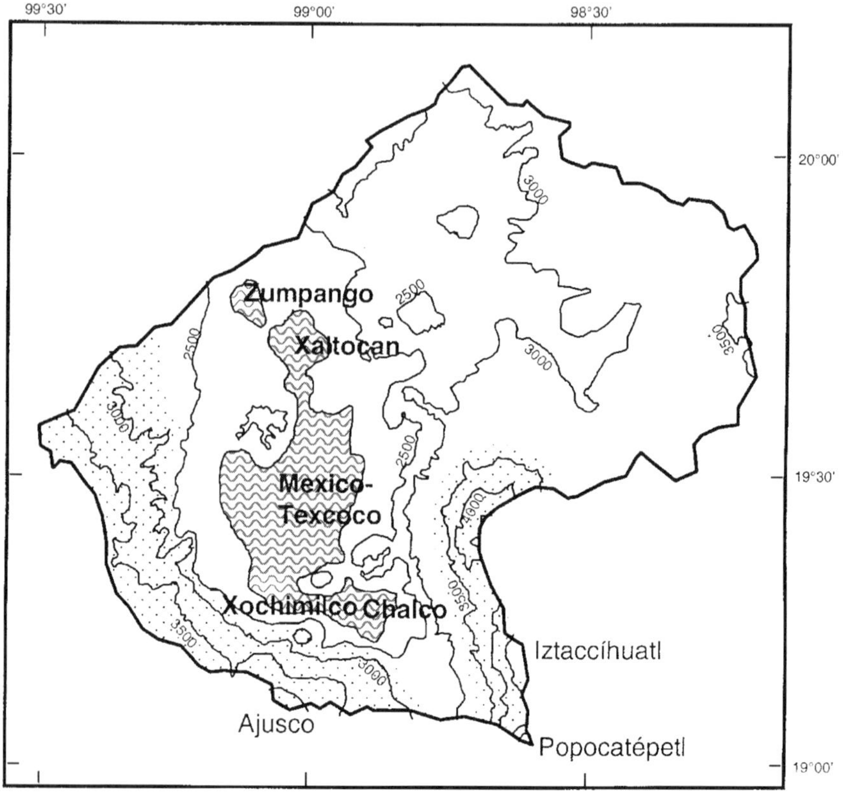

Figure 2.1. Map of the Basin of Mexico. The hatched area in the center of the Basin represents the original extension of the lakes at the time of the Spanish Conquest; the dotted areas in the southwest and southeast represent the forested areas of the Basin. The three highest volcanic peaks (Ajusco, Popocatépetl, and Iztaccíhuatl), which mark the boundary of the Basin, are indicate for references. (Modified from Ezcurra et al. 1999.)

Urban Growth and Social Distribution

Demographically, the Basin of Mexico concentrates almost 25% of Mexico's population, but its economic importance is even greater. In the 1940s, Mexico City constituted an economically optimal location for industrial development. Its extensive urban infrastructure, a concentrated market for industrial products, a variety of professional and financial services, and its abundant facilities for administrative transactions as seat of the federal government were all reinforced by a flow of government investments and fiscal incentives that stimulated and enhanced the concentration of economic development in the area.

In the early 1950s, industrial activity in the Basin became especially dynamic and, at a national level, reflected a clear trend towards concentration in the capital city. The dramatic industrial concentration in the capital during the present century has generated a situation in which the capital city produces almost half of the national industrial production. Recent efforts to redistribute industry elsewhere have apparently produced a relative decline in industrial growth in Mexico City and may have decreased the above percentages of concentration, but nevertheless the area remains heavily industrialized.

The city holds an enormous share of the major financial exchanges, private businesses, and central offices in Mexico. Mexico City also has the largest number of institutions of higher education and centers of culture. Indeed, compared with industry, services show an even higher tendency to concentrate in Mexico City. Many industrial firms that have relocated their manufacturing operations to outlying cities have kept their administrative headquarters in the capital.

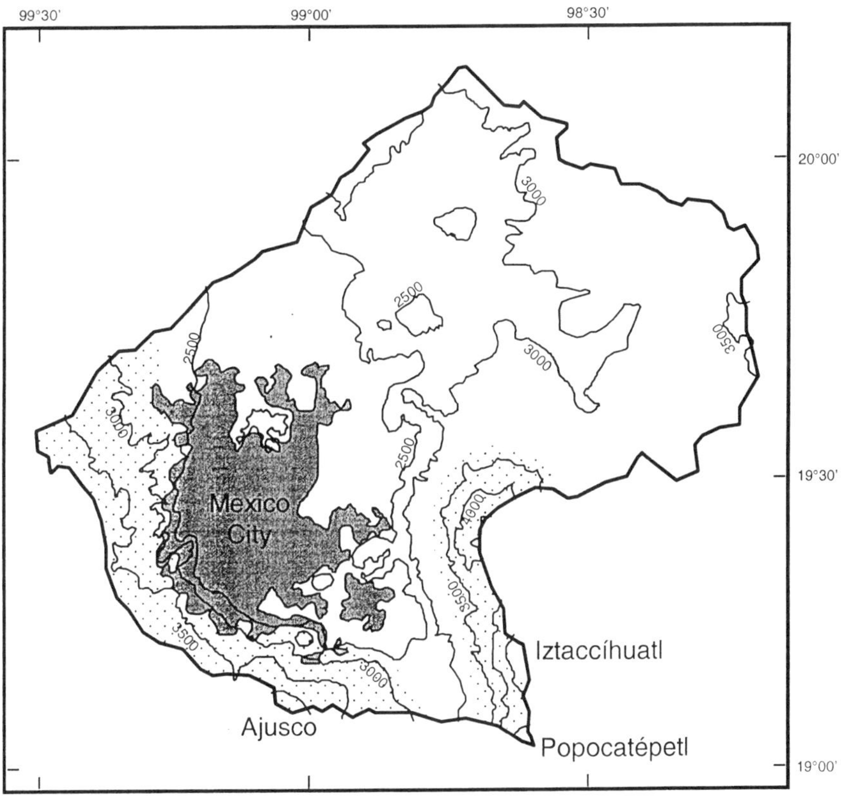

Figure 2.2. Map of the Basin of Mexico. The shaded area in the center represents the extension of the urbanized area in 1990; the dotted areas in the southwest and southeast represent the forested areas of the Basin (see for reference Fig. 2.1). Note the encroachment of the city's growth on the forested slopes of the southwest of the Basin. (Modified from Ezcurra et al. 1999.)

Urban Growth

With more than 15 million inhabitants in 1990, Mexico City is at present one of the largest cities in the world (Table 2.1). And despite a diminished population growth rate (Table 2.1, lower part), its total population is approaching 17 million. In the present century, the city has experienced a remarkable growth characterized by a great centralization of economic activities, notably industry, and high rates of population growth, driven to a large extent by migration of people from rural areas into the city. The population of the city reached one million inhabitants in 1930, more than 5 million in 1960, and more than 15 million in 1990.

However, the lure of the capital, a favorite destination for immigrants from rural or urban areas alike in 1940, gradually lost its appeal. The population began to notice that the concentration of economic, cultural, and political decisions was accompanied by negative side effects, perceived as a deterioration in the overall quality of life. In fact, the migration flows formerly destined for Mexico City now head for smaller metropolitan areas, where they represent an important component of rapid growth. Concomitantly, the capital city tends to export inhabitants, particularly of the middle class, into the periphery of the city or into other cities in the central region of the country, where the quality of life is perceived as more agreeable (Izazola and Marquette 1995).

Even so, the urban area still presents an active process of metropolitanization especially in neighboring territory of the State of Mexico. Although the growth

Table 2.1. Population in Mexico City 1519–2000, and growth rates 1950–2000

	Population (millions)		
Year	Federal District	State of Mexico	Total
1519 (Conquest)	0.3		0.3
1620 (Colony)	0.03		0.03
1810 (Independence)	0.1		0.1
1910 (Revolution)	0.7		0.5
1940 (Cardenist period)	1.8		1.8
1950	3.0		3.0
1960	4.8	0.4	5.2
1970	6.8	1.9	8.7
1980	8.8	5.0	13.8
1990	8.2	6.5	15.0
2000	8.6	9.1	16.7
	Mean annual growth rates per decade (%)		
1950–1960	4.7		5.5
1960–1970	3.5	15.6	5.1
1970–1980	2.6	9.7	4.6
1980–1990	–0.7	2.6	0.8
1990–2000	0.5	3.4	1.1

Source: DDF (1986), Ezcurra et al. (1999), DGE (1990), and INEGI (2000).
Pre-1950 dates have been chosen as approximate indicators and correspond with the important historical dates indicated in parentheses.

rate for the DF and, notably, the central city was negative over the years of 1980 to 1990, the municipalities in the neighboring State of Mexico are still growing at high rates (2.6% per year on average), and some of them have growth rates higher than 10% per year (Ezcurra et al. 1999). It is in these peripheral areas that urban growth is currently causing the stronger impacts on forests and natural ecosystems in terms of land occupation.

In spite of the lagging growth of the city itself, as other nearby metropolitan areas like Cuernavaca, Puebla, Tlaxcala, Pachuca, and Querétaro consolidate their functional integration to the country's capital, the urbanized areas around Mexico City will continue to grow (Garza 1987). In 1990, the metropolitan area concentrated 93% of the inhabitants of the Basin. In later years, however, population in the rest of the Basin has grown at a higher rate than in Mexico City itself (Ezcurra et al. 1999). Should this pattern persist, it may unfold a future characterized by more disperse settlements throughout the Basin. This more extensive pattern of land use may in turn put more pressure on the Basin's natural ecosystems and especially on the peripheral forests.

Physical Expansion

The spatial development of Mexico City until 1950 was characterized by a pattern of concentration in the four central *delegaciones* of the DF which contained around 70% of the urban area at that time. Thereafter the expansion of the city underwent rapid suburbanization that affected the surrounding *delegaciones* and finally produced a spill-over of population into the adjacent municipalities of other states. In this process, the city absorbed agricultural land. Land unsuitable for urbanization, such as the desiccated lake bed of Texcoco to the east, a group of abandoned sand quarries and open-cast mines to the west, and the forested mountain slopes in the south of the Basin, were also converted. To some extent, physical barriers in the south and west have contributed to the restriction of urban growth, because the large and flat valley of the Basin (approximately 2250 meters above sea level) is surrounded by volcanic mountain ranges that open only to the north and the southeast.

The Spanish established the colonial spatial structure of the city with their square grid layout for the central urban area organized around a central square (the *Zocalo*), which had formerly been the ceremonial center of the Aztec city. During colonial times, and even through most of the nineteenth century, this basic structure suffered only minor alterations. The nationalization of the large church landholdings in 1856, however, set important changes in motion. Land around the city was subdivided and sold off, and the upper classes moved to the periphery from their mansions in the city center. The downtown area, in turn, gradually deteriorated and became subdivided for low-income groups. The movement of the rich classes, largely towards the western forested slopes of the Basin, was accelerated by the purposeful construction of roads and avenues (Bataillon and Rivière d'Arc 1973).

The urban area of the Basin expanded dramatically from 360 km^2 in 1960 to 650 km^2 in 1970 (a 57% increase), and to 1114 km^2 in 1980 (an additional increase of 71%) (Graizbord and Salazar 1987). Accelerated urban growth led to the proliferation of illegal, usually poorly urbanized, settlements either in the State of Mexico or in the outskirts of the DF (Ward 1981). Between 1970 and 1975, some southern and western *delegaciones*, such as Tlalpan, Xochimilco, and Cuajimalpa, that harbor a large proportion of the Basin's forests, doubled or tripled their urban area (DDF 1986). This expansion has been in part stimulated by governmental decisions such as the construction of a paved highway to the Ajusco mountain range and the massive construction of large lower-middle-income housing units (mostly for governmental workers) in peripheral locations (DDF 1986). With these new urban expansions, Mexico City began to become connected across the mountain slopes with the states of México, Morelos, Puebla, and Hidalgo.

In its outward growth, the city is also changing internally. The central city is experiencing a depopulation process that started slowly in the 1970s and has continued at progressively higher negative growth rates. In a similar way, the inner ring of the city (but outside of the central city), after receiving a considerable population increase in the 1940 to 1960 period, has began to show the first signs of a depopulation process with negative growth rates after 1980. The same decade reveals particularly high growth rates in the intermediate and outer rings, with the former registering a growth rate of 12% and the latter a rate of 7%. Although these high rates decreased to between 4% and 6% in the following decade, the peripheral municipalities still have the fastest-growing populations of the megalopolis, enhancing the trend to incorporate contiguous municipalities (Ezcurra et al. 1999).

Changes in the Basin's Forests

Broad-Leafed Montane Forests

According to Sanders (1976b), by the time humans colonized the Basin of Mexico, particularly in the Texcoco region, a conifer forest covered the mountain ridges and the higher piedmont, whereas a moist broad-leafed (mesophyllous) forest, rich in oaks, probably covered the lower. Rzedowski (1969) reports that many of the moist ravines in the south and southeastern parts of the basin might have been occupied by this type of forest. Nowadays, the mesophyllous forest is best represented on the lower western slopes of Iztaccíhuatl, as well as on some of the eastern slopes of the Sierra de las Cruces at elevations between 2500 and 3000 m, but not more than 2 km^2 are occupied by this type of vegetation (Ezcurra 1990b). Near Amecameca, a relatively large town in the lower part of the Sierra Nevada, a small forested slope known as Sacromonte (the "Sacred Mount") rises from the alluvial plain. Though it has been deeply disturbed, some elements (e.g. *Quercus rugosa* and *Cupressus lindleyi*) of the original mesophyllous forest still grow on the slope. In contrast to most forests in the Basin of

Mexico, this one is rich in epiphytes and ferns. Rzedowski (1969) is of the opinion that this type of vegetation must have been better represented and more continuously distributed in the Basin of Mexico in the past. As a consequence of the deep disturbance it has suffered, probably many of the original elements have been lost.

Pine/Oak Forests

According to Rzedowski (1978) and Domínguez (1975), it is probable that many of the pine communities present in the Basin are of secondary origin, and their characteristic co-occurrence with tussocky grasses known as "*zacatón*" (*Muhlembergia* sp.) is a symptom of disturbance. Wood is cut in these forests for burning and building and also for producing paper. Although today oaks (*Quercus* spp.) cover only dispersed patches of different sizes, they were once very abundant in the Basin. Communities characterized by different species of *Quercus* covered the lower parts of the mountains and some patches of the volcanic substrate formed after the eruption of the Xitle Volcano (Álvarez del Villar 1971; Rzedowski 1975, 1978).

Conifer and oak forests have been deeply disturbed. Since the first settlements appeared in the Basin, these forests have been used as sources of wood for building and burning. Oak trees were also a main source for coal. Wood extraction had already been going for a long time when the Spaniards arrived in the Basin of Mexico. During the Colonial period, however, the rate of extraction increased significantly, because increasingly large amounts of wood were needed to build the new European-style houses of the conquerors. In the first half of the twentieth century, these forests were intensively utilized by three paper factories that established south and west of the city: San Rafael, Loreto, and Peña Pobre. Even zones officially designated as protected areas were given as concessions to the paper factories, as happened in San Rafael.

The Chapultepec Forest

Located to the west of the old downtown, Chapultepec Park is the largest forested section within the urbanized area of Mexico City. A recreational place as early as 1280, Chapultepec had, in Aztec times, a temple for the god Huitzilopochtli. Aztec emperors visited the area for its dense forest and its natural springs, which the ruling class used for bathing. At that time, the Chapultepec Forest was already planted with introduced, non-native plants. During the early colonial period, the Spanish conqueror of Mexico, Hernán Cortés, owned Chapultepec, but by 1530 the colonial government had designated it the public property of Mexico City and granted free entrance to everyone. Today, only 230 hectares of the 438 ha occupied by the original forest remain as a park, and a large proportion of this area has been invaded by exogenous species such as *Eucalyptus* spp.

The zone that surrounds what is now Chapultepec gained economic value towards the end of the Revolution (*circa* 1920), as people from the rural *haciendas* of the Mexican highlands fled from the violent struggle dominating the countryside and sought security in the Basin of Mexico. Many of the workers from the *haciendas* also fled the Revolution and crowded into the center of the city, prompting the upper classes to seek out and retreat to relatively secluded places. In fact, these migratory forces were one of the major factors giving rise to the incorporation into Mexico City of towns such as Coyoacán, Tacubaya, San Angel, Tizapán, and Tlalpan. By 1930, the pursuit of suburban plots in Mexico City had driven people into the green, mild zones west and south. One of the first manifestations of this preference was Lomas de Chapultepec, a trendy district developed in the 1920s by the new and wealthy postrevolutionary classes, built on lands surrounded or still occupied by the conifer forests that originally characterized the Chapultepec forest. As its name indicates, Lomas de Chapultepec emerged in the gently sloping hills where conifers such as sacred firs (*Abies religiosa*) and the native *ahuehuetes* (*Taxodium mucronatum*) originally grew (Suárez 1974).

Woody Shrub-Lands

Until the beginning of this century, the lava badlands of *El Pedregal*, in the southern part of the Basin, and their shrubby plant communities (see Rzedowski 1954) were considered a hostile environment, and no urban developments were attempted in them for many years. The lack of water flows or lakes, the high permeability of the rocky uneven substrate, and the abundance of dangerous animals, especially snakes, kept people away from *El Pedregal* for centuries. During colonial times, a few important *haciendas* (rural estates) were established in the parts where some soil had developed. The most important *haciendas* were located in zones with oaks and pines. The surrounding zones, like Coyoacán and San Angel, became wheat producers during the colonial period, and the occupation of these lands by large tenants forced the original inhabitants to move to the less favorable environments of *El Pedregal* (Carrillo 1995).

In the early 1950s, an affluent residential district, characterized by large houses of modern architecture with big gardens, began to emerge in *El Pedregal*, which still enjoyed a reputation as a secluded urban district, removed from the turmoils of the growing city. In fewer than 20 years, the original shrubby vegetation cover of the lava flow almost disappeared. Today fewer than 3 km^2 of the original *Pedregal* remain. Small patches can be found on some slopes and even in some gardens, but most of what is left is on the grounds of the campus of the National University (UNAM). The campus was built in the 1950s directly on *El Pedregal*, despite the protests of a small group of citizens that were already aware of the biological importance of this ecosystem. Paradoxically, it has proven crucial that the remnants of *El Pedregal* are on University grounds, because in 1983, a group of students and professors reacted firmly to a proposed urban development of this zone and succeeded in having it declared a protected area, only used for education and research (Rojo 1994). This is one of the few cases in the Basin of Mexico in which the construction of streets and commercial centers has been stopped or redesigned in order to protect a natural zone. Today, the natural reserve harbored by UNAM is the largest natural protected area to be found within an urban area in Mexico. Though small, and insufficient as a formal natural reserve, this area has become extremely important for the conservation and protection of the diversity that evolved around the Xitle lava flows (Rojo 1994). Six years after the area was protected, in 1989, a higher part of *El Pedregal*, in which deep disturbance had already taken place, was declared a protected area, the *Parque Ecológico de la Ciudad de México*. As a result, patches of open and dense oak forests, as well as some well-preserved patches of the characteristic *Sedum* shrubs, enjoy protection. Other patches, colonized by *Buddleia* or covered only by naked basalt and some rare rock vegetation, are also frequent in this region. In the Sierra de Guadalupe, a small representation of the natural xerophilous shrub-lands and some patches of *Eisenhardtia polystachia*, once an abundant dominant species, are still present. As in *El Pedregal*, this shrubby vegetation is rapidly disappearing as urban and industrial settlements appropriate the land.

Riparian and Lacustrine Forests

The lower regions of the Basin were covered by discontinuous vegetation types, including willows and other riparian trees along the borders of the lake, oak forests in the drier soils, and seasonally flooded patches occupied by aquatic species. Aquatic vegetation was once abundant in the Basin of Mexico; Rzedowzki (1975) presumes the existence of several endemic species that are now locally extinct. Some of these inferences are based on pollen profiles for the Teotihuacan Valley (Sanders et al. 1979), where the first large Mesoamerican urban settlement was established (Niederberger 1987a, 1987b). In these profiles, pollen of arboreal genera such as *Pinus*, *Abies*, *Quercus*, *Alnus*, and *Salix* was found together with that of herbaceous species. The abundance of pollen of sedge species (family Cyperaceae) in the older sediments is especially important, because it highlights the large original extension of the lake system.

However, the abundance of the pollen of aquatic species tends to diminish as corn records increase. This suggests that the shores of Lake Texcoco were gradually occupied by cornfields which replaced the riparian forests and the aquatic vegetation of the seasonally flooded shores. That is, the decline of riparian and lacustrine forests started far before the arrival of the Spaniards to Mexico. The ancestral springs and water-flows of Teotihuacan allowed intensive cultivation of the adjoining lands (Millon 1970; Sanders 1976a). The loss of these springs was probably related to the environmental degradation induced by human overuse of the natural resources, which seems to have played an important role in the decline of Teotihuacan (Ezcurra 1992, 1995; Sanders 1976a).

The Driving Forces of Environmental Change

The rapid rise and the enormous power of the Aztec state were based on the political control of much of Mesoamerica and on the subordination of hundreds of different groups that paid tribute to the emperor. Aztec wealth depended to some extent on the concentration of high-quality goods (e.g. metals, obsidian, tropical fruits, high-protein food) and labor collected as tribute from conquered groups. The Basin of Mexico, which initially allowed the rise of the Aztec culture through the appropriation and use of the highly productive *chinampa* technology, became a partly subsidized ecosystem receiving inputs of matter and energy from other areas.

This tradition, maintained under Spanish rule, has now reached overwhelming proportions. Few ecosystems in the world are so altered and so remote from self-sufficiency as the Basin of Mexico is at present. With much of the forests cut, most of the *chinampa* lands turned into urban developments, and practically all of the lakes dried up, the supply of raw materials and energy generated within the Basin does not suffice for even a small fraction of the almost 20 million residents. Thus, the Basin of Mexico imports vast amounts of food, energy, wood, water, building materials, and many other products from other ecosystems that, in effect, subsidize the water and energy flows of the Basin. With 22% of the population of the country, the Basin consumes around 27% of the country's oil and nearly one third of its electricity, both imported from outside the metropolitan region. Mexico City is at the hub of an immense system of concentration of power and work force, and also of appropriation of natural resources. The social domination of some groups by others is expressed in geographic and environmental terms, as a system of appropriation of natural resources from other regions and other social groups.

The ultimate cause behind the driving forces of environmental change is an immense system of subsidies, both in terms of natural resources (ecological subsidies) and in terms of financial resources (economic subsidies). We can, however, recognize a complex set of mechanisms that drive accelerated environmental change in the Basin of Mexico (Ezcurra et al. 1999). Of these, we shall discuss here two that are especially relevant for the problem of forest degradation: population change and accelerated economic growth.

Population Change

The last census reported a global population of 16.7 million for the whole megalopolis in 2000 (Table 2.1). While the DF is growing at a very slow rate, the *municipios* of the bordering State of Mexico are still growing at rates above 3%. There are, however, some uncertainties about the demographic data, especially for the DF. The official population count for the DF in 1980 was 8.8 million, while the count for 1990 was 8.2 million (DGE 1990). For those living in the city, however, this result does not seem plausible; during the period of 1980 to 1990, the global consumption of water and electricity by the city increased substantially, and the urbanized area grew considerably. Some recent governmental publications have now corrected the 1980 data, an action that suggests that the 1980 census was probably inflated. However, some researchers (e.g., Corona 1991) believe that the national 1990 census underestimated the total population of the country by some 2 to 6 million people. In spite of the potential errors in the censoring technique, we can conclude that the population of the metropolis is 17 million or more.

Between 1950 and 1980, the average annual growth rate of the population of Mexico City was 4.8%. However, the population has grown more quickly in the industrial zone of the State of Mexico, north of the DF, where the average rate of increase between 1950 and 1980 was 13.6%, compared to 3.3% in the DF. Much of the high growth rate of the City has been due to the continuous arrival of migrants from the economically depressed rural areas (Goldani 1977; Pick and Butler 1997; Stern 1977; Unikel 1974). In the period from 1950 to 1980, 5.43 million immigrants arrived in Mexico City. This influx accounts for 38% of the growth of the City (Partida 1987). In the one decade between 1970 and 1980, 3.25 million immigrants arrived in Mexico City (Calderón and Hernández 1987; Pick and Butler 1997). If this effect is taken into account, the intrinsic annual growth rate of the City for that period can be calculated as approximately 1.8%, considerably lower than the national average which was around 3.0% for the same period. It is immigration, more than reproductive growth, that has maintained the high rate of increase of the population of Mexico City (Goldani 1977). Between 1970 and 1980, approximately 650 babies were born every day, while some 1100 new immigrants arrived daily to stay in the Basin of Mexico. Of course, many of the newborn babies, were born to recently arrived immigrants, a phenomenon which further increased the growth rate of the population by elevating the natural birth rates. The growth rate for the period of 1980 to 1995 was around 1.0%, considerably lower than the rates for the previous decades. The population is still growing, but it is also quite clear that a demographic transition is in progress, and the growth rates seem to be slowing down.

The net population densities of Mexico City have stabilized since 1930 at around 14,500 persons/km^2. In the decade from 1920 to 1930, a dramatic decrease in population densities coupled with the introduction of inexpensive and efficient public transportation (trams and buses) into Mexico City changed the urban trace from a concentrated city that hinged around its downtown area into a more dispersed metropolitan area with a lower population density and a growing number of alternative commercial centers. This change in transportation technology was possibly the main factor leading to occupation of the forested slopes of the Basin of Mexico and the accelerated deforestation that ensued. In spite of this dispersement, the population densities of Mexico City are still high in comparison with other cities in the world (Ward 1991).

Severe environmental problems notwithstanding, the Mexican model of development has given priority to improving the quality of life in the large cities, where the social demand is more concentrated, at the expense of the rural areas, which have become comparatively poorer. From 1950 to 1980, the Basin experienced marked improvement in demographic and domestic indicators of quality of life, but the changes at the national level have been less dramatic. This difference in trends is, of course, much more pronounced if the Basin of Mexico is compared with the depressed rural areas from which most of the immigrants come. It is evident that such public services as education, drinking water, and drainage are rare in the poorer areas of central and southern Mexico that provide at a cheap price many of the products that are consumed in the Basin. These areas also generate most of the new immigrants that move into the area. In the Mixtec Region of Oaxaca, for example, the proportion of houses with drinking water is less than 4%, most of the towns do not have drainage, and the mean literacy and educational levels reach only elementary (primary school) level (Ezcurra et al. 1999).

In short, the population growth rate emerges as the main driving variable of environmental change. Between 1940 and 1980, with a growth rate of approximately 4% to 5%, the population doubled every 15 years, and with it the demand for water, electricity, transportation, housing, and other services. The number of cars has grown at a rate even higher than the population, doubling every decade. These accelerated growth rates have exacted a huge environmental toll, as more people necessarily use more resources and generate more waste and pollution.

Economic Development and Environmental Degradation

For nearly 2000 years, the Basin of Mexico has been one of the most densely populated areas of the world. During most of these 2000 years, the inhabitants of the Basin have used their political power to obtain advantages over other areas of the nation. The economic development of Mexico City after the Mexican Revolution (1930–1997) was promoted by governmental policies, local private enterprises, and foreign capital. The main objective has been massive industrial development, frequently at the expense of social equality. The direct driving force of urban growth has been the use of public resources allocated to the industrial and financial sectors. As a result of the rapid but highly unbalanced growth, the Basin now confronts an unsustainable pattern of urbanization and land use as well as an uneven and socially conflictive distribution of population and wealth.

Despite the existence of a wealthy social sector that includes government officials, private executives, public bureaucracy, and a large middle class with a high capacity to consume, most of the city consists of poor quarters inhabited by workers and underemployed people. These social groups consist mainly of immigrants that arrived in the metropolis in the decades between 1960 and 1980 and previously formed part of rural Mexico. They came to Mexico City in search of employment and some of the services and goods that the Basin seems to promise.

The concentration of wealth and the rapid development of the Basin was not accompanied by a concurrent emphasis on environmental protection. Decision-makers prioritized economic growth at all costs, while taking, in general, a complacent view in regard to environmental deterioration under the assumption that the technological development associated with rapid economic growth would eventually be able to restore degraded ecosystems and remediate environmental damage.

The Vulnerability of the Basin's Forests

Mexico City now spreads over approximately 1300 km^2 and has a population of around 17 million. The city now spills over the mountains into areas beyond the boundaries of the Basin, like the adjacent cities of Toluca and Querétaro and to a lesser extent Cuernavaca and Puebla, forming an immense conurbation complex. The accelerated peripheral expansion of the city has meant the deforestation of many of the surrounding areas which formerly acted as hydrological regulators for the severely disrupted water cycle of the basin.

Most (some 90%) of the megalopolis is occupied by built-up or paved urban spaces, whereas less than 10% of it is occupied by parks, private gardens, and public spaces. Taking into account the whole conurbation that forms this megacity, some 20 million people live in the Basin, enjoying only 5 m^2 of green area per person (Ezcurra et al. 1999). In some parts of Mexico City the situation is substantially worse; many inhabitants have less than 1 m^2 per person of open public spaces for recreational use. This is the case in Atzcapotzalco, a northern industrial quarter, and many other industrial areas. Mexico City has changed from the patchy mixture of urban and rural environments that was typical of the first half of the twentieth century to an overcrowded urban environment with little vegetation and open space for its inhabitants.

Unless new and more efficient wastewater treatment and recycling methods are adopted soon, meeting the growing demands for domestic water will require in the long term either a higher rate of pumping from the basin's aquifer (further increasing the current problem of subsidence) or, alternatively, the pumping of some 40 $m^3 s^{-1}$ of water from outside the Basin. Other external basins have been studied as potential water sources, but all these plans imply an enormous energy expense (on the order of 1.5 million MW-hour per year) to pump such vast volumes of water. Additionally, the extraction of water from other regions will necessarily generate significant ecological alterations and water shortages in the basins from where the water is to be pumped.

Factors Leading to Forest Endangerment

Most of the Basin's 20 national parks and protected areas were established between 1936 and 1939. Only one national park, Desierto de los Leones, had been established before this period (1917). Additionally, in 1983, a protected area was created in a small part of El Pedregal de San Angel, to protect the remnants of the vegetation and fauna that are endemic to the *pedregales* (lava flows) of the southern part of the Basin.

Other recently protected zones include part of the central region of the El Ajusco range, in the south of the metropolis in Lomas del Seminario. Illegal settlements had begun to be established there in the 1980s, endangering these humid southern ranges which are the most important source of water recharge into the Basin's aquifers, largely regulated by the existing remnants of the original vegetation. The area is now a reserve (*Parque Ecológico de la Ciudad de México*) with a center for research and educational activities. Other zones, such as El Cerro de Guadalupe, had also been colonized by squatters (*paracaidistas*), a process that in the past has been the spearhead of irregular urban expansion. The squatters have been moved, and the Cerro de Guadalupe is now being reestablished as a protected zone. The future of these reserves is not assured because population pressure on these areas is strong, and both economic and political interests could at some point acquiesce to their occupation by squatters. If this were to happen, the circle of unbridled growth and environmental destruction would recommence.

Most of the reserves in the Basin no longer represent true natural protected areas. Many are so deteriorated they can hardly be considered "green" areas anymore. Some of them have lost all their vegetation and been incorporated into the urbanized area. The deterioration of the national parks of the Basin of Mexico commenced almost immediately after their creation. For example, during the years of 1946 to 1952, one of Mexico's most important paper factories was officially allowed to log the forests in the protected parts of El Ajusco. Desierto de los Leones, Cumbres del Ajusco, Dinamos de Contreras, and Miguel Hidalgo are among the few zones that can still be considered effective protected zones, although they face the problem of increasing human use for recreation with increasing erosion from hiking trails and the additional problem of tree dieback owing to high levels of atmospheric pollution. The new reserves of *El Pedregal* and *Lomas del Seminario* have also managed to persist without being invaded or overly degraded by human use, although from a conservationist point of view, their size is much smaller than it should be for such a protected zone.

Most of the green areas in the Basin of Mexico are treated as recreational zones, with no ecological or conservationist purpose. Their management, design, and spatial distribution has been strongly dependent on recreational considerations. Green areas contain many nonindigenous species. This is the case with many of the most common tree species such as *Eucalyptus* sp., an Australian genus; *Schinus molle*, a South American tree; *Jacaranda mimosaefolia*, from Brazil; or *Ligustrum lucidum* from Asia. The original species of the Basin are only rarely used for planting, because the alien species frequently have higher growth rates, fewer predators, or both. These are important considerations, but this practice is accelerating the depletion of the original richness of environments and species in the Basin. The region is progressive losing its fauna because herbivores can no longer find the native plants from which they feed.

Summary

The urbanization process affects forests directly through the occupation and destruction of naturally vegetated areas. In the Basin of Mexico, air pollution, another consequence of urbanization, is also a debilitating stress on sensitive species of both natural and cultivated vegetation. Forest dieback in the surrounding mountains, triggered by atmospheric pollution, can in some places increase the degradation of the system of water recharge of the Basin's aquifer and hence its long-term hydrological sustainability.

The protection of these remaining reserves is of the utmost importance and must be achieved at all costs. Vegetated spaces are relevant not only to the quality of life of city dwellers but also to the hydrological cycle of the Basin. In the design of programs intended to rescue reserves invaded by urban settlements, the social factor merits consideration. People who are forced to leave these zones need an alternate place to settle without again damaging the environment. Their quality of life must also be improved with the move; otherwise, the occupation of the forested slopes of the Basin and the subsequent degradation of the Basin's forests will continue. It is evident that the protection of the few remaining forests in the Basin of Mexico and the restoration of as much as possible of those that have been affected by urban development should be a priority in any government program intended to rescue the Basin of Mexico from the ecological crisis it faces.

References

Aguilar, A.G., Ezcurra, E., García, T., Mazari-Hiriart, M., and Pisanty, I. 1995. The Basin of Mexico. In *Regions at Risk: Comparison of Threatened Environments*, eds. J.X. Kasperson, R.E. Kasperson, and B.L. Turner II, pp. 304–366. Tokyo: The United Nations University Press.

Álvarez del Villar, J. 1971. Panorama ecológico del Valle de México. In *IMERNAR, Mesas Redonadas Sobre Problemas de Ecología Humana en la Cuenca del Valle de México*, pp. 3–41. México, D.F.: Instituto Mexicano de Recursos Naturales Renovables.

Armillas, P. 1971. Gardens on swamps. *Science* 174:653–661.

Bataillon, C., and Rivière d'Arc, H. 1973. *La Ciudad de México*, Colección SepSetentas, no. 99, Translators C. Montemayor and J. Anaya. México, D.F.: Secretaría de Educación Pública.

Calderón, E., and Hernández, B. 1987. Crecimiento actual de la población de México. *Cienc. Desarrollo* 76:49–66.

Carrillo, C. 1995. *El Pedregal de San Ángel.* Mexico, D.F.: Coordinación de la Investigación Científica, Universidad Nacional Autónoma de México.

Coe, M. 1964. The chinampas of Mexico. *Sci. Am.* 211:90–98.

Cook, M. 1947. The interrelation of population, food supply, and building in pre-Conquest Central Mexico. *Am. Anti.* 8:45–52.

Corona, V.R. 1991. Confiabilidad de los resultados preliminares del XI Censo General de Población y Vivienda de 1990. *Est. Demográficos y Urbanos (El Colegio de México)* 6:33–68.

de Humboldt, A. 1811. *Essai Politique sur le Royaume de la Nouvelle–Espagne. (Avec un Atlas Physique et Géographique, Fondé sur des Observations Astronomiques, des Mesures Trigonométriques et des Nivellements Barométriques)*, 5 Vols. Paris: F. Schoell.

DDF. 1986. *Manual de Planeación, Diseño y Manejo de las Áreas Verdes Urbanas del Distrito Federal.* México, D.F.: Departamento del Distrito Federal.

DGE. (Dirección General de Estadística). 1990. *VI a XI Censo General de Población y Vivienda, Estados de Hidalgo, Mexico, Tlaxcala y el Distrito Federal, 1940–1990.* México, D.F.: Secretaría de Industria y Comercio e Instituto Nacional de Estadística, Geografía e Informática (INEGI).

Domínguez, V. 1975. *Estudio Ecológico del Volcán Popocatépetl, Estado de México.* Unpublished B.Sc. Thesis. México, D.F.: Facultad de Ciencias, Universidad Autónoma Nacional de México.
Ezcurra, E. 1990a. *De las Chinampas a la Megalópolis: El Medio Ambiente en la Cuenca de México*, Serie La Ciencia desde México. México, D.F.: Fondo de Cultura Económica.
Ezcurra, E. 1990b. Basin of Mexico. In *The Earth as Transformed by Human Action: Global, and Regional Changes in the Biosphere over the Past 300 Years*, eds. B.L. Turner II, W.C. Clark, R.W. Kates, J.F. Richards, J.T. Mathews, and W.B. Meyer, pp. 577–588. New York: Cambridge University Press with Clark University.
Ezcurra, E. 1992. Crecimiento y colapso en la Cuenca de México. *Ciencias* 25:13–27.
Ezcurra, E. 1995. Demographic, and resource changes in the Basin of Mexico. In *Global Land Use Change: A perspective from the Columbian Encounter*, eds. B.L. Turner II, A. Gómez Sal, F. González Bernáldez, and F. di Castri, pp. 377–396. Madrid: Consejo Superior de Investigaciones Científicas.
Ezcurra, E., and Mazari-Hiriart, M. 1996. Are megacities viable? A cautionary tale from Mexico City. *Environment* 38:6–15, 26–35.
Ezcurra, E., Mazari-Hiriart, M., Pisanty, I., and Aguilar, A.G. 1999. *The Basin of Mexico. Critical Environmental Issues and Sustainability.* Tokyo: The United Nations University Press.
Garza, G. 1987. Introducción. In *Atlas de la Ciudad de México*, ed. G. Garza, pp. 11–15. México, D.F.: Departamento del Distrito Federal—El Colegio de México.
Goldani, A.M. 1977. Impacto de los inmigrantes sobre la estructura y el crecimiento del área metropolitana. In *Migración y Desigualdad Social en la Ciudad de México*, eds. H. Muñoz, O. de Oliveira, and C. Stern, pp. 129–137. México, D.F.: Instituto de Investigaciones Sociales, Univ. Nac. Aut. de México, and El Colegio de México.
Graizbord, B., and Salazar, H. 1987. Expansión física de la Ciudad de México. In *Atlas de la Ciudad de México*, ed. G. Garza, pp. 120–125. México, D.F.: Departamento del Distrito Federal—El Colegio de México.
INEGI. 2000. *XII Censo General de Población y Vivienda, 2000.* Aguascalientes, México: Instituto Nacional de Estadística, Geografía e Informática (INEGI).
Izazola, H., and Marquette, C.M. 1995. Migration in response to the urban environment: Out-migration by middle-class women and their families from Mexico City after 1985. *Geog. Polonica* 64:225–256.
Millon, R. 1970. Teotihuacan: Completion of a map of the giant ancient city in the Valley of Mexico. *Science* 170:1077–1082.
Niederberger, C. 1987b. *Paléopaysages et Archéologie Pré-urbaine du Bassin de Mexico (Mexique)*, 2 Vols. Mexico, D.F.: Centre d'êtudes Mexicaines et Centraméricaines, Etudes Mésoaméricaines.
Niederberger, C. 1987a. De la prehistoria a los primeros asentamientos humanos en la Cuenca de México. In *Atlas de la Ciudad de México*, ed. G. Garza, pp. 40–43. México, D.F.: Departamento del Distrito Federal—El Colegio de México.
Parsons, J.R. 1976. Settlement, and population history of the Basin of Mexico. In *The Valley of Mexico: Studies in Prehispanic Ecology, and Society*, ed. E.R. Wolf, pp. 69–100. Albuquerque, New Mexico, USA: University of New Mexico Press.
Partida, V. 1987. El proceso de migración a la Ciudad de México. In *Atlas de la Ciudad de México*, ed. G. Garza, pp. 134–139. México, D.F.: Departamento del Distrito Federal—El Colegio de México.
Pick, J.B., and Butler, E.W. 1997. *Mexico Megacity.* Boulder, Colorado, USA: Westview Press.
Rojo, A. (ed.) 1994. *Reserva Ecológica "El Pedregal" de San Ángel: Ecología, Historia Natural y Manejo.* México, D.F.: Universidad Nacional Autónoma de México.
Rzedowski, J. 1954. Vegetación del Pedregal de San Ángel. *An. Esc. Nac. Cienc. Biol.* 8: 59–129.

Rzedowski, J. 1969. Notas sobre el bosque mesófilo de montaña en el Valle de México. *An. Esc. Nac. Cienc. Biol.* 18:91–106.

Rzedowski, J. 1975. Flora y vegetación en la cuenca del Valle de México. In *Memoria de las Obras del Sistema del Drenaje Profundo del Distrito Federal*, Vol. I, pp. 79–134. México, D.F.: Talleres Gráficos de la Nación.

Rzedowski, J. 1978. *Vegetación de México*. México, D.F.: Ed. Limusa.

Sala Catalá, J. 1986. La localización de la capital de Nueva España, como problema científico y tecnológico. *Quipu* 3:279–298.

Sanders, W.T. 1976a. The agricultural history of the Basin of Mexico. In *The Valley of Mexico: Studies in Prehispanic Ecology and Society*, ed. E.R. Wolf, pp. 101–159. Albuquerque, New Mexico, USA: University of New Mexico Press.

Sanders, W.T. 1976b. The natural environment of the Basin of Mexico. In *The Valley of Mexico: Studies in Prehispanic Ecology and Society*, ed. E.R. Wolf, pp. 59–67. Albuquerque, New Mexico, USA: University of New Mexico Press.

Sanders, W.T., Parsons, J.R., and Santley, R.S. 1979. *The Basin of Mexico: Ecological Processes in the Evolution of a Civilization*. New York: Academic Press.

Segura-Burciaga, S.G., and Martínez-Ramos, M. 1994. La introducción de especies a comunidades naturales: El caso de *Eucaliptus resinifera* Smith (Myrtaceae) en la reserva "El Pedregal" de San Ángel. In *Reserva Ecológica "El Pedregal" de San Ángel: Ecología, Historia Natural y Manejo*, ed. A. Rojo, pp. 177–186. México, D.F.: Universidad Nacional Autónoma de México.

Stern, C. 1977. Cambios en los volúmenes de migrantes provenientes de distintas zonas geoeconómicas. In *Migración y Desigualdad Social en la Ciudad de México*, eds. H. Muñoz, O. de Oliveira, and C. Stern, pp. 115–128. Mexico, D.F.: Universidad Nacional Autónoma de México, and El Colegio de México.

Suárez, L. 1974. *De Tenochtitlan a México*, Series Archivo del Fondo 16. México, D.F.: Fondo de Cultura Económica.

Unikel, L. 1974. *La dinámica del crecimiento de la Ciudad de México. Colección SEP–Setentas*. México, D.F.: Secretaría de Educación Pública.

Valverde C., and Aguilar, A.G. 1987. Localización geográfica de la Ciudad de México. In *Atlas de la Ciudad de México*, ed. G. Garza, pp. 19–22. México, D.F.: Departamento del Distrito Federal, and El Colegio de México.

Ward, P. 1981. Mexico City. In *Problems and Planning in Third World Cities*, ed. M. Pacione, pp. 124–147. London: Croom Helm.

Whitmore, T.M., and Turner II, B.L. 1986. *Population Reconstruction of the Basin of Mexico: 1150 B.C. to Present*. Technical Paper 1, Millennial Longwaves of Human Occupance Project. Worcester, Massachusetts: Clark University.

Whitmore, T.M., and Turner II, B.L. 1992. Landscapes of cultivation in Mesoamerica. In *The Americas before and after 1492: Current geographical research*, ed. K.W. Butzer. Special edition, Annals of the *Assoc. Am. Geog.* 82:401–425.

Whitmore, T.M., Turner II, B.L., Johnson, D.L., Kates, R.W., and Gottschang, T.R. 1990. Long-term population change. In *The Earth as Transformed by Human Action. Global, and Regional Changes in the Biosphere over the Past 300 Years*, eds. B.L. Turner II, W.C. Clark, R.W. Kates, J.F. Richards, J.T. Mathews, and W.B. Meyer, pp. 25–39. New York: Cambridge University Press with Clark University.

3. Physical Geography, Hydrogeology, and Forest Soils of the Basin of Mexico

Luis E. Marín, Oscar Escolero-Fuentes, and Antonio Trinidad-Santos

This chapter describes the physical, hydrogeological and edaphic characteristics of the Basin of Mexico and mountains which surround it to the east, south and west (Fig. 3.1). The description consists of the geographic position of the area, the elevation of the main points, geology, hydrogeological characteristics of the valley, and a description of soils found throughout the Basin with particular emphasis on forest soils.

The Valley of Mexico is located between 98°40′ to 99°25′ west longitude and from 19°05′ to 19°37′ north latitude. It lies within the Mexican Transvolcanic Belt (MTV), an area of approximately 105,000 km^2. Some of the highest peaks in Mexico are found within this area. The Valley of Mexico is located in an endoreic basin with a valley floor elevation ranging between 2240 and 2390 masl and an extension of 9600 km^2. It is limited to the west by the Sierra of Tepotzotlan, to the north by the Sierra of Tezontlalpan and the Sierra de Pachuca, to the south by the Sierra de Las Cruces, Zempoala, and Chichinautzin, and to the east by the Río Frío and the Sierra Nevada (Fig. 3.2). Mexico City is located in the lower part of the Basin and has a mean elevation of 2240 m above sea level (masl). The Popocatépetl Volcano rises to 5465 m, the highest elevation in the area, the Iztaccíhuatl Volcano rises to 5230 masl, and the Ajusco Volcano has an elevation of 3937 masl.

Within the MTV, uplift of the mountains and subsidence of adjacent areas formed closed basins. These basins were subsequently filled with alluvium and volcanic material from succeeding phases of erosion, sedimentation and volcan-

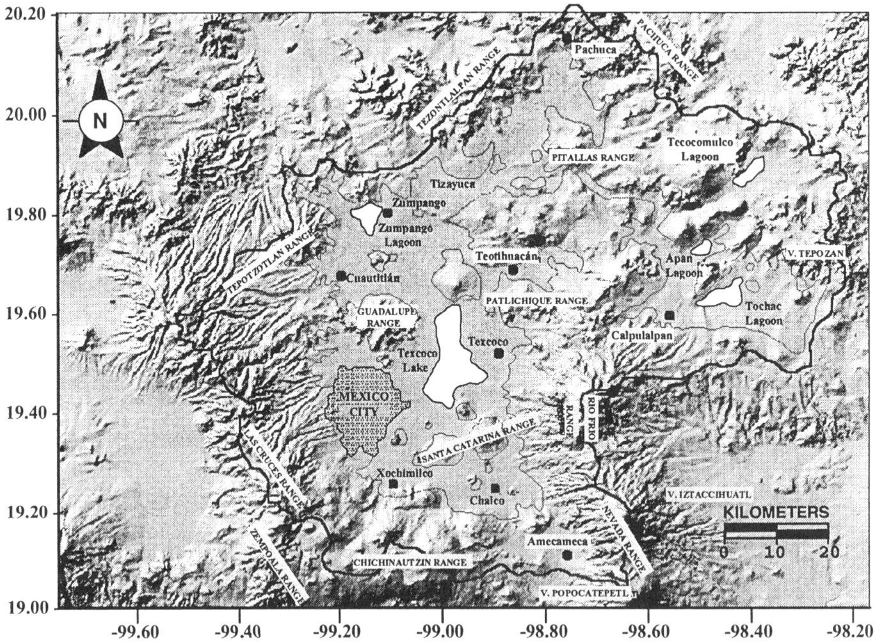

Figure 3.1. Three-dimensional figure of the Basin of Mexico.

ism. Drainage within the Basin of Mexico was closed due to subsequent volcanic extrusions. A large lake formed in the lower elevation of the valley, and great volumes of fine-grained sediments, interfingered with igneous rocks and alluvial materials, were deposited. Underneath Chalco (the southeastern part of the Basin of Mexico), for example, the approximate thickness of the clay deposits is on the order of 400 m (Ortega 1993).

The mountain ranges that surround Mexico City are composed primarily of extrusive igneous rocks. Within these rocks, Pliocene to Recent basalts are commonly found. The oldest igneous rocks that form the mountains are essentially impermeable; nevertheless, those that have been tectonically fractured or altered by weathering may have significant hydraulic conductivity values. Transient, unsaturated flow, a type of flow present primarily during the rainy season (May through September) is present in these rocks. As one approaches the piedmont, the flow regime changes to a more permanent regime, that is, quasi steady-state and saturated flow. Basalt flows, which are common in the mountains surrounding Mexico City, may also function as aquifers, particularly when they are fractured. Sánchez et al. (2000) report the following groundwater flow velocities for the different rock types found in the Basin of Mexico: lacustrine clays, 0.2 m/yr; alluvial deposits, 20 m/yr, and basalts alternating with pyroclasts, 750 m/yr.

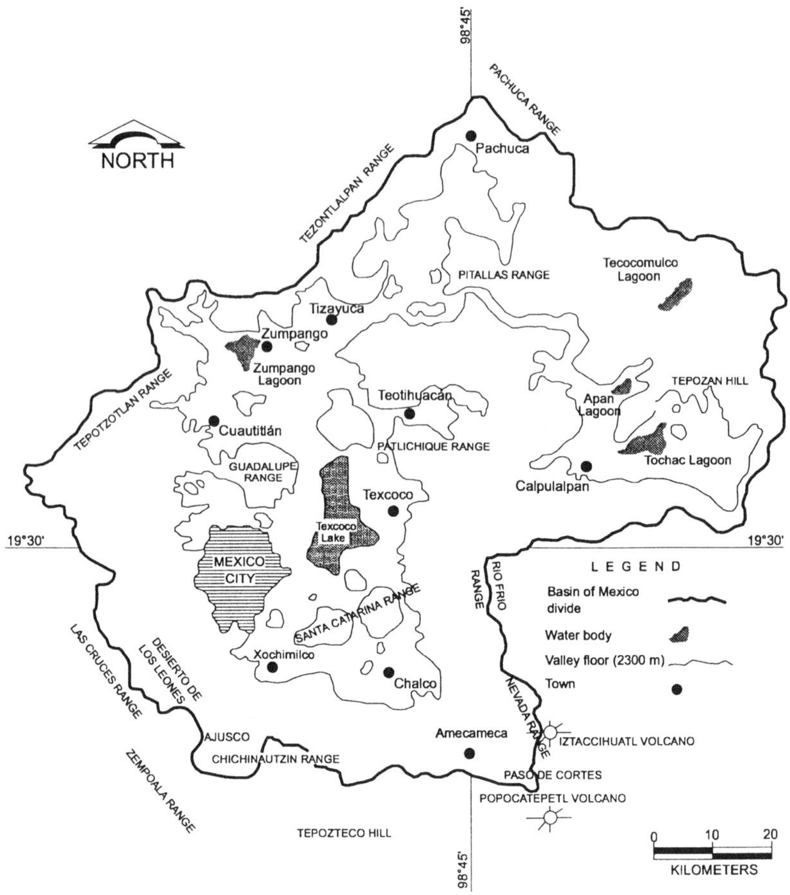

Figure 3.2. The Basin of Mexico, showing major mountain ranges (sierras) and water bodies (lakes and lagoons).

Hydrology

The lower part of the Basin is a flat lacustrine plain, that was covered by shallow lakes and swamps. Several of these lakes have dried up as a consequence of the human settlement of the Valley of Mexico from the time of the Aztecs through the present. Among these lakes, three important ones were the Lake of Mexico, the Lake of Xochimilco, and the Lake of Chalco. The latter one disappeared completely around the beginning of this century. Currently only two lakes remain within the Valley: Texcoco Lake and the Lagoon of Zumpango (Figs. 3.1 and 3.2). To the northwest of the basin, three depressions which have become shallow seasonal lagoons can be found: Apan, Tochac, and Tecocomulco (Figs. 3.1 and

3.2); these three disappear during the dry season (which typically runs from October through April).

There are no important rivers in the area which drain into the Valley of Mexico. There are, however, temporary streams and springs, particularly in the upper parts of the mountains, which carry water only during the rainy season (typically from May through September). These streams are more prevalent in the Desierto de los Leones (DL) National Park; some are also found at Los Dinamos, located between the DL and El Ajusco, and to a lesser extent in the Ajusco mountains. However, presently the water from these streams is intercepted and channeled into the water distribution system to supply drinking water for communities found within the mountain slopes. Thus, this water is no longer available for the forests that are found throughout this area.

Hydrogeology

Throughout the valley floor, there is a regional aquifer. Lesser et al. (1990) have subdivided the regional aquifer that underlies the Valley of Mexico into three subaquifers. These are (1) the granular aquifer found underneath the city, (2) the one found in the southern portion of the Valley, comprising the areas of Xochimilco–Chalco, and (3) the area around Texcoco Lake (Fig. 3.2). The regional aquifer is composed of fractured volcanic rocks which are covered by lacustrine and alluvial deposits of lower hydraulic conductivity values. For this reason, the aquifer within the Valley of Mexico is confined in some areas and semiconfined in others.

Recharge to the aquifer comes from the highest parts of the Basin and the hillsides of the three mountain ranges located to the east, west, and south (with the latter area providing the highest recharge to the aquifer). Historically, the whole valley floor has been the discharge zone for the Basin of Mexico.

At least two, and perhaps all three flow systems (local, intermediate, and regional), are present within the Valley of Mexico (Toth 1962). Local flow systems are common in the mountains (some are described by Cortes et al. (1997) and Morales (1998)), and intermediate flow systems occur along the mountain flanks (these currently recharge the regional aquifer). It is possible that water that recharges the intermediate flow systems may become part of the regional flow system. Marco Jacobo (personal communication) has suggested that recharge occurs at the top of the mountain ranges, and that this water discharges through springs in the valley floor. However, we were not able to substantiate this assertion. Furthermore, it seems unlikely that fractures and joints whose openings on the surface are at best tens of centimeters can be as deep as hundreds of meters.

Dominguez (1996) has described the general hydrostratigraphy of the Valley of Mexico. From the surface down it consists of:

1. Quaternary lacustrine deposits with a thickness that ranges between 0 and 400 m. These are semipermeable units that often have secondary permeability

from fracturing. These sediments alternate between clay layers (5–25 m thick) and volcanic sandstones, locally known as *capas duras* (hard layers).
2. Quaternary volcanic and pyroclastic material (0–2000 m). These rocks constitute the main aquifer in the southern part of the valley.
3. Quaternary alluvium deposits with a thickness that varies between less than 1 m to 10 m. These are found primarily along the mountain flanks. They are also the main fill of the valley and constitute the main aquifer system for the Valley of Mexico. Basement rocks consist of andesites and dacites.

Morales (1998) conducted a detailed hydrogeologic study in the Monte Alegre Valley, which is within the Ajusco range in the southwestern portion of the Valley of Mexico (Fig. 3.1). The soil cover in this area ranges from less than 1 m to more than 10 m. The rocks in this area are primarily fractured andesites. He mapped over 6000 joints and found that the predominant alignment of the joints follows the regional alignment, namely N 65° W. Morales (1998) described numerous springs which occur primarily along the contact between the soil and unfractured andesites (these are the local flow systems mentioned earlier). Other authors, such as Cortes et al. (1997), have described the presence of springs along fractures and in geologic contacts. Using stable isotope data, the latter authors also showed that the residence time of water in these springs is low, and that the water is of meteoric origin.

Water Quality

Cortes et al. (1997) report two different sources of water found in the subsurface of the Valley of Mexico based on stable isotope and major ion geochemistry. They were able to identify water of meteoric origin, which had a short residence time in the springs located along the mountain flanks that surround the Basin and water that had changed in chemical composition as it traveled through the rocks. Cortes et al. (1997) estimated that the average residence time for water in the Valley of Mexico is on the order of 50 years. Natural water quality is acceptable for human use (i.e., it typically has less than 500 mg/L TDS) except for the water found in the vicinity of Texcoco Lake, which typically have a TDS of 30,000 mg/L although concentrations as high as 130,000 mg/L TDS have been reported (Herrera 1995). Texcoco Lake has one of the lowest elevations of the Valley, and thus it is likely that groundwater discharge gravitates to this area. Evaporation of this water, with its high TDS load, leads to salt accumulation. Since the time of the Aztecs, this area has been mined for salt (Durazo and Farvolden 1989).

Few studies are available on water quality in the Basin of Mexico; however, Cardona and Hernández (1995) give some information on major ion chemistry. Most of the studies have dealt with physical hydrogeology issues (water balances, conceptual models, etc.). This is certainly an area that needs to be addressed in future investigations.

Land Subsidence

Due to the thick clay layers that are present throughout the Valley of Mexico, land subsidence became a major problem once groundwater extraction began on a regular basis in the middle of the nineteenth century. This problem (which continues today) became more acute in the 1940s when major ground water withdrawals from the regional aquifer were begun (these rates reached a maximum of 46 cm/yr in 1950–1951 [Herrera 1995]). In 1959, for example, the subsidence rate in the center of the Valley was on the order of 40 cm yr^{-1} (Durazo and Farvolden 1989). Birkle et al. (1998) report land subsidence greater than nine m in the Valley of Mexico as a result of groundwater withdrawals. Current land subsidence values for the Basin of Mexico range from 0 cm/yr (no subsidence) to more than 35 cm/yr in the Xochimilco area.

Hydrogeologic Evolution of the Valley

Mexico City, with a population of 8.5 million inhabitants within the city and approximately 20 million including the surrounding areas, obtains approximately 55% of its drinking water from groundwater (on the order of 19 m^3/s). As the population of Mexico City has continued to increase, so has the demand for groundwater. For example, as of 1988, Lesser et al. (1990) estimated that more than 33 million m^3 of groundwater were being withdrawn from storage annually, and that this volume is in excess of the recharge to the aquifer system. Arreguín-Mañon (1998) and Arreguín-Mañon and Terán (1994) discuss the recent hydrogeologic history of the basin.

Groundwater extractions began in 1847 with a well more than 50 m deep (Durazo and Farvolden 1989). Extraction increased dramatically in the 1940s, and consequently, land subsidence problems began to appear. As a result, groundwater extractions wells were moved from the center of the Basin to the southern portion of the valley. One of the immediate consequences of this new placement of the extractions wells, was that springs located along the flanks of the Chichinautzín mountains disappeared (Fig. 3.1), (Durazo and Farvolden 1989). Currently, the drawdown in the main aquifer is on the order of less than 1 to 3 m/yr (Herrera 1995). The unsaturated zone varies from 10 to more than 180 m from the surface.

Water Use in the Basin of Mexico

Initially, the whole floor of the Basin was the discharge zone. The calculated recharge rate is a maximum of 19 m^3/s in the Metropolitan Zone of the Basin of Mexico, whereas a recent estimate of the groundwater exploitation rate indicates that 51.4 m^3/s is being withdrawn from the Basin aquifer systems, resulting in a deficit of more than 30 m^3/s. Taking into account infiltration processes by leaking water supply systems, the calculated deficit is reduced to 20.5 m^3/s. Thus, ground-

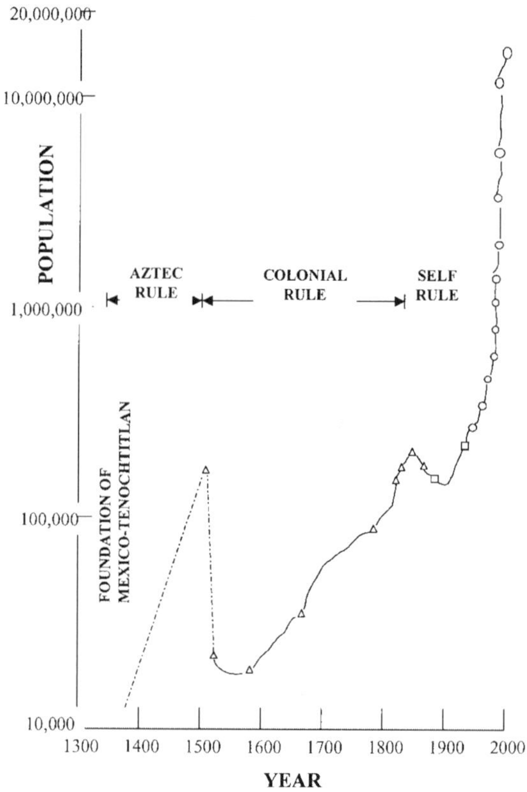

Figure 3.3. The approximate total population for the Basin of Mexico. (Modified from Durazo and Farvolden 1989.)

water is currently being mined at an approximate rate of 20.5 m^3/s (Birkle et al. 1998).

When the Spanish arrived in the 1500s, the Aztec population was approximately one million, with 30% living in the Valley of Mexico (Fig. 3.3). Large freshwater springs were common at the edge of the Valley where permeable aquifers pinch out. No hydrogeologic data is available from the time of the Aztecs, but we do know that numerous springs were present throughout the Valley. Durazo and Farvolden (1989) report potentiometric levels of 2.7 m above land surface for the center of the Valley in 1864 and a potentiometric surface of 3 to 7 m above land in 1954. Groundwater levels throughout the Valley have declined steadily, and current groundwater levels are declining on the order of 1 m/yr. One possible adverse effect on the forests located on the mountain flanks of the Basin is that water that perhaps would have flowed subhorizontally is now flowing vertically, reducing the availability of water for forest vegetation.

From the 1500s to the present, different types of engineering activities have taken place with the sole intention of draining the water from the Valley of Mexico to make the valley a more hospitable area to live in. Figure 3.3 shows how the population has grown from the 1500s through the present. Figure 3.4 shows how Mexico City has grown during the same time frame. Prol-Ledesma et al. (2002) report that the areal growth of Mexico City between 1970 and 1990 based on TM (Thematic Mapper) imagery has been on the order of 15%. One result of this uncontrolled growth on Mexico City, is that the mountain flanks, where the only remaining forests are, have become the prime residential areas. These are the areas where most of the recharge is concentrated (Robert Farvolden, personal communication). Thus, as the city continues to grow, it is both reducing the forest cover and the areas where most of the recharge is generated for the aquifer. Figure 3.5 shows this schematically. We suggest that a hydrogeologic reserve zone be established for the Valley of Mexico whereby strict land-use

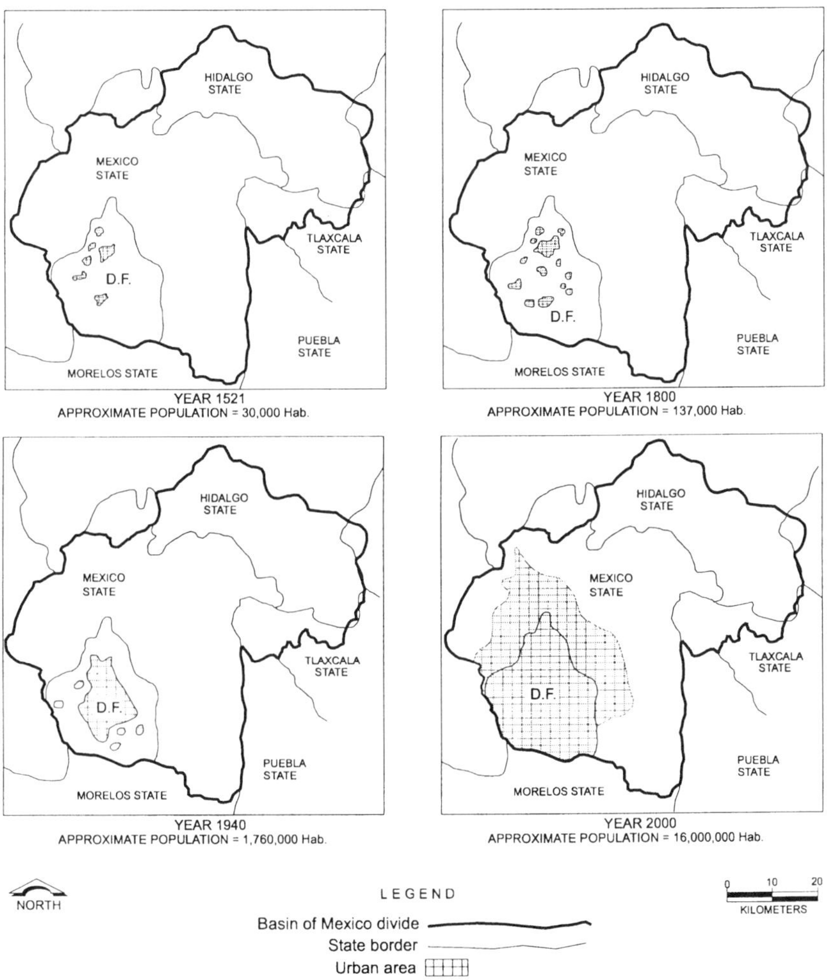

Figure 3.4. The approximate population for the Basin of Mexico for the years 1521, 1800, 1940, and 2000. (Modified from Herrera, 1995.)

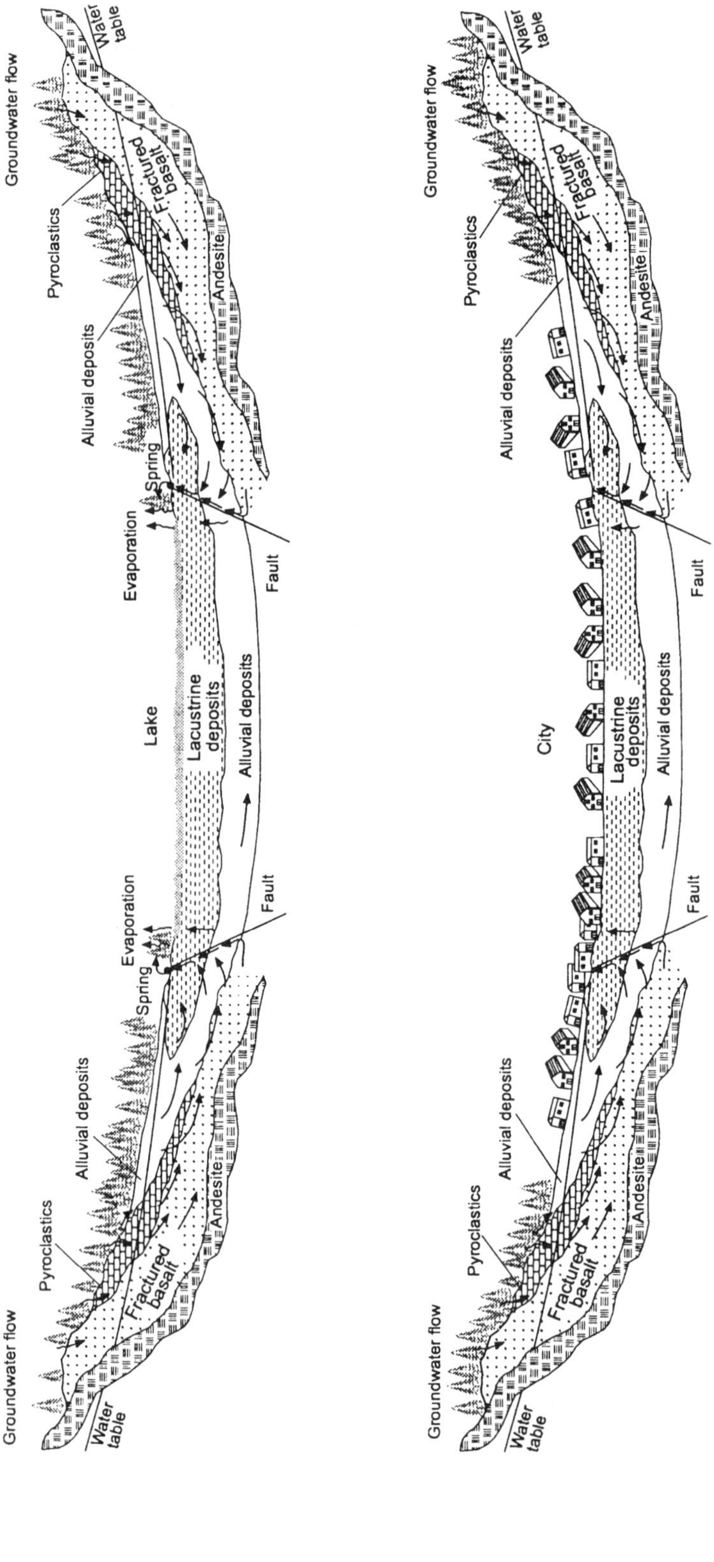

Figure 3.5. Schematic cross-section showing how as Mexico City continues to grow, the main recharge areas to the aquifer are being covered with asphalt and other artifacts of urbanization.

controls are enacted that would prohibit the destruction of any forested areas located along the mountain flanks of the basin. The last hydrogeologic reserve zone proposed in Mexico was the one proposed by Escolero et al. (2000) for the City of Merida, Yucatan. Furthermore, in areas where deforestation has occurred, it would help to replant trees, because this would reduce soil erosion, and greater soil cover would also increase infiltration capacity.

Soils

Different types of soils are found from the low-lying plains of the Basin of Mexico to the coniferous forests of the mountains in the eastern, southern, and western regions. In order to compare the major soil types in the Basin, the chemical and physical properties of soil profiles from representative locations of the plains and from montane sites of the Sierra Nevada and Sierra de las Cruces, and from DL and Zoquiapan National Parks are described (Tables 3.1–3.9).

In the lower part of the Basin of Mexico, the soils are identified as entisols. These soils contain amorphous minerals such as silica, kaolinite, smectite, and cristobalite in the clay fraction and normally have salinity problems and low drainage, as in the case of ancient Texcoco Lake soils (Gutierrez-Castorena 1997; Ortiz-Solorio et al. 2000).

The footslopes of the Sierra Nevada and the Sierra de Ajusco are covered by duripans (locally known as *Tepetates*) which develop after the soil, a eutric ustropept (Miehlich 1992; Soil Survey Staff 1975, 1990) has been eroded. In the forest area of the mountains, the soils are classified as inceptisols within the suborder of umbrepts. These are immature and shallow soils with high organic matter content in the A_1 horizon (8%–15%), low pH (5.5–6.5) and sandy loam or loamy sand textures (Gómez-Guerrero 1986). Sometimes in the mid-elevation montane regions (2800–3600 m) andisol soils are found.

Soils of the Low-Lying Plains of the Basin of Mexico

Soils located within the lower parts of the Basin of Mexico are formed from alluvial and colluvial materials and also from sediments accumulated on the ancient Texcoco, Chalco, and Xochimilco lakes. These soils have high salt and sodium concentrations as a result of long-term evaporation processes. The chemical and physical properties of soils near Texcoco Lake and Xochimilco Lake are presented in Table 3.1 (Espinoza-Tobar 1971; Gutierrez-Castorena 1997).

Soils in the Texcoco area are classified as fluvaquentic haplustolls and fluvaquentic endoaquolls (Gutiérrez-Castorena 1997; Soil Survey Staff 1975). The pH of these soils is above 8, the electrical conductivity is greater than 9 dS/m, and base saturation (BS) is above 85%. Amorphous minerals, as well as silica, kaolinite, smectite, or montmorillonite and cristobalite have been found within the clay fraction. Because of the high exchangeable sodium and salt content, these

Table 3.1. Soil characteristics of the lowlands of the Basin of Mexico[a]

Profile number[b]	Elevation (masl)	pH 1:2	Sand %	Silt %	Clay %	OM %	BD g/cc	EC dS/m	CEC $cmol_c\,kg^{-1}$	K $cmol_c\,kg^{-1}$	Ca $cmol_c\,kg^{-1}$	Mg $cmol_c\,kg^{-1}$	Na $cmol_c\,kg^{-1}$	BS (%)
16E	2239	9.4	25	45	30	7.0		28.6	42	9.5	2.4	0.4	37.5	118
43C	2259	8.5	42	36	22	1.5	1.1	9.4	24	4.4	11.7	4.3	5.6	95
44C	2259	8.7	32	49	19	1.4	1.3	14.4	25	2.4	16.1	4.8	6.3	97
45C	2250	9.5	34	33	33	2.5	1.1	14.1	36	5.6	4.7	5.0	15.9	92

[a] Data represent the average values for the soil profiles which ranged in depth from 130 to 175 cm, except for profile 16E which was 83 cm in depth. OM, organic matter; BD, bulk density; EC, electrical conductivity; CEC, cation exchange capacity; BS, base saturation.
[b] Profile 16E is from Xochimilco, and the other three profiles are from the former Lake Texcoco in Texcoco, State of Mexico.
Data from Gutierrez-Castorena (1997), Espinoza-Tobar (1971).

soils are classified as saline-sodic soils. Principal vegetation growing in these soils consists of *Distichlis spicata*, the dominant grass species which occurs in a patchy distribution, and planted species such as *Eucalyptus sp.* and *Tamarix sp.*

Soils located slightly higher than Texcoco Lake are typically nonsaline entisols classified as ustorthents. These are deep loamy and clayey soils with pHs from 7.0 to 7.5 and organic matter content from 1% to 2% in the plow layer. These Entisols are located between 2250 and 2270 masl, and with manure and fertilizer applications, they are very productive for crops such as corn, barley, wheat, alfalfa, and vegetables (Cachón et al. 1976).

Mid-Elevation Piedmont Soils

Piedmont soils are classified as eutric ustropepts (Miechlich 1992), vertic argiustolls and vertic haplustolls (Peña-Hernandez and Zebrowski 1992) and vertic haplustalfs (Ortiz et al. 2000). These soils are located in the piedmont of the Sierra Nevada from 2250 to 2800 masl. The slopes of the hills range from 7% to 15%. In some areas, the soils are cultivated with corn, barley, and beans, and others are used to grow fruit trees, including peaches, apricots, and plums. Several species of agave, cactus, and trees such as *Schinus molle* (Peruvian pepper tree or *pirul*) are also found in this region. One of the main problems with these cultivated soils and lands is that severe soil erosion occurs when vegetation cover is sparse, such as after the land is cleared of vegetation in preparation for agricultural use. It is common in these situations to have a total loss of the plow layer or A and B horizons, leaving only the hardened tepetate subsoil. Tepetate is formed from the subsoil of eutric ustropepts which is composed of older volcanic ash deposits. After the topsoil is eroded, the former subsoil (which normally is found at a depth of 40–60 cm) is exposed to wet and dry cycles that cause strong compaction and hardening of the clay/ash mixture, thus forming tepetates (Miehlich 1992). Large areas of the piedmont are covered with tepetate. Descriptive data for some of these piedmont soils are presented in Table 3.2.

Forest Soils

In the mountainous areas of the Basin of Mexico, forest soils are found from 2800 m to 4000 m. Sites at greater than 4000 m elevation are above the tree line, and understory vegetation is scarce. Andisols and Inceptisols (Soil Survey Staff 1990) are the dominant soil orders in forests of the Basin of Mexico. Andisols are located between 2800 and 3600 masl; within this order, typic haplustand, mollic haplustand, typic vitricryand, typic hapludand and typic and mollic ustivitrand can be found (Miehlich 1991). Inceptisols are found at higher elevation, some with a cambic B horizon and others without. The profile of these inceptisols is composed of only A_1, A_2, and R horizons. The amount of litter on the surface of these forest soils varies according to stand density. Pine (*Pinus*

Table 3.2. Characteristics of midelevation Piedmont soils from Sierra Nevada Tlaloc (SNT)[a]

Profile	Slope (%)	Parent material	Horiz	Depth (cm)	$CaCO_3$ (%)	pH 1:2	Sand %	Silt %	Clay %	Textural class	OM %	EC dS/m	CEC $cmol_c\,kg^{-1}$	K $cmol_c\,kg^{-1}$	Ca $cmol_c\,kg^{-1}$	Mg $cmol_c\,kg^{-1}$	Na $cmol_c\,kg^{-1}$	BS %
34 SNT	15	Volcanic	IT3[c]	Superficial	2.3	7.9					0.26	1.33	27.1	1.9	11.3	8.6	1.3	89
Tepetate		ash	IIT3	Superficial	1.9	7.9	61	30	9	Sandy loam	0.17	1.44	32.3	2.5	14.1	8.1	2.4	85
Samples[b]			IIIT3	Superficial	14.2	8.8	60	32	8	Sandy loam	0.28	1.32	27.6	1.2	15.3	11.9	1.3	107
			IT2	Superficial	1.6	7.5					0.27	1.47	19.5	1.2	7.7	6.2	0.9	82
			IIT2	Superficial	4.6	8.3					0.26	1.55	33.6	2.3	14.5	8.5	1.3	79
Typic	13	Colluvial	Ap	0–27	0.21	7.1	8	49	43	Silty clay	1.12	0.53	33.4	0.72	21.6	10.1	1.5	100
Haplustert			A	27–51	0.32	7.3	11	41	48	Silty clay	2.17	0.41	41.1	0.75	22.5	12.5	1.5	91
35 SNT[d]			C	51–61	0.21	7.3	20	47	33	Silty clay	0.49	0.41	29.3	0.61	22.5	6.5	1.7	105
			2Ck	61–72	0.74	7.8	28	46	26	Clay loam	0.49	0.53	29.7	0.43	17.0	8.9	2.1	96
			3Ck	72–91	1.59	8.1	49	33	18	Loam	0.07	1.18	38.1	1.17	26.7	5.4	3.9	98
			4C	91–108	0.21	8.1	51	32	17	Sandy clay loam	0.14	1.18	28.9	2.39	14.5	5.1	4.3	91
			5C	>108	0.21	8.1	54	31	15	Sandy loam	0.14	1.18	30.1	1.68	15.5	5.4	2.7	84
Vertic	12	Colluvial	Ap	0–27	0.5	6.7	46	39	15	Loam	1.50	0.40	18.9	0.3	9.5	4.0	0.7	77
Haplustalf			2Bss	27–62	0.4	5.1	23	30	47	Clay	1.00	0.20	17.0	0.2	10.3	4.8	0.1	90
36 SNT[e]			3Crt	62–90	0.4	5.8					0.80	0.20	33.5	0.6	16.3	7.8	0.3	74
			3Crtc	90–110	0.4	7.6					0.12	0.30	33.8	0.8	22.5	11.5	0.5	100
			3Crtc	110–140	0.2	7.9					0.12	0.30	39.0	0.9	23.5	16.3	0.5	100

[a] The peak of the Sierra Nevada Tlaloc site is located 25 km southeast of Texcoco. The study area (see also Tables 3.5 and 3.7), including the peak, encompasses western longitudes between 98° 42′ 51″ and 98° 48′ 12″ and northern latitudes between 19° 23′ 43″ and 19° 28′ 37″.

[b] Elevation is 2555 m, and site is devoid of vegetation.

[c] T1 to T3 refer to the age of the tepetates (hardpan layers) since the volcanic ash layers were deposited (from 1500–30,000 years ago). The Roman numerals preceding the letter "T" refer to the horizons in a vertical profile, with I being the surface layer, and II and III are the underlying layers.

[d] Elevation is 2295 m, and vegetation at this site is *Schinus molle*.

[e] Elevation is 2550 m, and vegetation at this site is wheat and planted trees.

Data from Gutiérrez-Castorena (1997), Ortiz et al. (2000), Peña and Zebrowski (1992), and Michlich (1992).

Table 3.3. The mean dry weights of different profiles of litter in the Sierra Nevada and Sierra de las Cruces

Elevation masl	Amount of litter kg/ha	Timber increment m^3/ha/year
2860	2200	1.0–1.5
2900	3250	3.0–5.0
3150	3080	6.0–8.0
3425	3000	1.0–2.0
3800	1000	1.0–2.0

Source: A. Trinidad and D. Martínez, unpublished data.

hartweggii), sacred fir (*Abies religiosa*), and hardwood species (*Quercus spp.*) are the main types of vegetation found in the Sierra Nevada and Sierra de las Cruces ranges. The more productive timber forests are found between 2800 and 3800 m. Timber production in this zone is related to litter production (Table 3.3). The litter, or organic layer, located above the A_1 horizon, is a major nutrient pool, but the pool size varies with the litter type (Table 3.4). The C:N ratio of litter also changes with forest type. Pine litter has a high C:N ratio (73:1) compared to hardwood (39:1) or sacred fir litter (37:1) (Table 3.4).

Soil depth, clay content, pH, cation exchange capacity (CEC), K, Ca, Mg, Na, and BS decrease with elevation in these mountainous forest soils (Table 3.5). However, the percentage of sand content tends to increase with elevation, favoring rapid drainage from the shallow soils of the upper mountain regions. Soil organic matter content increases with elevation (Table 3.5), perhaps as a result of the low temperatures, which impede organic matter decomposition. Notwithstanding the large nutrient pool in the litter of the forest floor and in the organic matter pool in the A_1 horizon (Table 3.6), these forests are commonly nutrient limited, because most of the nutrients are in organic form and not readily available for plant uptake (Gomez-Guerrero 1986). Thus tree growth increases (height, diameter, and timber production) have been observed in pine, cedar, and eucalyptus following N and P fertilization. On the tepetates, eucalyptus and pine respond to N and P application; in Andisols and Inceptisols, pines respond mostly to N (Saldivar-Castillo 1982), possibly because pines have a low P requirement (Leaf 1973; Zöttl 1971). A quadratic equation was used to predict the timber yield

Table 3.4. Nutrient concentrations in different types of forest litter

Source of leaf litter	N	P	K	Ca	Mg	C:N
Pine (*P. hartwegii*)	0.55	0.03	0.04	0.70	0.06	73
Sacred fir (*Abies religiosa*)	1.20	0.10	0.25	1.23	0.30	37
Hardwood species (*Quercus spp.*)	1.50	0.11	0.65	1.40	0.65	39

Source: A. Trinidad and D. Martínez, unpublished data.
Litter samples were collected at sites along an elevational transect in the Sierra Nevada between Tequesquináhuac and Tlaloc. *Quercus* litter was collected at the lowest elevation, *A. religiosa* at the intermediate site, and *P. hartwegii* at the highest elevation along the transect.
Units for nutrient concentration are percent dry weight.

Table 3.5. Soil profiles of the Sierra Nevada Tlaloc site[a]

Profile	Elev. (masl)	Slope %	Parent material	Vegetation	Horizon	Depth (cm)	pH	Sand %	Silt %
1 SNT	2830	25	Volcanic ash	*P. teocote*	O	0.0–0.5			
				P. leiophylla	A1	0.5–20	6.5	36	38
			Tepetate		A2	20–42	6.7	38	37
					A3	42–112	6.9	42	35
					B1	112–130	7.0	47	25
					IIB	130–X	7.2	32	26
						Average	6.9	39	32
2 SNT	2970	50	Andesite	*A. religiosa*	O	0–3.5			
			Volcanic ash		A1	3.5–21	6.7	42	36
					A2	21–46	7.0	44	39
					A3	46–84	6.8	40	30
					B1	84–156	6.2	42	30
					B2	156–180	7.1	52	27
					CR	180–X	7.7	48	28
						Average	6.9	45	32
3 SNT	3050	40	Granite	*A. religiosa*	O	0–6			
			Andesite		A1	0–26	6.3	52	35
			Volcanic ash		A2	26–80	6.4	58	27
					A3	80–113	6.3	62	25
					B1	113–175	6.6	50	19
					B2	175–210	7.1	44	32
					C	210–X	7.8	42	45
						Average	6.8	51	31
4 SNT	3170	35	Andesite	*A. religiosa*	O	0–0.3			
			Volcanic ash		A1	0.3–10	5.8	46	39
					A2	10–70	5.9	68	21
					A3	70–110	5.8	74	14
					B1	110–143	6.7	54	22
					B2	143–195	6.8	36	35
					C	195–X	6.4	46	42
						Average	6.2	54	29
5 SNT	3330	60	Granite	*A. religiosa*	O	0–2			
			Andesite		A1	2–16	6.0	58	37
			Volcanic ash		A2	17–43	6.1	58	36
					B	43–82	6.4	70	26
					C	82–X	6.5	74	21
						Average	6.3	65	30
6 SNT	3630	15	Volcanic ash	*P. hartwegii*	O	0–4			
					A1	4–16	6.1	60	34
					A2	16–49	6.1	48	47
					A3	49–80	6.2	52	41
					B1	80–111	6.4	72	23
					C	111–X	6.4	78	17
						Average	6.2	62	32
7 SNT	3680	20	Granite	*P. hartwegii*	O	0–3			
					A1	3–20	5.2	51	45
					A2	20–46	5.3	59	37
					RC	46–X	5.8	56	40
						Average	5.4	55	41
8 SNT	3780	15	Granite	*P. hartwegii*	O	0–2			
				Juniperus	A1	2–18	5.1	50	46
					A2	18–42	5.7	56	40
					RC	42–X	6.3	68	27
						Average	5.7	58	38

[a] The peak of the Sierra Nevada Tlaloc site is located 25 km southeast of Texcoco. The study area, including the peak, encompasses western longitudes between 98°42′51″ and 98°48′12″ and northern latitudes between 19°23′43″ and 19°28′37″.

Source: A. Trinidad and D. Martínez, unpublished data.

Table 3.5. *Continued*

Clay %	OM %	BD g/cc	EC dS/m	CEC $cmol_c$ kg^{-1}	K $cmol_c$ kg^{-1}	Ca $cmol_c$ kg^{-1}	Mg $cmol_c$ kg^{-1}	Na $cmol_c$ kg^{-1}	BS %
26	3.6	0.91	0.12	30.9	0.61	5.00	1.81	0.13	24.4
25	1.2	0.90	0.14	17.5	0.51	4.60	1.83	0.32	41.4
23	0.7	0.85	0.10	14.7	0.51	7.80	0.57	0.54	64.0
28	0.2	1.00	0.10	18.7	0.80	5.00	2.67	0.52	48.0
42	0.1	1.00	0.11	10.1	0.46	4.18	2.57	0.54	76.7
29	1.2	0.93	0.11	18.4	0.58	5.32	1.89	0.41	50.9
22	10.0	0.79	0.12	15.9	0.79	2.30	1.33	0.11	28.4
17	5.0	0.84	0.09	15.5	0.41	3.10	1.02	0.10	29.8
30	1.7	0.92	0.10	9.7	0.85	2.30	1.81	0.10	52.1
28	1.1	0.94	0.10	8.3	0.82	2.10	1.23	0.12	51.4
21	0.8	0.97	0.09	15.3	0.44	2.90	1.60	0.05	33.2
24	0.6	1.16	0.08	14.3	0.28	0.60	1.50	0.50	20.1
24	3.2	0.94	0.10	13.2	0.60	2.22	1.42	0.16	35.8
13	4.6	0.98	0.14	13.7	0.25	0.75	0.67	0.10	12.9
15	1.8	1.12	0.10	14.2	0.64	6.70	0.67	0.10	57.1
13	0.9	1.14	0.09	14.7	0.60	5.00	0.53	0.10	42.3
31	0.8	0.99	0.10	17.7	0.33	5.60	0.53	0.17	37.4
24	1.2	0.88	0.14	12.7	0.33	4.55	0.86	0.32	47.7
13	1.3	1.13	0.15	15.9	0.32	4.30	0.49	0.20	33.3
18	1.8	1.04	0.12	14.8	0.41	4.48	0.63	0.17	38.5
15	7.50	0.88	0.10	12.7	0.50	2.60	0.28	0.10	29.0
11	1.60	1.38	0.08	12.5	0.41	3.10	0.25	0.05	30.0
12	0.28	1.37	0.08	8.7	0.48	1.60	0.28	0.08	28.0
24	0.32	1.04	0.09	13.9	0.63	3.50	0.74	0.13	35.0
29	0.52	0.84	0.08	16.8	0.67	5.50	0.82	0.34	43.0
12	0.25	1.04	0.09	8.4	0.70	1.90	0.58	0.39	42.0
17	1.75	1.09	0.09	12.2	0.57	3.03	0.49	0.18	34.5
5	17.52	0.83	0.03	10.7	0.34	1.05	0.28	0.11	16.0
6	5.80	1.12	0.03	7.1	0.34	1.75	0.26	0.06	33.0
4	1.94	1.25	0.03	12.0	0.46	2.62	0.26	0.08	28.0
5	0.04	1.28	0.02	6.2	0.28	0.75	0.06	0.05	18.0
5	6.33	1.12	0.03	9.0	0.36	1.54	0.22	0.08	23.8
6	11.76	0.72	0.10	19.7	0.21	0.97	0.16	0.05	7.0
5	8.46	0.77	0.08	12.5	0.35	0.37	0.68	0.04	11.5
7	5.97	0.81	0.07	8.0	0.12	0.87	0.16	0.05	15.0
5	1.33	1.09	0.07	36.8	0.28	2.32	0.41	0.05	8.3
5	0.38	1.22	0.10	4.7	0.10	0.22	0.00	0.05	7.8
6	5.58	0.92	0.08	16.3	0.21	0.95	0.28	0.05	9.9
4	23.2	0.59	0.03	17	0.46	1.40	0.38	0.08	13.6
4	16.8	0.69	0.03	16	0.20	2.40	0.68	0.05	20.8
4	13.1	0.73	0.02	23	0.08	0.60	0.19	0.05	4.0
4	17.7	0.67	0.03	19	0.25	1.47	0.42	0.06	12.8
5	20.5	0.63	0.12	26	0.66	1.75	0.22	0.11	10.0
3	18.2	0.66	0.02	13	0.29	1.17	0.14	0.05	12.0
3	0.1	1.03	0.02	11	0.13	1.02	0.10	0.04	11.7
4	12.9	0.77	0.05	17	0.36	1.31	0.15	0.07	11.2

Table 3.6. Physical and chemical properties of the A_1 horizon of Sierra Nevada Tlaloc forest soils

Elevation (masl)	Vegetation	pH[a]	Sand %	Silt %	Clay %	Textural class	OM %	Total N %	C:N	P (Bray) ppm	K $cmol_c\ kg^{-1}$	Ca $cmol_c\ kg^{-1}$	Mg $cmol_c\ kg^{-1}$	Na $cmol_c\ kg^{-1}$
3000	*Abies religiosa*	6.7	34	42	24	Loam	12	0.41	29.3	6.9	1.4	29.9	3.5	0.60
3200	*Abies religiosa*	6.9	39	38	23	Clay loam	9	0.37	24.3	7.4	1.0	28.1	1.5	0.30
3400	*Abies religiosa*	6.4	38	36	26	Clay loam	6	0.33	18.2	3.7	0.7	8.7	1.3	0.20
3600	*Abies religiosa* and *Pinus hartwegii*	6.1	38	42	20	Loam	12	0.35	34.0	5.3	0.6	6.1	0.9	0.16
3800	*Pinus hartwegii*	5.6	40	35	25	Clay loam	13	0.37	35.0	5.7	0.2	2.8	0.3	0.12

[a] pH was determined with soil : water at a 1 : 1 ratio.
The peak of the Sierra Nevada Tlaloc site is located 25 km Southeast of Texcoco. The study area, including the peak, encompasses western longitudes between 98°42′51″ and 98°48′12″ and northern latitudes 19°23′43″ and 19°28′37″.
Source: Gomez-Guerrero (1986).

response curve for N fertilitation of 20-year-old pines (*Pinus michoacana*) in an Andisol soil (Comisión Forestal del Estado de Michoacán, unpublished data). The following equation with $R^2 = 0.92$ was obtained:

$$Y = 2.5 + 0.192N - 0.001N^2$$

where:

Y = timber yield m^3/ha/year
N = kg N/ha/3 years

The maximum yield was obtained with 96 kg N/ha/3years. On the straight line of the curve, each kg of N yielded 0.192 m^3 of timber.

Total N concentrations in soils at four *Pinus hartwegii* sites (Zoquiapan, Izta–Popo, and DL National Parks, and Ajusco) ranged from 3 to 6 g/kg, and C : N ratios were from 15 to 17 (Fenn et al. 1999, 2002b). Romero-Carmona (1986) reported total N concentrations of 3.7 to 6.1 g/kg and C : N ratios of 15.0 to 17.3 in forest soils in the DL (Table 3.7). The C : N ratios reported for the Sierra Nevada Tlaloc soils in Table 3.6 are higher, ranging from 18 to 35. It's not clear if C : N ratios are really higher at these sites or if differences are due to error or methodological differences. Soil C : N ratios less than 20 in forest soils are indicative of high N fertility (Fenn et al. 2002b), suggesting that the forest soils in the Basin of Mexico are relatively fertile. We are aware of only one published forest fertilization study in the Basin. Pine stands in the four sites mentioned above did not respond, at least in the short-term, to N fertilization with 300 kg N/ha (Fenn et al. 2002b). Nutrient analyses of an *Abies religiosa* plantation in the DL demonstrated that foliar N levels were adequate, P was deficient, but that K and Mn were the most deficient nutrients (López-López et al. 1998).

Soil pH at both the DL and Zoquiapan pine sites was 4.5 on average (Fenn et al. 1999). In Table 3.7, pH values in the A horizon ranged from 5.0 to 6.0 in soils of *P. hartwegii* and *Abies religiosa* stands in the DL. In the Sierra Nevada Tlaloc, soil pH values ranged from 5.6 to 6.7 under the same species. Data from several studies suggest a possible trend of lower soil pH in the higher elevation pine sites compared to fir sites in the Desierto de los Leones (Castro et al. 1997; Fenn et al. 2002a; López-López et al. 1998) (Table 3.6), but possibly in other sites as well. Further studies are needed to test this hypothesis. Forest soils in the Desierto de los Leones National Park show accumulation of lead, chiefly in the upper horizons (Table 3.7). Similarly, high concentrations of lead and other heavy metals were also measured in soil and vegetation from DL. These concentrations are a result of many years of heavy metal inputs from atmospheric deposition (Castro et al. 1997; Fenn et al. 2002a; Watmough and Hutchinson 1999). Additional data on forest soils in the Basin of Mexico are presented in Tables 3.8 and 3.9.

Summary

Even if the growth rate for Mexico City continues to decrease, the city will continue to grow, and so will the demand for water. One of the major problems that the city faces is that the predominant zones where recharge to the

Table 3.7. Soil profiles from the southwestern end of the Desierto de los Leones National Park

Profile[a]	Elevation (masl)	Slope (%)	Vegetation	Depth (cm)	pH 1:2	Sand	Silt %	Clay	Textural class	Total N g kg^{-1}	C:N	OM %	CEC cmol$_c$ kg^{-1}	K cmol$_c$ kg^{-1}	Ca cmol$_c$ kg^{-1}	Mg cmol$_c$ kg^{-1}	Na cmol$_c$ kg^{-1}	BS %	Pb ppm	P (Olsen) ppm
12A	3450	35	*P. hartwegii*	0–12	5.3	55	33	13	Sandy loam	10.8	14.0	26.4	55	0.20	10.40	0.90	0.28	22	5.1	36.7
			A. religiosa	12–50	5.0	51	37	13	Loam	6.3	18.4	20.0	46	0.15	5.90	0.56	0.21	15	0.0	26.2
				50–75	5.2	56	35	9	Sandy loam	5.7	14.4	14.2	43	0.10	3.65	0.62	0.20	11		16.3
				75–94	5.2	66	23	11	Sandy loam	2.6	11.4	5.1	44	0.07	1.00	0.20	0.46	4	2.0	26.2
				Average	5.2	57	32	12		6.4	14.6	16.4	47	0.13	5.23	0.59	0.29	13	1.8	26.4
13A	3480	90	*A. religiosa*	0–14	5.0	56	31	13	Sandy loam	8.3	18.6	26.5	64	0.20	4.95	0.46	0.06	9	2.4	35.6
				14–36	5.2	56	31	13	Sandy loam	8.7	15.7	23.5	71	0.17	3.00	0.39	0.24	5	1.1	20.2
				36–69	5.5	68	23	9	Sandy loam	8.5	15.1	22.2	65	0.07	3.65	0.60	0.09	7	0.4	24.0
				69–95	6.0	58	26	16	Sandy loam	6.3	14.3	21.0	62	0.10	2.65	0.28	0.18	5	0.2	22.4
				Average	5.4	60	28	13		8.0	15.9	23.3	66	0.14	3.56	0.43	0.14	6	1.0	25.6
14A	3610	90	*P. hartwegii*	0–19	5.5	56	31	13	Sandy loam	6.2	18.6	19.8	50	0.34	5.75	0.67	0.14	14	3.1	23.5
			Calamagrostis tolucensis	19–47	5.7	55	13	31	Sandy clay loam	5.9	16.8	16.0	54	0.14	2.20	0.63	0.12	6	0.5	46.0
				47–84	5.4	65	26	9	Sandy loam	7.7	13.6	18.1	66	0.11	5.60	0.65	0.52	11	0.4	13.6
				84–100	5.2	71	20	9	Sandy loam	8.4	15.3	5.2	65	0.06	9.00	1.72	0.00	17	0.9	18.0
				Average	5.5	62	23	16		7.1	16.1	14.8	58	0.16	5.64	0.92	0.33	12	1.2	25.3
15A	3640	40	*Calamagrostis*	0–20	6.0	73	17	11	Sandy loam	7.6	18.0	23.0	84	0.36	5.85	0.31	0.09	8	0.0	25.7
			tolucensis	20–37	6.3	65	28	7	Sandy loam	5.2	15.5	13.8	53	0.14	2.90	0.11	0.06	6	0.0	10.8
			and isolated	37–65	5.7	65	26	9	Sandy loam	3.8	16.7	10.9	41	0.09	3.05	0.19	0.05	8	0.7	14.7
			stands of *A.*	65–100	6.4	73	19	10	Sandy loam	2.0	17.8	6.0	39	0.06	1.00	0.19	0.00	3	0.0	21.3
			religiosa and *P. hartwegii*	Average	6.1	69	22	9		4.7	17.0	13.4	46	0.16	3.20	0.20	0.06	6	0.2	18.1

[a] The parent material for all these soils is andesite. The location and elevation for profiles 12A, 13A, 14A, and 15A are: Tlalpizayagua, 3450 m; Tlalpizayagua, 3480 m; Tlalpizayagua, 3610 m; and Los Hongos, 3640 m, respectively.

Source: Romero-Carmona (1986).

Table 3.8. Soil characteristics for forested sites in Paso de Cortés (Izta–Popo National Park)

Site	Location	Horizon	Depth (cm)	Sand %	Silt %	Clay %	Textural Class	pH	OM %	BS %	CEC $cmol_c\ kg^{-1}$	K $cmol_c\ kg^{-1}$	Ca $cmol_c\ kg^{-1}$	Mg $cmol_c\ kg^{-1}$	Na $cmol_c\ kg^{-1}$	P Olsen ppm
18	19°12′16″ N	Ap	0–8	48	32	20	Loam	4.7	5.9	<50	30.8	0.7	4.7	1.0	0.3	17.4
18	98°44′54″ W	A1	8–35	52	28	20	Sandy Clay Loam	4.8	1.0	<50	25.0	0.5	3.4	0.8	0.2	11.9
18		B21	35–64	52	30	18	Sandy Loam	5.6	1.2	<50	22.5	0.9	4.4	1.3	0.2	14.0
18		B22	64–100	58	26	16	Sandy Loam	5.6	1.2	<50	21.5	0.9	3.4	1.9	0.7	
39	19°08′47″ N	C1	0–27	64	22	14	Sandy Loam	5.5	1.2	<50	10.3	0.4	3.8	0.5	0.2	35.2
39	98°44′27″ W	C2	27–53	66	22	12	Sandy Loam	5.9	1.0	<50	9.9	0.1	4.1	0.3	0.1	39.8
39		C3	53–79	62	26	12	Sandy Loam	6.1	0.9	<50	10.5	0.2	4.7	0.9	0.1	
39		C4	79–100	56	26	18	Sandy Loam	6.1	1.2	>50	13.3	0.2	5.0	2.0	0.1	
43	19°08′22″ N	A1	0–31	66	36	8	Sandy Loam	5.8	1.5	<50	13.5	0.1	3.1	0.6	0.5	2.5
43	98°43′25″ W	A11b	31–50	54	32	14	Sandy Loam	6.1	3.3	<50	21.0	0.2	8.1	1.1	0.6	2.0
43		A12b	50–110	54	34	12	Sandy Loam	6.0	3.9	<50	23.5	0.2	7.2	1.8	0.6	
61	19°04′45″ N	C1	0–21	72	22	6	Sandy Loam	5.8	3.4	>50	9.3	0.1	5.3	0.6	0.1	34.1
61	98°40′08″ W	C2	21–54	74	20	6	Sandy Loam	5.8	1.6	>50	5.3	0.1	3.8	0.4	0.1	44.8
61		Ac	54–83	74	20	6	Sandy Loam	6.1	1.5	>50	6.5	0.1	3.8	0.3	0.1	
61		C1b	83–100	44	46	10	Loam	6.3	1.2	>50	6.8	0.1	4.4	0.3	0.1	
63	19°04′19″ N	C1	0–29	76	18	6	Loamy Sand	5.6	0.8	<50	4.3	0.2	1.3	0.3	0.1	54.1
63	98°42′17″ W	C2	29–71	78	16	6	Loamy Sand	5.6	0.3	<50	4.3	0.1	1.3	0.3	0.1	32.0
63		C3	71–100	68	22	10	Sandy Loam	6.0	0.5	>50	6.5	0.1	2.2	1.1	0.1	
64	19°04′19″ N	C1	0–21	72	22	6	Sandy Loam	5.7	1.5	>50	3.5	0.1	2.2	0.2	0.1	19.1
64	98°41′18″ W	C2	21–57	72	22	6	Sandy Loam	5.7	1.0	>50	3.8	0.1	1.9	0.2	0.1	19.3
64		C3	57–74	74	22	4	Loamy Sand	5.7	0.8	>50	3.0	0.1	1.9	0.1	0.1	
64		Ac1	74–100	54	40	6	Sandy Loam	5.6	0.8		4.0	0.1	1.6	0.2	0.1	

Data is from the published soil charts, Coordinación General (1982).

Table 3.9. Soil characteristics for forested sites in Zoquiapan

Site	Location	Horizon	Depth (cm)	Sand %	Silt %	Clay %	Textural Class	pH	OM %	BS %	CEC $cmol_c$ kg^{-1}	K $cmol_c$ kg^{-1}	Ca $cmol_c$ kg^{-1}	Mg $cmol_c$ kg^{-1}	Na $cmol_c$ kg^{-1}	P Olsen ppm
35	19°21′17″ N	A11	0–15	58	32	10	Sandy Loam	5.7	5.1	<50	19.8	0.4	7.5	1.1	0.1	14.3
35	98°40′44″ W	A12	15–24	66	28	6	Sandy Loam	5.9	2.6	<50	17.5	0.8	6.3	1.0	0.1	11.9
35		B21	24–73	64	24	12	Sandy Loam	6.1	0.7	>50	14.0	1.1	6.4	0.9	0.1	4.0
35		B22	73–100	82	14	4	Loamy Sand	6.1	0.6	>50	12.8	1.1	6.9	0.9	0.1	
62	19°17′50″ N	A	0–43	42	34	24	Loam	5.8	2.5	>50	14.0	1.0	6.3	2.1	0.3	21.9
62	98°44′42″ W	B	43–80	40	34	26	Loam	6.3	0.9	>50	18.5	1.2	7.5	4.2	0.7	
66	19°17′11″ N	A11	0–30	48	40	12	Loam	5.0	15.5	<50	32.8	0.2	6.6	0.9	0.1	5.1
66	98°40′36″ W	A12	30–45	54	32	14	Sandy Loam	5.2	10.2	<50	30.0	0.2	5.3	1.0	0.1	2.5
66		A13	45–75	52	32	16	Sandy Loam	4.9	8.1	<50	31.3	0.2	3.4	1.0	0.1	5.6
66		A14	75–110	60	28	12	Sandy Loam	4.9	8.4	<50	30.8	0.2	3.8	1.0	0.1	3.8
73	19°16′9″ N	A11	0–30	72	20	8	Sandy Loam	4.9	13.8	<50	30.0	0.2	4.1	0.7	0.1	3.9
73	98°42′3″ W	A12	30–45	60	30	10	Sandy Loam	5.0	6.9	<50	28.8	0.1	3.8	0.6	0.1	1.8
73		A13	45–60	56	36	8	Sandy Loam	5.1	5.4	<50	25.0	0.1	1.6	0.7	0.1	2.5
73		A14	60–90	60	34	6	Sandy Loam	5.4	4.4	<50	27.0	0.1	2.5	1.1	0.1	2.7

Data is from the published soil charts, Coordinación General (1983).

aquifer occurs are systematically being urbanized. The recharge zone is being paved over to build residential and commercial developments. We suggest that a hydrogeologic reserve zone be established immediately for the Basin of Mexico. If this same zone is protected, and trees are planted, this would also help to control soil erosion, and as more soil is retained, the recharge to the regional aquifer would also be increased. There is legislation concerning the establishment of groundwater reserve zones within the National Water Laws and Regulations (Ley Nacional de Aguas 1992; Ley Nacional de Aguas y su Reglamento 1994).

Different soil orders are found from the low-lying plains of the Basin of Mexico to the coniferous forests of the mountains in the eastern, southern, and western regions. In the lower part of the Basin, Entisols and Mollisols are found. Some of them have salinity problems, with low drainage and high salt and Na concentrations. Soils located slightly higher than Texcoco Lake are typically non-saline Entisols, and with good management they are productive for different crops. Inceptisols, Mollisols, and Alfisols are found in the piedmont region. In this area, when the topsoil is eroded, the former subsoil is exposed to wet and dry cycles causing strong compaction of the clay/ash mixture, thus forming duripans or tepetates. Large areas of the piedmont are covered with tepetates. Soils of the coniferous zone are classified as Andisols and Inceptisols. In these soils, the depth, clay content, pH, CEC, exchangeable cations, and BS decrease with elevation, but sand content increases with elevation, favoring rapid soil drainage. In the DL National Park the upper soil horizons contain high concentrations of Pb and other heavy metals.

References

Arreguín-Mañon, J.P. 1998. Aportes a la Historia de la Geohidrología en México 1890–1995. Mexico City: Asociación Mexicana de Geohidrología.

Arreguín-Mañon, J.P., and Terán, A. 1994. *Dos Testimonios sobre la Historia de los Aprovechamientos Hidráulicos en México*. Mexico City: Asociación Mexicana de Geohidrología.

Birkle, P., Torres-Rodríguez, V., and González-Partida, E. 1998. The water balance for the Basin of the Valley of Mexico and implications for future water consumption. *J. Hydrogeol.* 6:500–517.

Cachón-Ayora, H.E., Nery-Genes, H., and Cuanalo de la Cerda, H.E. 1976. *Los Suelos del Área de Influencia de Chapingo. Rama de Suelos*. Chapingo, México: Colegio de Postgraduados.

Cardona, A., and Hernández, N. 1995. Modelo geoquímico conceptual de la evolución del agua subterránea en el Valle de México. Ingeniería Hidráulica en México 10:71–90.

Castro-Servín, J.M., González-Kladiano, V., and Hernández-Tejeda, T. 1997. Contenido de metales pesados en algunos suelos y vegetación del Desierto de los Leones, Distrito Federal. *Rev. Cienc. Forestal México* 22:109–127.

Coordinación General de los Servicios Nacionales de Estadística, Geografía e Informática. 1982. *Dirección General de Geografía. Carta Edafológica. Zona Paso de Cortés.* Clasificación: AMECAMECA E14B41 (México, Distrito Federal y Morelos). Mexico City: Primera Edición.

Coordinación General de los Servicios Nacionales de Estadística, Geografía e Informática. 1983. *Dirección General de Geografía. Carta Edafológica. Zona Zoquiapan.* Clasificación: CHALCO E14B31 (México y Distrito Federal). Primera Edición. Mexico City.

Cortes, A., Durazo, J., and Farvolden, R.N. 1997. Studies of isotopic hydrology of the Basin of Mexico and vicinity: Annotated bibliography and interpretation. *J. Hydrol.* 198:346–376.

Dominguez, E. 1996. *Geoquímica del Agua Salina del Acuitardo Lacustre y Riesgo de Contaminación del Acuífero Subyacente que Abastece a la Ciudad de México*, M.Sc. Thesis. Mexico City, Mexico: División de Estudios de Posgrado, Facultad de Ingeniería, Universidad Nacional Autónoma de México.

Durazo, J., and Farvolden, R.N. 1989. The ground water regime of the Valley of Mexico from historic evidence and field observations. Journal of Hydrology 112:171–190.

Escolero, O.A., Marin, L.E., Steinich, B., and Pacheco, J. 2000. Delimitation of a hydrogeological reserve for a city within a karstic aquifer: the Merida, Yucatan, example. *Landscape Urban Plann.* 51:53–62.

Espinoza-Tobar, N. 1971. *Estudio Pedogenético de los Suelos de la Porción Suroeste de la Cuenca de México.* MSc. Thesis. Chapingo, México: Colegio de Postgraduados.

Fenn, M.E., Castro-Servín, J.M., and Hernández-Tejeda, T. 2002a. Heavy metals in forest soils, vegetation, and drainage waters in the Basin of Mexico. In *Urban Air Pollution and Forests: Resources at Risk in the Mexico City Air Basin*, Ecological Studies series, eds. M.E. Fenn, L.I. de Bauer, and T. Hernández–Tejeda, (this volume). New York: Springer-Verlag.

Fenn, M.E., de Bauer, L.I., Zeller, K., Quevedo, A., Perea, V., and Hernández, T. 2002b. Nitrogen and sulfur deposition in the Mexico City Air Basin: Impacts on forest nutrient status and nitrate levels in drainage waters. In *Urban Air Pollution and Forests: Resources at Risk in the Mexico City Air Basin*, Ecological Studies series, eds. M.E. Fenn, L.I. de Bauer, and T. Hernández-Tejeda, (this volume). New York: Springer-Verlag.

Gómez-Guerrero, A. 1986. *Relación Entre Algunas Propiedades Edáficas y el Crecimiento de Coníferas al Oeste del Tlaloc, Estado de México.* Tesis Profesional, Facultad de Agrobiología. Uruapan, Michoacán, México: Universidad Michoacana de San Nicolás de Hidalgo.

Gutiérrez-Castorena del, C.M. 1997. *Los Suelos de la Ribera Oriental del ex Lago de Texcoco (Macro y Micromorfología).* Ph.D. Thesis. Montecillo, México: Colegio de Postgraduados.

Herrera, I. (ed.). 1995. *El Agua y la Ciudad de México.* Mexico City, Mexico: Academia Mexicana de Ciencias.

Leaf, A.L. 1973. Plant analysis as an aid in fertilizing forests. In *Soil Testing and Plant Analysis*, eds. L.M. Walsh, and J.D. Beaton, pp. 427–454. Madison, Wisconsin, USA: Soil Science Society of America, Inc.

Lesser, J.M., Sánchez, L.F., and González-Posadas, D. 1990. Aspectos geohidrológicos de la Ciudad de México. *Rev. Ing. Hid. México* 5:52–60.

Ley Nacional de Aguas. 1992. Mexico City: Diario Oficial de la Federación, December 1, 1992.

Ley Nacional de Aguas y su Reglamento. 1994. Mexico City: Diario Oficial de la Federación, January 12, 1994.

López-López, M.A., Velázquez-Mendoza, J., Velázquez-Martínez, A., González-Romero, V., and Cetina-Alcalá, V.M. 1998. Estado nutrimental de *Abies religiosa* en una área con problemas de contaminación ambiental. *Agrociencia* 32:53–59.

Miehlich, G. 1991. *Chronosequences of Volcanic Ash Soils*, Hamburger Bodenkundliche Arbeiten 15. Hamburg, Germany: The University of Hamburg.

Miehlich, G. 1992. Formation and properties of tepetate in central highlands of Mexico. TERRA 10:137–144.

Morales, R. 1998. *Estudio Geohidrológico del Valle de Monte Alegre, Ajusco, Mexico*. M.Sc. Thesis. Montecillo, Texcoco, State of México, México: Colegio de Postgraduados.

Ortega, A., Cherry, J.A., and Rudolph, D.L. 1993. Large-scale aquitard consolidation near Mexico City. *Ground Water* 31:708–718.

Ortiz-Solorio, C.A., Gutiérrez-Castoreña del, C.M., López-Avila, G., Rodríguez-Tapia, S.A., and Segura-Castruita, M.A. 2000. *Guía de la Gira Técnica en el Municipio de Texcoco, México*. Montecillo, México: Congreso Universitario Internacional de Edafología "Nicolás Aguilera."

Peña-Hernández, D., and Zebrowski, C. 1992. *Estudio de los Suelos Volcánicos Endurecidos (Tepetates) de las Cuencas de México y Tlaxcala (México); Informe del Mapa Morfopedológico de la Vertice Occidental de la Sierra Nevada*. Montecillo, México: Colegio de Postgraduados.

Prol-Ledesma, R.M., Uribe-Alcántara, E.M., and Díaz-Molina, O. 2002. Use of cartographic data and landsat TM images to determine land use variations in the vicinity of Mexico City. International Journal of Remote Sensing (In Press).

Romero-Carmona, J. 1986. *Estudio de Reconocimiento de los Suelos Forestales del Parque Cultural y Recreativo Desierto de los Leones*. Tesis de Licenciatura de Biólogo. México City, D. F.: UNAM. Escuela Nacional de Estudios Profesionales Zaragoza.

Saldivar-Castillo, J.D. 1982. *Ensayo de Fertilización en una Plantación de Cinco Especies Forestales en Coatlinchan, México*. Tesis Profesional. Chapingo, State of México, Mexico: Departamento de Bosques, Universidad Autónoma de Chapingo.

Sánchez, L.F., Jacobo, M., and Cardona, A. 2000. *Estudio para la Recarga del Acuífero en el Suelo de Conservación del Distrito Federal*. Mexico City: Departamento de Ecología, Departamento del Distrito Federal.

Soil Survey Staff. 1975. Soil Taxonomy. *A Basic System of Soil Classification for Making and Interpreting Soil Surveys*. Agriculture Handbook No. 436. Washington, D.C.: Soil Conservation Service, U.S. Department of Agriculture.

Soil Survey Staff. 1990. *Keys to Soil Taxonomy*, 4th edn., Technical Monograph. No. 19. Blacksburg, Virginia, USA: Agency for International Development, U.S. Department of Agriculture, Soil Management Support Services and Virginia Polytechnic Institute and State University.

Toth, J. 1962. A theory of groundwater motion in small drainage basins in Central Alberta. *J. Geophys. Res.* 68:4375–4387.

Watmough, S.A., and Hutchinson, T.C. 1999. Change in the dendrochemistry of sacred fir close to Mexico City over the past 100 years. *Environ. Pollut.* 104:79–88.

Zottl, H.W., and Tschinkel, H. 1971. *Nutrición y Fertilización Forestal: Una Guía Práctica*. Medellín, Colombia: Centro de Publicaciones, Universidad Nacional de Colombia.

4. Forests in the Basin of Mexico: Types, Geographic Distribution, and Condition

Carlos Rodríguez-Franco

Most developing countries with high human concentrations in large urban areas have exhibited a notorious, accelerated reduction of their renewable natural resources. Forest resources, in particular, are heavily impacted by the strong pressures caused by new human settlements, the expansion of agriculture and animal husbandry activities, expanding urban infrastructure, and uncontrolled extraction of different products for consumption such as firewood, medicinal plants, timber for rural construction, food for humans and livestock, and water.

Forest resources are overexploited in the vicinity of large human populations as a matter of economic necessity to satisfy basic human needs of people who are struggling to survive and support their families. Under these conditions, conservation of natural resources or the environment is not a priority. This is exacerbated by the fact that a majority of those exploiting forest resources in a nonsustainable fashion are not aware of the value to society of managing these resources in an ecologically sustainable manner so that ecosystem services such as the capture of carbon dioxide, enhanced water infiltration for restoring aquifers, and the control of soil erosion can be sustained indefinitely.

The Basin of Mexico is no exception to the processes listed above. In the first place, since the foundation of Mexico City in the pre-Hispanic era, the Basin has contained large human concentrations. The impact of these settlements on the natural resources of this region, however, has evolved over time. Second, human values have changed over time; during the development of the area over the last 50 years, natural conservation has been accorded less importance. The consequences of this disregard for the natural resources of the Basin of Mexico, are now environmental matters of grave concern to government officials and others responsible for providing services to the ever-growing population (de la Lanza Espino and Garcia-Calderon 2002; Ezcurra et al. 2002; both in this volume).

The Basin of Mexico has a total area of 302,000 ha of land having the appropriate characteristics to grow forests, 40.3% of the total land area of the Basin. Much of this land was covered by forests at one time. Of this amount, 156,000 ha are now covered with trees. The forests in the Basin of Mexico have a wide diversity of vegetation including 14 pine species, one fir species, one species of white cedar, and several oak species. The main types of forest communities are pine forests, fir forests, oak forests, juniper forests, cloud forests, grasslands, xerophytic vegetation, halophytic vegetation, and scrub oak (Rodríguez and Magaña 1991).

The main factors affecting forest vegetation in the Basin of Mexico are high demographic pressure, illegal cuttings, forest fires, insect attack, soil erosion, uncontrolled grazing, and deforestation. These factors are responsible for the perturbation of approximately 13,000 ha of land on which some degree of soil erosion has occurred (Rodríguez and Magaña 1991; SARH, 1994). There are also an additional 29,500 ha of land with such severe erosion problems that restoration through reforestation activities and soil protection efforts will be needed. The objective of this chapter is to describe the condition, types, and extent of the forests surrounding Mexico City in the region known as the Basin of Mexico.

Location and Climate of Montane Areas in the Basin of Mexico

Rzedowski and Rzedowski (1979) define the Basin of Mexico as a closed watershed, with Mexico City located in the lower part. The Basin has a surface area of close to 7500 km^2 and its shape is slightly enlarged in the NNE–SSW direction. Its longer axis is 130 km in length and its maximum width is close to 90 km. The Mexico City Basin is part of the Transverse Volcanic mountain range, although its extensive flat surface represents a discontinuity in the mountain range system. The principal mountain ranges surrounding the watershed are oriented from north to south, but volcanic ranges on the east, south, and west partially enclose the Basin. To the north, the Basin is limited by a series of hills and low ranges. The most important mountains in the Basin of Mexico, those having the highest altitudes, are located at the edges of the Valley: to the east, the Sierra Nevada with the Popocatépetl Volcano (5452 m), Ixtaccíhuatl Volcano (5284 m), Telapon Hill (3996 m), and Tlaloc peak (4310 m); to the west, the Sierra de las Cruces and Monte Alto ranges with altitudes close to 3400 m. The Guadalupe mountain range in the center of the Basin of Mexico occupies almost 100 km^2 with altitudes slightly less than 3000 m. Finally, toward the south, the Santa Catarina mountain range (2750 m), which includes the Cerro del Pino peak (2800 m) is found (Rzedowski and Rzedowski 1979). Figures 3.1 and 3.2 in the previous chapter show the locations of these mountain ranges.

Much of the flat region of the Basin, the valley floor, covers the same area which originally formed the lake bottom. The elevation of this area varies from 2230 to 2250 m. Other parts of the flat region originated mostly from alluvial sediments, which are often characterized by gradual slopes. These slopes are located in the northern half of the Basin, where there is a large, continuous, gently sloped surface area oriented SW–NE, and extending from Zumpango and Los Reyes to Pachuca. In the central region of the Basin, there are flat areas forming a belt around the zone formed by the lake and in contact with the edges of the hills (Figs. 4.1 and 4.2). The largest area in this zone is close to Amecameca and located 2500 m above sea level (Rzedowski and Rzedowski 1979).

The climate of the Basin of Mexico is defined as tropical because of its latitude and altitude. The climate is similar to that of temperate cold regions based on values of the mean annual temperature, but the Basin has tropical weather characteristics because of the lack of well-defined thermal seasons. On the other hand, the rainy season is pronounced, and the Basin experiences wide temperature differences during the day. The mean temperature in most of the lower elevations of the Basin exhibits a difference of 5° to 7° C between the hottest and coldest months. In the montane regions, this difference is less than 5° C. The mean annual temperature at the lower elevations of the Basin ranges from 14° to 17° C, and that of the mountain ranges varies from 9° to 11° C. The mean annual rainfall ranges from 500 to 700 mm for two thirds of the Basin. Mean annual rainfall of 1000 to 1200 mm is restricted to the mountain ranges. Dominant winds in most of the Basin come from the northeast, although during the dry season (from the end of October to the end of May), there are frequent winds blowing from the northwest (Ezcurra et al. 1999; Rzedowski and Rzedowski 1979).

Forest Vegetation Types of the Basin of Mexico

A complete study of the vegetation of the Basin of Mexico has not been done, although according to Rzedowski and Rzedowski (1979), some important contributions are the work of Reiche (1914, 1923), Martinez (1948), Miranda (1963), and Rzedowski et al. (1964). There are also some studies describing the vegetation of certain locations within the Valley of Mexico (Anaya Lang 1962; Batalla and Ramírez Cantú 1939; Beaman 1962, 1965; Bopp 1956; Rzedowski 1954, 1957) and other studies related to plant communities (Cruz Cisneros 1969; Madrigal Sánchez 1967; Rzedowski and Rzedowski 1979; Sánchez 1968; Villegas 1970).

The distribution of the major vegetation types in the Basin of Mexico is shown in Figure 4.1 (see color plate). The main forest types are well represented in the national parks of the Basin. Together, the state of Mexico and Mexico City have 18 national parks covering 121,206 ha of protected area (McAllister 1991). Of these parks, 15 are located in the Basin of Mexico, encompassing 54,677 ha. The types of forest protected by these national parks are given in Table 4.1, and the

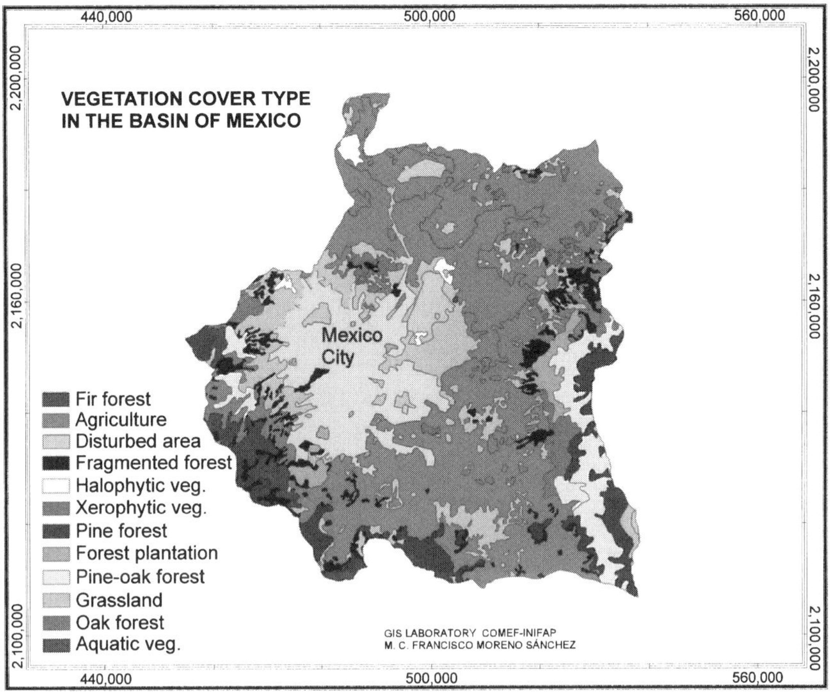

Figure 4.1. Location of major vegetation types within the Basin of Mexico. (See color plate.)

locations of the main national parks in the Basin are presented in Figure 4.2. The main forest types in the Basin of Mexico are described below in order of importance.

Pine Forest

Native pine stands represent 43% of the total forested land cover in the Basin of Mexico. They are found at altitudes from 2350 to 4000 m and grow in locations receiving from 700 to 1200 mm of annual precipitation with annual mean temperatures ranging from 12° to 18° C. They grow in both deep and shallow soil, and sometimes in rocky soils. The main pine species are: *Pinus hartwegii*, *P. montezumae*, *P. ayacahuite*, *P. leiophylla*, *P. rudis*, *P. teocote*, *P. patula* and *P. pseudostrobus*. The understories of these pine forests are relatively poor in brush but have an abundance of bunch grasses, commonly known as *zacaton* (*Muhlenbergia*, *Festuca* and others). These grasses are highly correlated with the occurrence of frequent forest fires. Most of the pine forests are concentrated in the southern part of the Basin of Mexico. They occupy 66,571 hectares in the Basin of Mexico (Maass et al. 1981; Rzedowski and Rzedowski 1979; SARH 1994).

P. leiophylla is the pine species that dominates between 2350 and 2600 m elevation. It forms mixed forests intermingling with several species of *Quercus* and sometimes other tree species. *P. leiophylla* trees measure from 8 to 12 m in height and do not grow as dense stands. They are found in the southern part of the Basin of Mexico in sites that have been disturbed, mainly by forest fires (Martinez 1948; Rzedowski and Rzedowski 1979; Sánchez 1968).

P. montezumae forests are found at altitudes between 2500 and 3100 m, almost always in the south central portion of the Basin. These forests are from 20 to 30 m in height with dense populations that grow mainly in pure stands, although they can be mixed with other species of *Pinus*, or with some species of *Quercus*, *Abies*, *Arbutus*, *Salix* and *Buddleia* (Maass et al. 1981; Rodríguez 1980; Rzedowski and Rzedowski 1979; Sánchez 1968).

P. rudis forests are found between 2700 to 3000 m in the northern and eastern mountains of the Basin. This is a medium sized tree that grows to 10 to 20 m in height. *Pinus rudis* is highly affected by forest fires because of the high resin content in its stem and branches. It also burns rapidly because of the drought conditions where it grows. This species is considered to be highly adapted to grow on land with disturbances such as forest fires. An important sign of this adaptation is its capacity to sprout from the stem in order to regenerate itself. *P. rudis* sometimes grows mixed with *Quercus*, *Alnus* or *Juniperus* trees. These forests

Table 4.1. National parks located in the Basin of Mexico

Name	County	Area (ha)	Main type of forest
El Sacromonte	Amecameca	45	Pine forest
Zoquiapan	Ixtapaluca	19,418	Pine forest, fir forest, and pine–oak forest
Molino de Flores	Texcoco	55	Hardwood forest
Los Remedios	Naucalpan	400	Pine forest
La Marquesa	Mexico State and Mexico City	1,750	Pine forest, fir forest
Iztaccíhuatl-Popocatépetl	Mexico, Puebla and Morelos States	25,679	Pine forest, fir forest
Molino de Belem	Miguel Hidalgo Delegation	17	Mixed forest and hardwood forest
Cerro de la Estrella	Iztapalapa Delegation	1,100	Hardwood forest
Desierto de los Leones	Cuajimalpa and Alvaro Obregón Delegations	1,866	Pine forest, fir forest, and oak forest
Cumbres del Ajusco	Tlalpan Delegation	920	Oak forest, pine forest, and fir forest
Fuentes Brotantes de Tlalpan	Tlalpan Delegation	129	Hardwood forest, mixed forest, and pine forest
Historico de Coyoacan	Coyoacán	584	Hardwood forest
Lomas de Padierna	Alvaro Obregón Delegation	670	Oak forest, pine forest, and fir forest
El Tepeyac	Gustavo A. Madero	294	Cloud forest
Miguel Hidalgo y Costilla	Cuajimalpa, Mexico City and Huixquilucan, Mexico state	1,750	Hardwood forest, and pine forest

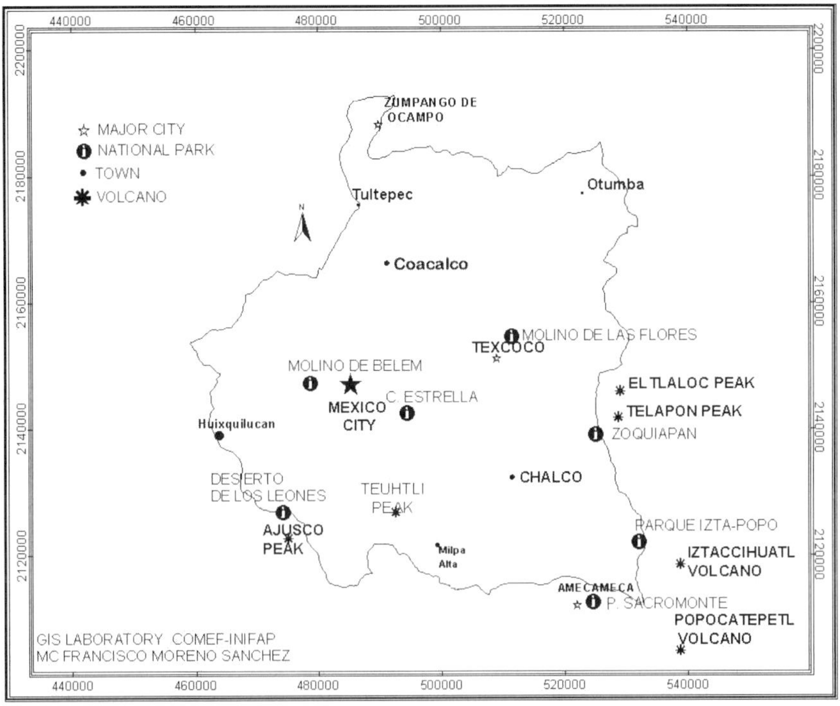

Figure 4.2. Map of the Basin of Mexico showing the location of national parks.

develop in shallow poor soils and are the pine forests that grow in drier conditions in the region (Martinez 1948; Rzedowski and Rzedowski 1979; Sánchez 1968).

P. hartwegii forests are the dominant species at higher altitudes. They are found from 2900 to 4000 m in pure stands, although sometimes they can be found mixed with small numbers of *Alnus firmifolia* trees. These forests are often found on steep slopes, with shallow, poor and rocky soils. They reach from 5 to 20 m in height, and stand density varies from moderate to open. The grass cover is sometimes greater than the tree cover (Maass et al. 1981; Martinez 1948; Rzedowski and Rzedowski 1979; Sánchez 1968). These forests grow near the limit of the arboreal vegetation in the Basin of Mexico. *P. hartwegii* is also the pine species in the Basin which is the most sensitive to ozone. Affected trees develop chlorotic mottle, premature senescence of the foliage, and thinned crowns (Miller et al. 2002).

Pine forests with more restricted distribution in the Basin of Mexico are formed by *P. teocote*, *P. ayacahuite*, *P. pseudostrobus*, and *P. patula*. The last two species are typically found in climates that are more humid than in the Basin of Mexico.

There are many herbaceous and shrub species associated with pine forests in the Basin of Mexico; these species are presented in Table 4.2 (Martinez 1948;

Table 4.2. Plant species associated with major forest vegetation types in the Basin of Mexico

Forest type	Associated shrub species	Associated herbaceous species
Pine forest	*Pernettya ciliaris, Ribes pringlei, Senecio angulifolius, Eupatorium glabratum*	*Alchemilla, Archibaccharis, Arenaria, Bidens, Eryngium, Eupatorium, Festuca, Geranium, Gnaphalium, Lupinus, Muhlenbergia, Penstemon, Ribes, Senecio, Stevia, Stipa*
Fir forest	*Symphoricarpos microphyllus, Pernettya ciliata, P. ciliaris, Baccharis conferta, Eupatorium glabratum, Ribes rugosum, Buddleja sessiliflora, Senecio angulifolium, S. platanifolius, S. barba-johannis, S. sanguisorbae*	*Acaena elongata, Fragaria indica, Asplenium monanthes, Brachypodium mexicanum, Sigesbeckia jorulensis, Alchemilla procumbens, Salvia elegans, Galium aschenbornii, Eupatorium patzcuarense, Cirsium pinetorum, Gaura coccinea, Geranium potentillaefolium, Castilleja tennuiflora, Penstemon gentianoides, Muhlenbergia sp., Thuindium delicatulum, Bryum procerum*
Oak forest	*Baccharis, Senecio, Eupatorium, Lamourouxia, Salvia, Symphoricarpos*	*Brickellia, Castilleia, Dahlia, Desmodium, Galium, Geranium, Muhlebergia, Penstemon, Stevia, Thalictrum, Valeriana*
Cloud forest	*Archibaccharis sescenticps, Cestrum terminale, Eupatorium aschenbornianum, Iresine ajuscana, Lamourouxia xalapensis, Lippia umbellate, Montanoa frutenscens, Salvia mocinnoi*	*Adiantum andicola, Bidens ostruthioides, Bromus dolichocarpus, Dryopteris parallelogramma, Peperomia hispandula, Phanerophlebia nobilis, Pteris cretica*
Scrub oak forest	*Dasylirion acrotriche, Nolina parviflora, Pithecellobium leptophyllum, Rhus standleyi*	*Brickellia, Desmodium, Stevia*
Grasslands		*Calamagrostis tolucensis, Festuca amplissima, F. livida, F. tolucensis, Muhlenbergia macroura, M. quadridentata, Stipa ichu, Pennisetum sp., Poa sp., Phacelia platycarpa, Gaura coccinea, Rumex acetosella, Castilleja tenuiflora, Polygonium acetosum* and *Platago sp.*
Cloud forest: epiphytic climber species	*Archibaccharis hirtella, Celastrus pringlei, Clematis dioica, Philadelphus mexicanus, Smilax moranensis, Solanum appendiculatum, Valeriana clematitis*	
Juniper forest	*Agave atrovirens, Quercus microphylla, Gymnosperma glutinosum, Brickellia veronicifolia, Eupatorium espinosarum, E. scorodonnioides, Stevia salicifolia*	

Rzedowski and Rzedowski 1979). More detailed botanical descriptions of Mexican pines species can be found in Look (1950), Mirov (1967), and Perry Jr. (1991).

The main insect pests of pine forests in the Basin of Mexico are bark beetle species including *Dendroctonus adjunctus* and *D. mexicanus*, foliar-feeding insects such as species of the genus *Neodiprion* and *Zadiprion*, and cone beetle species of the genus *Cydia* and *Conophthorus*. For a detailed description of the taxonomy, host ranges, damages caused, life cycles, habits, and importance of the forest insects of Mexico, see Cibrián-Tovar et al. (1995). A companion volume by Cibrián-Tovar et al. on the forest diseases of Mexico is in preparation as of this writing (2002).

Fir Forest

Abies religiosa is the only native fir species in the Basin of Mexico. Fir forests in the Basin are very well defined by their physiognomy. *A. religiosa* trees grow at altitudes ranging from 2700 to 3600 m and are almost always found on sites with high relative humidity throughout the year and with deep, well-drained soils with high organic matter content. The mean annual precipitation of *A. religiosa* sites varies from 1000 to 1400 mm and the mean annual temperature from 7.5° to 13.5° C. Fir forests are distributed in the highlands of the southern half of the Basin of Mexico covering 12,654 hectares (Benítez 1985; Maass et al. 1981; Rzedowski and Rzedowski 1979; SARH 1994).

This forest is evergreen and dense, and tree heights reach an average of from 20 to 40 m. The forest has one or two strata. The shrub and herbaceous strata are scarce under natural conditions, but they increase substantially when the forest has been disturbed. The forest soil is often covered almost totally by mosses, and lichens are common as epiphytic organisms (Maass et al. 1981; Rodríguez 1982; Rzedowski and Rzedowski 1979).

The dominant tree species in the main stratum is *Abies religiosa*, although sometimes other tree species, such as *Alnus firmifolia*, *Cupressus lindleyi*, *Quercus laurina*, *Salix oxylepis*, *Pseudostsuga macrolepis*, *Garrya laurifolia*, and *Prunus serotina* spp. *capuli*, are found mixed in with the firs. The most common species in the shrub and herbaceous strata are presented in Table 4.2. (Maass et al. 1981; Rzedowski and Rzedowski 1979).

Oak Forest

Oak tree species grow in ecological habitats that are very similar to pine forests within an elevational range from 2350 to 3100 m. Soils vary from deep to shallow, and this forest group occurs in areas that have from 700 to 1200 mm of annual precipitation. There are several types of oak forests in the Basin, and they are classified according to the dominant oak species. Most of the oak forests in the

Basin of Mexico are small in size, reaching heights from 5 to 12 m, with moderate stand density. Many oak species lose their leaves for several weeks; others are evergreen, or practically evergreen. At elevations below 2500 m, the dominant oak species are *Quercus laeta*, *Q. deserticola*, *Q. crassipes* and *Q. obtusata*. Frequently, several oak species are found growing together, and *Pinus leiophylla* often occurs in this association as well. Pine–oak forests occupy 2352 hectares in the Basin of Mexico (Martínez and Matuda 1979; Rzedowski and Rzedowski 1979; SARH 1994).

Q. rugosa forests grow between 2500 and 2800 m. This oak species has large and moderately rigid leaves. *Q. rugosa* can be found in pure stands, but is most frequently associated with Q. *mexicana* or *Q. crassipes*, *Arbutus xalapensis* and some species of *Pinus*. *Cupressus*, *Garrya* and *Clethra* can also be found occasionally in oak forests. In some locations close to Amecameca, *Q. rugosa* stands reach a height of 25 m, and in the Pedregal of San Angel, and in certain locations in the northern part of the Basin, this species grows as a shrub (Martínez and Matuda 1979; Rzedowski and Rzedowski 1979).

Q. laurina grows at higher elevations, between 2800 to 3100 m. It is the most common oak forest in the Basin of Mexico. This species has thin leaves and is small in height, ranging from 8 to 12 meters. Sometimes it is found mixed with *Q. crassifolia*, *Q. rugosa*, *Abies*, *Arbutus*, *Juniperus* and some species of pine trees (Martínez and Matuda 1979; Rzedowski and Rzedowski 1979).

In drier locations, mainly in the northern part of the Basin, *Q. microphylla*, *Q. greggii* and *Q. mexicana* are dominant species in oak forests. These species form very low forests that reach from 3 to 5 m in height. They also quite often form open stands. The shrub and herbaceous genera of the oak forests are presented in Table 4.2. (Martínez and Matuda 1979; Rzedowski and Rzedowski 1979).

Cloud Forest

This type of vegetation, also known as mesophyllous forests, grows in a small area no larger than 2 km^2 on steep hillsides and in narrow canyons located at the base of the Ixtaccíhuatl Volcano and in the Las Cruces mountain range. It is found between 2500 and 2800 m of elevation in sites protected from excessive insolation and winds where the mean annual precipitation is more than 1000 mm, and the mean annual temperature varies from 12° to 14° C. The soils are generally deep, rich in organic matter, and high in water content almost throughout the year (Rzedowski and Rzedowski 1979).

The forest varies from 10 to 25 m in height and is dense, and most of the species found are evergreen, although some are deciduous. These forests are characterized by an abundance of epiphytes, mostly mosses and ferns, and woody climbers that cover the trunks and branches of the dominant trees (Rzedowski and Rzedowski 1979). The dominant tree species in this type of forest are *Clethra mexicana*, *Cornus disciflora*, *Garrya laurifolia*, *Ilex tolucana*, *Meliosoma dentada*, *Prunus brachyobotrya*, and *Quercus laurina*. Other tree species that can sometimes be found are *Abies religiosa*, *Alnus arguta*, *Buddleia cordata*, *Cornus excelsa*, *Cupressus lindleyi*, *Eupatorium mairetianum*, *Pinus ayacahuite*, *P. patula*, *P. pseudostrobus*, *Prunus serotina spp*, *capuli*, *Quercus rugosa*, *Sambucus mexicana*, and *Viburnum stenocalyx*. Some of the shrubs, herbaceous species and epiphytic climber plant species associated with this type of forest are listed in Table 4.2.

Juniper Forest

This kind of forest is open and low in stature. It occupies relatively extensive areas on hillsides and in semiarid plains in the northern and eastern parts of the Basin, at elevations ranging from 2450 to 2800 m. The dominant species in this forest is the Junipero (*Juniperus deppeana*), a small, evergreen tree that reaches a height of from 3 to 6 m and has small scaly leaves (Rzedowski and Rzedowski 1979). The mean annual temperature varies from 11° to 14°C, and the mean annual precipitation ranges from 600 to 800 mm.

Juniper forest in the Basin of Mexico is a secondary succession ecosystem that appears after the destruction of pine and oak forests. The other two species of *Juniperus* in the Basin are *J. monticola* and *J. flaccida*. The first species sometimes grows as a dense bush. It is a successional stage after the establishment of *Abies* forest, and is sometimes present as a creeping shrub above the limit of arboreal vegetation. The second species only grows as isolated individuals (Rzedowski and Rzedowski 1979). These juniper forests have much open space favoring the growth of shrubs and herbaceous plants that are listed in Table 4.2.

Scrub Oak Forest

This shrubby community is dominated by *Quercus frutex*. This species grows in a dense cover due to its root-based vegetative propagation. It is frequently found in the northeastern part of the Basin, but is also found in the north, east, west, and center of the Basin (Rzedowski and Rzedowski 1979). It grows at elevations from 2350 to 3000 m in shallow soils with a mean annual precipitation varying from 700 to 900 mm, and a mean annual temperature from 9° to 13°C. The height of this bush community varies between 20 and 100 cm but is more commonly between 40 to 80 cm. The dominant species is deciduous, although it is without foliage for only a brief period. The underbrush of *Quercus* in the Basin seems to be an induced community maintained by fire, because it grows in sites where pine and oak forests were previously found. The species associated with this type of forest are listed in Table 4.2. (Rzedowski and Rzedowski 1979).

Grasslands

Rzedowski and Rzedowski (1979) indicated that there are at least five types of grass communities growing from 2250 to 4300 m in the Basin of Mexico. The grassland of *Hilaria cenchroides* is the most important because of the amount of surface area that it occupies. It prospers mainly on hillsides of moderate slopes between 2300 and 2700 m with a mean annual precipitation of 600 to 750 mm (Rzedowski and Rzedowski 1979). This grassland covers large areas of land in the northwest of the Basin, in the region of Huehuetoca, Tepotzotlán, and Tlalnepantla. It is found in lowlands and along the foothills of the Monte Alto and Sierra Nevada ranges. The species found with *Hilaria cenchroides* are *Abildgaardia mexicana*, B*outeloua radicosa*, *B. hirsuta* and *Stevia serrata* (Rzedowski and Rzedowski 1979).

Another kind of rangeland is that formed by *Buchloe dactyloides*. This is a very small grass, which forms a dense covering due to vegetative reproduction by root sprouts. This species is found in small patches at altitudes from 2250 to 2800 m. The most extensive patch is located along the east side of Zumpango Lake (Rzedowski and Rzedowski 1979).

A frequent and extensive grassland is a mixed grass in which annual grasses such as *Aristida adscensionis* and *Bouteloua simplex* often prevail. *Lycurus phleoides*, *Hilaria cenchroides*, *Erioneuron spp.*, *Bouteloua spp.*, *Enneapogon desvauxii* and other species can also be found (Rzedowski and Rzedowski 1979, 1990).

Potentilla candicans (Sínfito) grows in the Basin of Mexico at altitudes ranging from 2900 to 3500 m occupying clearings in the *Abies* and *Pinus* forests. It grows in places where soil drainage is low. In the dry season, this species is found as a ground cover; in the rainy season a group of species from gramineous and cyperaceous families form a dense carpet of grasses and sedges dominated by *Muhlenbergia repens*, *Festuca myuros*, *Deschampsia pringlei*, *Cyperus seslerioides*, and *Carex peucophila*. At higher elevations (4000 m or more), above the timberline of the *Pinus hartwegii* forests, are alpine meadows of *Muhlenbergia* and *Festuca*. These are bunchy tussock grasses with clusters of tough erect leaves 60 to 120 cm high. Forest fires are an important ecological factor in the existence, dynamics, and utilization of these grasslands. One of the main characteristics of this grassland is that, together with *Pinus hartwegii* forests, they are the habitat for the endangered species of volcano rabbit (*Romerolagus diazi*). The common species of these associations of grasses are listed in Table 4.2. (Ezcurra et al. 1999; Maass et al. 1981; Rzedowski and Rzedowski 1979).

Forest Plantations in the Basin of Mexico

In the Mexico City area there are 3201 hectares planted as forest plantations, using mainly native species. Little data is available, however, on their characteristics (SARH 1994). Since 1970, several efforts to establish forest plantations have been attempted through formal government programs. In 1973, the Ministry of Agriculture and Water Resources established forest plantations in the eastern slopes surrounding the former Lake Texcoco with the aim of recovering and protecting soil in the area. From 1973 to 1976 there were 1,766,733 seedlings of nine forest species planted; four of them were exotic species (*Pinus radiata*, *P. halepensis*, *Casuarina sp.*, and *Eucalyptus sp.*) and five were native species (*Pinus michoacana*, *P. montezumae*, *P. patula*, *Cupressus sp.*, and *Acacia sp.*). These forest plantations covered 2651 ha near the towns of San Pablo Ixayoc, Tequexquinahuac, Huexotla, and Coatlinchan in Texcoco County and near the town of Coatepec in Chimalhuacan County (Pedraza 1983).

A forest inventory, carried out in 1982 when the species were nine years old, showed that *Eucalyptus* was the most prospering hardwood species as indicated by its fast diameter and height growth. The mean height of the *Eucalyptus* trees was 4.34 m, and the mean diameter was 4.23 cm (Pedraza and Rodríguez 1985). Eleven years after the seedlings were planted, Pedraza, Rodríguez, and Fierros (1987) also reported that *Eucalyptus* was the fastest growing species in these forest plantations. The average number of trees of this species per hectare was 246 with an average height of 4.5 m and average diameter at breast height (dbh) of 4.5 cm. According to these authors, the types of soil where these forest plantations were growing were yellow, gray, and red hardpan soils and soils without hardpan. The fastest growth rates were found in *Eucalyptus* growing in red hardpan soils with an average height of 6.06 m and a dbh of 6.29 cm (Pedraza et al. 1987).

Factors Affecting Forests in the Basin of Mexico

Because of the high demographic pressure in this region, the effect of natural and anthropogenic disturbances is notorious. There are a high number of forest fire events each year. These fires affect more than 11,000 ha/yr of grassland, shrubs, forest, and regeneration areas. The highest occurrence of forest fires in Mexico occurs within the Basin of Mexico. Every year, there are approximately 7600 forest fires in the country, of which 3298 (2215 in the State of Mexico and 1083 in Mexico City) or 43% of the total number of forest fires per year occur in the Basin. Approximately 14,000 ha are burned per year in the Basin, accounting for 12% of the total forested area burned in Mexico (SEMARNAP 1999). The average land area affected by each fire in the forested area of Mexico City is 2 ha, compared with an average of 5 ha per fire in forests of the State of Mexico. Other important states in which considerable amounts of forest land are burned, although they experience a much lower number of forest fires, include Chiapas, Durango, and Chihuahua. In Chiapas, where approximately 335 forest fires occur each year affecting an average of 63 ha per burn, a total of 21,000 ha burn per year. Durango experiences 316 forest fires per year affecting an average of 74 ha burned per event amounting to 23,500 ha affected per year. In Chihuahua, approximately 513 forest fires occur per year affecting an average of 34 ha per event, and around 17,300 total ha (SEMARNAP 1999). Forest fires in the region are the result of uncontrolled grazing in forest areas. Areas of grass are intentionally burned during the dry season with the aim of creating a more vigorous resprouting of the grass during the next rainy season. These fires often get out of control and spread to other areas. Uncontrolled grazing in forest areas is also responsible for problems related to soil compaction, soil erosion, decreased water infiltration, high mortality of trees in stand regeneration areas, and deformation of the growth of young seedlings (Rodríguez and Magaña 1991).

Forests in the Basin of Mexico produce around 300,000 m^3 of pine and fir wood (not including illegal cuttings) annually. However, in addition, due to the high demand for firewood needed for domestic and small industry use, illegal harvesting in the forests of the Basin of Mexico is causing an overexploitation that, in turn, causes soil erosion, loss of biodiversity, and stressed or injured trees predisposed to insect and disease attack. The loss of forested area due to changes in land use from forest to agriculture, urbanization, industrial development, and road expansion is also of grave concern. According to the National Forest Inventory, there are 13,000 ha impacted by these disturbance factors (Rodríguez and Magaña 1991; SARH 1994).

The Protectora de Bosques (PROBOSQUE, an agency) of the government of the state of Mexico, has the most recent data on forest distribution and stand characteristics. They have vegetation maps based on aerial photography (scale of 1 : 20,000) and land ownership maps with forest stands delimited by watersheds for each county around the Basin of Mexico. The cartography was the basis of a state forest inventory and estimate of the parameters of the forests of the Basin of Mexico (PROBOSQUE 1990). The study provided forestry data for eight of the 21 municipalities in the Basin of Mexico: Huixquilucan, Jalatlaco, Juchitepec, Lerma, Naucalpan, Ocoyoacac, Ocuilan de Arteaga, and Tianguistenco. These eight municipalities cover a total area of 141,908 ha, with a forested area of 34,287 ha. By comparison, the total forested area in the Basin of Mexico is 156,000 ha (Rodríguez and Magaña 1991).

The counties closest to Mexico City had less tree cover. For example, at Naucalpan, located 11.3 km from the main square of Mexico City, the stand density was 210 trees/ha with a wood volume of 102.9 m^3/ha. In those counties farther away from Mexico City, the situation is reversed. For example, Ocuilan de Arteaga and Juchitepec, located 17.1 and 18.5 km from the main square of Mexico City, had stand densities of 410 and 417 trees/ha and wood volumes of 282.9 and 278.4 m^3/ha (Table 4.3). This trend provides further evidence of demographic pressure exerted on the forests closest to urban areas. The locations of these counties with respect to Mexico City are presented in Figure 4.3.

The values for tree height, dbh, stand density, and volume varied widely (Table 4.3.), indicating highly irregular stand structures in the forests in the Basin of Mexico. The irregular stand structure is associated with the frequent occurrence of noncatastrophic disturbances which result in a new age class of trees in the stand, thus affecting the average values of the stand and giving these extreme

Table 4.3. Average stand characteristics of pine forests in the Basin of Mexico

County	Average tree and stand characteristics				Distance from downtown Mexico City km
	Height (m)	dbh (cm)	Volume (m^3/ha)	Trees/ha	
Naucalpan	16.5 (5.6–32.3)	27.5 (12.3–65.2)	102.9 (93.1–115.2)	210 (190–235)	11.3
Huixquilucan	16.2 (5.6–31.8)	25.5 (12.5–43.1)	139.8 (97.9–181.7)	267 (187–347)	11.7
Ocoyoacac	13.9 (5.8–30.0)	26.1 (12.0–56.1)	120.2 (97.3–143.1)	231 (188–275)	12.5
Jalatlaco	12.1 (11.0–13.0)	19.2 (17.6–20.7)	151.9 (139.7–164.0)	280 (257–302)	12.6
Lerma	16.2 (6.0–34.0)	28.4 (11.1–67.5)	165.1 (140.3–189.9)	271 (230–311)	14.1
Tianguistenco	17.7 (5.2–29.6)	23.4 (7.5–39.4)	129.5 (112.6–146.9)	243 (222–265)	15.0
Ocuilan de Arteaga	20.2 (8.2–31.5)	31.2 (16.7–43.7)	282.9 (268.7–297.0)	410 (378–443)	17.1
Juchitepec	16.5 (8.5–32.0)	29.3 (12.1–48.4)	278.35 (253.3–303.4)	417 (355–466)	18.5

Values in parentheses are ranges.

values. The main forest disturbances that are the sources of this situation are forest fires, uncontrolled grazing, illegal cutting, silvicultural treatments, insect attack, disease, and air pollution. Except for air pollution which is a chronic stress in forests SSW of the city, these disturbances occur almost every year in the same general area, but not in the same forest stand.

These disturbance factors, together with the lack of appropriate forest-management practices needed to protect these watersheds, are the main causes of the severe problems related to water supply for human consumption in the vicinity of large cities such as Mexico City and its suburbs and also near Guadalajara and Monterrey. The scarcity of water in these cities is due to the decreasing water level of the aquifers which is no longer enough to meet the local demand for water. Therefore, it has been necessary to import great amounts of water from distant watersheds, greatly increasing the cost of water.

The Government of the State of Mexico recently allocated additional funds to increase the amount of water available to cover the demand in Mexico City and its surrounding 17 counties. If this trend continues, future water demand will entail even higher costs and cause greater environmental and technological challenges. The severity of this problem was evidenced when the Federal District Water Authority announced in 1998 that water service would be suspended for one million people per day during the drought period.

In order to reverse this problem, it will be necessary to modify land use in some areas, increase forest cover, reforest with fast growing species, effectively manage forest plantations, and create private water-production companies to

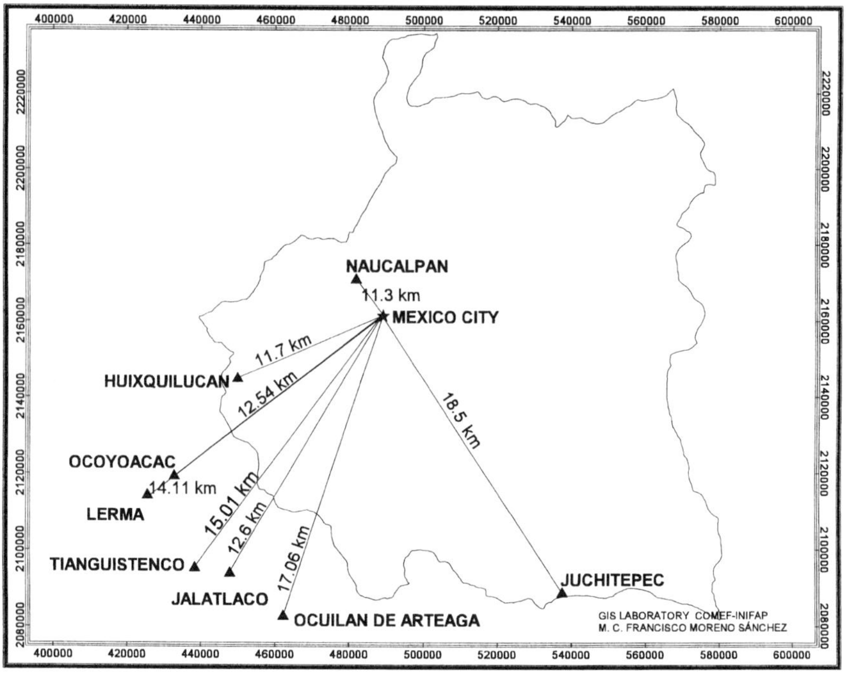

Figure 4.3. Location of eight sites within the Basin of Mexico where forest inventory data was collected in 1994 by the National Research Institute of Forestry, Agriculture, and Animal Husbandry (INIFAP) under a memorandum of understanding with PROBOSQUE (Protectora de Bosques, State of Mexico).

supply urban water demand. It will also be necessary to establish an efficient system of valuation and payment for the owners of forests and forest plantations, a system based on water production, recharge of aquifers, soil protection, biodiversity, and long distance water transport. In addition, it will be necessary to carry out environmental education programs at all levels with the aim of highlighting the importance of water as a natural resource and principles for its efficient use and conservation. Expansion of rain collection systems in urban and forested areas is also needed to optimize water use and management.

Summary

Most developing countries with high demographic concentrations in large urban areas share a notorious deterioration of their natural resources around these cities. This is due to the many anthropogenic stresses on natural resources due to growth of the cities, changes in land use from forest to urban, agricultural and industrial activities, and the expansion of infrastructure such as electric power lines, water storage, and conveyance systems and roads. Other causes of deterioration are illegal cutting in forests, wildfires, extraction of wood products for construction and firewood, and harvesting of nonwood products for human consumption and medicinal uses. As a result, forests surrounding population centers, for example, many of the forests surrounding the Basin of Mexico, are highly disturbed and stressed and susceptible to insect, disease, and air pollution impacts.

The main recommendations derived from this review are:

1. Conduct a forest inventory of natural resources in the Basin of Mexico, including forest vegetation (trees, shrubs and herbs), wildlife, hydrology, insect and disease incidence, soils, and the effects of disturbances to provide currently lacking data on the state of resources in the basin.
2. Increase the extent of protected areas.
3. Increase the amount of land area used for forest plantations in order to recover disturbed natural forests, prevent further erosion of soils, and improve areas with high potential to grow trees.
4. Based on the results of this forest inventory, develop a master plan for land use, taking into account urban growth, natural areas and vegetation to be protected, growth of industry, and other uses.

References

Anaya Lang, M.L. 1962. *Estudio de las Relaciones Entre la Vegetación, el Suelo y Algunos Factores Climáticos en Seis Sitios del Declive Occidental del Ixtaccíhuatl.* Tesis Profesional. Facultad de Ciencias. Universidad Nacional Autónoma de México. México.

Batalla, M.A., and Ramírez Cantú, D. 1939. Contribución al estudio florístico del Valle de México. Estudio cualitativo de las Sinecias que habitan los cerros del norte del Valle de México. *An. Inst. Biol. (México)* 10:227–267.

Beaman, J.H. 1962. The timberlines of Ixtaccíhuatl and Popocatépetl, México. *Ecology* 43: 377–385.

Beaman, J.H. 1965. A preliminary ecological study of the alpine flora of Popocatépetl and Ixtaccíhuatl. *Bol. Soc. Bot. Méx.* 29:63–75.

Benitez, B.G. 1985. *Arboles y flores del Ajusco.* Mexico City: Instituto de Ecología—Museo de Historia Natural de la Ciudad de México.

Bopp, O.M. 1956. *Contribución al Estudio de la Flora Fanerogámica de los Cerros Situados al Norte de la Ciudad de México: Sierra de Gualdalupe, Cerros Chiquihuite, Ticomán y Zacatenco.* Tesis Profesional. Mexico City: Facultad de Ciencias, Universidad Nacional Autónoma de México.

Cibrián-Tovar, D., Méndez-Montiel, J.T., Campos-Bolaños, R., Tates III, H.O., and Flores-Lara, J.E. 1995. *Forest Insects of Mexico.* Chapingo, Mexico: Universidad Autónoma Chapingo.

Cruz-Cisneros, R. 1969. *Contribución al Conocimiento de la Ecología de los Pastizales en el Valle de México.* Tesis Profesional. Mexico City: Escuela Nacional de Ciencias Biológicas.

de la Lanza Espino, G., and García-Calderon, J.L. 2002. Historical summary of the geology, climate, hydrology, culture, and natural resource utilization in the Basin of Mexico. In *Urban Air Pollution and Forests: Resources at Risk in the Mexico City Air Basin*, Ecol. Stud. Series, eds. M.E. Fenn, L.I. de Bauer, and T. Hernández-Tejeda. New York: Springer-Verlag.

Ezcurra, E., Mazari-Hiriart, M., Pisanti, I., and Aguilar, A.G. 1999. *The Basin of Mexico: Critical Environmental Issues and Sustainability.* New York: United Nations University Press.

Ezcurra, E., Mazari-Hiriart, M., Pisanty, I., and Aguilar, A.G. 2002. Socioeconomic change and its impact on forest resources in the Basin of Mexico. In *Urban Air Pollution and Forests: Resources at Risk in the Mexico City Air Basin*, Ecol. Stud. Series, eds. M.E. Fenn, L.I. de Bauer, and T. Hernández-Tejeda. New York: Springer-Verlag.

Loock, E.E.M. 1950. *The Pines of Mexico and British Honduras: A Report on a Reconnaisance of Mexico and British Honduras during 1947.* Bulletin no. 35. Pretoria, South Africa: Union of South Africa, Department of Forestry.

Maass, J., Rafael, P., Suarez, A., Blanco, S., Cevallos, G., Galindo, C., and Pescador, A. 1981. *Ecología de la Estación Experimental Zoquiapan. Colección Cuadernos Universitarios.* Chapingo, México: Universidad Autónoma Chapingo.

Madrigal-Sánchez, X. 1967. *Contribución al Conocimiento de la Ecología de los Bosques de Oyamel (Abies religiosa (H.B.K.) Schl. Et Cham.) en el Valle de México*, Bol. Téc. 18. Mexico City: Instituto Nacional de Investigaciones Forestales.

Martinez, M. 1948. Los Pinos Mexicanos, 2nd edn. Mexico, D.F.: Editorial Botas.

Martinez, M., and Matuda, E. 1979. *Flora del Estado de México. Edición Facsimilar de los Fascículos Publicados en los Años de 1953 a 1972*, Tomo I, Biblioteca Enciclopédica del Estado de México. Toluca, State of Mexico, Mexico: Gobierno del Estado de México, Fondo Nacional para las Artes (FONAPAS), Instituto Nacional de Bellas Artes.

McAllister, H.M. 1991. Recreación y esparcimiento. In *Proceedings of I Simposio Sobre el Papel de los Bosques del Valle de México en el Siglo XXI*, ed. J.M. Torres Rojo, pp. 23–29, 30 de Mayo de 1991. Chapingo, Mexico: Centro de Genética Forestal A.C.—Instituto Nacional de Investigaciones Forestales Agrícolas y Pecuarias–Protectora de Bosques.

Miller, P.R., de Bauer, L.I., and Hernández-Tejeda, T. 2002. Oxidant exposure and effects on pines in forests in the Mexico City and Los Angeles, California air basins. In *Urban Air Pollution and Forests: Resources at Risk in the Mexico City Air Basin*, Ecol. Stud. Series, eds. M.E. Fenn, L.I. de Bauer, and T. Hernández-Tejeda. New York: Springer-Verlag.

Miranda, F. 1963. Commentary. In *Mesas Redondas Sobre Problemas del Valle de México*, pp. 189–199. Mexico City: Ed. Instituto Mexicano de Recursos Naturales Renovables.

Mirov, N.T. 1967. *The Genus Pinus.* New York: The Ronald Press.

Pedraza, C.E. 1983. *Evaluación Dasométrica de Plantaciones Forestales en la Zona Oriente de la Cuenca del Valle de México.* Tesis profesional. Departamento de Enseñanza, Investigación y Servicio en Bosques. México.

Pedraza, C.E., and Rodríguez, F.C. 1985. Evaluación de plantaciones para protección y recuperación de suelos de la cuenca oriental del Ex Lago de Texcoco. In *Memoria de III Reunión Sobre Plantaciones Forestales*, Pub. Esp. 48, pp. 769–803. México City: Secretaria de Agricultura y Recursos Hidraulicos (SARH).

Pedraza, C.E., Rodríguez, F.C., and Fierros, G.A.M. 1987. Establecimiento de cinco especies forestales en tres tipos de tepetate de la zona de Tequexquinahuac a Coatepec, Estado de México. In *Uso y Manejo de los Tepetates par el Desarrollo Rural*, ed. J.F. Ruiz Figueroa, pp. 128–140. Chapingo, Mexico: Departamento de Enseñanza, Investigación y Servicio en Bosques; Universidad Autónoma de Chapingo.

Perry Jr., J.P. 1991. The Pines of Mexico and Central America. Portland, Oregon, USA: Timber Press.

PROBOSQUE. 1990. *Proceedings of the Segundo Estudio Dasonómico del Estado de Mexico (Sedemex)*, Vols I and II. Estado de Mexico: Secretaria de Desarrollo Agropecuario del Estado de Mexico—Protectora de Bosques.

Reiche, C. 1914. *La Vegetación de los Alrededores de la Capital de México.* México City: Tipografía Económica.

Reiche, C. 1923. Die vegetationsverhaeltnisse in der umgebung der hauptstadt von Mexico. *Bot. Jahrb.* 58 (Beibl. 129):1–116.

Rodríguez, F.C. 1980. *Estudio Epidométrico de Pinus montezumae Lamb., Mediante Análisis Troncales en el C.E.F. de San Juan Tetla, Puebla.* Tesis profesional. Chapingo, Mexico: Universidad Autónoma Chapingo.

Rodríguez, F.C. 1982. *Determinación de la Calidad de Estación de Pinus montezumae Lamb. A Través de Análisis Troncal en el C.E.F. de San Juan Tetla, Puebla.* Tesis de maestría. Chapingo, México: Colegio de Postgraduados, Centro de Genetica–Programa Forestal.

Rodríguez, F.C., and Magaña, T.O.S. 1991. La situación actual de los bosques del Valle de México. In *Proceedings I Simposio Sobre el Papel de los Bosques del Valle de México en el Siglo XXI*, ed. J.M. Torres Rojo, pp. 1–13. May 30. Chapingo, Mexico: Centro de Genética Forestal A.C.—Instituto Nacional de Investigaciones Forestales Agrícolas y Pecuarias—Protectora de Bosques.

Rzedowski, J. 1954. Vegetación del Pedregal de San Angel (Distrito Federal, México). *An. Esc. Nac. Cienc. Biol. México* 8:59–129.

Rzedowski, J. 1957. Algunas asociaciones vegetales de los terrenos del Lago de Texcoco. *Bol. Soc. Bot. México.* 21:19–33.

Rzedowski, J., and Rzedowsky, G.C. 1979. *Flora Fanerogámica del Valle de México. Generalidades, Gymnospermae, Dicotiledoneae*, Vol. I. Mexico City: Editorial CEC S.A.

Rzedowski, J., and Rzedowsky, G.C. 1990. *Flora Fanerogámica del Valle de México. Monocotyledoneae*, Vol. III. Patzcuaro, Michoacan, Mexico: Instituto de Ecología. Centro Regional del Bajío.

Rzedowski, J., Guzmán, G., Hernández-Corzo, A., and Muñiz, R. 1964. Cartografía de los principales tipos de vegetación de la mitad septentrional del Valle de México. *An. Esc. Nac. Cienc. Biol. México.* 13:31–57.

Sánchez, S.O. 1968. *La Flora del Valle de México*, 1st edn. Mexico, D.F.: Editorial Herrero. S.A.

SARH (Secretaria de Agricultura y Recursos Hidraulicos). 1994. *Inventario Nacional Forestal Periódico. Memoria Nacional.* Mexico City: Subsecretaría Forestal y de Fauna Silvestre, SARH.

SEMARNAP (Secretaria del Medio Ambiente, Recursos Naturales y Pesca). 1999. *Atlas Forestal de México.* Mexico City: Dirección General Forestal, SEMARNAP.

Villegas, D.M. 1970. Estudio florístico y ecológico de las plantas arvenses de la parte meridional de la Cuenca de México. *An. Esc. Nac. Cienc. Biol. México* 18:17–89.

5. The Climate of the Mexico City Air Basin: Its Effects on the Formation and Transport of Pollutants

Ernesto Jáuregui

Mexico City is located in the southwestern portion of an elevated basin in the central region of the country and covers an area of 7500 km^2. Volcanic ranges surround the lacustrine plains on the east, south, and west. To the north, the basin is bounded by discontinuous ranges. The lowest portion of the lacustrine plain has an average elevation of 2250 meters above sea level (masl). The main forests occupy the upper piedmont of the ranges above 2500 m to the west (Serranía de las Cruces), south (Serranía del Ajusco), and east (Sierra Nevada) and harbor temperate plant communities of pines, fir, and junipers (Sanders et al. 1979). Figure 5.1 shows the locations of these geographical features.

Data Sources

Temperature and precipitation data taken from the book of climatological normals of the National Meteorological Service for the period of 1951 through 1980 were used to draw the climatological maps presented in this chapter (locations of stations used are shown in Fig. 5.2). Unfortunately, after this period, due to the economic crisis in the country, most stations were closed. Climatic conditions prevailing in the basin after 1980 are discussed later in this chapter. Hourly wind and ozone (O_3) data used to construct tables, pollution roses, and surface wind patterns correspond to the meteorological and air quality network of stations operated by the Ministry of the Environment of the Federal District (RAMA).

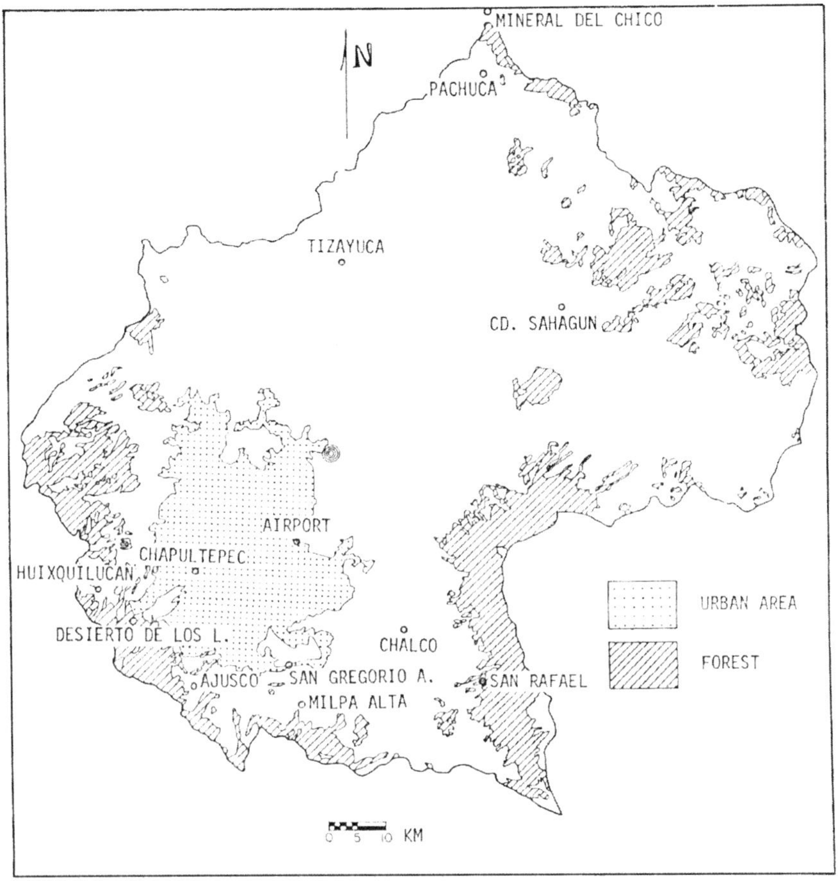

Figure 5.1. Location of forests in the Basin of Mexico.

The General Climate

The climate of the Mexico City Air Basin (MCAB) is determined by two large-scale alternating regional currents: the predominantly dry Westerlies that sweep over Mexico during winter and spring, and the northeastern moist trade wind current that brings the rains from May to October.

The Dry Season

During the cool period (November to April), modified polar air masses, coming from North America, periodically affect the Basin of Mexico, sometimes causing a drop of temperature, sparse rain, and occasionally (every five or more years),

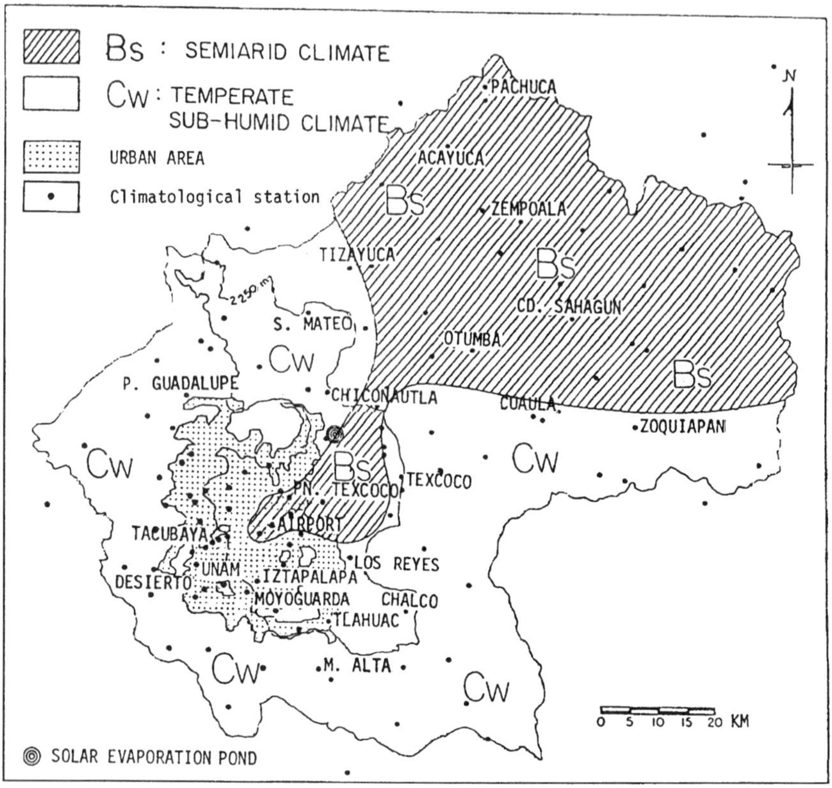

Figure 5.2. The climates of the Basin of Mexico according to Koeppen's classification and location of climatological stations used.

snow over the high mountain ranges to the south and east. After such polar fronts pass, anticyclonic weather systems (e.g., clear skies and calm conditions) prevail favoring the development of strong surface inversions that enable high concentrations of pollutants in the Air Basin.

Since these polar air masses coming from the North American continent are relatively dry, the nocturnal radiational cooling of the surface air is more pronounced, and therefore the cool season is also the smog season in the Valley of Mexico. This is in contrast with subtropical coastal regions with a Mediterranean type of climate (e.g., Los Angeles; Santiago, Chile), where the smog season is shifted to the summer season in which the stability of the surface (moist) layer is enhanced by the presence of a cool marine current rather than by surface radiational cooling. At the end of the dry season (February to April), turbulent winds associated with the last frontal passages and/or the first manifestation of convective activity, produce wind erosion of the dried-up soils in the plains, resulting, at times, in severe dust storms (Jáuregui 1989).

The Wet Season

As the Westerlies recede northwards by the end of the dry season, they give way to the moist easterly trade-wind current that brings convective rains from May to October. Simultaneously the low-pressure trough lying between the trade-wind regions of the two hemispheres (known as the Intertropical Convergence Zone) moves along the northeastern Pacific toward the north. Approaching the Pacific littoral of Southern Mexico, it favors the onset of the rainy season in most of the central and southern parts of the country, including the MCAB. Using mean 5-day outgoing long-wave radiation (OLR) data for the period of 1975 to 1987 and for the tropical Pacific region of Mexico and Central America, Cortes (1999) has described the annual cycle of convective activity over Mexico. Other regional systems modulating the rainfall regime in the MCAB are the tropical disturbances from the Caribbean known as easterly waves, which reach Mexican coasts (Jauregui 1967, 1995a), and to a large degree, the El Niño phenomenon (Magaña et al. 1999a; Mosiño and Morales 1995).

The bimodal distribution in the annual cycle of precipitation over central Mexico is discussed by Magaña et al. (1999b). These authors conclude that the so called midsummer drought observed in the region (and in the MCAB) is not associated with the meridional migration of the intertropical convergence zone (ITCZ) but rather with fluctuations in its intensity and location.

Climate Classification

According to a universal climatic classification (Koeppen 1923), the climate of the basin is semiarid (BS) in a large portion to the northeast and in the central plains, whereas in the western and southern areas of the Basin, the temperate subhumid climate (Cw) prevails (Fig. 5.2).

The Thermal Climate

Increasing urbanization and deforestation are causing a gradual increase in temperature in the MCAB. As would be expected, the increase is not spatially uniform across the Basin because it is modulated by the specific land use. In the case of central Mexico City, the mean annual temperature has changed from 14.5° C (in 1900) to 17.0° C in 1995 (Jáuregui 1995b), an increase several times larger than the expected global temperature change. In contrast, suburban/rural areas located downwind of the Mexico City smog plume are showing signs of declining afternoon temperatures on the order of 0.4° C/decade (Jáuregui and Luyando 1999).

Annual Temperatures

Figure 5.3 shows the distribution of the mean annual temperature in the MCAB. As would be expected, the lowest mean annual temperatures (those in the range

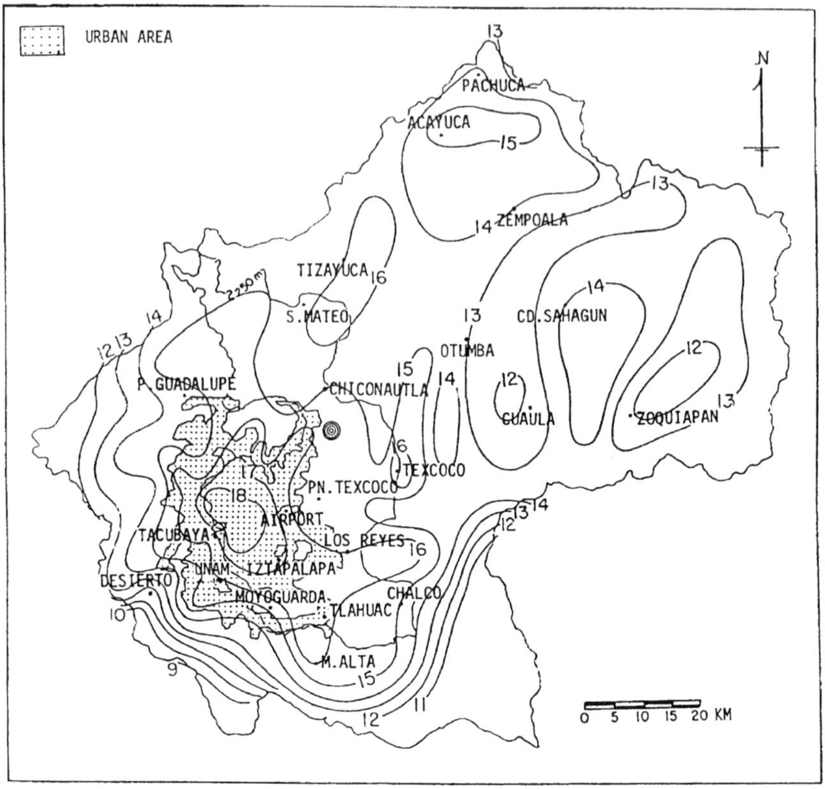

Figure 5.3. Mean annual temperature (°C): Period 1951–1980.

of 9°C to 13°C) are found in higher elevations to the east, south, and west, where most of the remaining forests in the Basin are located (Fig. 5.1). On the other hand, the warmer areas are situated on the valley plains around the center of the city. The higher mean annual temperatures (around 18°C) reflect the heat island phenomenon (see Jáuregui 1986, 1993, 1997).

Mean Extreme Temperatures

Figures 5.4 and 5.5 show the distribution of average minimum and maximum temperatures for January (the coldest month) and April (the warmest month) for the period of 1951 to 1980. Again, the coldest areas in the Basin are found over the mountains, but also in the semiarid plains to the north around Ciudad Sahagun and Tizayuca, where nocturnal radiational cooling to a dry atmosphere brings the thermometer down to temperatures below freezing. The higher temperatures in the MCAB are caused by the heat island in the capital city and also in the semi-arid north due to the high levels of insolation (Fig. 5.5).

Absolute Extreme Temperatures

Figures 5.6 and 5.7 show the distribution of absolute extreme temperatures that occurred in the period of 1951 to 1980. The values displayed on these maps were observed only once or on a few days. While they represent infrequent (not necessarily simultaneous) events, they are useful in showing how far temperature extremes have reached in the MCAB.

The coldest spot in the MCAB, where temperatures as low as 10° C to 12° C below zero have been registered, are the dry plains of the center (Texcoco) and north (Tizayuca), as well as the Chalco valley (Fig. 5.6). Although temperatures in the forests of the Basin are not as low as in the lower plains, the coldest temperatures are in the range of 3° C to 6° C below freezing. Less extreme low temperatures in the forests are probably due to the higher humidity provided by evapotranspiration of the forests, which acts as a screen to reduce radiational cooling at night (Fig. 5.6). As would be expected, extremes of high temperatures (up to 34° C) occur in the same region where the lowest temperatures are observed,

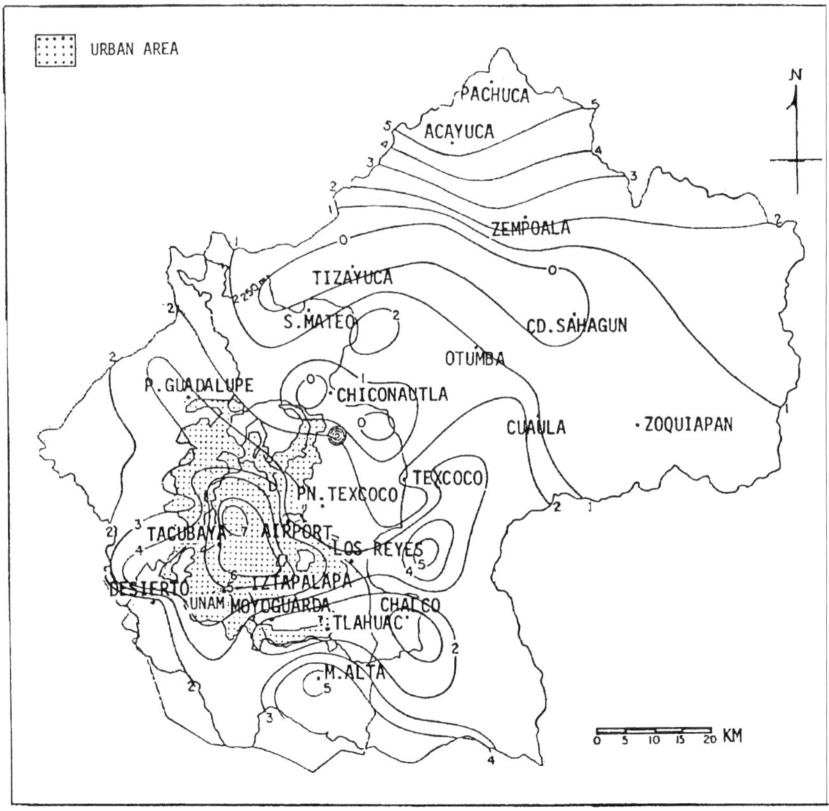

Figure 5.4. Average minimum temperature (° C) in January: Period 1951–1980.

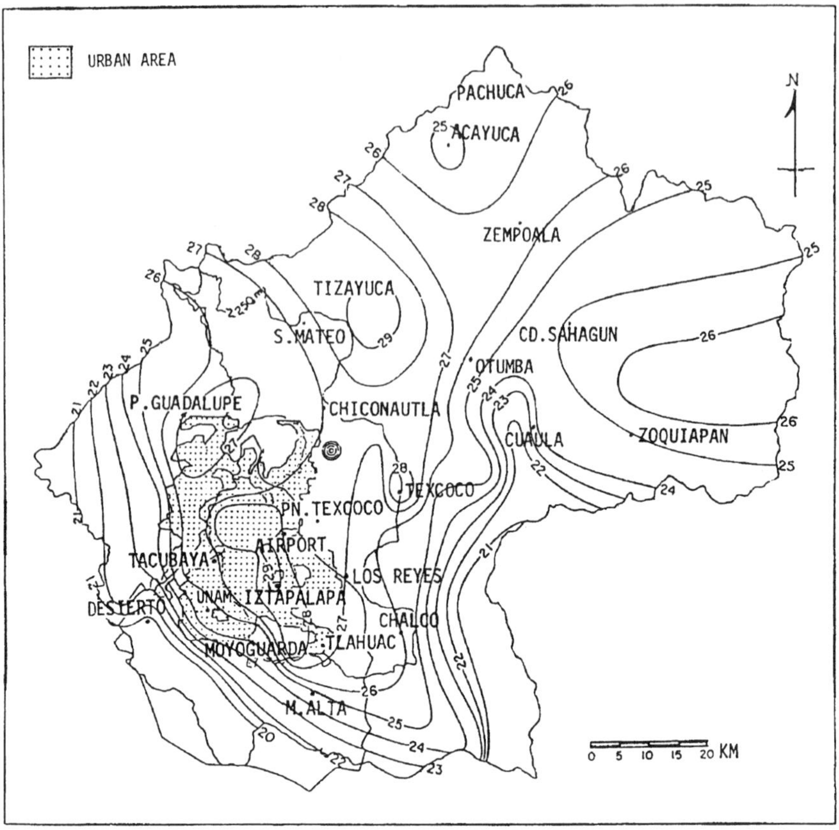

Figure 5.5. Average maximum temperature (°C) in April: Period 1950–1980.

in the northern portion of the Basin, due to the prevailing dry conditions. The other hot spot in the Basin is situated in the Mexico City area where urbanization favors the formation of a dome of warmer air over the urban area (Fig. 5.7).

Frequency of Freezing Days

Figure 5.8 shows that the MCAB location with the least number of days with subzero temperatures lies within the urban area of the capital city, in contrast with the northern portion of the Basin, where the number of freezing days per annum is greater than 70.

Long-Term Effects of Urbanization on the Thermal Climate of the MCAB

It is well established that the accelerated urbanization process in the capital city has caused a gradual increase of temperature during most of the twentieth century and until the early 1980s in the central districts of Mexico City (see Jáuregui

1986). The climatological maps presented in this chapter correspond to the period of 1951 to 1980. As mentioned earlier, only a few stations (e.g. two observatories) have continued in operation after 1980. The question of whether climatic conditions have stayed within the range of natural variability during the last two decades of the twentieth century or whether the climate has been affected by land-use changes remains unanswered.

During the second half of the twentieth century, the Basin of Mexico experienced a widespread conversion of natural vegetation to urban and agricultural land. Deforestation and urbanization have caused a considerable reduction in sources of moisture in the Basin. Therefore the impact would be toward a warmer and drier climate.

Comparison of temperature trends for the periods of 1951 to 1980 and 1981 to 1999 for the urban Tacubaya station (see Figure 5.1 for its location) shows that while the rate of temperature increase during the first period was 0.2°C per decade, it increased to 0.7°C per decade during the second period (1981–1999). This suggests that as the capital city continued its accelerated growth, the heat island effect increased the rate of change in the average temperature, because it is well established that heat islands increase in intensity in proportion with city

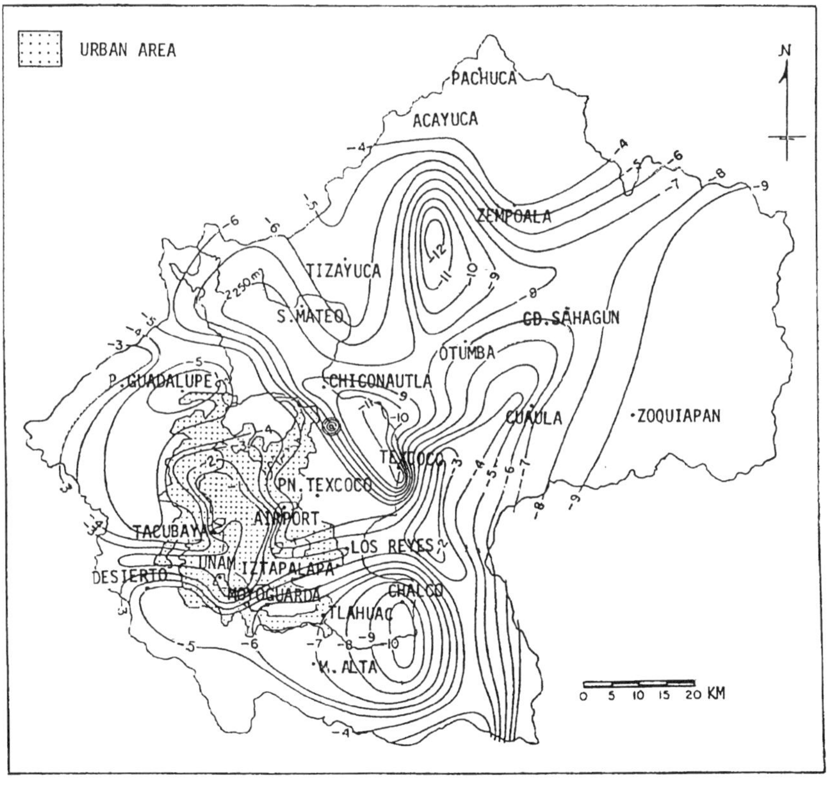

Figure 5.6. Extreme minimum temperature (°C) in January: Period 1951–1980.

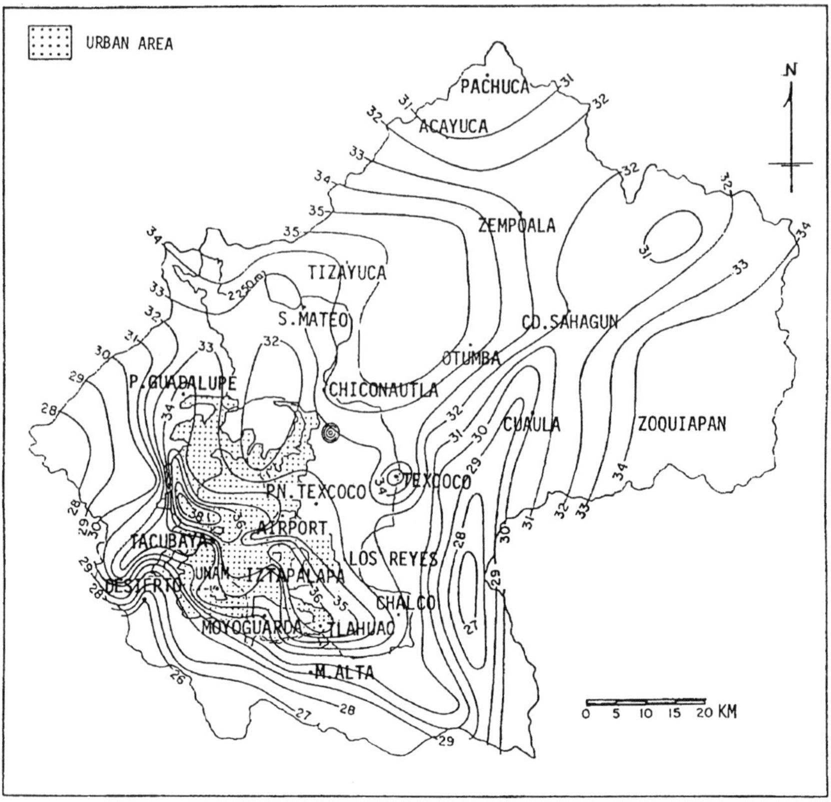

Figure 5.7. Extreme maximum temperature (°C) in April: Period. 1951–1980.

size. However, given the various sizes of urban settlements, this result may not be extended to the whole Basin. For example, in the case of Mexico City and its region, a previous study illustrates that while the northern part of the city shows a trend toward warmer temperatures, the afternoon (maximum) temperatures in the southern (downwind) portion of the urban area and in areas toward the mountain slopes (including the mountain stations Ajusco and Desierto) have experienced a significant ($p = 0.01$–0.1) decrease up to the mid-1980s. It is hypothesized that this is due to the dominant radiational cooling by the ever increasing aerosol particles in the mixed layer (Jáuregui and Luyando 1999).

The Rainfall Regime

Annual Rainfall

The location of the MCAB in the tropics determines the convective character of precipitation during summer. This precipitation is in the form of rain showers usually accompanied by gusty winds, lightning, and occasional hail. Contrasting with these afternoon showers of mostly short duration, the scant precipitation that occurs during the dry season is usually in the form of continuous rain associated with frontal passages. These provide less than 10% of the annual rainfall totals. Appendix A gives more details of the precipitation and thermal regime prevailing in the forests. In addition, tables and climograms of the mean monthly values of precipitation (mm) and mean maximum–minimum temperatures for some stations located near or in the vicinity of the forests of the Basin and at higher altitudes are included. The mean annual precipitation in the higher elevations to the west, south, and east, enhanced by orographic lifting, is on the order of 1000 to 1300 mm, whereas the rather flat northern half of the study area receives less than 600 mm of precipitation (Fig. 5.9). Another location with lower rainfall due to subsiding air lies to the east of Mexico City in the ancient lake-bed plains. This location also constitutes the main source of dust storms during the dry period (see Jáuregui 1989).

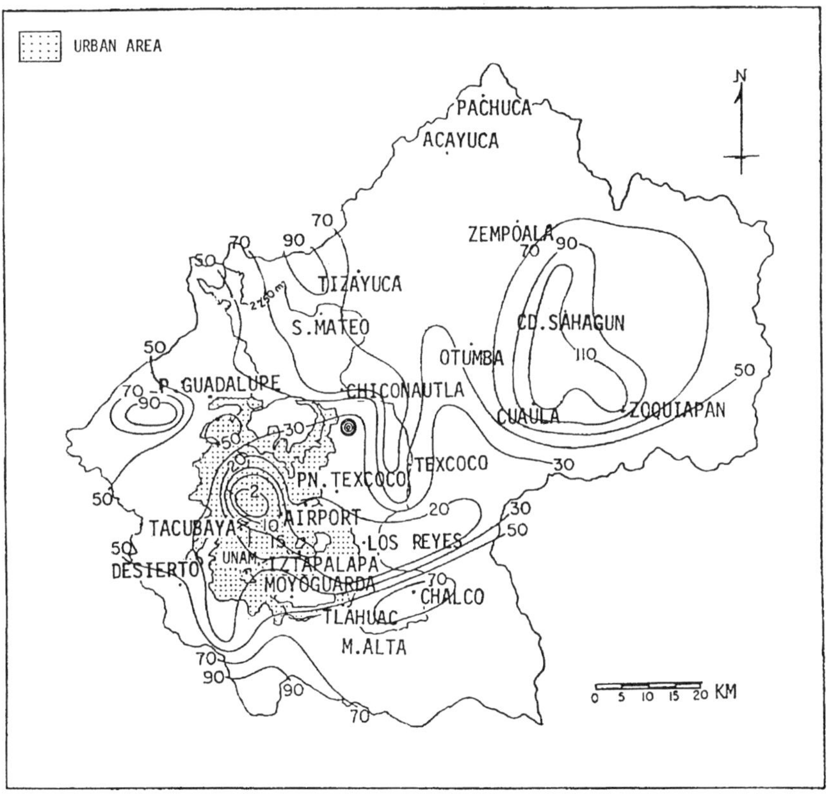

Figure 5.8. Mean frequency of freezing days/year: Period 1951–1980.

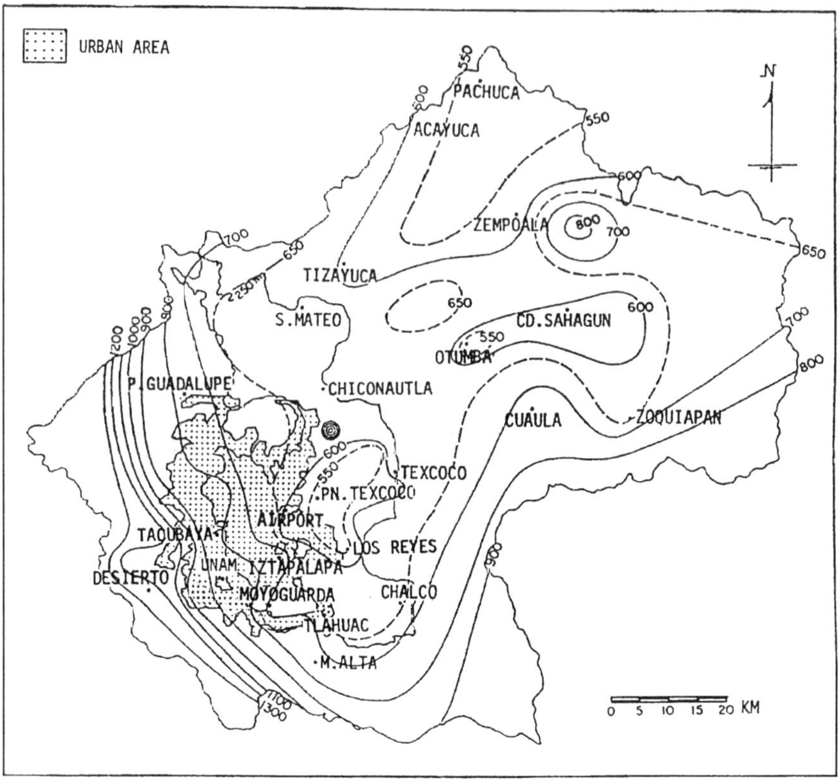

Figure 5.9. Mean annual precipitation (mm): Period 1951–1980.

Evidence of urban-induced increases in convective precipitation has been found in the daily and seasonal precipitation in Mexico City. A long-term significant increase in wet-season precipitation during the period of 1940 to 1990 has been found by Jáuregui and Romales (1996) for the urban station Tacubaya to the west of the downtown area. In Mexico City, intense thunderstorms occur mainly during the afternoon and early nighttime hours. Severe weather conditions, such as strong winds, frequent lightning, and heavy rainfall that results in street flooding (Jáuregui and Romales 1996), are related to these thundershowers.

The Number of Days with Rain

Figure 5.10 shows the distribution of the number of days with appreciable rain. In general the forests to the west, south, and east receive rain from 90 to 120 days per year in the form of convective showers. The mean intensity of this precipitation ranges between 70 and 100 mm per 24-hour period (Fig. 5.11).

The Solar Radiation Climate

Global Radiation

The radiation climate in the MCAB is modulated mainly by clouds, aerosols, and other local pollutant gases and particles. Figure 5.12 shows the seasonal variation of solar radiation components observed at the Institute of Geophysics solar radiation station located at the university (Universidad Nacional Autónoma de México, UNAM) in the southern suburbs of the capital city. The instrument is an EPPLEY S/N 28281 pyranometer with shade ring for measuring total incoming and diffuse shortwave radiation with a spectral range of 293.9 to 383.9 nm. Global radiation (diffuse plus direct) reaches maximum intensity at the end of the dry season, during March, April, and May, when the noon sun is already high in the clear skies. Once the moist Trades arrive in the valley, the increasing cloud cover reduces global radiation.

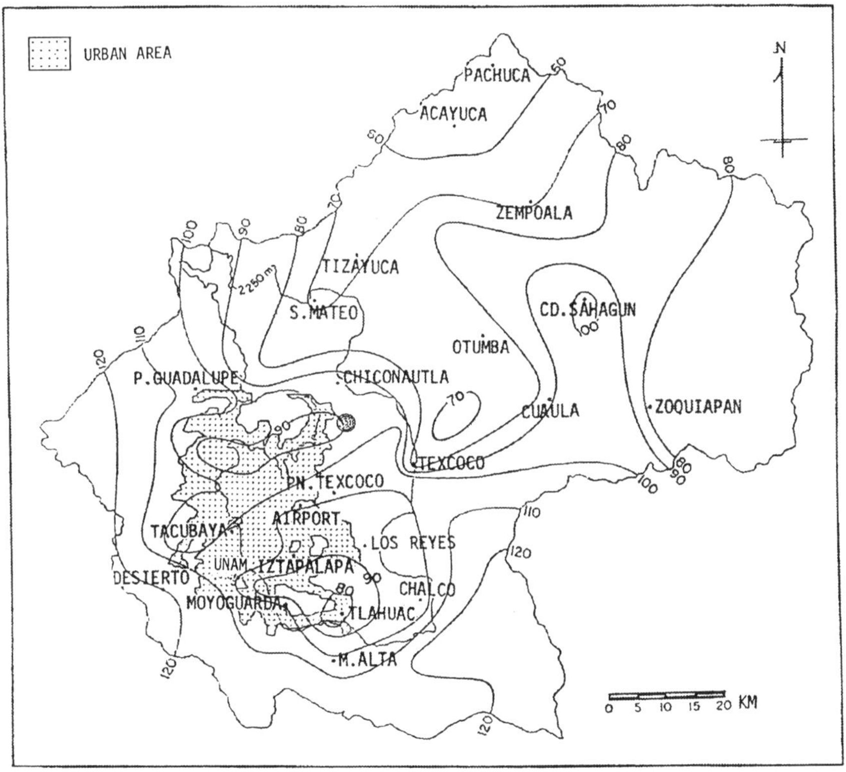

Figure 5.10. Mean number of days with rain/year: Period 1951–1980.

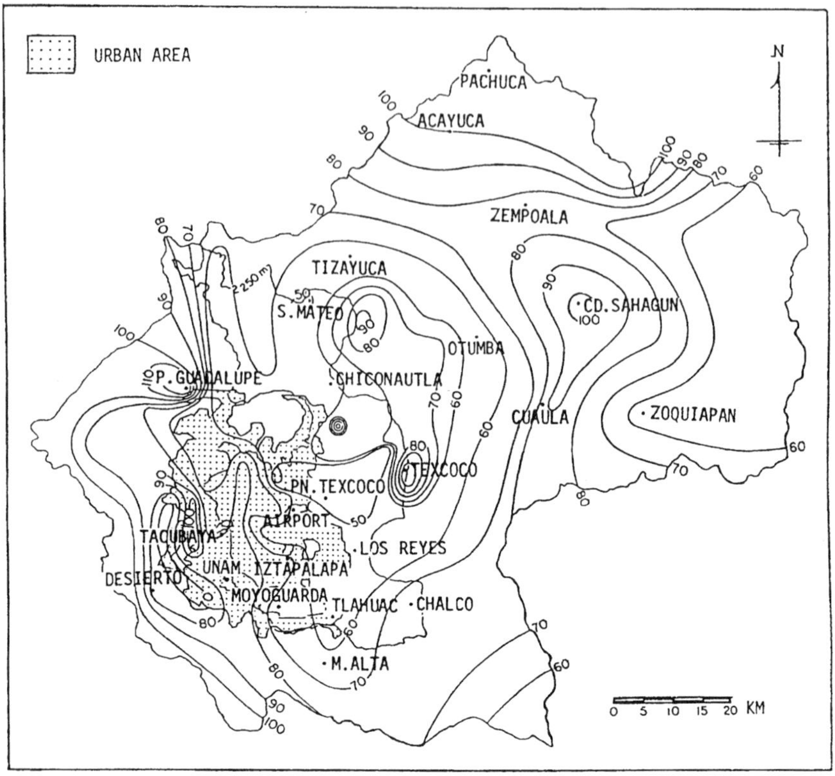

Figure 5.11. Mean maximum 24-hour rainfall (mm): Period 1951–1980.

Diffuse Radiation

Conversely, it is during most of the rainy season that diffuse radiation reaches higher intensities because of the presence of abundant convective clouds that scatter a large portion of the incoming radiation. This effect is enhanced by the urban layer of aerosols.

Ultraviolet-B Radiation

The ultraviolet (UV) portion of the solar spectrum displays a smaller range of seasonal variation (Fig. 5.13). UV-B radiation, like global radiation, peaks during March, April, and May and has lower monthly values, as expected, during the winter months.

If, as anticipated, stratospheric O_3 levels decrease, the resulting increase in UV-B near the earth surface should increase the chemical reactivity of the troposphere. As noted by Tevini (1993), plant responses to enhanced UV-B radiation vary markedly within species and among species. Damaging effects of enhanced UV-B radiation on vegetation may be reduced leaf area, stem length, and photo-

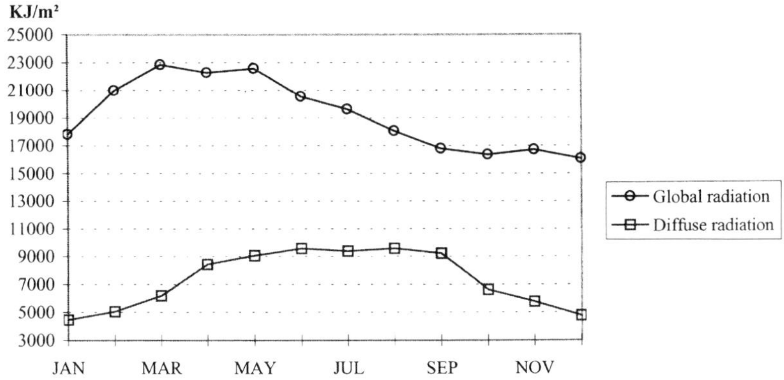

Figure 5.12. Mean seasonal variation of global and diffuse radiation at the University solar radiation observatory: Period 1995–1998. (*Source*: Institute of Geophysics, Universidad Nacional Autónoma de México.)

synthetic activity. However, as Tevini pointed out, most research in this area has been done with temperate-latitude agricultural species, and little is known regarding how native plants and natural forests in the tropics would cope with the predicted increase in solar UV-B radiation that would occur with O_3 depletion.

While UV-B radiation is bound to increase with the thinning of stratospheric O_3, particulate matter in the pollution layer would tend to counterbalance this effect. Mean global radiation attenuation by pollutants in an urban central site in Mexico City, where pollution levels are densest, was as high as 21.6% on clear days of the dry season (Jáuregui and Luyando 1999). These authors suggest that while particulate pollutant levels tend to be lower during the rainy season (because of the removal of a considerable amount of ambient aerosols by rainout

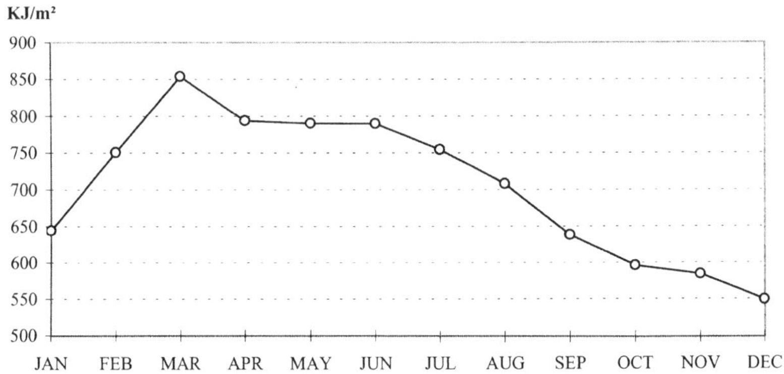

Figure 5.13. Mean seasonal variation of UV-B (W/m^2) in Mexico City (University grounds): Period 1995–1998. (*Source*: Institute of Geophysics Solar Radiation Bulletin, Universidad Nacional Autónoma de México.)

and the wash-out effect), the higher prevailing humidity favors an increase in turbidity. This dimming effect of aerosols decreases with distance from the urban source; nevertheless, it is likely to remain important in the forested mountains of the MCAB, thus helping to reduce any impact of increases in UV-B radiation levels by stratospheric ozone depletion. Since UV-B radiation increases with altitude, the pollution layer would also have the effect of bringing the UV-B level down to that corresponding to a site of lower altitude.

A series of simultaneous UV-B measurements were taken at the university radiation observatory in southern Mexico City and at a lower altitude (1500 masl) smog-free rural site in the adjacent Cuernavaca basin 150 km to the south (Temixco). UV-B values at the latter site were not lower (as expected due to its lower altitude), but higher by 5% to 30% than those observed at the MCAB (A. Muhlia and J. L. Bravo, unpublished data). This result would mean that the observed UV-B levels of radiation in the MCAB correspond, on the average, to those at much lower elevation because of the attenuation effect of the smog layer. It might be concluded that ground-level O_3 pollution in the MCAB has a mitigating effect on UV exposures for both humans and vegetation. This attenuation effect is likely to be more marked in the early afternoon when ozone concentrations peak and the low solar zenith angle contributes to high actinic flux.

Insolation

The mean number of hours of insolation at the university Solar Radiation Station is illustrated in Figure 5.14. Once the rainy season is established in June, the duration of isolation is reduced by interception of incoming radiation by clouds and reaches a minimum in September.

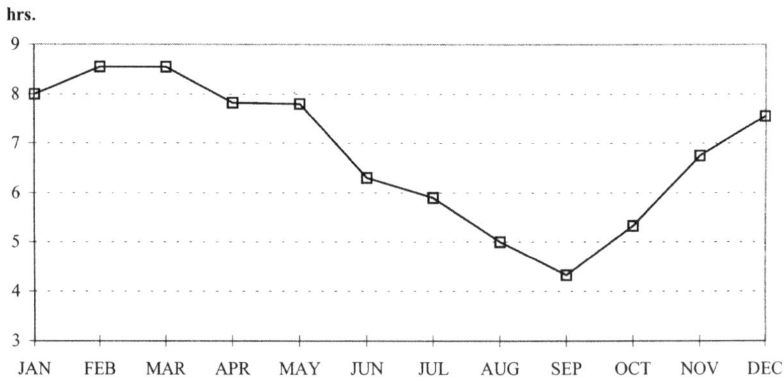

Figure 5.14. Mean monthly variation of insolation (hours) in Mexico City (University): Period 1995–1998. (*Source*: Institute of Geophysics Solar Radiation Bulletin, Universidad Nacional Autónoma de México.)

Wind Circulation

Surface wind circulation in the MCAB is the result of the interaction of regional winds, that is, winds associated with large-scale synoptic situations (frontal passage, anticyclonic condition, etc.), and local winds determined by the topography. When the pressure gradient in the region is weak, then the mountain and valley circulation of winds prevails. This is the case on many days in the MCAB when anticyclonic weather dominates, especially during the dry season.

Mountain Winds

In this condition, down-slope mountain winds (or katabatic winds) converge toward the valley plains as illustrated in Figure 5.15 for January, 1997, from mid-

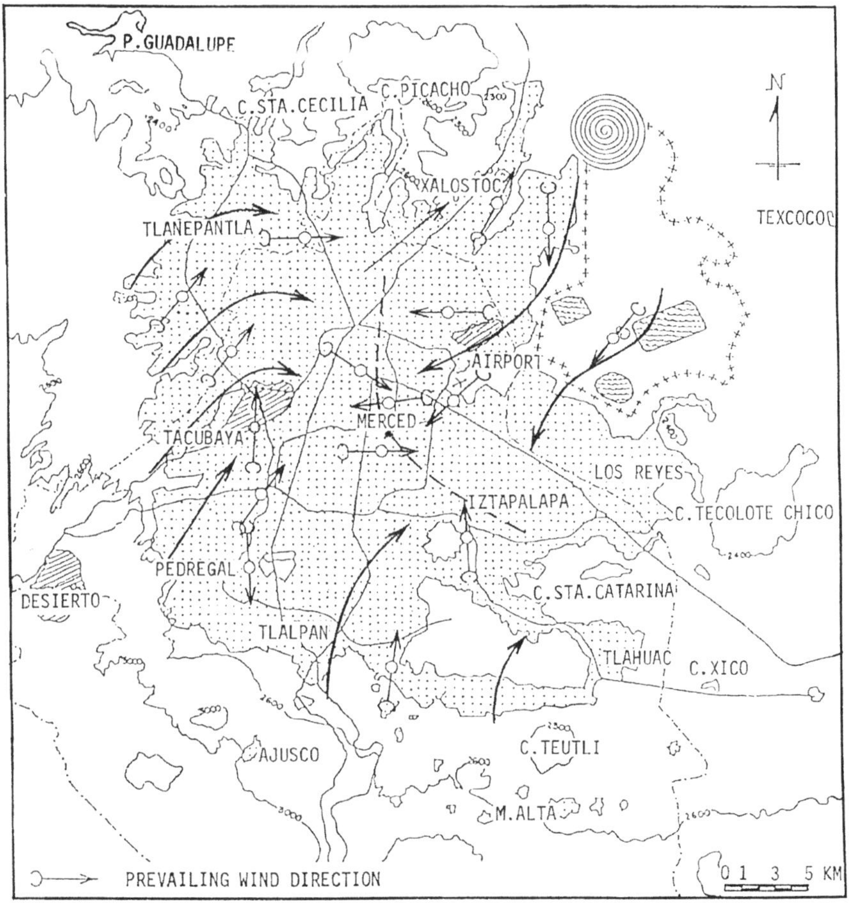

Figure 5.15. Surface wind-flow pattern for January, 1997: Period 0:00 to 8:00 hours.

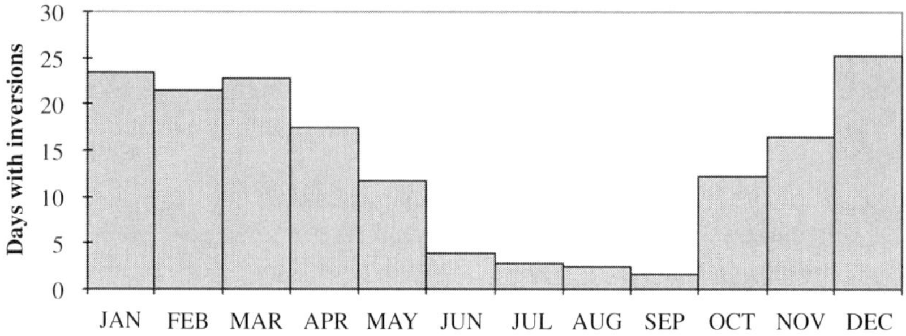

Figure 5.16. Frequency of surface inversions: Period 1994–1997.

night to 8 a.m. Arrows indicate the prevailing wind observed hourly in the anemometric stations operated by the local automatic environmental monitoring agency, RAMA. During the second half of the night, the converging winds tend to accumulate pollutants laterally. This confluence of pollutants is reinforced by the centripetal circulation induced by the heat island (Jáuregui 1988).

On the other hand, the katabatic winds, being cooled from below over sloping terrain at night, produce a shallow blanket of colder air. This heavy air exchanges little energy with the overlying air above the valley plains. This strong stable condition, called a surface inversion, is frequent in the MCAB, especially during the dry season when clear calm nights promote intense radiational cooling (Fig. 5.16). The depth of this large pool of cold air often reaches 400 to 500 m over the plains, about the height of the mountain passes. Its upper limit is clearly visible as one leaves the city by car or plane.

From the late evening until around midmorning, this stable air condition promotes the accumulation of pollutants, mainly primary air pollutants, in the inversion layer. This increases the concentration of such pollutants as particulate matter (including the fine fraction of less than 10 microns), sulfur dioxide (SO_2), and nitrogen oxides (NOx), some of which may be highly toxic to humans and vegetation.

Valley Winds

Shortly after sunrise, strong solar heating of the ground induces turbulent mixing that erodes the morning inversion creating a deep boundary layer after about noon. Maximum mixing heights obtained by the indirect Holtzworth (1964) method were estimated by Jáuregui (1983) to be around 2500 to 3000 m during the warm season (March–May) decreasing to 2000 m during the wet season. Recent measurements of the mixed layer, made with an aircraft, give more detail and a range of 300 m to up to 4000 m (Nickerson et al. 1992). Once the valley winds are well established, around noon, they carry all the pollutants, including NOx and hydrocarbons, that when exposed to the abundant solar radiation, generate the oxidant atmosphere characteristic of the MCAB. The afternoon weak valley winds blow across the capital city (average speed: 0.5–1.5 m/s, see Table 5.1) and up the slopes of the surrounding mountains. This is illustrated in Figure 5.17 for the month of June, 1997, which was uncharacteristically dry. In this case, the valley wind circulation was being reinforced by the regional trade-wind current from the northeast. These wind-flow patterns are clearly an oversimplification, because it is difficult to characterize typical near-surface flow patterns in the MCAB, as noted by Fast and Zhong (1998).

Even though O_3 precursors are emitted close to the surface, it is believed that the presence of absorbing particles within the boundary layer causes a gradient in the photo-dissociation rates such that the production of O_3 is more efficient aloft (Raga et al. 1997). This hypothesis has been substantiated by aircraft observations, taken in February of 1991, that show a peak of both O_3 and fine particle concentrations at the top of the mixed layer (Pérez-Vidal and Raga 1998). This condition of the vertical O_3 structure, with higher concentrations at the top of the mixed layer, implies that a large portion of the forests around the Basin are likely to sometimes be exposed to O_3 concentrations higher than those observed in the city.

Daily O_3 observations taken in April and May of 1990 at the Desierto de los Leones Park (see location in Fig. 5.1) showed maximum values of 0.162 ppm near midday when the maximum temperature was 28°C. This was closely associated with upslope valley winds from the north and east. In contrast, minimum O_3 concentrations were observed during the night (0.027 ppm) when the minimum temperature was 12°C and katabatic or downslope winds from the southwest were blowing (Jáuregui et al. 1990). These authors concluded that the high O_3 levels observed at the Desierto reserve were carried by the upslope thermal wind circulation. In a recent pilot project undertaken in the Ajusco reserve, Raga et al. (1999) have confirmed that the diurnal upslope winds observed at the southwestern corner of the valley are also responsible for the transport of pollutants toward the Ajusco reserve.

Ozone pollution impact is particularly severe in those forests close to the urban emission sources and in those located directly downwind of the gigantic urban plume of warm (from the heat island) polluted air generated by the city. This is the case in the Ajusco range and in the Desierto reserve where ozone injury to *Pinus hartwegii* Lindl. and *P. montezumae* has been reported since the 1980s (Hernández-Tejeda and Bauer 1984; Miller et al. 1994). During the wet season, the presence of moist turbulent trade winds reduces the frequency of strong surface-based inversions (see Fig. 5.16). In addition, the wash-out of pollutants by convective rainfall should contribute in part to better air quality in the valley, especially for particulate matter. However, during this season, morning skies are usually clear, and the strong solar radiation leads to photochemical generation of oxidants.

Table 5.1. Hourly frequency of afternoon (13:00–18:00 hr) surface winds related to pollution concentrations of O_3 (ppm) for five sites in Mexico City: Period November–February 1998–1999

Station: Cerro de la Estrella (SE)								
				O_3		O_3		
Direction	Frequency	%	m/s	<0.11	%	>0.11	%	max. O_3
N	164	27.8	0.5	118	72.0	46	28.0	0.21
NE	144	24.4	0.5	110	76.4	34	23.6	0.174
E	59	10.0	0.6	49	83.1	10	16.9	0.149
SE	26	4.4	0.5	24	92.3	2	7.7	0.187
S	38	6.4	0.8	33	86.8	5	13.2	0.151
SW	31	5.3	0.8	21	67.7	10	32.3	0.181
W	45	7.6	0.5	29	64.4	16	35.6	0.226
NW	83	14.1	0.4	51	61.4	32	38.6	0.239
Total	590	100		435		155		
Station: Merced (downtown)								
				O_3		O_3		
Direction	Frequency	%	m/s	<0.11	%	>0.11	%	max. O_3
N	100	17.6	1.1	85	85.0	15	15.0	0.175
NE	168	29.5	0.9	121	72.0	47	28.0	0.194
E	111	19.5	0.8	76	68.5	35	31.5	0.226
SE	62	10.9	0.9	38	61.3	24	38.7	0.208
S	35	6.2	1.1	33	94.3	2	5.7	0.186
SW	6	1.1	1.1	4	66.7	2	33.3	0.168
W	24	4.2	0.9	19	79.2	5	20.8	0.133
NW	63	11.1	1.1	57	90.5	6	9.5	0.146
Total	569	100		433		136		
Station: Pedregal (SW)								
				O_3		O_3		
Direction	Frequency	%	m/s	<0.11	%	>0.11	%	max. O_3
N	219	36.5	1.6	90	41.1	129	58.9	0.282
NE	194	32.3	1.3	23	11.9	171	88.1	0.309
E	31	5.2	1.0	12	38.7	25	80.6	0.236
SE	19	3.2	0.7	11	57.9	8	42.1	0.225
S	13	2.2	1.2	7	53.8	6	46.2	0.237
SW	6	1.0	0.4	3	50.0	3	50.0	0.199
W	22	3.7	1.4	17	77.3	5	22.7	0.151
NW	96	16.0	1.6	50	52.1	46	47.9	0.203
Total	600	100		213		393		

Transport and Dispersion of Relevant Air Pollutants

Transport of some primary pollutants has been the object of a study based on ground level data. Applying cross correlation analysis to hourly SO_2 series from

Table 5.1. *Continued*

Station: Tlanepantla (NW)								
				O_3		O_3		
Direction	Frequency	%	m/s	<0.11	%	>0.11	%	max. O_3
N	281	47.8	2.2	252	89.7	29	10.3	0.188
NE	88	15.0	1.5	64	72.7	24	27.3	0.185
E	21	3.6	0.8	9	42.9	12	57.1	0.208
SE	74	12.6	1.5	39	52.7	35	47.3	0.212
S	46	7.8	1.0	13	28.3	33	71.7	0.25
SW	14	2.4	1.0	6	42.9	8	57.1	0.246
W	10	1.7	1.0	7	70.0	3	30.0	0.151
NW	54	9.2	1.6	47	87.0	7	13.0	0.2
Total	588	100		437		151		

Station: Xalostoc (NE)								
				O_3		O_3		
Direction	Frequency	%	m/s	<0.11	%	>0.11	%	max. O_3
N	139	23.3	1.7	134	96.4	5	3.6	0.127
NE	150	25.1	1.8	139	92.7	11	7.3	0.147
E	76	12.7	1.0	53	69.7	23	30.3	0.212
SE	124	20.8	1.2	88	71.0	36	29.0	0.195
S	57	9.5	1.2	37	64.9	20	35.1	0.203
SW	19	3.2	0.8	12	63.2	7	36.8	0.185
W	13	2.2	0.7	9	69.2	4	30.8	0.156
NW	19	3.2	1.1	17	89.5	2	10.5	0.148
Total	597	100		489		108		

an air quality network, Jáuregui et al. (1981) found a distinct peak of correlation, at one-hour and two-hour intervals, indicating a southward flow (from NNW to SSE) of this contaminant along the foothills of the mountains to the west of the city, shifting toward the southeast. This suggests a cyclonic circulation around the low pressure induced by the heat-island phenomenon hovering over the center of the city. Since this study was made, SO_2 emissions have been markedly reduced since the early 1990s because of a series of actions taken by the city authorities, among them:

A shift from sulphur-rich petroleum to natural gas in 365 industries and in two large power stations located in the north of the basin
The shut down of an oil refinery in the northern sector of the city
The introduction of low-sulphur diesel fuel in 1993

As a result of these measures, SO_2 concentrations have consistently remained below the standard (0.11 ppm) after 1993 (Fig. 5.18). Although much less pronounced, and therefore less significant, NOx (the precursors for O_3 formation)

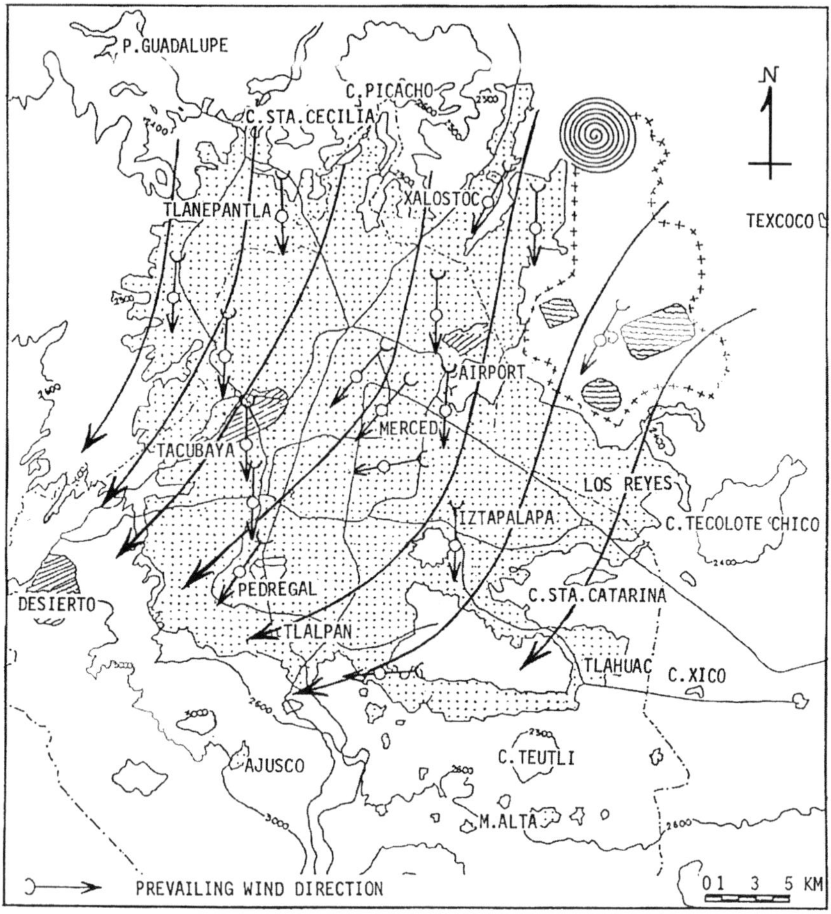

Figure 5.17. Surface wind-flow pattern for June, 1997: Period 13:00 to 18:00 hours.

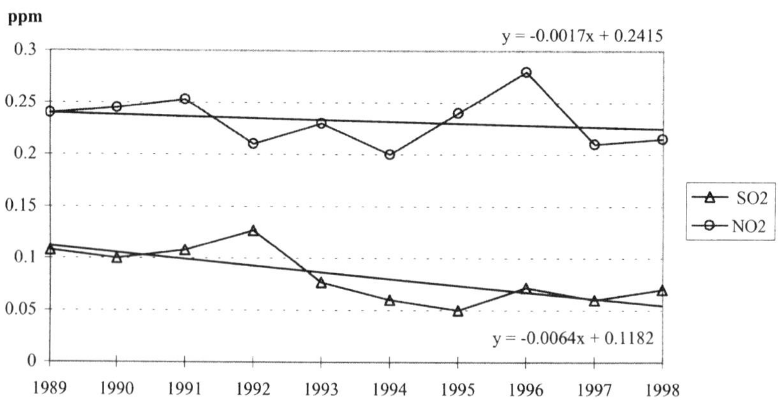

Figure 5.18. Mean annual maximum values of SO_2 and NO_2 in the Mexico City region: Period 1989–1998. (*Source*: Ministry of the Environment, Federal District (RAMA) Annual Report 1998. Ministry of Environment.)

also seem to be declining, as shown in this figure. More recently, also applying linear correlation analysis but with a larger network of sampling stations, Raga and LeMoine (1996) arrived at a similar conclusion regarding the north to south transport of pollutants.

Air Pollution Roses

Table 5.1 shows the frequency of afternoon winds (hourly values from 13:00 to 17:00 hours) for five stations located at the center and four corners of the urban area of Mexico City (see Fig. 5.1 for locations) for four months of the dry season (November, 1998, to February, 1999). Also included in the table are the frequencies for each direction and for winds linked to values of O_3 above and below the standard (0.11 ppm). At the right end of each subtable are the maximum values of O_3 observed for each direction. At the two northern stations, Tlalnepantla and Xalostoc, as well as at the Cerro de la Estrella station to the southeast, afternoon winds blow mostly from the northern sector. These directions are mostly linked to below-standard concentrations (Fig. 5.19). Conversely, the prevailing northerly winds at station Pedregal located in the southwest sector, the one closest to the Desierto and Ajusco forests, are nearly always associated with above-standard O_3 concentrations. Maximum values of O_3 observed at this station are the highest (almost three times the standard for winds from the northeast direction) of the five observation sites (Fig. 5.19).

Morning Pollution Wind Roses

Contrasting with the afternoon surface wind roses, the morning roses (07:00 to 11:00 hours) for the same four months showed more variable and light winds associated with below-standard O_3 concentrations, as would be expected. The only exception was the Tlalnepantla station to the northwest where a small proportion of the total wind directions were linked to above-standard concentrations of O_3 (not shown).

Wind Storms

In the preceding sections, the thermally driven mesoscale wind circulation in the MCAB has been examined in relation to dilution and transport of pollutants from the urban area and their potential impact on the mountain forests surrounding the plains. There are times, however, when this local wind circulation is obliterated by regional winds associated with the arrival of modified polar masses of dry continental air, especially at the end of the dry season. Strong gusty winds may then reach up to 70 to 80 km/hr, causing the development of dust storms both in the plains and in deforested areas of the mountain slopes in the southern portion of the basin. Due to a foehn effect, these dry winds become even drier in descending over the lee (ENE facing) side of the ranges, which on occasions induces forest fires. Strong gusty winds may also occur during the wet season in connection with cumulonimbus thunder clouds. This may also result in forest fires initiated by cloud to ground lightning (Table 5.2).

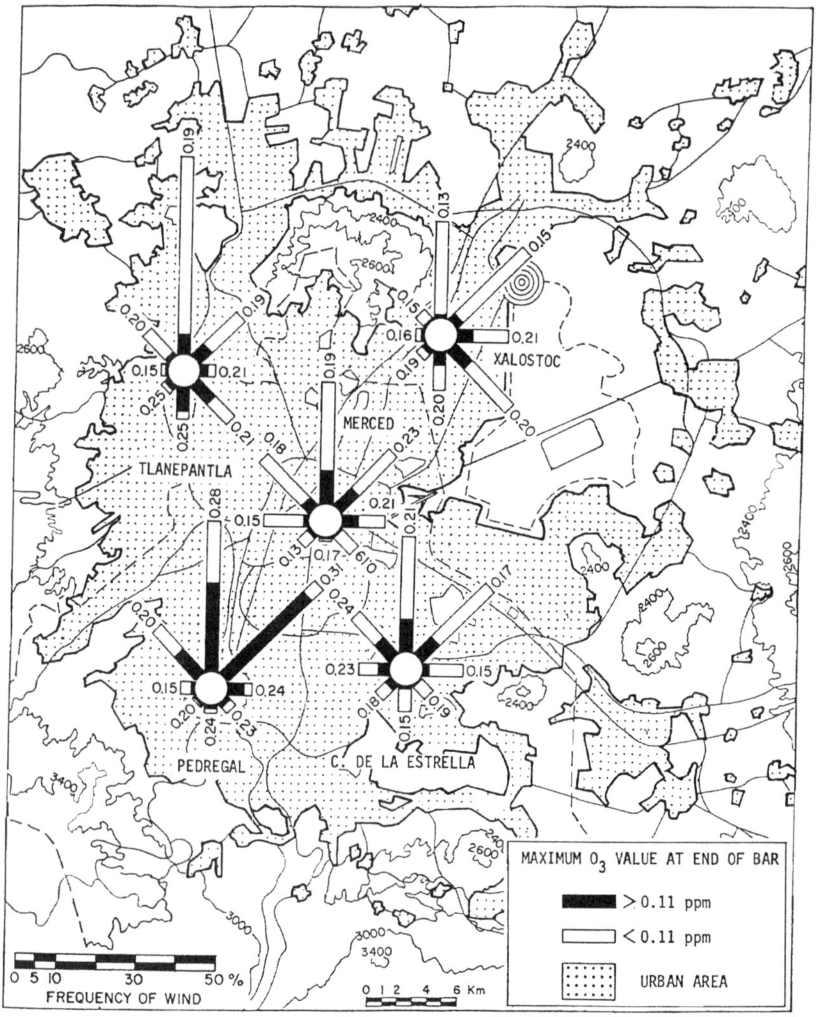

Figure 5.19. Air pollution wind roses for 5 locations in Mexico City: Period November–December 1998 and January–February 1999, from 13:00 to 17:00 hrs.

Climate Change and Forests in the Basin of Mexico

Climatic change (including elevated carbon dioxide (CO_2) levels) could enhance the supply of nutrients to plants in the forests. However, more important impacts are likely to be nutrient leaching and soil erosion wherever tree cover is lost either because of drought, fire, or removal (by logging, clearing or grazing), especially where high rainfall occurs, such as in the forests of the south and west of the Basin. Villiers and Trejo (1997) have related vegetation patterns to present

Table 5.2. Maximum winds (km/h) at Tacubaya Observatory: Period 1994–1998

	JAN	FEB	MAR	APR	MAY	JUN	JUL	AUG	SEP	OCT	NOV	DEC
1994												
Wind Velocity	44.3	44.3	52.6	60.5	83.2	26.6	59.4	72.7	56.5	54.0	38.9	40.3
Direction	NW	WSW	WNW	WNW	SSW	S	ENE	E	S	SE	NW	W
1995												
Wind Velocity	36.4	52.2	46.1	64.1	72.4	56.5	52.6	46.4	49.3	39.2	50.0	49.0
Direction	NW	NE	SW	N	SW	N	SE	ESE	N	N	N	W
1996												
Wind Velocity	39.2	53.3	48.2	79.6	62.6	50.0	48.2	52.9	45.4	44.3	41.4	37.1
Direction	W	S	S	W	ENE	WNN	E	S	N	SSW	N	N
1997												
Wind Velocity	65.5	42.5	65.5	55.4	59.4	46.4	52.6	51.8	49.3	36.7	46.4	49.3
Direction	S	W	SS	W	W	N	NNE	WNW	N	N	SW	SSW
1998												
Wind Velocity	42.5	62.6	51.5	44.3	51.5	51.5	43.2	40.3	42.5	36.4	28.8	32.4
Direction	WSW	SSW	SW	SSW	WSW	N	N	SSE	SSW	WNW	SE	NNW

climate conditions in Mexico, and subsequently they have generated climate change scenarios in order to evaluate the changes in climate and vegetation resulting from global climate change. For the mountain forests of central Mexico, using a model-derived scenario in which an increase of 2° C and a 10% decrease in precipitation is considered, they have estimated that the impact on conifer and oak forests like those encircling the Basin of Mexico would be that 65% of the forests would suffer from the modeled climate change.

Since water availability is likely to be a limiting factor (it already is), it may be concluded that forest ecosystems in the Basin of Mexico will experience a critical stress situation in future decades. This will be aggravated by continual population increase and subsequent deforestation, especially in the forests to the south and west closest to the current urban area. It is to be hoped that at least the forests where rainfall is abundant will be protected to foment infiltration for recharge of the aquifers and thus help preserve this important source of water for the capital city.

Summary

This chapter dealt with a general description of the climate of the MCAB. A brief discussion was made of the main features determining the regional climate such as the cold-air-mass type of weather during the winter, and the prevailing easterly trade winds associated with the rainy season in summer. The chapter also focused on the thermal and precipitation regimes, illustrated with maps in which emphasis is placed on the climatic conditions in the forested areas of the surrounding mountains.

Maps of typical surface wind-flow patterns are discussed in the context of dispersion and transport of air pollutants emitted in the capital city. Particular attention is given to climatic conditions leading to high concentration of pollutants (e.g., surface-based inversions) occurring especially during the dry (winter) season.

Finally, air pollution roses are examined at five sites in the urban area in order to determine which wind sectors are linked to above-regulatory-standard O_3 concentrations. At three stations located in the north (Tlalnepantla and Xalostoc) and east (Cerro de la Estrella), afternoon wind directions from the north sector are mostly linked to below-standard levels. In contrast, at the Pedregal station (in the southwest), the station closest to the Desierto and Ajusco reserves, northerly winds (e.g., upslope winds) are likely to be associated with above-standard O_3 concentrations.

It is clear from the above analysis that while all forested areas of the mountains in the Basin of Mexico are subject to the impact of air pollutants, those in the south and western portions (e.g., Desierto and Ajusco ranges) are the most directly affected in their growth because of their closeness to the emission sources and to the transport of pollutants by the upslope afternoon mesoscale circulation. Because these forests are being gradually decimated by illegal felling and urban developments, it is important to protect them so that precipitation can infiltrate and recharge the aquifer. The forests also play a role in controlling runoff and modulating the local climate by evapotranspiration. Climatological data at forest sites are important for studying photosynthesis and forest growth, the forest water cycle, and forest damage. The available climatological stations are located (with exceptions) mostly in rural areas or small settlements somewhat distant from the mountains. Moreover, these stations were planned to provide temperature and rainfall data with minimum instrumentation. Therefore, while their information is useful for some purposes, it is not representative of meteorological conditions in the forests. It would be desirable for a small network of fully equipped automatic stations to be installed in strategic places in the forests themselves to better understand their climatic conditions.

Acknowledgments. The author is grateful to Elda Luyando and Mario Casasola for doing the pollution wind roses and climograms, to Alfonso Estrada for drawing the maps and to Liliana Cisneros for typing the text. Thanks are also extended to the Ministry of the Environment of the Federal District (RAMA) for providing the meteorology and air-quality data. Thanks are also extended to the Meteorological Service for providing data for the climatological maps. Solar radiation data were taken from the annual reports of the Solar Radiation Observatory, Geophysics Department, Universidad Nacional Autónoma de México.

References

Cortes, M. 1999. Annual cycle of convective activity in Mexico. *Atmósfera* 12:101–110.

Fast, J.D., and Zhong, S.Y. 1998. Meteorological factors associated with inhomogeneous ozone concentrations within the Mexico City basin. *J. Geophys. Res.—Atmos.* 103: 18927–18946.

Hernández-Tejeda, T., and de Bauer, Ma.L.I. 1984. Evolución de daño por gases oxidantes en *Pinus hartwegii* y *P. montezumae* en el Ajusco. *Agrociencia* 56:183–194.

Holtzworth, G.C. 1964. Estimates of mean maximum mixing depths in the contiguous *U.S. Monthly Weather Rev.* 92:235–242.

Jáuregui, E. 1967. Easterly waves and tropical cyclones in Mexico. *Ing. Hid. México.* 22: 197–208.

Jáuregui, E. 1983. Estimates of atmospheric diffusion in Mexico. *Bol.* Institute of *Geogr. UNAM* 13:9–52.

Jáuregui, E. 1986. The urban climate of Mexico City. In *Proceedings Urban Climatology and its Applications with Special Regard to Tropical Areas* ed. T.R. Oke, pp. 63–86. WMO No. 652. Geneva, Switzerland: World Meteorological Organization.

Jáuregui, E. 1988. Local wind and air pollution interaction in the Mexico Basin. *Atmósfera* 1:131–140.

Jáuregui, E. 1989. The dust storms of Mexico City. *Int. J. Climatol.* 9:169–180.

Jáuregui, E. 1993. Mexico City's heat island revisited. *Erdkunde, Band* 47:185–195.

Jáuregui, E. 1995a. Rainfall fluctuations and tropical storm activity in Mexico. *Erdkunde, Band* 49:39–48.

Jáuregui, E. 1995b. Some long-term changes in the climate of Mexico City due to urbanization. *Bull.* Investigaciones *Geogr. Inst. Geogr.* 31:9–44.

Jáuregui, E. 1997. Heat island development in Mexico City. *Atmos. Environ.* 31: 3821–3831.

Jáuregui, E., and Luyando, E. 1999. Global radiation attenuation by air pollution and its effects on the thermal climate of Mexico City. *Int. J. Climatol.* 19:683–694.

Jáuregui, E., and Romales, E. 1996. Urban effects on convective precipitation in Mexico City. *Atmos. Environ.* 30:3383–3389.

Jáuregui, E., Klaus, D., and Lauer, W. 1981. An estimation of SO_2 transport in Mexico City. *Geofís. Int.* 20:55–79.

Jáuregui, E., Bauer, L.I., Quevedo, A., Miller, P.R., and Alvarado, A. 1990. Ozone characterization in a mountain forest near Mexico City. In *Proceedings Symposium on Urban Climate, Air Pollution and Planning*, ed. E. Jáuregui, pp. 7. Guadalajara, Mexico: World Meteorological Organization/Organización Mexicana de Meteorología.

Koeppen, W. 1923. *Die Klimate der Erde. Grundkriss der Klimakunde.* Berlin: Degruyter.

Magaña, V., Amador, J., and Medina, S. 1999b. The midsummer drought over Mexico and Central America. *J. Climate* 12:1577–1588.

Magaña, V., Pérez, J.L., Vázquez, J.L., Carrisoza, E., and Pérez, J. 1999a. El Niño and the climate in Mexico. In *Impacts of El Niño in Mexico*, ed. V. Magaña, pp. 23–66. Mexico City: Universidad Nacional Autónoma de México (UNAM), Secretaría de Educación Pública (SEP)–Consejo Nacional de Ciencia y Tecnología (CONACYT), Instituto Interamericano para la Investigación del Cambio Global del Clima (IAI).

Miller, P.R., Bauer, M.L., Quevedo-Nolasco, A., and Hernández-Tejeda, T. 1994. Comparison of ozone exposure characteristics in forested regions near Mexico City and Los Angeles. *Atmos. Environ.* 28:141–148.

Mosiño, P., and Morales, T. 1988. Tropical cyclones, El Niño and rainfall in Mexico City. *Geof. Int.* 27:61–82.

Nickerson, C.E., Sosa, G., Hochstein, H., McCasline, P., Luke, W., and Shanot, A. 1992. Measurements of Mexico City air pollution by aircraft. *Atmos. Environ.* 26B:445–451.

Pérez-Vidal, H., and Raga, G.B. 1998. On the vertical distribution of pollutants in Mexico City. *Atmósfera* 11:95–108.

Raga, G.B., and LeMoine, L. 1996. On the nature of air pollution dynamics in Mexico City–I non-linear analysis. *Atmos. Environ.* 30:3987–3993.

Raga, G.B., Raga, A.C., and Ruiz Suárez, L.G. 1997. The interaction of particles and gases with solar radiation in Mexico City. In *Proceedings Conference on Visual Air Quality, Aerosols, and Global Radiation Balance*, ed. Air and Waste Management Association, American Geophysical Union, pp. 459–468. Pittsburth, Pennsylvania, USA: Air and Waste Management Association and the American Geophysical Union.

Raga, G.B., Baumgardner, D., Kok, G., and Rosas, I. 1999. Some aspects of boundary layer evolution in Mexico City. *Atmos. Environ.* 33:5013–5021.

Sanders, W.T., Parsons, J.R., and Stanley, R.S. 1979. *The Basin of Mexico: Ecological Processes in the Evolution of a Civilization.* New York: Academic Press.

Tevini, M. 1993. Effects of enhanced UV-B radiation on terrestrial plants. In *UV-B Radiation and Ozone Depletion: Effects on Humans, Animals, Plants, Microorganisms, and Materials*, ed. M. Tevini, pp. 125–153. Boca Raton, Florida, USA: Lewis Publishers.

Villiers, L., and Trejo, I. 1997. Assessment of the vulnerability of forest ecosystems to climate change in Mexico. *Climate Res.* 9:87–93.

Appendix A: Climograms

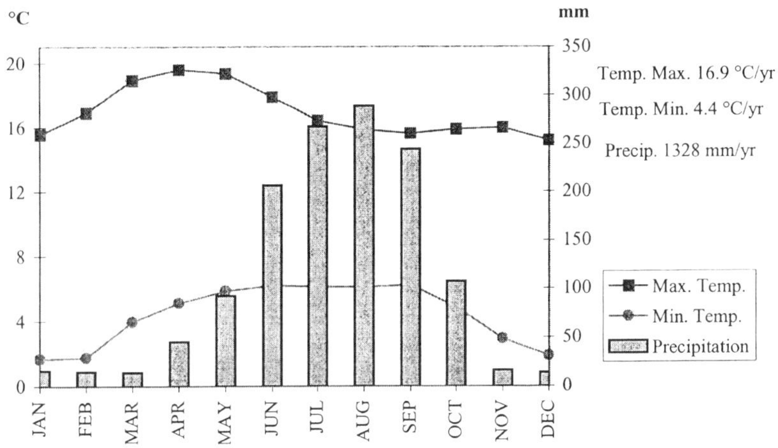

Figure 5A.1. Desierto de los Leones: Period 1961–1982, Altitude 3220 m.

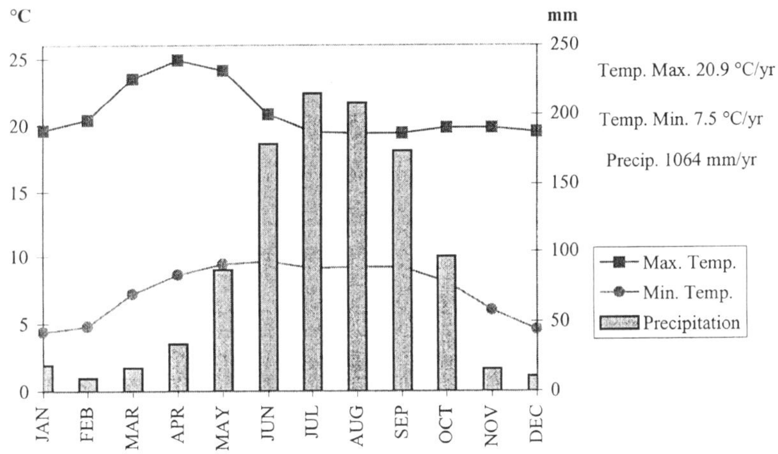

Figure 5A.2. San Rafael: Period 1951–1980, Altitude 2530 m.

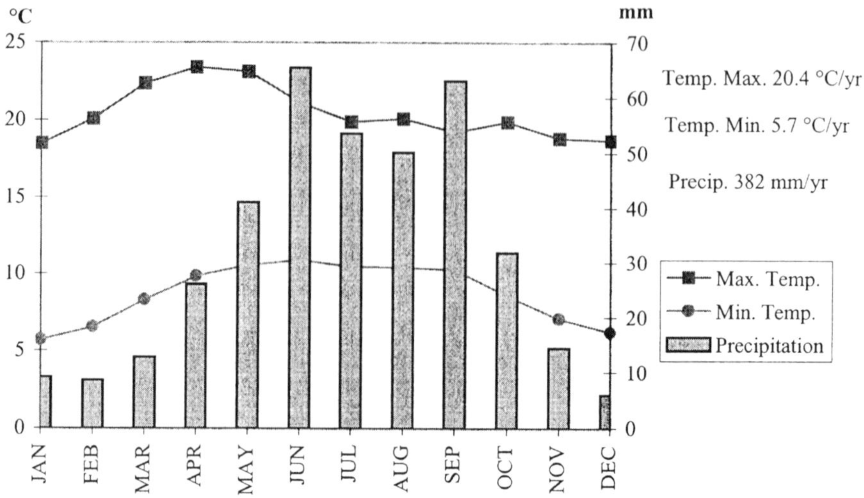

Figure 5A.3. Pachuca: Period 1921–1995, Altitude 2435 m.

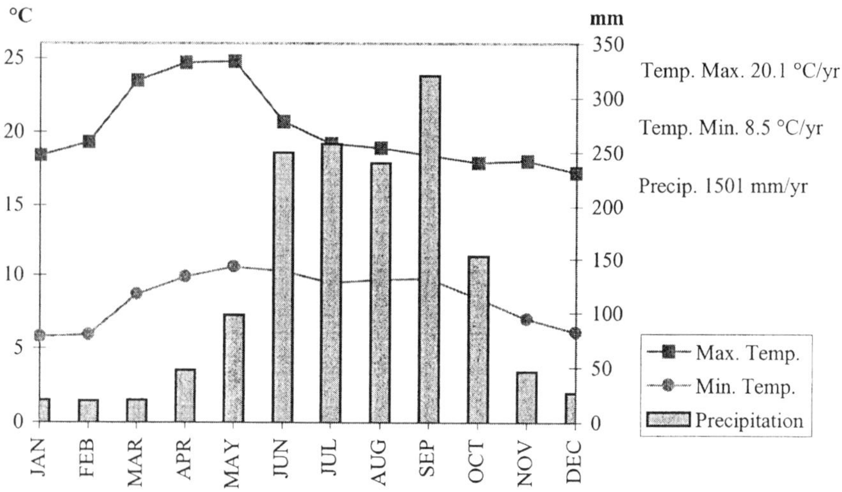

Figure 5A.4. Mineral del Chico: Period 1951–1980, Altitude 2320 m.

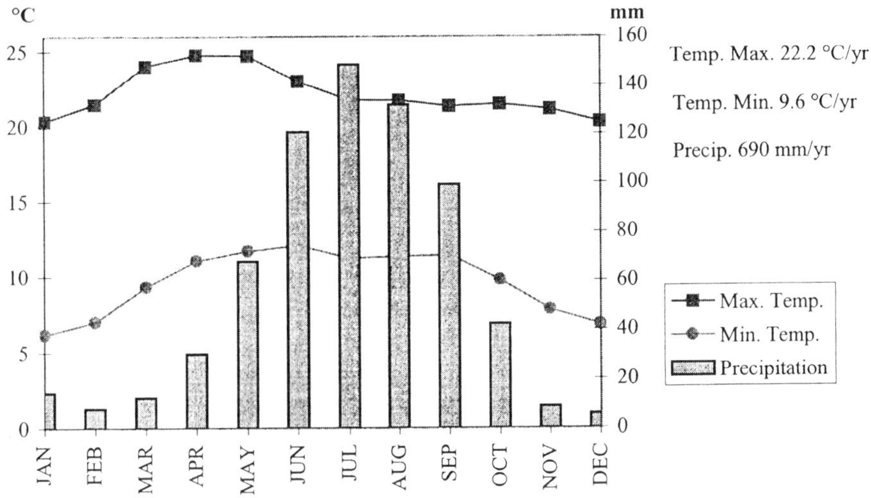

Figure 5A.5. Milpa Alta: Period 1963–1986, Altitude 2420 m.

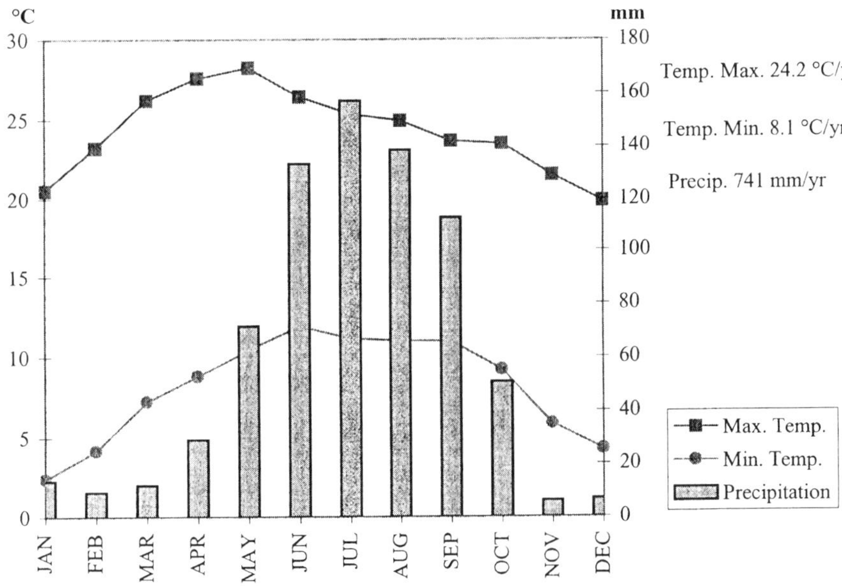

Figure 5A.6. San Gregorio Atlapulco: Period 1961–1985, Altitude 2259 m.

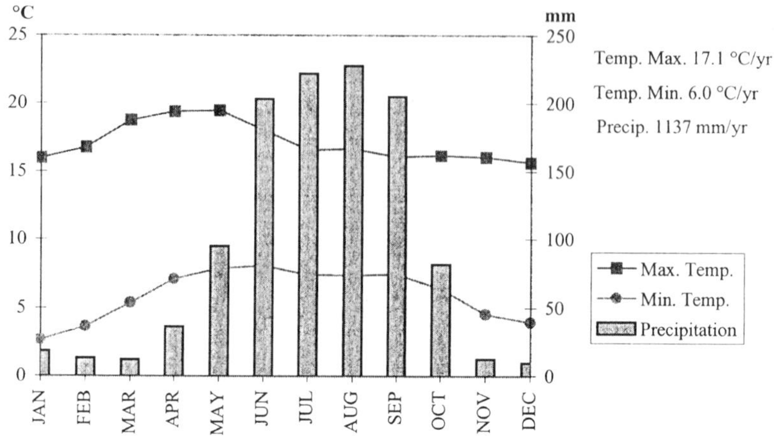

Figure 5A.7. Ajusco: Period 1961–1980, Altitude 2839 m.

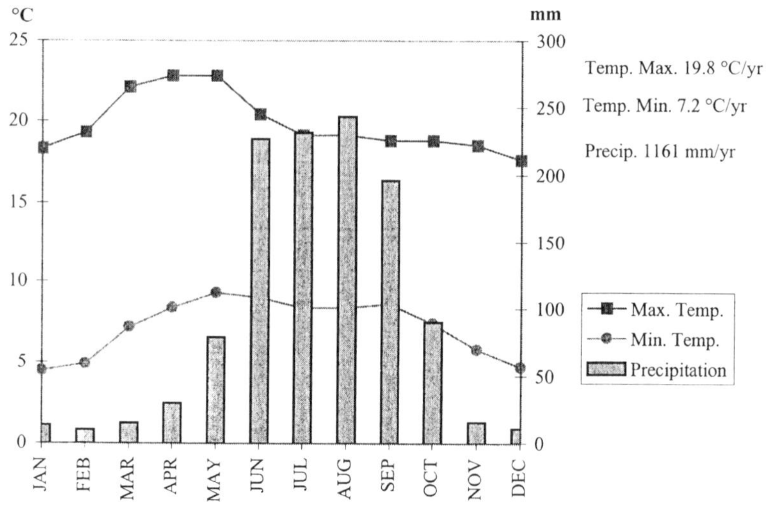

Figure 5A.8. Huixquilucan: Period 1951–1980, Altitude 2732 m.

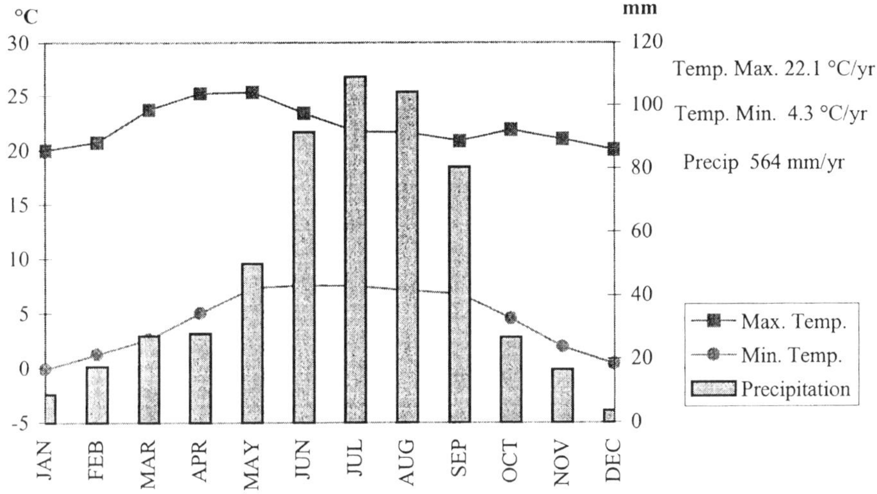

Figure 5A.9. Zoquiapan: Period 1978–1983, Altitude 2450 m.

2. Air Pollution in the Mexico City Air Basin

6. Air Pollution Levels and Trends in the Mexico City Metropolitan Area

Humberto Bravo-Alvarez and Ricardo Torres-Jardón

On many sunny days, the residents of the Mexico City Metropolitan Area (MCMA) note that a thick gray haze builds up within this atmospheric basin impairing the visibility and the quality of the air. MCMA air quality has been recognized as among the worst in the world (WHO 1997). The MCMA is located at the southern end of a semiclosed basin (19°25′N, 99°10′W), at a mean altitude of some 2200 m. The average height of the surrounding mountain ranges is about 3200 m, with some peaks, including two volcanoes, exceeding 5000 m. Because of this irregular topography and the relatively light winds, ventilation of the basin is poor. In addition, surface as well as upper air temperature inversions occur frequently, trapping pollutant emissions close to the surface. During winter, inversions occur up to 25 days per month (DDF 1991). These climatological and topographical factors combine to provide plentiful sunshine which, in comparison to more northern latitudes, is enhanced in the ultraviolet (UV) radiation that drives atmospheric photochemistry to produce secondary pollutants such as ozone (O_3) and fine particles (Streit and Guzman 1996).

Urban expansion in the MCMA has been the result of high rates of population growth and rural to urban migration combined with strong traditions of centralization in the city. The MCMA is formed by the Federal District and some counties of the State of Mexico. The 2000 census (INEGI 2000) recorded that 16.9 million people lived in the MCMA. This represented more than one fifth of Mexico's total population. From the 1930s to the 1970s, Mexico has experienced a strong sustained industrial growth, a large part of which occurred in the MCMA.

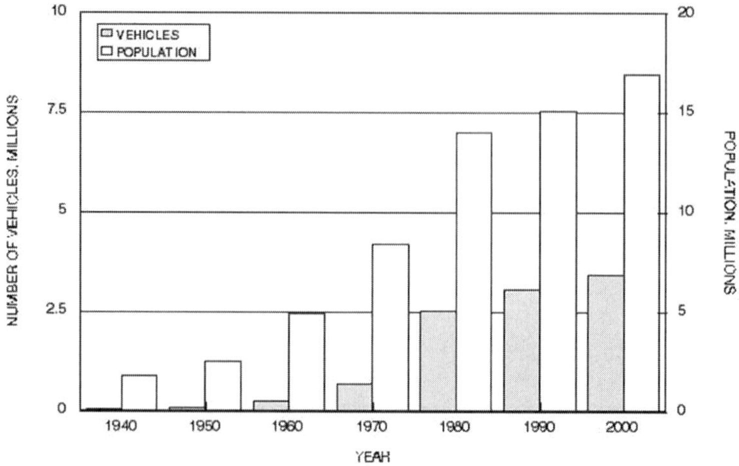

Figure 6.1. Trend in population and number of vehicles in the MCMA, 1940–2000. (Data from Benitez (1984), INEGI (2000), and GDF (1998).)

However, economic difficulties through the 1980s led to a decline in the number of industrial establishments. More than 30,000 industries, of all types and sizes, and 12,000 commercial/service facilities are now located in the MCMA. While practically all the commercial and service facilities (restaurants, bakeries, hospitals, hotels, public baths, etc.) utilize only combustion processes, nearly 4000 industrial establishments use combustion and/or transformation processes (Streit and Guzman 1996). It is estimated that current industry and other facilities consume 50,940 m^3 of low-sulfur fuel oil and 9.6 million m^3 of natural gas each day. There are also, a large number of noncombustion pollutant sources such as dry cleaning, printing, and solvent use.

In contrast with the decline in the industrial sector, since 1950, the MCMA vehicle fleet has grown to almost forty times its 1950 numbers, making the transportation sector the main source of air pollutants (Fig. 6.1). By 1999, motor vehicles consumed an estimated volume of 20 million L of gasoline and diesel each day. The high level of transport emissions is attributable to the large number of road vehicles in use, poor vehicle maintenance and illegal tuning practices, unfavorable fuel properties, unpaved roads in the peripheral areas, and localized traffic congestion (resulting in polluting "stop-and-go" driving conditions). In many instances, the air quality has worsened as a consequence of the implementation of counter-productive regulatory and preventive measures (Bravo and Torres 2000; Ezcurra and Mazari-Hiriart 1996). Table 6.1 shows a summary of the main air pollution control measures implemented to date.

Table 6.1. Main actions taken by the Mexican authorities to control air pollution in the Mexico City Metropolitan Area

Action	Year												
	86	87	88	89	90	91	92	93	94	95	96	97	98
Change in gasoline formulation and reduction in lead content of gasoline	■												
Temporary switching of oil-fired power plants to natural gas	■												
Implementation of a vehicle inspection and maintenance program for hydrocarbons (HC) and CO emissions			■										
Addition of 5% methyl–t–butyl ether (MTBE) to gasoline to reduce HC and CO emissions				■									
"Hoy no circula", a rotating one day/week driving ban				■									
Closing of the "18 de Marzo" Pemex oil refinery					■								
Introduction of unleaded reformulated gasoline for cars equipped with catalytic converters					■								
Adoption of new vehicle emissions standards for HC and CO						■		■	■				
Obligatory use of catalytic converters in new-model cars						■							
Switching industrial combustion from heavy fuel oil to gas-oil							■						
Definitive switching of oil-fired power plants to natural gas							■						
Change in gasoline formulation in both low-leaded and unleaded gasolines										■			
Implementation of HC vapor recovery plants for gasoline distribution terminals											■		
Installation of HC vapor recovery systems in gasoline service stations												■	
Adoption of tighter vehicle emissions for HC, CO and NOx												■	
Introduction of new reformulated unleaded gasolines for cars with and without catalytic converters													■

Air Quality Monitoring in the MCMA

Air pollution monitoring in the MCMA began in the late 1950s with some exploratory monitoring efforts. Bravo and coworkers measured settled dust, suspended particulate matter (SPM) as total suspended particles (TSP), organic fractions in SPM, smoke haze, sulfur dioxide (SO_2), nitrogen dioxide (NO_2), formaldehyde (HCHO), and ammonia (NH_3) at a limited number of sites in the city (Bravo and Viniegra 1958; Bravo 1960). Official monitoring activity began in the late 1960s on a more systematic basis, when the Pan American Health Organization supported the operation of 14 monitoring stations to measure smoke, TSP (using a nonspecific methodology), NO_2, and SO_2 (Bravo 1987; PAHO 1982).

After this initial effort, air pollution monitoring in the MCMA became of more interest to both academic and governmental groups. In the early 1970s, the Mexican authorities and the United Nations Environment Programme (UNEP) developed a program to install a manual network of 22 stations for monitoring TSP and SO_2. The stations were augmented in mid-1976 with additional automated equipment (Marquez 1977). However, the automated stations were difficult to operate, and by 1980 the automated network was put out of service (Legorreta 1989).

A new automated official air quality monitoring system was installed in 1986 replacing part of the equipment of the earlier manual system. The network of 25 automatic monitoring sites, known as the Red Automática, covered a large part of the MCMA. At five of the automatic stations, one in each urban quadrant or zone and one in the urban center, SO_2, carbon monoxide (CO), O_3, oxides of nitrogen (NOx), and nonmethane hydrocarbons (NMHC) were measured. In the other sites, a subset of these pollutants were measured. In addition, 16 stations for measuring SO_2 and SPM from the older manual network were operating for five additional years. Finally, after receiving strong support from the World Bank in 1992, the automated air quality monitoring system (RAMA, which stands for Red Automática de Monitoreo Atmosférico or Automated Network for Atmospheric Monitoring) was expanded to 32 stations in which several, if not all of the following air pollutants are measured: O_3, SO_2, NOx, and CO. Figure 6.2 shows the location of the present monitoring stations in the MCMA. In several of these stations, meteorological parameters such as wind speed, wind direction, temperature, precipitation, are also measured. At the same time RAMA was expanded (1992), the former manual network was updated and expanded to 19 sites equipped with new semiautomated samplers to measure TSP and particles of less than 10 μm diameter (PM_{10}). Since 1997, this manual network has had only 12 stations. In many instances, the manual stations are located in the same place as the automatic stations. The monitoring sites for both networks and the pollutants measured at each site as of 1997 are listed in the following web site: *http://www.ine.gob.mx/dggia/cal_aire/libros/segu_info/mexico.html.*

Furthermore, since the early 1980s, more academic and government-supported research groups have become involved with the study of the occurrence, chemistry, and fate of air pollutants in the Mexico City Air Basin. In addition

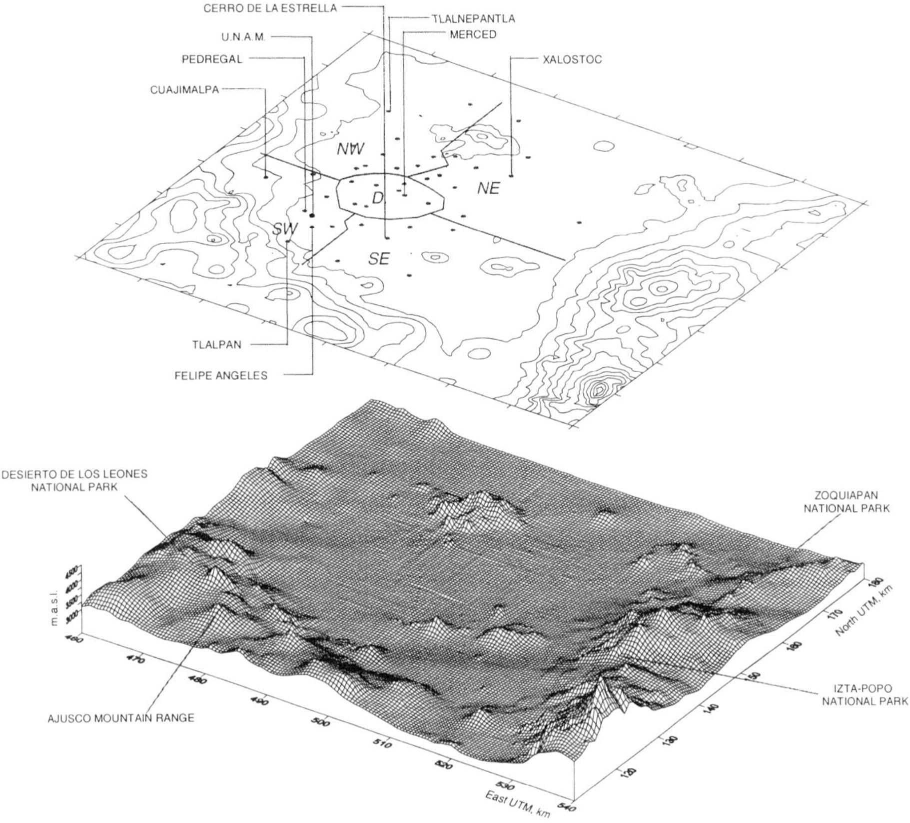

Figure 6.2. Location of the air quality monitoring stations in the different sectors of the MCMA and the main forested areas in the Basin. Several of the marked sites share both automatic and manual stations. The circular region labeled *D* represents the downtown sector. The perspective of the 3-D map has been modified to facilitate the viewing of the Basin. (Courtesy of Alfonso Salas, Centro de Ciencias de la Atmósfera, UNAM.) Note that the North UTM values given on the Y-axis are 2000 km less than the actual value (e.g., 120 km north is actually 2120 km north).

to several academic research groups at the National Autonomous University of Mexico (UNAM) and the Metropolitan Autonomous University (UAM), institutions such as the Japanese International Cooperation Agency (JICA) and the Mexican Petroleum Institute (IMP), among others, have produced a number of air pollution reports (Corey et al. 1989; Edgerton et al. 1999; Ezcurra and Mazari-Hiriart 1996; MARI 1994; Streit and Guzmán 1996). However, the dependence on official budgetary supporters of many of these studies has resulted only in the improvement of the public image of several federal organizations.

Since the beginning of air quality monitoring in the MCMA, two kinds of reports on air pollution have been released: official and academic reports. In the former, the apparent interest has been to smooth public perception of the air pollution problem, while in the latter, the aim has been to highlight the inability of the authorities to solve the problem. As a result, comprehensive long-term measurements of most of the air pollutants are incomplete, and available research data on air pollutant control strategies are so biased that they often confound meaningful analysis and demonstrate the lack of seriousness of the institutions responsible for such investigations (see comments of Stocksdale and Steinberg 2000).

On the other hand, because in Mexico, "right to know" laws do not exist as they do in other countries, the release to the public of raw data on air quality and of specialized research studies sponsored by federal agencies or other government institutions have been banned for many years. Only recently did the Government of Mexico City open the air quality data from the RAMA to academic groups and the public. However, other government-related research groups still do not disclose valuable information to either academia or the general public. For the above mentioned reasons, in-depth analysis of air quality trends, including quality assurance of the measurements, is not included in this report.

In this chapter, air quality progress is measured by comparing ambient air pollution levels with the appropriate primary air quality standards or with recommended international guidelines (EPA 1985). Based on this statement, the chapter is focused on the evaluation of air quality improvement in the MCMA according to all of the available information including official and academic reports which have already been statistically filtered by their respective authors (reference INE 1999 as an example). It is not the intention of this work to carry out a critical evaluation of the validity of the collected data.

Meteorology and Air Quality

Daytime winds in the Basin most often blow to the south and the southwest (Fig. 6.3). Since most of the heavy industrial areas and the most populated areas are in the northeast and northwest sectors of the MCMA, the wind carries air masses rich in oxidant precursor (NO_x and volatile organic compounds) emissions into the downwind populated and forested areas. The blocking effect of the high mountains on the southern edge of the Basin, results in some recirculation of polluted air carried to these areas during the daytime. However, downslope wind flows during the night contribute to the return of O_3 and NO_x to the lower elevations of the Basin (Bravo et al. 1999; Raga et al. 1999; Torres 1997). Mesoscale modeling analyses of the three-dimensional meteorology in the Basin show that, in fact, air mass circulations in the vicinity of the MCMA are highly complex (Fast and Zhong 1998). Wind circulation within the Basin seems to be further

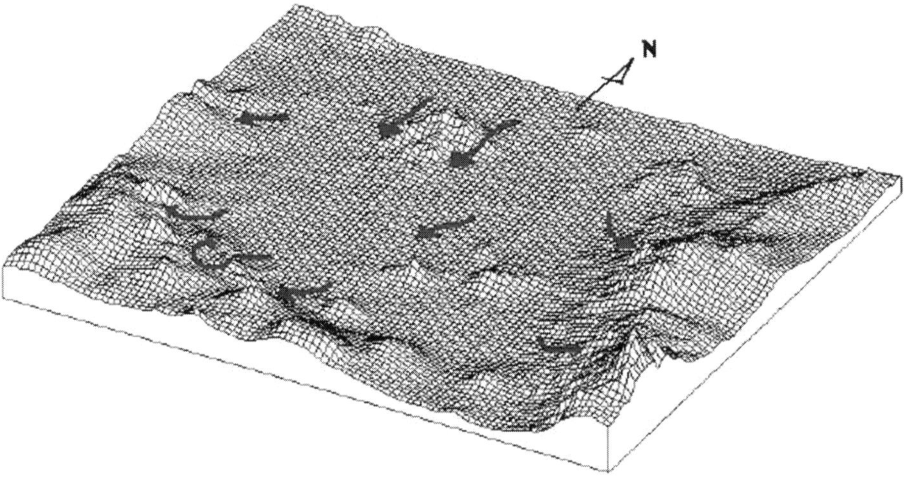

Figure 6.3. Idealized typical diurnal flow of winds in the MCMA region.

exacerbated by the frequent occurrence of nocturnal thermal inversions, which occur every month and are especially severe and prolonged during the "winter" dry season from November to March. In an inversion, a layer of cold dense air near ground level in the Basin is thermally stable, preventing pollutants emitted at ground level from dispersing upward. Above the nocturnal or stable layer, O_3 concentrations from the previous day remain high, even though the surface O_3 concentration can be quite low during the night. Research studies performed on the vertical distribution of O_3 in several sites in Mexico City confirm the occurrence of nocturnal trapping of "aged" O_3 aloft of the inversion layer (Bravo et al. 1999; Bravo-Cabrera et al. 1999; MARI 1994; Porch et al. 1992; Raga et al. 1999). The monthly average depth of the inversion layer has been estimated to range between 200 and 400 m (Jauregui 1983).

After sunrise, as the daytime convective boundary layer grows, the O_3-rich air is entrained into the mixing layer and transported to the surface. This may produce some increase in morning surface O_3 concentrations, but more importantly, it adds a very photochemically reactive component to the fresh morning emissions. When the inversions coincide with other mesoscale meteorological phenomena, such as descending high-pressure air masses, pollutant concentrations frequently reach critical levels. Although it has been suggested that contributions from air pollutants resulting from the previous day are probably small (Fast and Zhong 1998; Gaffney et al. 1999), there is evidence that such contributions are important for the air pollution levels of the following day (MARI 1994; Raga et al. 1999; Torres 1997). The combination of high ground-level emissions and poor dispersion leads to high pollutant concentrations, especially during the winter months.

Primary Air Pollutants

Sulfur Dioxide

Sulfur dioxide is produced by the combustion of sulfur-containing fuels and, to a much lesser degree, from smelting ores and other processes such as petroleum refining and fertilizer production. In the atmosphere, SO_2 reacts with O_3 and other photochemical oxidants to produce sulfur trioxide (SO_3), which in turn, rapidly combines with water to form sulfuric acid and with various atmospheric cations to form sulfate aerosols. The effects of exposure to short-term high concentrations include irritation of the respiratory tract, bronchitis, and bronchoconstriction in asthmatics. Sulfur dioxide is one of the man-made pollutants causing economic damage to vegetation. The effects of SO_2 on plants can be visible, invisible, acute, or chronic (APTI 1976; EPA 1973).

There are two Mexican air quality standards for SO_2 (MSO_2AQS): an annual arithmetic mean of 0.03 ppm and a 24-hr average level of 0.13 ppm. Both standards are health-related standards. The annual standard is not to be exceeded, while the short-term standard is not to be exceeded more than once per year.

Until the early 1990s, SO_2 was a very serious air pollution problem in the MCMA. The combustion of bunker and diesel oil fuels rich in S (around 3.6% and 1% by weight respectively) by practically all the industrial and commercial facilities, as well as emissions from the operation of the presently closed PEMEX refinery, led to high emissions of this pollutant. PEMEX is the Mexican governmental oil company. Bravo (1960) measured SO_2 at two sites: Tacuba (very close to the PEMEX refinery in the northwest) and at the UNAM campus (in the southwest). The respective 6-month average levels were 0.15 and 0.03 ppm while the 24-hour maximum levels were 1.28 and 0.42 ppm. The latter two values were well above the SO_2 short-term air quality standard, although the former could be influenced by emissions from an industrial boiler stack located very close to the sampling site at that time.

In the 1970s, SO_2 levels in the Mexico City Air Basin showed some increase due to urban development, the expansion of industrial activities, and increases in the vehicle fleet (Bravo and Magaña 1979). In 1986, the Federal Commission of Electricity (CFE), began a program to switch its power plants located in the northern sector of the MCMA to run on natural gas rather than the usual fuel oil during the winter season. This action was implemented to prevent SO_2 from reaching the high levels typically associated with the cold months. Nonetheless, several reports on the SO_2 situation in the MCMA during the late 1980s showed that no clear-cut trends in the data were evident from 1986 to 1991 (Bravo et al. 1990; Lacasaña-Navarro et al. 1999; WHO/UNEP 1994). Apparently the switching of fuel oil in the CFE power plants was not implemented as expected.

However, in March 1991, a presidential mandate ordered the immediate closing of the PEMEX refinery located in the northeast of the MCMA. This mandate rapidly brought down the SO_2 levels typically observed in this area. By the end of 1991, the Mexican authorities had dictated the definitive switching from fuel oil to natural gas in the two CFE power plants and the mandatory use of low-sulfur diesel fuel in all industrial and commercial combustion processes in the Basin (PEMEX 1996; WHO/UNEP 1994). As expected, areawide atmospheric SO_2 levels diminished.

Figure 6.4a shows the trend in annual average SO_2 concentrations for several representative sites in the MCMA from 1977 to 1997 after a compilation of data from Gavin (1982); the former Secretariat of Ecology and Urban Development (SEDUE 1986), and Lacasaña-Navarro et al. (1999). 1991 was the critical year for change in SO_2 levels. Although the long-term average levels of SO_2 have shown an important decrease, in 1997, the annual average levels of this pollutant were still not in compliance with the annual MSO_2AQS. No official data are available for the years 1998 and 1999. Because of the current relatively high content of S in Mexican gasolines, it is expected that the annual average levels of SO_2 will remain above the respective air quality standard for several more years.

Figure 6.4b summarizes maximum SO_2 levels and the number of days above the short-term 24-hour MSO_2AQS per year between 1988 and 1999 in the whole Basin (GDF 2000; INE/SEMARNAP 1997). From this figure, it is clear that since 1993, there has been an important reduction in short-term SO_2 levels. Values are now well below the 24-hour air quality standard.

Historically, the highest SO_2 levels have been registered close to the industrial areas located at the northeast and northwest of the Basin. Because of the meteorological characteristics of the Basin, a declining spatial gradient in SO_2 concentrations occurs towards the south (Bravo et al. 1990; SEDUE 1986). There is strong evidence that, due to the strong atmospheric photochemical activity in the Basin, most of the SO_2 emitted upwind is transformed to sulfate aerosols and transported to the southwest. These aerosols are subsequently brought to the ground by atmospheric deposition processes (Bravo et al. 1988; Fenn et al. 1999).

Nitrogen Dioxide

Nitrogen dioxide, a yellowish-brown gas, plays a key role in the photochemical production of O_3. It is also a toxic and irritant gas, and the precursor to the formation of nitric acid and nitrate particles, which are significant contributors to acid deposition and PM_{10} levels. Some NO_2 is directly emitted as a result of combustion. However, most of the NOx emissions (NOx = NO + NO_2) from combustion sources is in the less toxic form of nitric oxide (NO). This NO is subsequently oxidized to NO_2 as part of the same photochemical cycle that produces O_3. In the MCMA, major sources of NOx emissions are gasoline and diesel vehicles, fuel combustion in power plants and industrial burners, and liquefied petroleum gas (LPG) for domestic use. Early in the morning on a typical weekday, the concentrations of NO and NO_2 remain relatively constant. As urban activity increases from 7 to 9 a.m., the concentrations of primary pollutants such as NO increase dramatically. Then, in response to increasing solar UV radiation, the amount of NO_2 increases as NO is converted to NO_2. As the NO concentration falls to very low levels (less than 0.1 ppm), photochemical oxidants begin to

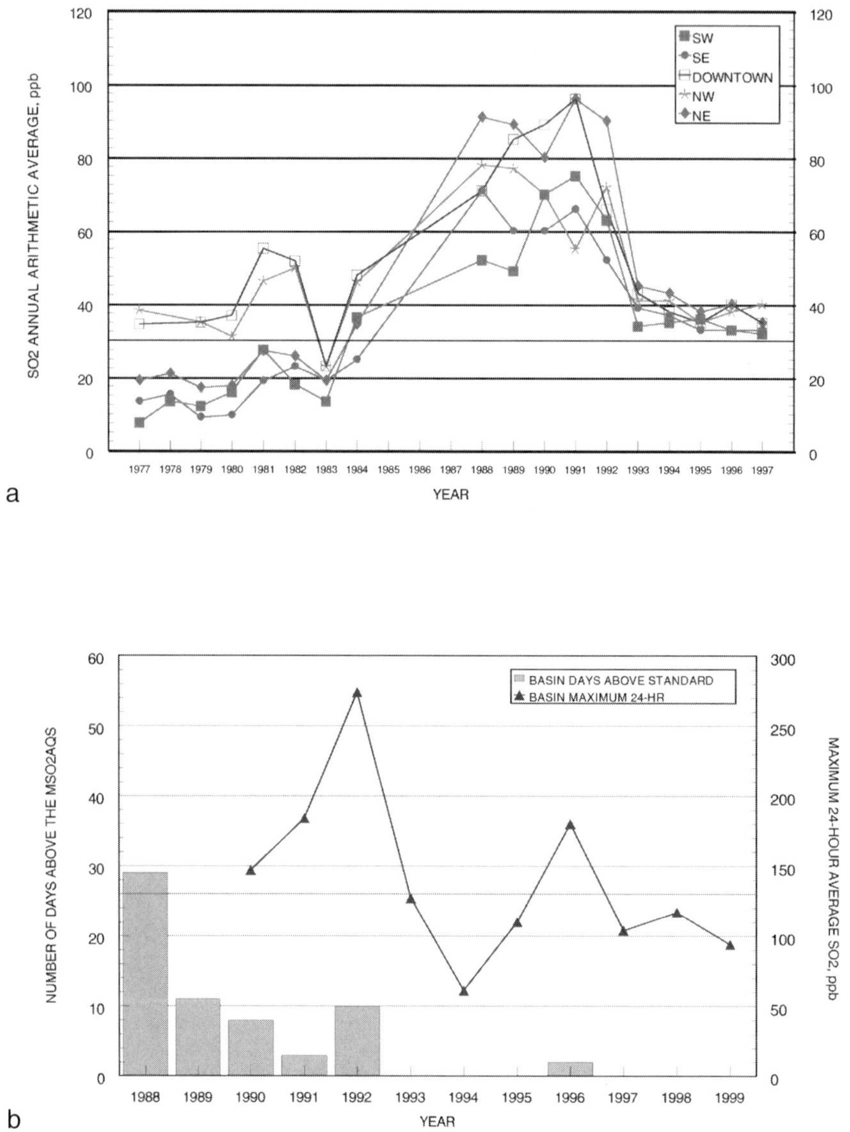

Figure 6.4. (a) Trend in annual average sulfur dioxide (SO_2) concentrations at five representative sites of the MCMA, 1977–1997. (b) Number of exceedances of the 24-hour Mexican air quality standards for SO_2 (MSO_2AQS) and the trend in maximum 24-hour average SO_2 concentrations, 1988–1999.

accumulate and reach a peak about midday. The increase in automobile traffic in the late afternoon and evening causes an increase in the NO concentration. Even in the absence of sunlight, NO_2 continues to be formed from NO by O_3 until the O_3 supply is exhausted. Therefore, the highest one-hour average concentrations of NO_2 tend to be very close to the areas of greatest NOx emissions.

Short-term exposure to high levels of NO_2 causes lung edema and damage to lung cells and increases susceptibility to bronchial infections among other health effects. Nitrogen dioxide can also adversely affect plants, either directly or indirectly. Direct effects result mainly from NO_2 exposures and occur in localized areas. Indirect effects result from the key role of NOx in forming highly phytotoxic photochemical oxidants, such as O_3 and peroxyacetyl nitrate (PAN) (APTI 1976). Further indirect effects of deposition of atmospheric nitrogen from NOx include enrichment of terrestrial and aquatic ecosystems with elevated levels of biologically-available forms of nitrogen. The result of chronic atmospheric nitrogen deposition is the eutrophication of these systems and a cascade of deleterious environmental and ecological effects (Fenn et al. 1998; Vitousek et al. 1997). The Mexican air quality standard for nitrogen dioxide (MNO_2AQS) is set at a maximum daily 1-hour average NO_2 concentration of 0.21 ppm, not to be exceeded more than once a year. By comparison, the World Health Organization (WHO) recommends two air quality guidelines for NO_2 concentrations: a 24-hr average concentration of 0.0795 ppm and an annual average of 0.053 ppm.

In the late 1950s, NO_2 levels were measured for the first time using a wet-sampling method (Bravo 1960). By 1968, the Pan American Health Organization (PAHO) manual network included the sampling and measurement of NO_2 in several of their stations in the MCMA although with a number of operating problems. Bravo and Lowe (1976) took the yearly averages reported by PAHO for the years 1968 and 1969 for a station located in the southwest of the MCMA and compared these averages with the NO_2 annual means they themselves had found for the years 1975 and 1976 at the University of Mexico Monitoring Station (UMMS). The UMMS is also located in the southwest of the urban area. The PAHO annual averages for the years 1968 and 1969 were 0.017 and 0.03 ppm, and the UMMS means for the years 1975 and 1976 were 0.014 and 0.05 ppm respectively. Apparently, there was no clear change in the annual average NO_2 levels between 1968 and 1975 in the southwest of the Basin. The annual WHO air quality guideline was not exceeded, at least during these years. However, Bravo and Lowe found that the maximum 24-hour NO_2 average levels for the years 1975 and 1976, 0.076 and 0.083 ppm respectively, were close to the annual WHO guideline for this pollutant. Thus, short-term levels of NO_2 began to appear as an air pollution problem in the MCMA in the mid-1970s.

Bravo and Torres (1986), monitored NOx concentrations in 1984 and 1985 at the UMMS in order to study the role of NOx in determining the O_3 concentrations registered in the MCMA. The major conclusion of this study was that the southwest area of the Basin was the principal smog receptor for the MCMA. Bravo and Torres (1986) also found that in 1985, average NOx concentrations ranged between 0.01 and 0.27 ppm with a median of 0.07 ppm during the morning rush from 6 to 9 a.m. This range was slightly greater than the corresponding morning rush-hour values reported for several cities of the United States. The United States values varied between 0.018 ppm in Texas City to 0.114 ppm in Cleveland (Baugues 1986). Clearly, by 1985, morning concentrations of NOx in the MCMA represented a serious air quality problem.

Since 1985, continuous ambient NO_2 levels have been recorded by the official network using automatic analyzers. The trend in the days per year in excess of the NO_2 air quality standard at any of the stations in the basin (basin-days above the MNO_2AQS), and the trend in the Basinwide annual hourly maximum NO_2 concentrations from 1987 to 1999 are shown in Figure 6.5a (GDF 2000; INE/SEMARNAP 1997; Legorreta and Flores 1992; Romieu et al. 1991). According to this figure, the MNO_2AQS has been continuously exceeded during at least the last 13 years. Although the daily maximum seems to be going down, it is clear that this air pollutant is still a problem in the MCMA.

The large number of days in which the MNO_2AQS was exceeded in the Basin between 1993 and 1998 may be attributed to the exhaust emissions of a large number of cars whose engines had been deliberately tuned to pass the newly enforced emission standards. Prior to 1998, vehicle emission standards only considered the emissions of hydrocarbons (HCs) and CO but not NO_x. Because lean fuel-air mixtures in the engine produce lower HC and CO emissions, the tuning-up of old cars at this fuel-air range was a recurrent practice among mechanics in the MCMA. However, lean fuel-air mixtures led to high NOx emissions. An emission standard for NOx was finally implemented in 1998.

The long-term trend in NO_2 levels in the MCMA is best observed by reviewing the annual mean data for representative monitoring sites. Average annual NO_2 levels from 1988 to 1997 confirm that the long-term levels of this air pollutant have remained practically constant in the MCMA (Fig. 6.5b). As previously mentioned, there is no long-term air quality standard for NO_2 in Mexico. However, the comparison of mean annual NO_2 levels with the annual WHO guideline (0.053 ppm) clearly shows that through 1997, the long-term Basinwide air quality was poor with regards to this pollutant.

The main origin of this pollutant is indicated by the spatial distribution of NO_2 surface concentrations, which is highest in those zones of the MCMA with severe traffic, such as the downtown and the northeast areas. NO_2 levels toward the southwest of the Basin are lower because the photochemical plume travels away from the high NOx emission zones. Most of the NO_2 formed during the morning period is transformed to other photochemical products (i.e., peroxyacetyl nitrate and nitrate aerosols) after promoting O_3 formation. For this reason, it is expected that high concentrations of secondary nitrogenated pollutants are reaching the southern areas of the basin, once the photochemical plume has moved away from the high NOx emission zones. There is evidence that some of the NO_2 transported during the daytime to the southern mountains remains nearly unaffected until nocturnal catabatic winds return these air masses to the valley (Raga et al. 1999; Torres 1997).

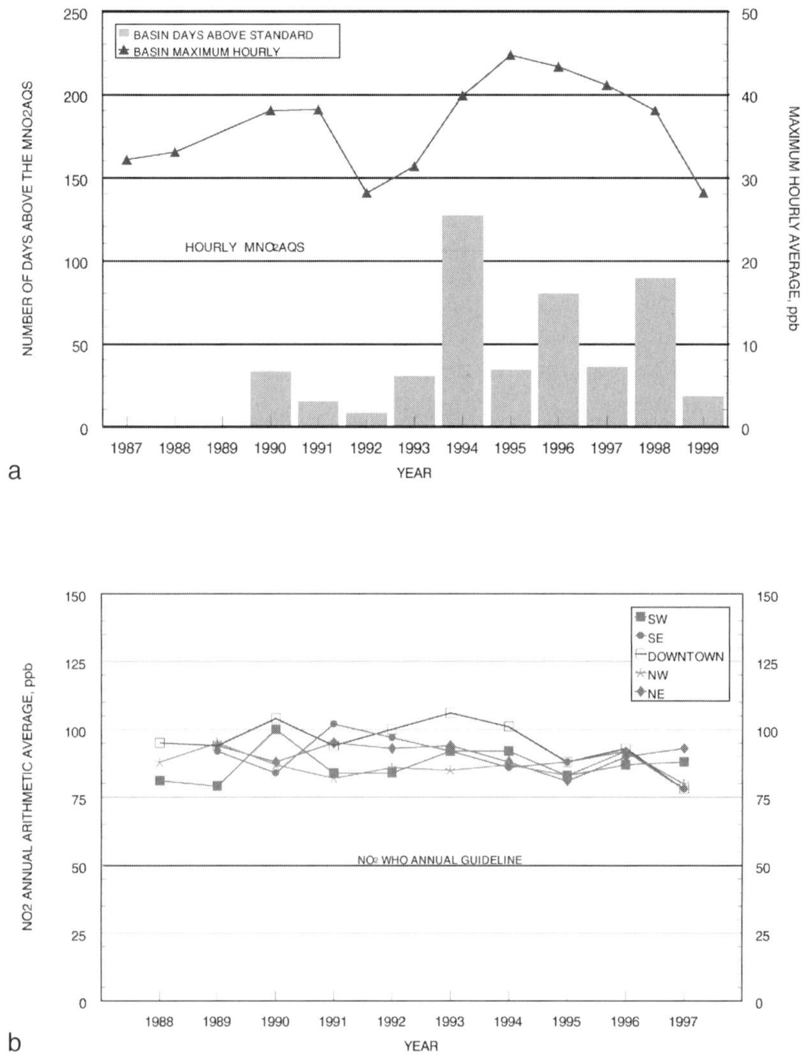

Figure 6.5. (a) Number of daily exceedances of the 1-hour Mexican air quality standards for nitrogen dioxide (MNO_2AQS) and trend in the maximum 1-hour nitrogen dioxide concentration at all monitoring sites, 1987–1999. (b) Trend in the annual average NO_2 concentrations at five representative sites of the Mexico City Metropolitan Area, 1988–1997.

Carbon Monoxide

Carbon monoxide is formed by the incomplete combustion of carbon-containing fuels. Almost all CO emissions are produced by gasoline vehicles. Carbon monoxide levels peak on weekdays between 7:00 and 9:00 a.m. when low temperatures, atmospheric stability (thermal inversions), and heavy traffic occur simultaneously. There is another peak in the evening, but this is usually lower than the morning one. In urban areas, total mass emissions of CO far outweigh all other pollutants combined. When inhaled, CO binds to hemoglobin in the blood to form carboxyhemoglobin. This deactivates the hemoglobin molecule, preventing the transport of oxygen through the body and reducing oxygen supplies to sensitive organs. The results of moderate CO toxicity include shortness of breath, increased blood pressure, headaches, and difficulty in concentration. There is just one short-term Mexican carbon monoxide air quality standard (MCOAQS). This standard specifies an 8-hour average of 11 ppm that should not be exceeded for more than an 8-hour continuous period in a year. The MCOAQS is around 20% less stringent than the respective Environmental Protection Agency (EPA) recommendation of 9 ppm for 8 hrs and the WHO guideline of 8.75 ppm for the same period. Prior to the 1980s, vehicular traffic was not a problem in Mexico City; therefore, neither were ambient levels of CO. Since the late 1970s, however, as motor vehicles became an increasingly important source of air pollutants, CO gained importance as a potential deteriorating factor for air quality in the MCMA.

Besides some exploratory monitoring work by Bravo and Corona (1969) during the Olympic Games held in Mexico City in 1968, there were no other measurements of CO until the end of the 1970s, when the official air quality monitoring network included several CO monitors in their stations. Nevertheless, little reliable data resulted from this first stage of monitoring. According to the former Undersecretariat for Environmental Improvement, the maximum 8-hour average CO concentrations from November, 1977, through June, 1978, ranged between 3.4 and 22.4 ppm (SMA 1978). Gavin (1982) reports that except for the southwest area of the MCMA, levels of CO increased over a large part of the city from 1978 to 1981 to almost reach the higher 8-hour average levels observed in the downtown area. These levels ranged between 19.1 and 17.8 ppm.

The CO monitoring network was improved and enlarged by the mid-1980s. However, an exposure assessment study made in the early 1990s, showed that former and current CO levels measured at these monitoring stations underestimate the actual exposures of pedestrians and commuters (Fernandez-Bremauntz 1991). Although air quality standards must be achieved at all spatial scales, experience has shown that areas of greater public exposure are more likely to experience standard violations than are areas that contain few people. For the above reason, the US EPA has recommended since the first criteria were established for selecting sites for CO monitoring in the 1970s, that middle-scale stations be located close to roadways or traffic corridors with air sampling inlets at 3 m height (EPA 1975).

Figure 6.6 shows the trend in Basin-days in excess to the MCOAQS and the Basin-maximum moving 8-hour average CO concentrations from 1988 to 1999 (GDF 2000; INE/SEMARNAP 1997). In spite of PEMEX's introduction of methyl–*t*–butyl ether (MTBE) into the low-leaded gasolines sold in the MCMA since 1989, the number of days above the MCOAQS increased notoriously.

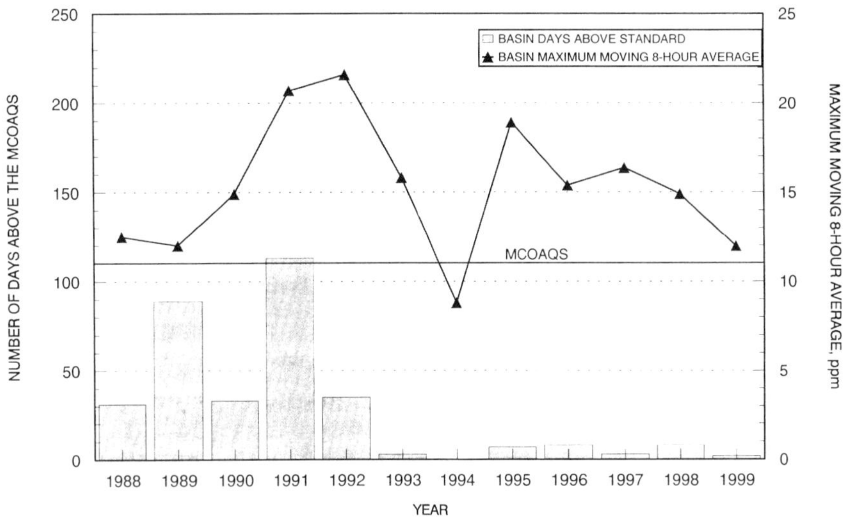

Figure 6.6. Number of daily exceedances of the Mexican air quality standards for carbon monoxide (CO) (MCOAQS) and trend in the maximum 8-hour average CO concentration from all monitoring sites of the Mexico City Metropolitan Area, 1988–1999. An 8-hour "moving" average CO concentration is calculated every hour as an average of the preceding 8 hours.

PEMEX had declared that such an oxygenated additive would help to improve engine efficiencies and to reduce the levels of HC and CO. There was no reduction in ambient CO levels until 1992, when the federal government required catalytic converters on all new model cars and set more stringent CO and HC emission standards for old cars.

The persistence of poor maintenance and tuning practices in a fraction of the automotive fleet in the MCMA is revealed by the trend in the daily maximum CO levels shown in Figure 6.6. Although the number of days above the 8-hour standard has decreased, this guideline is still exceeded in some isolated points of the city. The highest 8-hour average levels are currently monitored near heavily congested roads and problematic street intersections in the downtown and northeast areas.

Total Suspended Particulate Matter

All solid or liquid particles (aerosols) that can remain suspended in ambient air because of their physical properties such as size and weight are identified as SPM. SPM can be either primary or secondary in origin. It is produced naturally (e.g., pollen, spores, salt spray, and soil erosion), by human activity (e.g., soot, fly ash, and cement dust), and by atmospheric photochemical reactions (secondary aerosols) and occurs in a broad range of sizes. In broad terms, SPM tends to cluster in two principal size groups: coarse particles, generally larger than 2 to 3 μm in diameter; and fine particles, generally smaller than 2 to 3 μm in diameter.

Fine fractions of the SPM can have effects on both health and welfare. Particles may be deposited in (1) the extrathoracic airways, (2) airways of the tracheobronchial region, and (3) the alveolar region. SPM also has an effect on visibility and speeds the deterioration of many man-made materials. Through the years, SPM has been measured as total suspended particles (TSP). Total suspended particles is a more specific term used to define the ambient SPM levels based on the sampling method to measure them. The TSP monitoring method is supposed to be effective in sampling all size ranges up to about 100 μm (EPA 1996), but does not provide information with regards to the fractions of the total SPM. The Mexican TSP air quality standards (MTSPAQS) are 260 μg/m^3 as a 24-hour average not to be exceeded more than once per year and 75 μg/m^3 as an annual arithmetic mean. The reference method is the "high-volume" sampler (Diario Oficial 1993). Measurements of TSP in the MCMA using high-volume samplers began in the late 1950s; however a number of problems, related to the maintenance and support of the official network between the early 1970s and mid 1980s, led to a loss of continuity and credibility in the data.

Early measurements of TSP were made during 1959 in a limited number of sites in the MCMA (Bravo 1960). In this study the highest TSP levels were found in the northwest sector of the city. The 24-hour values ranged from 40 to 360 μg/m^3 with an annual average of 120 μg/m^3. In the late 1960s and early 1970s, the mean 24-hour TSP levels measured with a wider sampling network ranged from 60 to 150 μg/m^3 (PAHO 1982). The corresponding maximum daily values varied between 200 and 1000 μg/m^3 at a site located in the northeast area of the MCMA. According to these data, there was a definitive increase in TSP levels during the 1960s. Although there are not sufficient and reliable TSP data on maximum 24-hour concentrations between 1975 and 1985, several academic studies concluded that the TSP increased remarkably during this 10-year period (Bravo 1987; García-Gutiérrez 1990; Espinoza and Zúñiga 1982).

The evaluation of the historical trend of 24-hour TSP levels in the Mexico City Air Basin after 1985 has also been a matter of confusion. While the Department of the Federal District reports maximum TSP daily values from 1100 to 1300 μg/m^3 from 1988 to 1995 (DDF 1997), the National Institute of Ecology states that for the same period, the maximum 24-hour concentrations varied from 1494 to 2913 μg/m^3 (INE 1999). In spite of the discrepancies in the data, the seriousness of the TSP in the air quality of the MCMA is obvious.

With regard to the annual TSP arithmetic means, the official reports are more consistent. According to these reports, from 1976 to 1986, the annual average levels increased markedly as a direct reflection of the growing population. The Basinwide annual arithmetic TSP means ranged from 120 to 412 μg/m^3 in 1976 and from 182 to 590 μg/m^3 in 1987 (García-Gutierrez 1990; Romieu et al. 1991). According to the Metropolitan Commission (1993) and Lacasaña–Navarro et al. (1999), from 1988 to 1997, annual TSP means show a decreasing trend, although Basinwide values in 1997 were still clearly above the annual MTSPAQS of 75 μg/m^3 (Fig. 6.7).

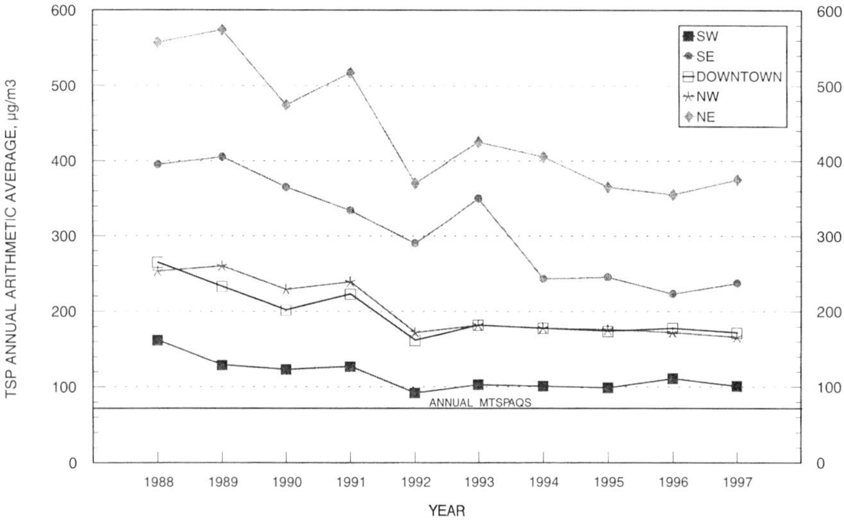

Figure 6.7. Trend in the annual arithmetic mean for total suspended particulates at five representative sites of the Mexico City Metropolitan Area.

Historically, the area with the most unhealthy levels of TSP has been the northeast sector, because of the locally high particulate matter emissions from numerous industries, including a power plant, and many unpaved roads. As a consequence of the combination of the spatial distribution of particulate emissions and the prevalent wind direction in the Mexico City atmospheric basin, there is a gradient of TSP concentrations decreasing towards the southwestern mountains of the Basin (Fuentes 1990).

The major constituents of suspended particles are sulfates, nitrates and other inorganic ions, carbonaceous compounds, metals, and materials of crustal origin. Some chemical forms of sulfates and nitrates have the potential to affect humans, plants, and materials. Both are secondary particles or aerosols. Sulfate and nitrate aerosols are considered to be the main constituents of the fine fraction of suspended particles. However, they are discussed only briefly herein because historically they were measured as a part of the TSP.

Sulfate aerosols (mainly sulfate salts and some acidic aerosols) are the result of the oxidation of SO_2 through complicated atmospheric reactions (EPA 1996). For this reason, sulfate aerosols are good indicators of historical trends for SO_2 in industrialized regions and urban areas. Since the late 1960s, several researchers have followed the evolution of sulfates in the MCMA (Bravo 1960; Bravo et al. 1972, 1988; Espinoza and Zúñiga 1982; Falcón et al. 1990). The most important change in the ambient levels of sulfates in TSP occurred between 1985 and 1992. Before the switch from bunker oil to natural gas in the early 1990s, the sulfate content in TSP was relatively high. The annual average sulfate concentrations ranged between 12.5 and 17.0 $\mu g/m^3$ from 1981 to 1987 in the southwest of the Basin (UNAM), while in the west of the city (Tacubaya), the sulfate levels ranged between 8.5 and 12.5 μg/m^3 (Bravo et al. 1988). However, after the closing of the PEMEX refinery and the official mandate to lower S fuels in the MCMA in 1991, the sulfate content in TSP began to decrease. A study of the sulfate content in TSP performed during the first semester of 1992 at the UNAM reported an average sulfate concentration of 5.2 μg/m^3 (Metropolitan Commission 1993).

Nitrate aerosols (such as ammonium nitrate and nitric acid) are the result of the occurrence of terminal atmospheric chemical reactions of photochemical products such as NO_2 and PAN among others (EPA 1996). Although the nitrate content in TSP has not been studied with the same intensity as the sulfate content, the available data show a strong relationship between the concentrations of these compounds and the photochemical activity of the Mexico City atmospheric Basin. Data from Bravo and coworkers (1988) and the Metropolitan Commission (1993) indicate that the nitrate fraction in TSP had increased steadily from 1981 to 1989, with nitrate concentrations ranging from 4 to 8.5 μg/m^3. Nevertheless, data on nitrate content in TSP measured after 1991 show that nitrate levels have been decreasing. According to the Metropolitan Commission (1993), the average nitrate concentration for the first semester of 1992 was 7 μg/m^3 for a site in the southwest of the MCMA.

Typically the highest sulfate and nitrate levels in TSP have been observed in the southern areas of the atmospheric basin as a consequence of the combination of photochemical processes and atmospheric transport of air masses. Because new sampling devices for secondary aerosols, devices different from the original high-volume samplers, have been used since the early 1990s, data on sulfate and nitrate content in TSP are not available for recent years.

Elemental carbon or soot, another important fraction in TSP, is largely responsible for visibility reduction in urban areas. Soot is generated by poorly controlled combustion processes. In urban areas, buses and trucks are the main sources. There are few studies related to soot content in TSP in the MCMA (Bravo et al. 1986; Nava 1991). According to studies before the 1990s, around 20 percent of the TSP was elemental carbon. The quarterly average concentrations of soot (January through March) from 1982 to 1987 in TSP filters sampled in the west, downtown, and southwest areas of the MCMA (Tacubaya, Casco de Santo Tomás and UNAM, respectively) ranged between 16 and 85 μg/m^3. The highest concentrations were found in the downtown area, and the lowest values in the southwest zone. No correlation was found between soot content and TSP concentrations. More than 30% of the visibility reduction in the Mexico City atmospheric basin was attributable to the soot content in TSP (Bravo et al. 1986).

Lead

Lead (Pb) is commonly present in the SPM of urban atmospheres. The main source is the combustion of Pb antiknock compounds in gasoline. Lead is a cumulative, systematic toxin with effects on blood pressure regulation and on the nervous system. The Mexican air quality standard for Pb (MPbAQS) is 1.5 μg/m^3 on a quarterly average basis. Before 1986, Pb was probably the most harmful

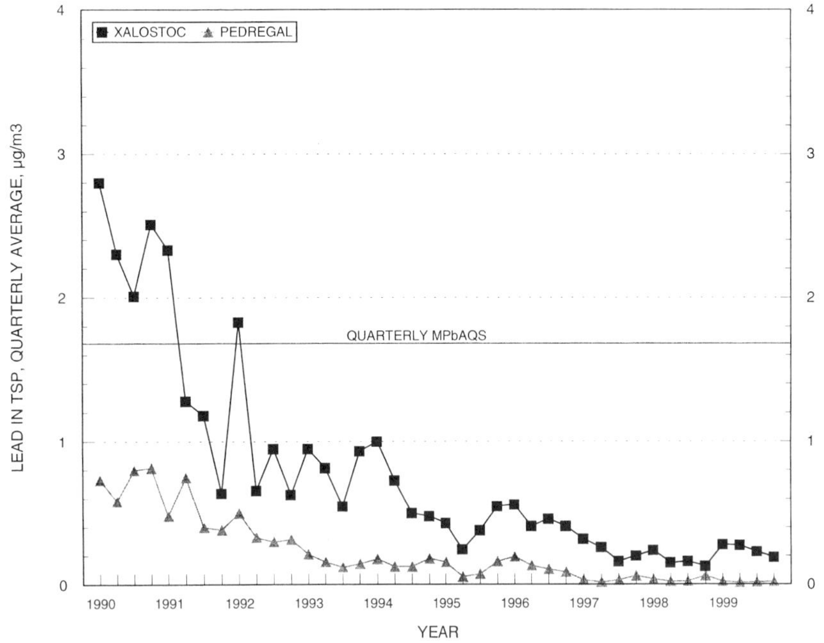

Figure 6.8. Trend in the quarterly arithmetic means for levels of lead at two representative sites of the Mexico City Metropolitan Area (Xalostoc in the northeast and Pedregal in the southwest), 1990–1999. MPbAQS, Mexican air quality standard for lead.

pollutant in the Mexico City Air Basin. Until that date, only leaded gasoline was sold in the MCMA, and the concentration of Pb in the air increased steadily with the number of cars until average Pb values of almost three times the MPbAQS were reached. The reduction by PEMEX of the maximum Pb content of gasoline, from 3 ml/gal of tetra-ethyl-lead to less than 1 ml/gal in 1986, undoubtedly led to a decrease in both Pb emissions and atmospheric concentration of Pb in the air. However, at the same time an unexpected increase of photochemical smog (Bravo and Torres 2000) also occurred. The reduction of tetra-ethyl-lead in gasoline continued steadily until 1998, when PEMEX finally phased out this additive. The chemical composition of the new additive required to enhance the octane number of gasoline used by cars without catalytic converters is still not available to the public.

Figure 6.8 shows the trend of quarterly average concentrations of lead in TSP in the Pedregal (southwest residential area) and Xalostoc (northwest industrial area) sampling stations, two representative sites of the MCMA, during the period 1980 to 1999 (GDF 2000). Historically, the highest concentrations have been recorded in the northeast and in the downtown areas of the MCMA as a result of the combination of emissions of Pb smelters and industrial activities in the area in addition to emissions from vehicular sources.

Secondary Air Pollutants

Ozone

Ozone is a "secondary" pollutant, formed by complex photochemical reactions of "primary" directly-emitted pollutants in the atmosphere. It is the marker chemical for a family of photochemical oxidants produced by reactions between their precursors, NOx, and reactive volatile organic compounds (VOC) in the presence of sunlight (Kley et al. 1999). The high levels of NOx and VOC emissions in the MCMA and the intense sunlight resulting from its high altitude and tropical latitude combine to produce O_3 concentrations which are among the highest measured in any city (WHO 1997).

Because of its photochemical nature, peak concentrations of O_3 are typically reached during the afternoon in the downwind southwest sector of the Basin. Since O_3 is highly reactive, its concentration at ground level drops markedly in the evening, mainly because it is scavenged by fresh NO emitted during the night. On the other hand, O_3 can remain at elevated concentrations in the ambient air above the mixing layer. As previously mentioned, this elevated reservoir of "aged" O_3 can then contribute to elevated ground level O_3 on the following day as air mixing increases. Other related chemical reactions involving NOx, VOC and O_3 result in the production of secondary pollutants such as organic, sulfate and nitrate aerosols, gaseous nitric acid and PAN, among others (EPA 1986).

The high levels of O_3 and other oxidants that occur in the downwind southwest sector of the Basin depend upon many interrelated factors. Simple advection of photochemically reactive air masses, local winds, diurnal wind cycles, and contributions from retained "aged" O_3 aloft appear to be the main factors involved in pollution transport in the Mexico City Air Basin (Torres 1997). As a consequence of the time lag involved in O_3 formation, the highest O_3 levels are registered in the afternoon on sunny days towards the southwest of the Basin, in the direction of the prevailing diurnal winds.

Short- and long-term exposure to high ambient O_3 concentrations causes irritation and inflammation of the eyes and respiratory tract, increased mucous production, and impaired lung function, among other effects (Lippmann 1989). It has been suggested that the health of most of the inhabitants in the MCMA has been affected to some degree by the high levels of O_3 and high dosage of exposure currently experienced in this urban area (Calderón-Garcidueñas et al. 1992; Lacasaña-Navarro et al. 1999). The short-term Mexican O_3 air quality standard (MO_3AQS) of 0.11 ppm maximum daily 1-hour average not to be exceeded more than once in three years, is comparable to the US EPA O_3 air quality standard of 0.12 ppm and the WHO guideline of 0.10 ppm. In addition to its effects on human health and well-being, high ambient O_3 concentrations adversely affect vegetation, rubber and plastics, and surface coating such as paints (Kley et al. 1999).

Before 1985, O_3 levels in the MCMA occasionally exceeded the MO_3AQS (Bravo et al. 1989). However, after PEMEX replaced high-leaded gasolines sold in this urban area with low-lead-content fuels in mid-1986, O_3 pollution became a serious problem. While atmospheric concentration of Pb did indeed fall, photochemical smog increased because of the change in reactivity of the exhaust emissions (Bravo and Torres 2000). In order to restore the octane number lost due to the reduction of lead-alkyl additives formerly in the gasoline, PEMEX reformulated the "new" fuels with more iso-paraffins, naphtenes, and alkylaromatics. PEMEX did not take in account that a relatively high content of olefins and aromatics in gasoline could be important in terms of photoreactivity. In addition, in 1986, the car fleet of Mexico was not equipped with catalytic converters which could destroy the new balance of reactive VOC and NOx in the exhaust. Low-leaded gasoline sales in the MCMA have shown a strong correlation with violations of the MO_3AQS (Bravo and Torres 2000; Riveros 1996).

Exploratory O_3 measurements in the MCMA were first made during the mid 1970s and early 1980s by the University of Mexico (Bravo et al. 1978; Bravo and Torres 1985, 1986). There was also some official O_3 monitoring activity during this same period, but the results were very questionable (Legorreta 1989). Continuous O_3 air quality monitoring began in 1986 when the old network was totally replaced with new monitoring equipment. Although the monitoring objectives of the academic and official groups have been different through the years, both coincide in recognizing that O_3 levels in the MCMA have been exceptionally high at least during the last 15 years. Figure 6.9a shows the maximum hourly O_3 concentrations in the Basin from 1985 to 1999 (GDF 2000; INE 1999; MARI 1994). The seriousness of the O_3 air quality problem in the MCMA is better understood by reviewing the annual number of exceedances of the MO_3AQS. Figure 6.9b shows the number of Basin-hours and the number of hourly exceedances to the MO_3AQS registered by the official monitoring network and the University of Mexico monitoring station from 1986 to 1999 respectively (Bravo and Torres 2000; GDF 2000; INE 1999). The MO_3AQS specifies a maximum 1-hour concentration of 0.11 ppm not to be exceeded more than once in three years. However as a gross average, the O_3 air quality standard is exceeded by more than 4 to 5 hours per day for at least 300 days per year.

Since 1991, new model cars are equipped with catalytic converters, but their proportion in the MCMA car fleet is less than 50%. Although the effect of catalytic converters is now reflected in the downward trend in O_3 concentrations, it is clear that O_3 is still a major air pollution problem for the authorities and the inhabitants of the MCMA.

Ozone air quality impacts in this urban area are not restricted only to the health of the people. Due to topographical and meteorological characteristics of the Basin, air mass transport carries high levels of photochemical products to the forested mountains to the south. Evidence for this mesoscale effect was first presented by Krupa and de Bauer in the late 1970s after they conducted a series of surveys of pollutant-induced visible plant injury in the Ajusco mountains (Krupa and de Bauer 1976). This phenomenon was confirmed lately by Bravo and Torres (1985), who observed air mass transport to and accumulation of O_3 in the high altitude sectors in the southwest of the Basin. Subsequent studies monitoring O_3

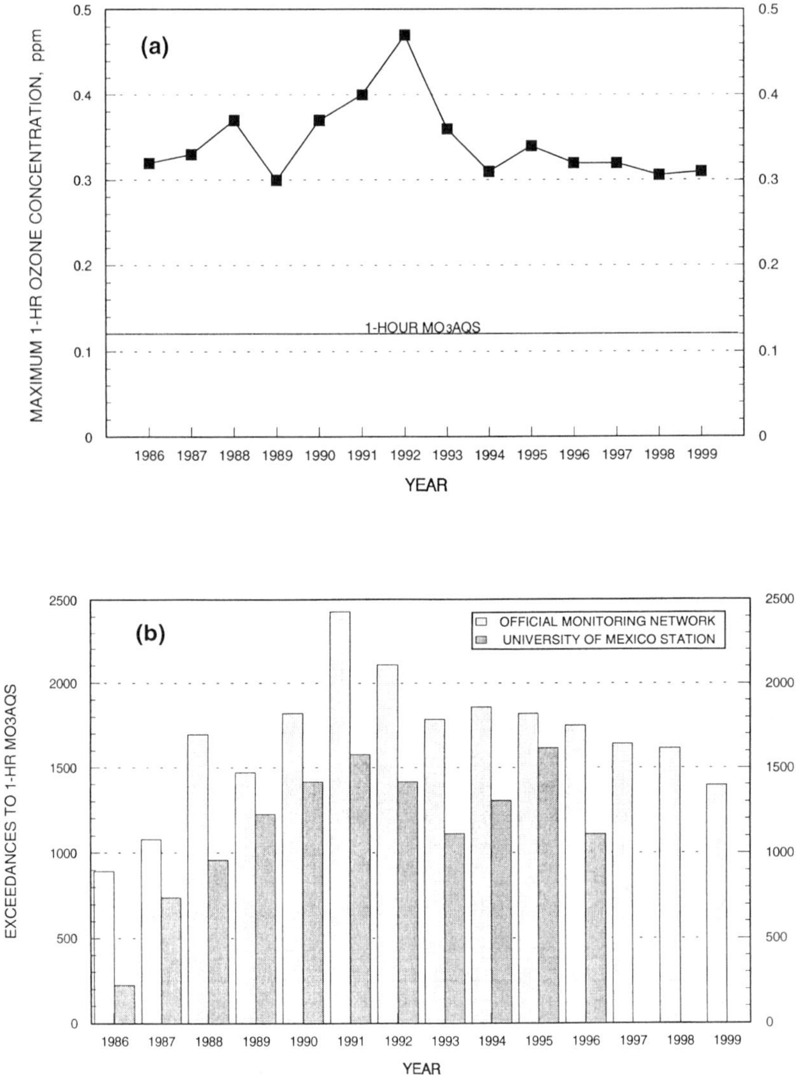

Figure 6.9. (a) Trend in the daily maximum 1-hour ozone concentration from all monitoring sites in the MCMA, 1986–1999. (b) Number of annual exceedances of the Mexican air quality standard for ozone (MO_3AQS) at both the University of Mexico (1986–1996) and all official monitoring stations (1986–1999).

in the Ajusco and the Desierto de los Leones mountains have demonstrated that high O_3 levels are common in these forested areas (Bravo et al. 1999; Miller et al. 1994; Raga et al. 1999).

For several months in 1992, Bravo and coworkers (1999) measured O_3 concentrations on the smog receptor north-facing slope of the Ajusco at a site located

440 m above the average surface level of the valley. They found that the daily peak O_3 concentrations occurred a short time later than at the closest upwind station located at UNAM. They also found that the average duration of the periods when O_3 concentrations exceeded 0.11 ppm were longer than those recorded in the valley. The location of the nocturnal observations of high O_3 levels coincided with the zone where the inversion layer most often contacted the mountain slope. Because of the absence of O_3 scavenging by nocturnal ambient NO at the Ajusco site, nocturnal O_3 levels at this location were higher than those measured at the same hour in the valley. Figure 6.10 shows the hourly average measurements of O_3 at the Ajusco and UNAM sites for the period June 4 through June 9. It is likely that O_3 remains aloft of the stable mixing layer which develops on the valley after sunset. Measurements of vertical O_3 profiles in different sites of the MCMA during the day have supported this theory (Bravo-Cabrera 1999; Porch et al. 1992). It was concluded that in some cases, the nocturnal inversion acts as a reservoir for O_3 transported to these downwind locations. Raga et al. (1999) arrived at similar conclusions after conducting a field monitoring campaign in 1998 at the Ajusco site.

Respirable Particulate Matter

As in other urban polluted areas, particulate matter less than 10 μm in diameter, PM_{10}, makes up a very important fraction of the SPM in the MCMA. Major types of PM_{10} are those produced by combustion (diesel and stationary sources), process systems (vapors and fine dusts), and aerosol particles (organic, sulfate, nitrate, and ammonium) formed by atmospheric photochemical reactions. Because of its small size, PM_{10} has greater potential to enter the lower portions of the lungs than larger particles (>10 μm). Acute exposures to high short-term levels of respirable SPM has been linked to increases in illness and death from respiratory causes, especially when the particulate includes acid aerosols such as sulfate and nitrate particles (EPA 1996). The short-term Mexican PM_{10} air quality standard ($MPM_{10}AQS$) is 150 μg/m^3 for a 24-hour average concentration, and the respective long-term air quality standard is 50 μg/m^3 as an annual arithmetic mean. The WHO short-term guideline for PM_{10} is 70 μg/m^3 (24-hour).

Besides its health effects, PM_{10} is also associated with dry acidic deposition, soiling, and reduced visibility. The most important of these is acidic deposition, which causes damage to plants, materials, and buildings (EPA 1996).

The first measurements of fine fractions of SPM in the MCMA were done during the late 1970s and the early 1980s (Bravo et al. 1981, 1984, 1988; Sigler et al. 1982; SMA 1978). The concentrations of fine SPM were strongly linked to TSP levels and ranged from 20 to 320 μg/m^3 (30%–40% of TSP) during the period of 1976 to 1977, 79 to 100 μg/m^3 (54%–67% of TSP) between 1980 and 1981, and 74 to 191 μg/m^3 (35%–95% of TSP) from 1981 to 1987.

In 1988, the Mexican environmental authorities included new PM_{10} samplers in several of their manual monitoring stations. Figure 6.11a shows the trend in Basin-days above the 24-hour $MPM_{10}AQS$ as well as the trend in maximum

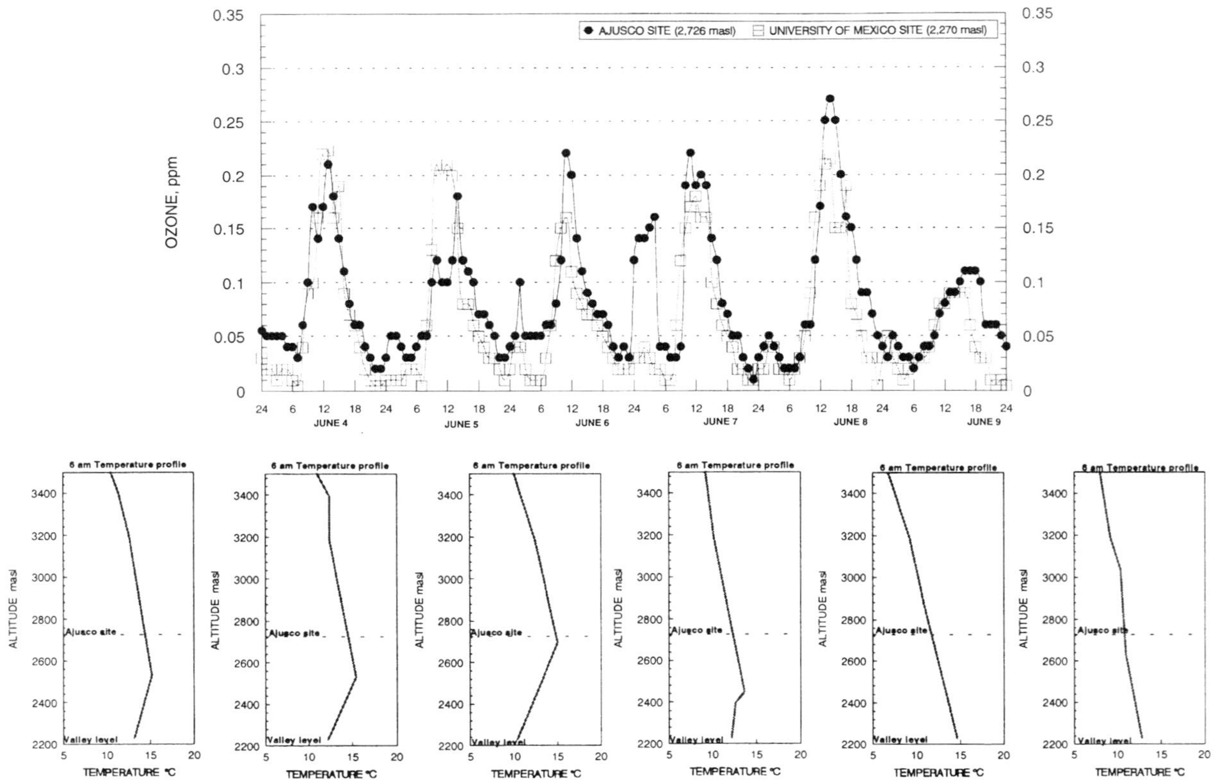

Figure 6.10. Diurnal variation of ozone at the Ajusco and the University of Mexico monitoring station and morning vertical temperature profile at the International Airport in Mexico City, June 4 to 9, 1991.

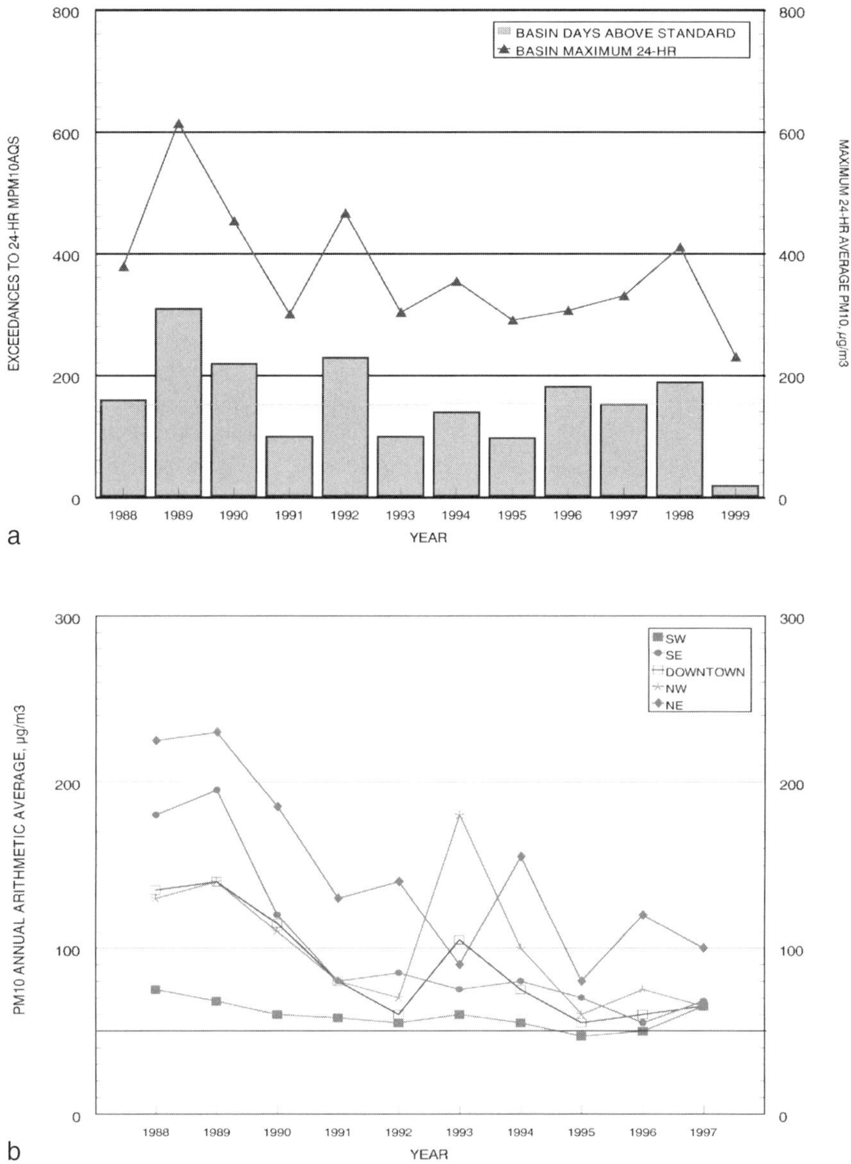

Figure 6.11. (a) Number of 24-hour exceedances of the Mexican air quality standards for particles of less than 10µm diameter ($MPM_{10}AQS$) and trend in the maximum 24-hour average PM_{10} concentrations from all monitoring sites. (b) Trend in the annual arithmetic mean PM_{10} concentrations at five representative sites of the MCMA.

24-hour PM_{10} concentrations from 1988 to 1999 (GDF 2000; INE/SEMARNAP 1997). The gross trend for the PM_{10} annual arithmetic mean for five representative stations in the MCMA is shown in Figure 6.11b. Although a downward trend has been observed, the levels are still above the respective PM_{10} long-term air quality standard.

According to studies on the composition of PM_{10} collected in several areas of the MCMA during the 1980s, on average 30% to 50% of the fine particles were organic material, 6% to 20% was soot, and sulfate and nitrate ions made up 18% to 60%. Ordinary dust accounted for 3% to 11% of the PM_{10}, and metals, mainly Pb and Cd, constituted 3% to 24% (Barbiaux 1990; Bravo et al. 1984; Cahill et al. 1996; Falcón 1992).

Several research groups have done more detailed studies on the composition of the ultra-fine fraction of SPM in the MCMA since the 1980s (Cahill et al. 1996; Edgerton et al. 1999; Falcón 1992; Melgar and Ruíz 2000; Metropolitan Commission 1993; Miranda et al. 1992, 1996). Prior to the 1990s, the "coarse fraction" (CF) of the PM_{10} (diameter between 2.5 and 10 µm) was much greater than the "ultra-fine fraction" (less than 2.5 µm or $PM_{2.5}$) in both concentration and relative abundance (Table 6.2). In recent years, these two fractions are of similar proportions.

Compared with relatively old reports on $PM_{2.5}$ concentrations for urban sites in North America, the fine particulate matter in the MCMA is relatively high. The mean concentration of $PM_{2.5}$ for 14 cities of Canada was 13.9 µg/m^3 between 1986 and 1994 (Brook et al. 1997), and the range of averages of "ultra-fine" particles in five cities of the United States was between 11.2 and 29.6 µg/m^3 between 1977 and 1987 (Schwartz et al. 1996). $PM_{2.5}$ concentrations in the MCMA in 1997 ranged from 39 to 103 µg/m^3 (Edgerton et al. 1999).

The composition of the $PM_{2.5}$ present in the MCMA has also been studied (Cahill et al. 1996; Edgerton et al. 1999; Melgar and Ruíz 2000). According to Melgar and Ruíz (2000), the percentage composition of $PM_{2.5}$ collected in a site located in the southeast of the Basin (Cerro de la Estrella) between February and March of 1997 was: organic carbon, 35% to 40%; soot, 12% to 15%; sulfates, 8% to 12%; nitrates, 5% to 10%; soil, 8% to 12%; ammonium, 4% to 6%; salt, <2%; and others, 8%.

Other Noncriteria Air Pollutants

Volatile Organic Compounds

Volatile organic compounds (VOCs) are noncriteria air pollutants; thus there are no particular air quality standards for them. However, some of them are important because of their toxicity to humans and their potential for O_3 formation. Because of the technical difficulties in collecting and analyzing VOC, there are few studies of this air pollution problem in the MCMA. The first study on this issue was conducted between 1992 and 1993 by the IMP and the US EPA (MARI

Table 6.2. Means of particles less than 10 μm diameter (PM_{10}), CF (coarse fraction) and particles less than 2.5 μm diameter ($PM_{2.5}$) reported by several research studies performed in representative sites of the MCMA from 1984 to 1997

Area	Period	PM_{10} μg/m^3	CF μg/m^3	$PM_{2.5}$ μg/m^3	Reference
Northwest (Metropolitan Autonomous University-Azcapotzalco)	Winter 1984	237	208 (87%)	29 (13%)	Falcón 1992
East (Museo Tecnologico)	Average, 1987–1989	133	85 (64%)	48 (36%)	Cahill et al. 1996
Southwest/North (University City/ Mexican Petroleum Institute)	January–June 1992	190	110 (58%)	80 (42%)	Metropolitan Commission 1993
Southwest (University City)	Fall 1993	106	56 (53%)	50 (47%)	Miranda et al. 1996
Northeast (Xalostoc)	February–March 1997	57	21 (37%)	36 (63%)	Edgerton et al. 1999
Northwest (Tlalnepantla)	February–March 1997	103	59 (57%)	44 (43%)	Edgerton et al. 1999
Downtown (Merced)	February–March 1997	58	28 (48%)	30 (52%)	Edgerton et al. 1999
Southeast (Cerro Estrella)	February–March 1997	78	45 (58%)	33 (42%)	Edgerton et al. 1999
Southwest (Pedregal)	February–March 1997	39	18 (46%)	21 (54%)	Edgerton et al. 1999

The numbers in parentheses represent the percentage of total PM_{10} that the particular fraction represents.

1994). Air samples collected between 6 a.m. and 9 a.m. in March 1992, in several sites of the basin, showed that mean concentrations of nonmethane organic compounds (NMOC) in the MCMA ranged from 1979 to 4805 ppbC with values as high as 7170 ppbC. Mean concentrations obtained in March 1993 ranged from 1300 to 6800 ppbC with a high value of 7700 ppbC. These were very high levels, approximately two to four times the peak levels currently experienced in southern California, a region with strong photochemical air pollution. The highest NMOC concentrations were observed at the northeast site of Xalostoc and the downtown site of Merced, and the lowest were registered at the southwest site of Pedregal.

In 1993, Blake and Rowland (1995) carried out an independent study on the VOC composition in the MCMA and its potential for O_3 formation. They concluded that urban leakage of liquefied petroleum gas (LPG) and incomplete combustion of this gas were the main sources of O_3 precursors in the MCMA. They based their hypothesis on the high levels and persistent ratios of propane and butane observed in their VOC data. However, careful analysis of the concentrations and relationships of vehicular indicators, performed later on, showed that mobile sources are the most important sources of O_3 precursors (Bravo and Torres 1999; DDF 1996; Guzmán et al. 1996).

Bravo and Torres (1999) analyzed the reactivity potential of the mix of VOCs present in the MCMA based on the concept of the maximum incremental reactivity potential (MIR) developed by Carter (1994). They used the original VOC data collected by the US EPA in 1992 and made a comparison between the reactivity potential of atmospheric mixtures of VOC from different urban areas with strong photochemical pollution problems, including the MCMA. They found that the combined potential of the top 38 most abundant organic compounds found in the MCMA was almost twice the respective potential for Athens and Osaka City and nearly three times the respective potentials for Los Angeles and Sydney. Figure 6.12 shows this comparison in terms of the contribution of each organic family to the total reactivity. Ethylene was the compound with the highest individual reactivity potential. Other important contributors to the reactivity of the atmosphere of MCMA were alkyl-aromatics and other olefins. The latter two are emitted by vehicular sources. Propane and butane, the main components of LPG, have a mild reactivity potential.

Measurements of VOC carried out in March, 1997, showed that the NMOC levels have not changed as expected in response to the several VOC emissions control that have been implemented in the MCMA since 1991 (Edgerton et al. 1999). The mean average NMOC concentration reported in this study was 4106 ppbC, with levels ranging from 500 to 6700 ppbC.

According to the original data of the collaborative study between the IMP and the US EPA in 1992, the three-hour morning average concentration of benzene, a carcinogen typical of urban areas, ranged from 30 to 60 ppbC. The maximum benzene concentrations, 86 ppbC, were observed at Merced and Xalostoc. Other abundant hydrocarbon species found were acetylene and ethylene. The average concentration of the sum of these compounds ranged between 65 and 374 ppbC.

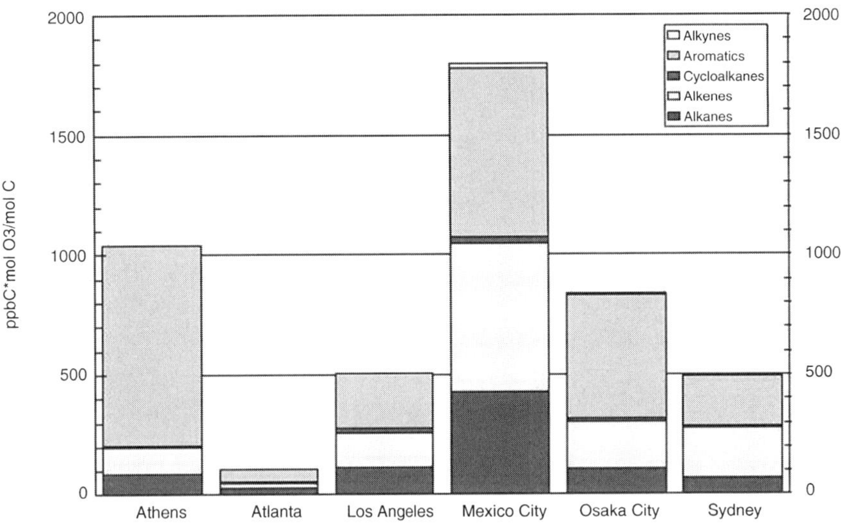

Figure 6.12. The contribution of hydrocarbon families to the total reactivity potential for ozone production of six urban areas including the MCMA.

Ethylene is recognized as one of the few VOCs from the family of the alkenes that pose a direct threat as air pollutants to vegetation (APTI 1976).

Formaldehyde (HCHO), an organic compound from the family of the carbonyls, is also present in the atmosphere of the MCMA. Formaldehyde is a nonselective toxic that may pose a threat to urban, agricultural, and natural vegetation with increased atmospheric concentrations (Barker et al. 1991). This carbonyl has two origins. First, it can be emitted in the exhaust of gasoline and diesel powered vehicles, and second, it can be generated as a byproduct of atmospheric photochemical activity. Ambient levels of HCHO in the Mexico City Air Basin increased as a consequence of the introduction of MTBE in the gasolines sold in the MCMA since 1989 (Bravo et al. 1991; Gaffney et al. 1999). Several studies on ambient HCHO levels in different sites of the MCMA indicate that for this carbonyl, the concentrations found depend on the site and period of sampling (Baez et al. 1995; MARI 1994; Mugica et al. 1998). The reported average morning levels of HCHO ranged from 12 to 37 ppb between 1991 and 1996. The maximum HCHO concentration (110 ppb) was found in the spring of 1993 in a site located at the southwest of the MCMA (Baez et al. 1995). In general, the ambient HCHO levels present in the Mexico City Air Basin during the 1990s were among the highest levels reported in the literature for the last 20 years.

Peroxyacetyl nitrate (PAN) is another secondary air pollutant of importance in the MCMA. PAN is a constituent of photochemical smog, although at a given location, its concentration and correlation with O_3 differ as a function of time. PAN is an unstable compound resulting from the reaction between organic acyl radicals and NO_2 (EPA 1986). The chief property of interest for PAN is its oxidizing ability. PAN is phytotoxic and is a lachrymator agent (EPA 1986; Taylor 1969).

PAN concentrations measured near the downtown area of the MCMA during 1997 were in the range of 10 to 35 ppb (Gaffney et al. 1999). PAN levels greater than 30 ppb have not been reported in the literature since similar levels were observed in Los Angeles in the 1970s. These high MCMA PAN concentrations may be explained by the significant levels of PAN precursors (higher aldehydes and olefins) in the Mexico City Air Basin. PAN concentrations exhibited a strong diurnal variation with the peak around noon (Gaffney et al. 1999). The sampling site was located just north of the central district of Mexico City; even higher levels of PAN are likely to occur in southern sites of the Basin.

Acidic Precipitation

Atmospheric pollution also has a considerable influence on the quality of rainwater. Precipitation is one of the principal removal mechanisms by which the atmosphere cleanses itself. Without such removal, the concentrations of air pollutants would continue to increase as a result of more or less continuous emissions, and toxic levels would have been reached long ago. Acids in precipitation result mainly from sulfuric acid and nitric acid being either absorbed directly into precipitation or formed in the aqueous phase from precursor compounds. The major sources of acid rain precursors are emissions of SO_2 and NOx. The normal pH of rain, as a result of the chemical equilibrium between atmospheric CO_2 and water, is approximately 5.6. Acidity problems develop because there are insufficient neutralizing alkaline substances in the air and in the lakes, streams, and soils on which deposition occurs. The potential adverse effects of excess acidity include loss of aquatic life in poorly buffered lakes and damage to impacted forests (NAPAP 1991).

Studies of acid precipitation in the MCMA have been conducted since the early 1980s (Bravo 1987; Paramo et al. 1987). Bulk deposition collectors (wet and dry deposition together) were used in the first stage of those studies; thus no clear appreciation of acidic values was observed. Since 1987, studies of acid rain have been carried out with automatic wet/dry collectors. Analysis of wet-only samples has allowed more definitive monitoring of wet deposition trends.

According to data from the official acid-rain network, the Basinwide frequency of acid-rain events has decreased. In 1987, more than 80% of the precipitation samples had pH values below 5.6, but by 1995, less than 20% of the rain samples had pH levels below 5.6 (Secretaría del Medio Ambiente 1996). However, when pH data for specific stations in the Basin are analyzed, such a decreasing trend is observed only in northern sampling sites. The volume-weighted pH values of wet deposition samples collected at the southwest smog receptor site of the UMMS between 1984 and 1998 have remained nearly constant (Fig. 6.13a) (Humberto-Bravo and Ma. Isabel-Saavedra, unpublished data). Acid-rain events with pH values as low as 3.5 were still common in 1998. Data from the official acid-rain network also show that the highest acidity values are commonly regis-

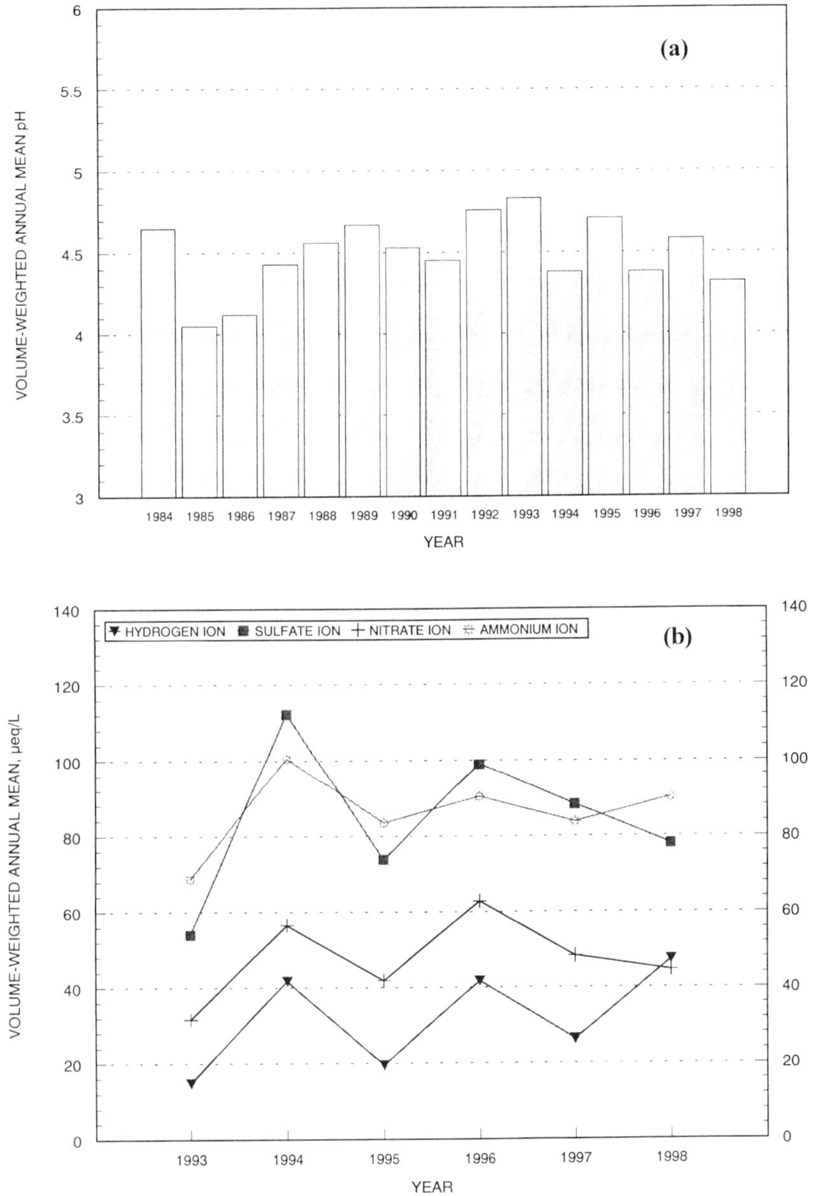

Figure 6.13. (a) Volume-weighted annual mean pH of wet-only precipitation samples collected at the University of Mexico monitoring station (UMMS; 1984–1998). (b) Trend in volume-weighted annual mean concentrations of ammonium, nitrate, sulfate, and hydrogen ions ($\mu eq\ L^{-1}$) in wet-only precipitation samples collected at the UMMS (1993–1998).

tered in the southwest forested areas of the Basin (GDF/SMA 2000). Figure 6.14 shows the spatial distribution of the volume-weighted pH values in wet precipitation collected during 1998 at all rain sampling sites located in the Mexico City region (16 from the official network and one from the UMMS). The information confirms that the spatial distribution of wet acidic deposition in the MCMA is a direct consequence of atmospheric processes in the Basin, namely wind transport and washout of air pollutants by precipitation.

The volume-weighted annual mean concentrations for hydrogen ion, sulfate, nitrate, and ammonium measured between 1993 and 1998 in wet precipitation samples collected at the UMMS are shown in Figure 6.13b. In comparison with the highest ion concentrations reported for the United States in the same year (NADP 2000), volume-weighted concentrations at the UMMS in 1998, were nearly equal for pH, approximately 1.3 times higher for sulfate, and nitrate, and almost two times higher for ammonium. The relatively high levels of ammonium observed in wet precipitation are the consequence of the availability of gaseous ammonia (NH_3) in sufficient quantities to neutralize the acid aerosols present in the Mexico City atmospheric basin. A preliminary NH_3 emission inventory for the MCMA (Gasca and Osnaya 2000), reports that 75% of the NH_3 emissions

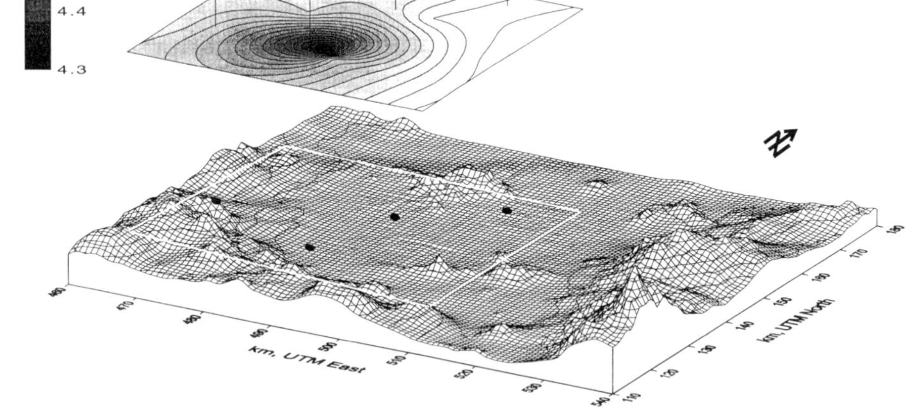

Figure 6.14. Idealized projection of the spatial distribution of the volume-weighted annual mean pH of wet-only precipitation in 1998 over the Mexico City region. Dots on the three-dimensional map indicate the locations of the four reference sites shown on the upper plane. Note that the North UTM values given on the Y-axis are 2000 km less than the actual value (e.g., 120 km north is actually 2120 km north).

come from human activities and associated domestic animals, and the origins for the remaining 25% are combustion processes, industrial refrigeration, sewage, and other sources. The biological decomposition of fecal matter from open fecalism of low-income people and from unchecked dogs contributes to a very important air pollution and health threat problem in the MCMA (Ferriss 1999; Rosas et al. 1995).

Summary

Over the past 35 years, the number of mobile and stationary sources of air pollutants has increased phenomenally in the MCMA. The review of the data and trends of the different air pollutants in the last 20 years, indicate that problems with air quality in this region reached critical proportions in the late 1970s and continued until the late 1990s. Emissions reductions in the recent past have led to a slow but gradual decrease in air pollutant levels. Although very high peaks have been reduced, levels of some pollutants, specifically those of photochemical origin, are still high. In the early 2000s, the critical air pollutants are O_3, total and fine suspended particulate matter, and to a lesser extent, SO_2, NO_2, and CO. Of all the criteria pollutants, O_3 is by far the most important air pollutant in terms of time of persistence, frequency, spatial distribution, and health and welfare impacts, despite efforts to control air pollution. Ozone levels currently exceed the Mexican O_3 air quality standard on more than 300 days per year. Total suspended particles still represent a serious problem in the northern sector of the MCMA. Although the short-term PM_{10} and SO_2 levels are just within the respective air quality standards, these pollutants are in excess of their respective long-term Mexican air quality standards. In a similar way, short-term concentrations of NO_2 are lower than in previous years, although Basinwide annual averages have been consistently well above the recommended WHO guideline for the last ten years. Carbon monoxide is currently in compliance with the air quality standard; however, the data on concentrations of this pollutant do not represent the levels to which pedestrians are exposed because of problems with the location of the monitoring stations. Lead has been in compliance with the air quality standard since 1991.

Benzene, formaldehyde, and PAN, three noncriteria air pollutants, are important contributors to the poor air quality of the MCMA. Even though there are no ambient air quality standards for these air pollutants, the relatively high levels of these air toxics observed during the 1990s suggest that there exists a significant health risk potential for the exposed population. The impact of introducing inappropriate formulations of unleaded gasoline in the MCMA has been a very high potential for O_3 formation. The reactivity of the most abundant volatile organic compounds in the Mexico City Air Basin is among the highest in the world.

Geographic and climatic conditions in the Mexico City Air Basin play an important role in the occurrence of high air-pollution levels. Daytime winds in the Basin most often blow to the southwest. Since a large proportion of the heavy industrial and residential areas are located north and northwest of the city, winds carry both industrial and vehicular emissions towards residential and forested areas in the southwest of the Basin. Ozone and other photochemical products reach their maximum concentrations downwind of the source areas for the main precursors. Meteorological transport of secondary air pollutants in the Basin is leading to the deposition of or exposure to high levels of air pollutants in the rural and forested areas to the south and southwest. Some of these pollutants (e.g., O_3 and PAN) are phytotoxic, and other pollutants are plant nutrients. Wet deposition of pollutants in the southwest of the MCMA was greater than the highest values reported for the United States, suggesting the likelihood of ecological effects in this part of the basin.

More research is needed [illegible] arify the role of atmospheric processes in the occurrence of h[illegible] levels and therefore, to justify the adoption of air pollution co[illegible] r the medium and long term. Because of the present economi[illegible] l situation of Mexico, the abatement of air pollution levels in t[illegible] particular, of photochemical air pollutant concentrations, will [illegible] han five years, if strict and perhaps unpopular control measures [illegible] ese measures should include requirements to provide fuels form[illegible] the emissions of photochemical precursors by a vehicular fleet c[illegible] proportion of older vehicles.

[illegible] eferences

APTI 1976. *Diagnos*[illegible] *ry Caused by Air Pollution*. Research Triangle Park, North Caroli[illegible] aining Institute. U.S. Environmental Protection Agency.

Báez, A.P., Belmont, [illegible] H. 1995. Measurements of formaldehyde and acetaldehyde in the [illegible] xico City. *Environ. Pollut.* 89:163–167.

Barbiaux, M. 1990. *C*[illegible] *Respirable Particulate Matter in Mexico City, Mexico*. M.S. Thesi[illegible] llinois, Chicago.

Barker, J.R., Lantis, J.M., and Tingey, D.T. 1991. Atmospheric formaldehyde and the potential for plant effects. Presented at the 78th Annual Meeting and Exhibition of the Air Pollution Control Association. June 16–21, Vancouver, B.C. Paper 91-77.3.

Baugues, K. 1986. A Review of NMOC, NOx and NMOC/NOx Ratios Measured in 1984 and 1985. Research Triangle Park, NC: U.S. Environmental Protection Agency. EPA-450/4-86-015.

Benitez, F. 1984. *Historia de la Ciudad de México*. México D.F.: Salvat Mexicana de Ediciones.

Blake, D.R., and Rowland, F.S. 1995. Urban leakage of liquefied petroleum gas and its impact on Mexico City air quality. *Science* 269:953–956.

Bravo, A.H. 1960. Variation of different pollutants in the atmosphere of Mexico City. *J. Air Pollut. Control Assoc.* 10:447–449.

Bravo, A.H. 1987. *La Contaminación del Aire en México*. México D.F.: Ed. Fundación Universo Veintiuno.

Bravo, A.H., and Corona, T.L. 1969. *La Contaminación Atmosférica y su Relación con el Flujo de Vehículos en la Ciudad de México*. Technical Report. UNAM. México: Instituto de Ingeniería.

Bravo, A.H., and Lowe, A. 1976. Air pollution station operating at Ciudad Universitaria. Presented at the 69th Annual Meeting of the Air Pollution Control Association. Portland, Oregon, June 1976.

Bravo, A.H., and Magaña, R. 1979. The actual air pollution situation in Mexico City. *Staub–Reinhalt* 39:427–428.

Bravo, A.H., and Torres, J.R. 1985. Ozone monitoring and night concentration events at the campus of the University of Mexico. Presented at the 78th Annual Meeting and Exhibition of the Air Pollution Control Association. June 16–21, Vancouver, B.C. Paper 85-59B.4.

Bravo, A.H., and Torres, J.R. 1986. *Relationship of Nitrogen Oxides to Oxidants in the Southwest Mexico City Atmosphere during 1985*. Technical Report. UNAM, México: Sección de Contaminación Ambiental, Centro de Ciencias de la Atmósfera.

Bravo, A.H., and Torres, J.R. 1999. *Comparative Assessment of the Hydrocarbon Potential for Ozone Formation in Mexico City Using Different Reactivity Schemes*. Technical Report. UNAM, México: Sección de Contaminación Ambiental, Centro de Ciencias de la Atmósfera.

Bravo, A.H., and Torres, J.R. 2000. The usefulness of air quality monitoring and air quality impact studies before the introduction of reformulated gasolines in developing countries. Mexico City: a real case study. *Atmos. Environ.* 34:499–506.

Bravo, A.H., and Viniegra, O.G. 1958. Informe Preliminar Acerca de la Polución Atmosférica en la Ciudad de México. In *Memorias de la XII Reunión Anual de la Sociedad Mexicana de Higiene Industrial*. November. Puebla, Puebla, Mexico.

Bravo, A.H., Monkman, J.L., and Lowe, A. 1972. A comparison of sulphur compounds (H_2SO_4, F.S.S. and T.S.S.) found in the atmosphere of Mexico City, Mexico; Montreal and Ottawa, Canada. Presented before the Division of Water, Air and Waste Chemistry, American Chemical Society. August 27–September 1, New York.

Bravo, A.H., Sosa, E.R., and Torres, J.R. 1990. Study of the horizontal sulfur dioxide concentration on the Metropolitan Zone of the Mexico City. Presented at the 83rd Annual Meeting and Exhibition of the Air Pollution Control Association. June 24–29, Pittsburgh, PA U.S.A. Paper 90-135.4.

Bravo, A.H., Tirado, S.D., and Torres, J.R. 1984. Determinación del comportamiento de las partículas suspendidas totales y la fracción respirable en la zona céntrica del Casco de Sto. Tomás, D.F. In *Memorias del IV Congreso Nacional de Ingeniería Sanitaria y Ambiental*, pp. 495–501. October 18–29, Morelia, Michoacán.

Bravo, A.H., Torres, J.R., and Aguilar, A.S. 1981. Determinación de la fracción respirable de PST en el Área Metropolitana de la Ciudad de México. Presented at the 2ndo. Congreso Nacional de la Asociación Nacional de Egresados de la Facultad de Medicina, UNAM, A.C., Jun. México D.F.

Bravo, A.H., Magaña, Z.R., Lowe, A.C., Ayanegui, J.S., and Sigler, A.E. 1978. Preliminary study of the photochemical process in the air pollution of Mexico City. In *Proceedings of the International Clean Air Conference*, pp. 183–191. May 15–19, Brisbane, Australia. The Clean Air Society of Australia and New Zealand.

Bravo, A.H., Saavedra, R.M.I., Torres, J.R., Lomas, A.G., Nava, T.D., and Tirado, S.D. 1986. Particulate carbon: A significant contributor to the visibility reduction in Mexico City. In *Proceedings of the International Conference on Particulate Matter and Fugitive Dusts*, pp. 113–127. October, Tucson, Arizona U.S.A. Air Pollution Control Association.

Bravo, A.H., Camacho, C.R., Saavedra, R.M.I., Sosa, E.R., and Torres, J.R. 1988. NO_3^- y $SO_4^=$ en partículas (PST) y respirables en la Cd. de México, como indicadores de estrategias ambientales (1981–1987). In *Memorias del VI Congreso Nacional de la Sociedad Mexicana de Ingeniaría Sanitaria y Ambiental, A.C.*, Aug 10–13, Querétaro, Querétaro, Mexico.

Bravo, A.H., Perrin, G.F., Sosa, E.R., and Torres, J.R. 1989. Results of an air pollution strategy (lead reduction in gasoline) on the air quality of Mexico City. In *Proceedings of the 8th World Clean Air Congress*, Vol. IV, pp. 31–37. September 11–15, The Hague, the Netherlands.

Bravo, A.H., Camacho, C.R., Roy-Ocotla, G., Sosa, E.R., and Torres, J.R. 1991. Analysis of the change in atmospheric urban formaldehyde and photochemistry activity as a result of using MTBE as an additive in gasolines of the Metropolitan Area of Mexico City. *Atmos. Environ.* 25B:285–288.

Bravo, A.H., Torres, J.R., Sánchez, A.P., and Coutiño, A.M. 1999. Estudio preliminar de ozono a dos nivels de altitud en la zona suroeste de la Ciudad de México. In *Memorias del 4°. Congreso de Meteorología de la CFE*, pp. 190–192. April 29–30, México D.F.

Bravo-Cabrera, J.L., Valdés, B.M., and Alatorre, M.N. 1999. Distribución Vertical del Ozono en la Ciudad de México. Presented at the 3er. Simposio de Contaminación Atmosférica. Universidad Autónoma Metropolitana/El Colegio Nacional. Feb. 17–19, México D.F.

Brook, R.J. Dann, F.T., and Burnett, T.R. 1997. The relationship among TSP, PM_{10}, $PM_{2.5}$ and inorganic constituents of atmospheric particulate matter at multiple Canadian locations. *J. Air Waste Manage. Assoc.* 47:2–19.

Cahill, T.A., Morales, R., and Miranda, J. 1996. Comparative aerosol studies of Pacific Rim cities—Santiago, Chile (1987); Mexico City, Mexico (1987–1990); and Los Angeles, U.S.A. (1973 and 1987). *Atmos. Environ.* 30:747–750.

Calderón-Garcidueñas, A.L., Velázquez, O.A., Bravo-Alvarez, H., Delgado-Chávez, R., and Barrios-Márquez, R. 1992. Histopathologic changes of the nasal mucosa in southwest Metropolitan Mexico City inhabitants. *Am. J. Path.* 140:225–232.

Carter, W.P.L. 1994. Development of ozone reactivity scales of volatile organic compounds. *J. Air Waste Manage. Assoc.* 44:881–899.

Corey, G., Lima, J.L., Aguilar, C., Albert, L., and Palomares, G.L. 1989. *Bibliografía Sobre Contaminación Ambiental en México*. Metepec, Edo. de México: Centro Panamericano de Ecología Humana y Salud, Programa de Salud Ambiental, Organización Mundial de la Salud.

DDF. 1991. *Comprehensive Pollution Control Program for the Mexico City Metropolitan Zone*. April. México D.F.: Departamento del Distrito Federal.

DDF. 1996. *Programa para Mejorar la Calidad del Aire en el Valle de México 1995–2000*. México D.F.: Departamento del Distrito Federal.

DDF. 1997. *Datos de Vigilancia de la Calidad del Aire para la Ciudad de México, 1995*. México D.F.: Departamento del Distrito Federal.

Diario Oficial. 1993. Norma Oficial Mexicana NOM-CCAM-002-ECOL/1993. Método de medición para partículas suspendidas totales, Secretaría de Desarrollo Social. In *Diario Oficial de la Federación*. October 18. Gobierno Constitucional de los Estados Unidos Mexicanos, Mexico City.

Edgerton, S.A., Bian, X., Doran, J.C., Fast, J.D., Hubbe, J.M., Malone, E.L., et al. 1999. Particulate air pollution in Mexico City: a collaborative research project, *J. Air Waste Manage. Assoc.* 49:1221–1229.

EPA. 1973. *Effects of Sulfur Oxides in the Atmosphere on Vegetation*. Research Triangle Park, N.C.: U.S. Environmental Protection Agency. EPA/RS/74-030.

EPA. 1975. *Selecting Sites For Carbon Monoxide Monitoring*. Research Triangle Park, N.C.: U.S. Environmental Protection Agency. EPA-450/3-75-077.

EPA. 1985. *National Air Quality and Emissions Trend Report, 1983*. Research Triangle Park, N.C.: U.S. Environmental Protection Agency. EPA-450/4-84-029.

EPA. 1986. *Air Quality Criteria for Ozone and Other Photochemical Oxidants*, Vol. II. Research Triangle Park, NC.: U.S. Environmental Protection Agency. EPA-600/8-84/020bF.

EPA. 1996. *Air Quality Criteria Document for Particulate Matter*. Washington D.C.: U.S. Environmental Protection Agency, U.S. Government Printing Office. EPA/600/1-95/001/1af.

Espinoza, R.E.M., and Zúñiga, M.A. 1982. Partículas y compuestos de azufre en el norte de la Ciudad de México. In *Memorias del III Congreso Nacional de Ingeniería Sanitaria y Ambiental*. Acapulco, Guerrero.

Ezcurra, E., and Mazari-Hiriart, M. 1996. Are megacities viable? A cautionary tale from Mexico City. *Environment* 38:6–15, 26–35.

Falcón, I.Y., Ramírez, R.C., and Vega, G.S. 1990. México City aerosol study: sulfates and nitrates in PM-10 and TSP. Presented at the 83rd Annual Meeting and Exhibition of the Air Pollution Control Association. June 24–29, Pittsburgh, PA. Paper 90-71.3.

Falcón, Y. 1992. La contaminación por partículas suspendidas totales en la atmósfera. In *La Contaminación Atmosférica en México*, coordinator I. Restrepo, pp. 147–166. México D.F.: Comisión Nacional de Derechos Humanos.

Fast, J.D., and Zhong, S. 1998. Meteorological factors associated with inhomogeneous ozone concentrations within the Mexico City basin. *J. Geophys. Res.* 103 (D15): 18,927–18,946.

Fenn, M.E., Poth, M.A., Aber, J.D., Baron, J.S., Bormann, B.T., Johnson, D.W., et al. 1998. Nitrogen excess in North American ecosystems: Predisposing factors, ecosystem responses, and management strategies. *Ecol. Appl.* 8:706–733.

Fenn, M.E., De Bauer, L.I., Quevedo-Nolasco, A., and Rodríguez-Frausto, C. 1999. Nitrogen and sulfur deposition and forest nutrient status in the Valley of Mexico. *Water Air Soil Pollut.* 113:155–174.

Fernández-Bremauntz, X. 1991. *Commuters Exposure to Carbon Monoxide in the Metropolitan Area of Mexico City*. Ph.D. Dissertation. London: Imperial College, University of London.

Ferriss, S. 1999. Mexico City urges residents to scoop dog poop, a health threat. Cox News Service. *http://www.coxnews.com/washington/MEXDOGS.HTM.*

Fuentes, G.V. 1990. Tendencias y distribución de las partículas suspendidas y el plomo en la atmósfera del Valle de México. In *Memorias del VII Congreso Nacional de la Sociedad Mexicana de Ingeniería Sanitaria y Ambiental, A.C.*, pp. D.21–D.24. September 19–21, Oaxaca, Oaxaca.

Gaffney, J.S., Marley, N.A., Cumingham, M.M., and Doskey, P.V. 1999. Measurements of peroxyacetyl nitrates (PANs) in Mexico City: implications for megacity air quality impacts on regional scales. *Atmos. Environ.* 33:5003–5012.

García-Gutierrez, A. 1990. Tendencias de las concentraciones atmosféricas de partículas suspendidas totales en la Ciudad de México entre 1976 y 1988. In *Memorias del VII Congreso Nacional de la Sociedad Mexicana de Ingeniería Sanitaria y Ambiental*, A.C., pp. D.25–D.29. September 19–21, Oaxaca, Oaxaca.

Gasca, J., and Osnaya, P. 2000. Mexico City emission inventory. Presented at Tropospheric Aerosols: Science and Decisions in an International Community. A NARSTO Technical Symposium on Aerosol Science. October 23–26, Querétaro, México.

Gavin, J. 1982. *Air Pollution in Mexico City: An Update*, U.S. Embassy in Mexico Reporting Program, pp. 10–12. July.

GDF. 1998. *Dirección General de Estadística, Dirección de Contabilidad Nacional y Estadísticas Económicas*, Anuario Estadístico del DDF. México D.F.: Gobierno del Distrito Federal.

GDF. 2000. 1999. *El Mejor Año Ambiental de la Década*. Secretaría del Medio Ambiente, México D.F.: Gobierno del Distrito Federal.

GDF/SMA. 2000. Precipitaciones Acidas en la ZMCM, Informe Anual 1998. México D.F.: Secretaría del Medio Ambiente, Gobierno del Distrito Federal.

Guzmán, F., Ruíz, M.E., and Vega, E. 1996. Air quality in Mexico City. *Science Lett.* 271:1040–1041.

INE. 1999. Calidad del aire en la zona metropolitana del Valle de México. In *Segundo Informe Sobre la Calidad del Aire en Ciudades Mexicanas 1997*. Instituto Nacional de Ecología. *http://www.ine.gob.mx/dggia/cal_aire/libros/segu_info/mexico.html.*

INE/SEMARNAP. 1997. Calidad del aire. In *Dirección General de Gestión e Información Ambiental*. México D.F.: InstitutoNacional de Ecología, Secretaría de Medio Ambiente Recursos Naturales y Pesca.

INEGI. 2000. Instituto Nacional de Estadística Geografía e Informática. México. *http://www.inegi.gob.mx*.

Jáuregui, E.O. 1983. Una primera estimación de las condiciones de difusión atmosférica en la República Mexicana. In *Boletín del Instituto de Geografía, UNAM*, No. 13, pp. 9–51. México, D.F.: UNAM.

Kley, D., Kleinmann, M., Sanderman, H., and Krupa, S. 1999. Photochemical oxidants: state of the science. *Environ. Pollut.* 100:19–42.

Krupa, S.V., and De Bauer, L.I. 1976. La ciudad daña los pinos del Ajusco. *Panagfa* 4:5–7.

Lacasaña-Navarro, M., Aguilar-Garduño, C., and Romieu, I. 1999. Evolución de la contaminación del aire e impacto de los programas de control en tres megaciudades de América Latina. *Salud Pública México* 41:203–215.

Legorreta, J. 1989. *Transporte y Contaminación en la Cd. de México*. México D.F.: Centro de Ecodesarrollo.

Legorreta, J., and Flores, A. 1992. La contaminación atmosférica en el Valle de México. In *La Contaminación Atmosférica en México*, coordinator I. Restrepo, pp. 61–97. México D.F.: Comisión Nacional de Derechos Humanos.

Lippmann, M. 1989. Health effects of ozone. A critical review. *J. Air Waste Manage. Assoc.* 39:672–695.

MARI. 1994. *Mexico City Air Quality Research Initiative (MARI)*. Los Alamos, New Mexico, U.S.A.: Los Alamos National Laboratory and the Mexican Petroleum Institute. Los Alamos Report # LA-12699.

Márquez, M.E. 1977. *Informe Sobre la Calidad del Aire en Algunas Ciudades del País*. México D.F.: Proyecto Mexico–PNUD, Subsecretaría del Mejoramiento del Ambiente, SSA.

Melgar, M.P., and Ruíz, S.M.E. 2000. Composición de las partículas suspendidas en la Zona Metropolitana de la Ciudad de México y sus efectos en la calidad del aire. *Rev. Indust. Tecnofórmula Ambiental* 1:10–11.

Metropolitan Commission. 1993. *Partículas Suspendidas. Situación Actual en la Zona Metropolitana de la Ciudad de México*. México D.F.: Comisión Metropolitana para la Prevención y Control de la Contaminación Ambiental en el Valle de México.

Miller, P.R., de Bauer, M.L., Quevedo, N.A., and Hernández, T.T. 1994. Comparison of ozone exposure characteristics in forested regions near Mexico City and Los Angeles. *Atmos. Environ.* 28:141–148.

Miranda, J., Andrade, E., López-Suárez, A., and Ledesma, R. 1996. A receptor model for atmospheric aerosols from a southwestern site in Mexico City. *Atmos. Environ.* 30: 3471–3479.

Miranda, J., Morales, J.R., Cahill, T.A., Aldape, F., and Flores, J. 1992. A study of elemental contents in atmospheric aerosols in Mexico City. *Atmósfera* 5:95–108.

Múgica, V., Vega, E., Arriaga, J.L., and Ruíz, M.E. 1998. Determination of motor vehicle profiles for non-methane organic compounds in the Mexico City Metropolitan Area. *J. Air Waste Manage. Assoc.* 48:1060–1068.

NADP. 2000. *National Atmospheric Deposition Program (NRSP-3)/National Trends Network*. Champaign, Illinois, D.S.A. NADP Program Office, Illinois State Water Survey.

NAPAP. 1991. *1990 Integrated Assessment Report. The U.S. National Acid Precipitation Assessment Program*. Washington D.C.: The NAPAP Office of the Director.

Nava, T.D. 1991. *Determinación de Partículas Suspendidas Totales y su Contenido de Carbón, Plomo, Cobre y Cadmio como Evaluación de la Contaminación Atmosférica de la Zona Metropolitana de la Cd. de México*. B.S. Thesis. México D.F.: Facultad de Química, UNAM.

PAHO. 1982. *Red Panamericana de Muestreo de la Contaminación del Aire: Informe Final 1967–1980*. Lima, Perú: Pan American Health Organization. CEPIS 23.

Páramo, V.H., Guerrero, M.A., Morales, M.A., Morales, R.E., and Contreras, D.B. 1987. Acidez de las precipitaciones en el Distrito Federal. *Cienc. Desarrollo* 72:59–66.

PEMEX. 1996. *Calidad de Combustibles y Proyectos Ambientales. Technical Presentation*. México D.F.: Petróleos Mexicanos.

Porch, W.M., Clements, W.E., and Herwehe, J.A. 1992. *Analysis of Tethered Balloon-Borne Measurements in Mexico City. September 1990 and February 1991*. Los Alamos, New Mexico, U.S.A.: Los Alamos National Laboratory. Technical Report LA-UR-92-1295.

Raga, G.B., Baumgardner, D., Kok, G., and Rosas, I. 1999. Some aspects of boundary layer evolution in Mexico City. *Atmos. Environ*. 33:5013–5021.

Riveros, G.H. 1996. Contaminación en la Zona Metropolitana de la Ciudad de México. Hacia el Tercer Milenio Marcos Mazari coordinator, pp. 243–267. México: El Colegio de México.

Romieu, I., Weitznfeld, H., and Finkelman, J. 1991. Urban air pollution in Latin America and the Caribbean. *J. Air Waste Manage. Assooc*. 41:1160–1171.

Rosas, I., Salinas, E., Arreguín, R., and Rodríguez-Romero, A. 1995. *Escherichia coli* in settled dust and air samples collected in Mexico City. Applied and Environmental Microbiology 63:4093–4095.

Schwartz, J., Dockery, W.D., and Neas, M.L. 1996. Is daily mortality associated specifically with fine particles? *J. Air Waste Manage. Assoc*. 46:927–939.

Secretaría del Medio Ambiente. 1996. *Programa: Precipitaciones Ácidas en la Zona Metropolitana de la Ciudad de México. Informe Anual 1995*. Dirección General de Prevención y Control de la Contaminación/RAMA, Secretaría de Medio Ambiente, Departamento del Distrito Federal, México.

SEDUE. 1986. *Informe Sobre el Estado del Medio Ambiente en México*. México D.F.: Subsecretaria de Ecología, Secretaría de Desarrollo Urbano y Ecología.

Sigler, A.E., Fuentes, G.V., and Vargas, A.C. 1982. Análisis de la contaminación del aire por partículas en Ciudad Universitaria. In *Memorias del III Congreso Nacional de Ingeniería Sanitaria y Ambiental*, Acapulco, Guerrero.

SMA. 1978. *Situación Actual de la Contaminación Atmosférica en el Área Metropolitana de la Cd. de México*. México D.F.: Subsecretaría de Mejoramiento del Ambiente, SSA. Memorándum Técnico SMA/DGS At/04-78.

Stocksdale, T., and Steinberg, M. 2000. Letters to the Editor. *J. Air Waste Manage. Assoc*. 50:919–921.

Streit, G.E., and Guzmán, F. 1996. Mexico City air quality: Progress of an international collaborative project to define air quality management options. *Atmos. Environ*. 30: 723–733.

Taylor, O.C. 1969. Importance of peroxyacetyl nitrate (PAN) as a phytotoxic air pollutant. *J. Air Pollut. Control Assoc*. 19:347–351.

Torres, J.R. 1997. *Construcción y Uso de Trayectorias de Retroceso de Parcelas de Aire para el Estudio de la Contaminación Atmosférica Fotoquímica en la Z.M.C.M*. M.S. Thesis. México D.F.: Facultad de Ingeniería, UNAM.

Vitousek, P.M., Aber, J.D., Howarth, R.W., Likens, G.E., Matson, P.A., Schindler, D.W., et al. 1997. Human alteration of the global nitrogen cycle: sources and consequences. *Ecol. Appl*. 7:737–750.

WHO. 1997. *WHO Guidelines for Air Quality*. December. WHO Information Fact Sheet no. 187.

WHO/UNEP (World Health Organization/United Nations Environment Programme). 1994. Air pollution in the world's megacities. *Environment* 56:4–37.

7. Emissions Trends and Pollution Control Strategies in the Mexico City Metropolitan Area

M. Esther Ruiz and Jorge Gasca

There were a few preliminary air pollution studies in Mexico City in the late 1950s and 1960s, but only in the early 1970s did air quality in Mexico City begin to be more widely recognized as a serious problem by scientists, the general population, and government authorities. A chronicle of air quality (AQ) monitoring in Mexico City can be found in *http://sma.df.gob.mx/sma/gaa/monitoreo/o/antecedentes.htm*. This concern has been the primary motivation for taking action to manage AQ in the Mexico City Metropolitan Area (MCMA). Along with other methods, the integration of emission inventories (EI) emerged as a very important tool for designing appropriate pollution-control strategies. Emission inventories are a snapshot of the emissions arising from the different activities performed by the population, services, and industry. They also include the contribution of natural phenomena such as soil erosion, wildfires, or emissions from vegetation.

The Mexican government's attentiveness towards the environment resulted in the establishment of the Environmental Improvement Office (Subsecretaría de Mejoramiento del Ambiente) at the undersecretariat level within the Health and Welfare Secretariat (SSA, Secretaría de Salud y Asistencia Pública) in 1971. In the early 1980s, that office was raised to the first level of importance and renamed the Urban Development and Ecology Secretariat (SEDUE, Secretaría de Desarrollo Urbano y Ecología). The SEDUE established the National Ecology Institute to manage environmental issues nationwide. Environmental issues are currently administered by the Secretariat of Environment, Natural Resources, and Fisheries (SEMARNAP, Secretaría de Medio Ambiente, Recursos Naturales y Pesca), created in 1994.

In 1988, the SEDUE instituted the General Law for Ecological Equilibrium and Environmental Protection. This law introduced environmental standards and concepts for environmental policy. Some of these concepts serve as the basis for the establishment and protection of natural areas, environmental risk and impact assessments, and for support of research and education. The law also emphasizes the importance of information, pollution monitoring, and enforcement of environmental policies.

Before the law was published, several Air Quality Programs were implemented for the MCMA. From 1979 to 1982, the Coordinated Program to Improve Air Quality in the Valley of Mexico (PCMCA, Programa Coordinado para Mejorar la Calidad del Aire en el Valle de México) was in effect. Regulatory actions continued in the 1980s with additional programs based on the experiences of other countries. Two of these remain controversial: "A day without a car" (Hoy No Circula), first implemented in 1989 as a voluntary program, which became mandatory in 1990; and the "Compulsory Emission Verification" (Verificación), a local version of an inspection and maintenance program (I/M). The I/M was voluntary between 1982 and 1988 and was made compulsory in 1989. The Comprehensive Program against Atmospheric Pollution (PICCA, Programa Integral Contra la Contaminación Atmosférica de la Zona Metropolitana de la Ciudad de México), implemented from 1990 to 1995, was followed by the Program to Improve the Air Quality in the Valley of Mexico 1995–2000, also known as Pro Aire. All these programs took into account physical, chemical, technical, and socioeconomic aspects of air pollution as well as pollution control options. However, the environmental policies consisted mostly of direct control measures, with less attention paid to economic instruments, though these are more frequently used in some other countries (Belausteguigoitia, 1994).

A preliminary version of the program for the next ten years (ProAire 2000) can be found in *http://sma.df.gob.mx/publicaciones/aire/prog_cal_aire00_10/indice.htm* both in Spanish and English. ProAire 2000 maintains some strategies of the previous programs but includes new chapters devoted to power-plant modernization, climate change, economic incentives, and international cooperation. The World Bank Clean Air Initiative for Latin American cities is also included.

The pioneer program, PCMCA, was mainly a short-term plan to prevent and deal with atmospheric contingencies. It was focused on transportation and vehicular pollution, industrial pollution control, natural sources, research, and education. A voluntary inspection and maintenance program for in-use vehicles was begun, reformulated gasoline and diesel fuels were introduced, and two power plants in the MCMA were converted to natural gas (NG).

PICCA was a more ambitious program created with the assistance of several international agencies including the US Environmental Protection Agency (USEPA), the Japanese International Cooperation Agency (JICA), the German Technical Control Agency (GTZ), and the World Bank. The main goals of this program were to improve the quality of fuels, restructure public and private transport, modernize industrial technologies, relocate the most polluting industries (such as refineries) outside of the MCMA, and protect and reestablish natural areas. This program also proposed specific options for solid-waste management, education, communication, and citizen participation. It included the use of lead-free gasoline and catalytic converters in new vehicles, low sulfur diesel, and some evaporative emissions control. PICCA also included intensive reforestation. The reorganization of public transportation was also considered. Microbuses with catalytic converters were introduced to replace combis and microbuses without catalytic converters (combis are minivans, vans, or other vehicles licensed to carry about nine fare-paying passengers; microbuses have a capacity of about 20 passengers).

The Metropolitan Environmental Commission (CAM, Comisión Ambiental Metropolitana) was established in 1996 to coordinate all pollution-control programs within the MCMA with the leadership of the city government, the State of Mexico, and federal environmental authorities. The CAM includes representatives from the highest levels of government. Other government entities and academic and nongovernmental organizations have been added to this Commission. In 1996, the CAM established the Pro Aire Program (DDF et al. 1996) based on recommendations of the Organization for Economic Cooperation and Development (OECD). Pro Aire continued some of the former program's agenda and established broader objectives including urban policies and transport policies along with emission-control regulations. This program identified specific goals for air quality, as well as a few economic incentives. The general goals included reduced emissions from industry and vehicles and improvements in public transportation to reduce emissions (e.g., the Metro).

Special attention has been given to ecological recovery in the MCMA in order to enhance regeneration of the native flora and fauna, restore the hydroclimatic equilibrium of the Basin, and improve environmental conditions. This includes programs to reinstate the natural marshy conditions in places like Tláhuac, Texcoco, Zumpango, and Xochimilco, the creation of new artificial water bodies and green parks, and reforestation of urban, suburban, and rural sites.

In this chapter, we present an introduction to EI methodology along with an analysis of emission trends and pollution-control options included in the various MCMA AQ programs. NOx control scenarios for the year 2010 are analyzed, and the basis for the integration and selection of control strategies is also discussed. The 1994 Emission Inventory (94–EI) is used as the base-year inventory for comparison with subsequent EIs. Some analyses and tables presented here are based on a detailed analysis of and application of diagnostics to the EIs of the MCMA carried out jointly by the Integrated Program on Urban, Regional, and Global Air Pollution, based at the Massachusetts Institute of Technology, and the Mexican Petroleum Institute (Molina et al. 2000).

Emissions Inventories

As an essential tool for environmental management, especially for heavily polluted regions, EIs require continuous improvement in order to take into account changes in the economy, population growth, or societal activities in addition to the effect of pollution control options. For instance, fuel consumption, which is closely related to the levels of emissions, decreased drastically in 1995 because of the Mexican economic crisis. As the country recovered economically, fuel consumption also increased. However in 1996, fuel consumption was still 3% lower than in 1994. Such a reduction influences the amount of emissions and must be accounted for in the corresponding EI. During the same period, the composition of emissions changed due to the shift from leaded gasoline to unleaded regular gasoline in the MCMA, among other reasons (Table 7.1). Other examples of changes to be taken into account in the EIs are the improvements in emission control devices in industry and the renewal of the vehicular fleet.

The EI evaluates the emissions of 'criteria air pollutants', which are those pollutants that adversely affect public health or welfare, and for which a "safe ambient criteria concentration" has been documented. The pollutants currently considered in the local EI are carbon monoxide (CO), nitrogen oxides (NOx), hydrocarbons (HC), volatile organic compounds (VOC), sulfur dioxide (SO_2), total suspended particulate matter (TSP) and particulate matter less than ten microns (PM_{10}).

Inventories are calculated by allocation of emissions according to the source; consequently, EIs contain uncertainties introduced by the estimation methodologies, particularly at the more detailed source category level. There are six common techniques used to estimate emissions:

1. **Source Sampling**: direct measurement of pollutant concentrations in a known volume of gas and flow rate. It is most commonly used for combustion emission sources.

2. **Emissions Models**: equations developed to estimate emissions not directly related to a single parameter. Emissions models seek to describe complex emissions processes, like those from mobile sources. The input may consist of parameters such as temperature, altitude, fuels properties, or vehicular fleet characteristics. The outputs may consist of composite emissions factors (e.g., vehicular fleet g/km) or the emission inventories by vehicular category. The models most commonly used for the integration of MCMA EIs are the MOBILE5–MCMA.a3 (Gobierno del Distrito Federal-RADIAN 1996), BEIS2, Biogenic Emission Inventory System, (USEPA 1995b), TANKS v. 3.0 (USEPA 1997), LANDFILL (USEPA 1994), SIMS (USEPA 1990), and FAEED2.1 (Federal Aviation Administration 1998). Such models allow the estimation of emissions of a number of pollutants. Some of the models (e.g., BEIS2 or MOBILE5) contain assumptions tied to the geographical location or local conditions of the area to be modeled. To avoid uncertainties in applying such models to different areas, they should be adapted to the new conditions. For instance, MOBILE5–MCMA was adapted (Gobierno del Distrito Federal–RADIAN 1996) for its application in the MCMA to take into account the differences in vehicular fleet between the United States and Mexico City. In the case of the BEIS2 model, the differences in vegetation and the corresponding emission factors should be also taken into account. The application of both models to the MCMA is under revision by CAM. Emission factors from MOBILE5–MCMA will be replaced with experimental Federal Test Procedure (FTP) emission factors. The

Table 7.1. Mexico City Metropolitan Area fuel consumption (million m^3/year)

	Year			Year		
Fuel	1994	1996	Percent change	1996	1998	Percent change
Natural gas	3050.9	3525.4	15.55	3525.4	Not available	
LPG	3.61	3.35	−7.25	3.35	3.82	13.98
Nova gasoline[a]	4.08	3.04	−25.58	3.04	0.00	−100.00
Magna gasoline	2.61	3.18	21.83	3.18	6.03	89.68
Premium gasoline	0.00	0.0005	100.00	0.0005	0.40	8106.13
Total gasoline	6.69	6.22	−7.02	6.22	6.43	3.41
Diesel Sin	1.18	1.21	1.95	1.21	1.32	9.25
Diesel special 1	0.35	0.28	−18.41	0.28	0.17	−39.05
Industrial gas oil	0.26	0.31	17.94	0.31	0.38	22.62
Total intermediate distillates	1.79	1.79	0.34	1.79	1.86	3.94

[a] The leaded gasoline Nova, was discontinued in August, 1997. Magna and Premium are unleaded gasolines.
Source: CAM (1999). Data presented as in Molina et al. (2000).

EPA FTP describes emission and fuel economy testing of light-duty vehicles and heavy-duty engines to certify compliance with US federal vehicle emission standards for CO, unburned HC, and NOx, (USEPA 1991). Emission factors for BEIS2 will be replaced with the few biogenic emission factors available for the MCMA.

3. **Emission Factors**: ratios that relate the quantity of a pollutant emitted to a single activity unit, which may be process data (i.e., throughput, hours of operation, surface area) or census data (i.e., population, employment). USEPA AP-42 is the best example of a comprehensive set of emission factors (USEPA 1995a). It is important that the emission factors not be directly applied without considering the uncertainties introduced by the differences in technology and economic activity patterns between the United States and Mexico. Although the application of emission factors from AP-42 to the MCMA took into account those differences, future work must be done to compile actual MCMA emission factors.

4. **Surveying Questionnaires**: data collected at point sources or integrated from representative sampling of a given source area category.

5. **Material Balance**: a method most commonly used for sources with no data to support the other approaches.

6. **Extrapolation**: a method based on scaling parameters and emissions previously determined for a known source of the same type but a different size (e.g., production quantity, land area, or number of employees).

Frequently, a combination of approaches is used to develop the inventories; however, it should be stressed that each one of them carries uncertainties related to methodologies, efficiency of the processes, actual emissions, or underlying assumptions in the calculation of the inventory.

The first published emission inventory for the MCMA (JICA 1988) reported that total emissions were just above 2 million tons/yr. This inventory was limited to three major pollutants, NOx, SO_2, and CO and was limited to the Federal District jurisdiction not including nearby municipalities. Since then, several EIs have been done (Table 7.2). In the 1989 PICCA emission inventory (Secretariado Técnico Intergubernamental 1990), total emissions were estimated as 4 million tons/yr. Industrial and services emissions represented only 8.4% of the total, and the largest amount (77%) was attributed to mobile sources. The large decrease in

Table 7.2. Mexico City Metropolitan Area emissions inventories (tons/year)

	Pollutant			
Year	NOx	SO_2	HC	CO
1988	65,749	105,282	Not quantified	1,050,221
1989	177,390	205,860	572,320	2,951,025
1994	129,007	45,468	1,025,760	2,358,141
1996	174,020	28,600	569,099	1,947,767
1998	184,351	22,687	554,972	2,033,973

NOx, nitrogen oxides; HC, hydrocarbons.

NOx and SO_2 emissions and the large increase in CO emissions, reported between 1988 and 1989 (Table 7.2), are more likely due to differences between methodologies rather than to actual differences in emissions between years.

The 1994 EI (DDF et al. 1996) cannot be compared with previous ones, because different assumptions and methodologies were used. The 1996 inventory (CAM 1999) updates the data and was the basis for some of the emission-control options contained in the Pro Aire Program. Since the methodology applied was similar to that used in 1994, it is possible, in principle, to compare the emission trends of the two.

The 96–EI was developed and used:

- to support some of the options to control and prevent air pollution;
- to assess how well the environmental regulations were met;
- to evaluate impacts from the introduction of selected control technologies; and
- to follow emission trends.

It is important to emphasize the usefulness of the EI in AQ modeling. Along with meteorological variables and initial conditions, EI results are essential as input data. AQ modeling allows assessment of AQ improvement as a result of different control strategies. In general, AQ models address the question of what happens if a specific control strategy is implemented.

A cooperative research effort between the Mexican Petroleum Institute (IMP) and the Los Alamos National Laboratory (LANL) in New Mexico (United States) resulted in a major advance in understanding the problem of air pollution in Mexico City with immediate applicability for decision makers. One product of this effort was to provide appropriate models for the MCMA that are the basis for the analysis of air-quality improvement options (IMP–LANL, 1994a). To use EIs for evaluation of AQ improvement options, additional temporal and spatial desegregation of the inventories is required. Several Mexican institutions are developing a common basis for the modeling and analysis of control options and strategies (CAM 2000).

Particulate matter (PM) was included in the 1989 and 1994 EIs as TSP and was attributed mainly to soil erosion (95%). A preliminary estimation for PM_{10} from mobile sources was presented in 1992 (World Bank 1992). In the 1994 EI, particle emissions from sources other than erosion were considered for the first time. Primary PM_{10} emissions from the most important sources, including the industrial sector, were considered in the 1996 EI. The latest EI, based on actualized economic activity data for 1998, is presented in the Federal District Government web page (GDF 2000a). This last inventory does not report emission estimates for TSP, but contains estimates for primary sources of PM_{10} and particles of less than 2.5 μm diameter ($PM_{2.5}$).

Source Characterization

The broad classifications of sources for the 1994, 1996 and 1998 EI are as follows:

- **Point Sources, Industrial Sector**: This type of source is defined as any established facility which carries out industrial processes, performs commercial activities, provides services, or carries out any other activity that generates large air-pollutant emissions.
- **Mobile Sources, Transportation Sector**: This category includes all on-road motor vehicles on public roads, that is, cars, trucks, buses and other vehicles.
- **Area Sources; Residential, Commercial, and Services Sectors**: This category takes into account the emissions from sources that are too numerous and dispersed to be efficiently included in a point-source inventory. To ensure completeness, such sources are considered collectively in the emissions inventories. Non-road mobile sources are also included in this category.
- **Natural Sources, Vegetation, and Soil Erosion**: Emissions arising from natural phenomena such as meteorological, geological, or metabolic processes, including soil and vegetation emissions.

Geographic Limits of the MCMA Emission Inventory

The MCMA EIs include the 16 delegations of the Federal District and 18 adjacent municipalities of the State of Mexico. The total area, 4681 km^2, contains more than 16 million inhabitants and about 44,580 industrial, commercial, and services facilities. In the last five years, the population growth in the Federal District was 0.5%, whereas for the municipalities in the State of Mexico, growth was 3.3%. Currently 51% of the MCMA population lives in the Federal District and 49% in the suburbs (CAM 1999). The vehicular fleet for this area is estimated at 3.2 million vehicles.

In the analysis of biogenic emissions trends and in comparisons with emissions from other sources it is important to take into account the base conditions for the calculations. Comparisons and interpretation of the results are easier when base conditions for the various sources remain constant (e.g., when the land area under consideration is the same for all sources). In the 1996 and 1998 EIs the environmental authorities increased the total area for estimating biogenic VOC emissions from 4681 km^2 (the value used in 1994, approximately equalling the area covered by urban settlement), to 20,000 km^2. This increase was considered to include emissions from forests and crops surrounding the MCMA based on satellite images. The larger area includes the area between the Chichinautzin, Las Cruces, Santa Catarina, Nevada and Zempoala mountain ranges and corresponds to the area enclosed between latitudes 19° and 20° north and between longitudes 98° and 100° west. The political entities in this area are the Federal District, approximately 90% of the State of Mexico, 23 municipalities of the State of Puebla, 13 municipalities of the State of Hidalgo, and 4 municipalities of the State of Morelos, together with most of the State of Tlaxcala, including 62 important villages. Soil NOx emissions were estimated from a core area of 5600 km^2.

Table 7.3. Mexico City Metropolitan Area fuel consumption (percent of total energy units, PJ)

	Year		
Fuel	1994	1996	1998
Liquid petroleum gas	18.65	17.66	19.15
Natural gas	24.87	28.34	26.96[a]
Industrial gas oil	1.86	2.23	2.60
Special diesel	2.47	2.05	1.19
Diesel sin	8.46	8.06	9.12
Nova gasoline	26.58	20.34	0.00
Magna gasoline	17.11	21.28	38.40
Premium gasoline	0.00	0.03	2.58
Total	100.00	100.00	100.00

[a] NG consumption in 1998 was estimated from 1996 figures.
Source: CAM (1999). Presented partially in Molina et al. (2000).

Fuel Consumption

Analyses of fuel consumption provide a picture of the different components of the economic and industrial activity of a certain area. They reflect not only the effect of economic enterprises but the introduction of new technologies using cleaner fuels. Such technological changes affect the amount and quality of emissions. As mentioned above, from 1994 to 1996, gasoline and liquid petroleum gas (LPG) consumption decreased due to the post-1994 economic crisis (Table 7.1). Total gasoline consumption, which represents more than 40% of the fuel consumption in the MCMA, increased steadily from 1988 to 1994. However annual sales in 1995 decreased by 4% compared with 1994, and an additional 3% decrease was observed in 1996. In 1997, gasoline consumption increased again by about 3%. By 1999, annual gasoline consumption had increased to an average of 40,531,800 barrels per month, approximately 77,851 more than in 1996 (PEMEX 1999).

Because it was required for new model cars with catalytic converters, unleaded Magna and Premium gasoline consumption increased the share of these fuel types in total fuel consumption from 1994 to 1996. At the same time, consumption of Nova leaded gasoline decreased (Table 7.3). The decrease in CO emissions as a consequence of changes in emissions from mobile sources is noticeable in the EIs (Table 7.2).

Summary of 1994 Emissions

The 94–EI (Table 7.4) can be considered as the first EI integrated on appropriate grounds. It includes the whole MCMA geographical area, and the sources are categorized by sectors. 49 major source categories were included in the 94–EI, 24 of which were from the service sector, 15 were from industry, eight were related to transportation, and two were natural sources. Besides the different calculation methodologies compared with previous inventories (Table 7.2), the figures in the 94–EI also demonstrate the effect of some of the pollution-control measures. In 1989, a vehicular fleet of 2,720,000 vehicles was reported (Table 7.5), 71% of which were private cars. Private vehicles remained the largest component (73%) of the total vehicular fleet in 1996. The introduction of catalytic converters in new vehicles for private use in 1991 resulted in a reduction of CO emissions in excess of 500,000 tons/year in 1994 compared with 1989 (Table 7.2). An important reduction in SO_2 emissions is also evident when the 94–EI is compared with former inventories. This is a result of the specification for sulfur content in fuel being reduced from 0.2% to 0.1% by weight in gasoline, from 0.5% to 0.05% in diesel, and from 4% to 2% in gas oil. The actual average composition of these fuels is reported in Table 7.6; the new fuel specifications (the standards published in ProAire) are fully satisfied. New source categories were included for HCs, mainly for HCs related to solvent use and evaporation of fuel from gas stations and mobile sources. The major source of HC, obtained from mass-balance calculations, was adjudged to be fugitive LPG emissions. In the 94–EI, particle emissions (as TSP) from sources besides soil erosion were considered for the first time (Table 7.4).

Biannual Emissions Inventory Trends from 1994 to 1998

The detailed emission inventories for 1994, 1996, and 1998, are presented in Tables 7.4, 7.7, and 7.8, respectively. These inventories show the different pollution sources and suggest the relative importance of each one. With the exception of NOx, a reduction was observed for each pollutant from 1994 to 1996. It is, however, premature to conclude that this is a trend, and efforts should be made to understand changes in emissions estimates. For example, the decrease in HC emissions could be attributed to methodological changes in calculating the apportionment of LPG (CAM 1999; PEMEX–Gas y Petroquímica Básica 1997) and to the incorporation of estimates for evaporative emissions from mobile sources (DDF et al. 1996) as well as to the effect of control measures. In addition, the 96–EI includes local emission factors for graphic arts, industrial coating, architectural surface coating, automotive painting, and road and street painting (CAM 1999). The increase in NOx emissions is probably overestimated in the 96 and 98 inventories. The uncertainty in NOx emissions is associated with cargo trucks and heavy-duty diesel vehicles, for which the estimated NOx emissions increased by 228% and 750% from 1994 to 1996, whereas the whole vehicular fleet increased only 21% from 1989 to 1996 (Table 7.5). These figures are still under revision.

Other Important Pollutants

Besides the criteria pollutants, many other pollutants, such as the greenhouse gases CO_2, CH_4, N_2O, CFC's, and HFC's, as well as NH_3 and $PM_{2.5}$, should be included in the EIs. The latter two are important for the implementation of

Table 7.4. 1994 Mexico City Metropolitan Area emission inventory (tons/year)

		Pollutant					
Category	Source	TSP	PM_{10}	SO2	CO	NOx	HC
Industry	Electric power generation	163	NE[a]	19	1,291	17,855	97
Industry	Several consumer products	67	NE	109	74	678	304
Industry	Printing products	776	NE	19	15	14	5,015
Industry	Metallic products	197	NE	559	653	468	1,548
Industry	Medium-life consumer products	99	NE	38	101	70	599
Industry	Long-life consumer products	93	NE	172	524	196	2,959
Industry	Others	3	NE	5	8	4	121
Industry	Refining/ petrochemical	7	NE	85	5	28	158
Industry	Chemical industry	974	NE	3,443	2,601	2,477	7,198
Industry	Metal processing	550	NE	622	1,458	553	461
Industry	Nonmetallic minerals	1,675	NE	11,711	323	4,934	3,168
Industry	Vegetables and animal products	111	NE	842	40	260	239
Industry	Wood and derivatives	384	NE	3,912	463	1,822	1,442
Industry	Food consumer industry	799	NE	2,111	406	1,069	397
Industry	Clothing Industry	460	NE	2,405	734	1,091	605
Services	Cleaning and degreasing	0	NE	0	0	0	29,044
Services	Hospital sterilization	0	NE	0	0	0	20
Services	Hospital incineration	0	NE	0	1	1	0
Services	Asphalt paving	0	NE	0	0	0	19,095
Services	Sewage plants	0	NE	0	0	0	56
Services	Airplane fuel provision	0	NE	0	0	0	543
Services	Residential combustion	372	NE	1,483	730	3,808	290
Services	Commercial combustion	705	NE	5,734	218	1,531	64
Services	Graphic arts	0	NE	0	0	0	8,788

Table 7.4. *Continued*

Category	Source	Pollutant					
		TSP	PM_{10}	SO2	CO	NOx	HC
Services	Gasoline storage	0	NE	0	0	0	543
Services	Industrial coating	NE	NE	NE	NE	NE	NE
Services	Solvent use	0	NE	0	0	0	42,005
Services	Sanitary landfills	NE	NE	NE	NE	NE	NE
Services	Open waste disposal	NE	NE	NE	NE	NE	NE
Services	Foreign railroad	39	NE	26	51	414	17
Services	Local railroad	25	NE	37	52	294	30
Services	Airports	0	NE	0	1,583	1,295	523
Services	Gasoline distribution	0	NE	0	0	0	20,127
Services	Liquid petroleum gas distribution	0	NE	0	0	0	242,272
Services	Dry cleaning	0	NE	0	0	0	12,213
Services	Architectural surface coating	0	NE	0	0	0	1,598
Services	Bakery	0	NE	0	0	0	2,291
Services	Automotive painting	0	NE	0	0	0	5,976
Services	Road and street painting	0	NE	0	0	0	3,381
Soil	Erosion	425,337	NE	0	0	0	0
Transport	Private cars	10,321	NE	6,062	1,044,008	31,913	253,866
Transport	Pickup	1,409	NE	354	73,419	2,675	19,374
Transport	Combis and microbuses[b]	439	NE	1,478	359,032	14,314	101,582
Transport	Taxis[b]	613	NE	3,073	529,530	15,983	126,575
Transport	Ruta-100 Buses	1,900	NE	366	5,655	6,751	2,337
Transport	State of Mexico buses	2,825	NE	502	59,110	5,437	2,837
Transport	Light-duty trucks	360	NE	37	271,321	5,868	46,100
Transport	Heavy-duty trucks	1,902	NE	266	4,736	7,205	2,080
Vegetation	Vegetation	0	NE	0	0	0	38,909
	Total	452,604	NE	45,468	2,358,141	129,007	1,006,845

[a] NE, not estimated.

[b] Taxis are cars with taxi license plates and generally with distinctive paint schemes. Combis are minivans, vans, or other vehicles licensed to carry fare-paying passengers (about nine). Microbuses have a capacity of about 20 passengers. The local authorities regulate the number of license plates for each of these vehicle types.

Table 7.5. Mexico City Metropolitan Area fleet size and distribution[a]

Type of vehicle	1996 Fleet Emission Inventory 96	1994 Fleet COMETRAVI (1999)	1989 Fleet World Bank (1992)	1989 Fleet Legorreta (1989)
Private cars	2,301,445	2,865,624	2,210,000	1,942,400
Taxis	91,765	26,555	56,950	145,800
Microbuses[b]	52,158[c]	16,589	11,080	
Combis[b]		104,039	72,000	
Pickups	48,507			
Cargo trucks	463,962	378,833	196,218	480,600
Ruta-100	2,794	2,440	3,500	
Suburban buses		1,284		
Private buses		4,013		
Buses	5,297		4,000	
Heavy-duty diesel trucks	191,946		49,055	
Passenger buses				51,300
Others				99,900
Motorcycles		39,040		
Foreign buses		164,900		
Total vehicles	3,157,874	3,603,317	2,602,803	2,720,000

[a] Data presented as in Molina et al. (2000).
[b] Taxis are cars with taxi license plates and generally with distinctive paint schemes. Combis are minivans, vans, or other vehicles licensed to carry fare-paying passengers (about nine). Microbuses have a capacity of about 20 passengers. The local authorities regulate the number of license plates for each of these vehicle types.
[c] Value includes both microbuses and combis.

particulate matter control strategies because they are precursors to secondary aerosol formation. The integration of these pollutants into the inventory is the starting point for correlating emission trends, as reflected in lower ambient air measurements, with the effectiveness of control strategies.

A second stage in EI integration and formulation of control strategies is the speciation (identification of the individual components) of the volatile and semi-volatile organic compounds. The nature of the individual species determines its reactivity in the atmosphere and its participation in the formation of secondary pollutants. For instance, small olefins are highly reactive in the formation of ozone (O_3), and polyaromatic hydrocarbons along with NOx are precursors for secondary aerosol formation (Finlayson–Pitts and Pitts 1986).

Emission Inventory Uncertainties

The determination of uncertainties in the EIs is not a straightforward procedure. The accuracy of the numbers in the EIs is based on the uncertainties contained in the emission factors and the activity data and on the validity of all of the quantitative factors used in the calculations. For example, the precision of the

Table 7.6. Options identified for improving air quality in Mexico City

Options	Done	Comment
1. Produce gasoline conforming to international standards	Yes	Current gasoline contains 27.2% aromatics, 10.3% olefins, 8.1 pounds per square inch guage Reid Vapor Pressure, 1% Benzene, 0.048% sulfur, and 1% oxygen. PEMEX invested $1.3 billion dollars to produce better gasoline, diesel, and gas oil. Leaded gasoline was phased out in August, 1997.
2. Produce low-sulfur diesel	Yes	Current diesel has a 55.4 Cetane Index, 0.04% sulfur and 26.6% aromatics. Implemented in 1993.
3. Produce low sulfur fuel oil and gas oil (combination of fuel oil and diesel)	Yes	Current industrial gas oil has 2% sulfur. Implemented in 1998.
4. Expand Ruta-100 (Mexico City's public bus system)	No	Ruta-100 declared bankruptcy.
5. Authorize expanded bus routes	No	Mexico City Metropolitan Area main transport system consists of minibuses.
6. Continue the 'Day-without-a-Car' (Hoy no Circula) program	Yes	Implemented in 1989.
7. Expand the verification program for gasoline and diesel vehicles	Yes	Made compulsory in 1989 for gasoline vehicles. Not yet implemented for diesel vehicles.
8. Convert public vehicles and delivery trucks to liquid propane gas and install catalytic converters	Yes	Partial, some companies converted their delivery trucks to LPG. Minibuses conversion program failed.
9. Retrofit catalytic converters on minibuses	Yes	Minibuses with catalytic converters were introduced in Mexico City in 1993. State of Mexico minibuses require retrofit.
10. Require recent model taxis (newer than 1984) and combis (newer than 1980)[a]	Yes	Taxis were renewed in 1992.
11. Substitute fuel oil for natural gas in industry	Yes	Implemented in 1992.
12. Clean and/or relocate the foundries in the Valley of Mexico	Yes	
13. Improve combustion and install control equipment on small boilers	No	
14. Substitute fuel oil for natural gas in power plants in Mexico City and the Valley of Mexico	Yes	Implemented in 1992.
15. Implement a reforestation program in Mexico City and the Valley of Mexico	Yes	40 million trees were planted between 1998 and 2000.
16. Remove gross-polluting cars from circulation and prevent them from entering the city	Yes	There is an inspection and maintenance program. New vehicles can circulate on all days. Pre-1991 vehicles can't circulate one day per week. Scrapping programs failed.

(Continued)

Table 7.6. *Continued*

Options	Done	Comment
17. Prohibit all open burning	Yes	
18. Limit use of paints and finishes in industries that don't control vapors	Yes	
19. Limit operation of closed trash burners to optimum time of the day	No	
20. Relocate polluting industries	Yes	
21. Install vapor recovery systems in filling stations (both storage and delivery)	Yes	90% of gasoline stations have vapor recovery systems.
22. Prohibit parking on selected streets	No	
23. Eliminate restrictions on operating hours for some businesses	No	
24. Extend the operating hours of buses	Partial	There is a recently introduced night program in the Federal District.
25. Encourage nighttime deliveries	No	
26. Require private enterprises to maintain their fleets	Yes	
27. Require private enterprises to promote environmental programs among employees	Yes	Partial
28. Promote the use of bicycles	No	
29. Coordinate traffic lights to speed flow of traffic	Yes	Partial
30. Implement information system on traffic conditions	Yes	Partial
31. Construct and police parking lots next to public transportation stations	Yes	Partial
32. Improve taxi efficiencies by use of taxi stands	No	
33. Create toll streets and roads for single-passenger vehicles	No	
34. Control and increase fees on parking lots	Yes	
35. Construct Line 8 of the Metro	Yes	Line 8 was finished in 2000.
36. Improve electric transport (trolleys)	Yes	
37. Improve the image of public transportation	No	
38. Carry out a complete assessment of the advantages of public transportation	No	
39. Better organization of traffic and parking in the Central Historical District	No	
40. Encourage political support for greater financing of public transportation	No	
41. Reduce emissions from metal cleaning and degreasing	Yes	

(*Continued*)

Table 7.6. *Continued*

Options	Done	Comment
42. Require radial tires on light-duty trucks	No	
43. Reduce emissions from dry cleaning	Yes	
44. Control emissions from commercial charbroiling	No	
45. Control emissions from woodworking operations	Yes	
46. Control emissions from off-road mobile sources	No	
47. Control emissions from large commercial bakeries	Yes	
48. Control emissions from livestock waste	No	
49. Control fugitive emissions from construction of roads and buildings	No	
50. Use of "feebate" system for taxing new cars	No	There is a law initiative in the Congress to tax new cars according to their NOx emissions. Less-polluting cars will pay less tax.
51. Purchase of old polluting cars	No	Scrapping program failed for financial reasons.
52. Continue to require catalytic converters on automobiles	Yes	Implemented in 1993 for all new vehicles.
53. Convert gasoline trucks to compressed natural gas (CNG)	No	There is a recent program to convert microbuses[a] to compressed natural gas (CNG). To date, 200 microbuses use CNG.
54. Require gasoline trucks to conform to 1993 standards	Yes	
55. Replace gasoline trucks not suitable for conversion to CNG or liquefied petroleum gas	No	
56 Pave roads	Yes	Partial

Measures Only in Place When There Is an Air Pollution Emergency	
57. Suspend the use of 50% of government vehicles	Yes
58. Reduce electricity production from the "Valley of Mexico" power plant by 50%	Yes
59. Suspend services such as public baths, dry cleaning, painting, and others	Yes
60. Use policemen to speed traffic flow	Yes
61. Suspend street maintenance and government-funded construction	Yes
62. Suspend asphalt production	Yes
63. Reinforce the elimination from circulation of high-polluting vehicles	Yes
64. Suspend activities in nonessential commerce and entertainment	Yes
65. Add a second day without a car	Yes
66. Suspend all construction and demolition activities and transport of construction material	Yes

[a] Taxis are cars with taxi license plates and generally with distinctive paint schemes. Combis are minivans, vans, or other vehicles licensed to carry fare-paying passengers (about nine). Microbuses have a capacity of about 20 passengers. The local authorities regulate the number of license plates for each of these vehicle types.

Table 7.7. 1996 Mexico City Metropolitan Area emission inventory (tons/year)

		Pollutant					
Category	Source	TSP	PM-10	SO2	CO	NO_x	HC
Industry	Electric power generation[a]	294	294	103	797	12,507	41
Industry	Several consumer products	75	70	110	71	91	808
Industry	Printing products	45	37	18	22	53	3,214
Industry	Metallic products	278	251	749	1,155	2,717	2,194
Industry	Medium-life consumer products	163	149	73	196	564	747
Industry	Long-life consumer products	396	309	277	809	611	2,346
Industry	Others	53	51	111	22	78	88
Industry	Refining/ petrochemical	2	1	5	2	3	7
Industry	Chemical industry	936	743	2,488	2,033	2,212	4,323
Industry	Metal processing	567	513	691	1,589	770	278
Industry	Nonmetallic minerals	2,383	1,923	2,340	1,103	4,379	706
Industry	Vegetables and animal products	48	44	248	34	103	14
Industry	Wood and derivatives	435	377	4,924	379	1,833	641
Industry	Food consumer industry	583	558	1,366	829	1,408	605
Industry	Clothing industry	435	381	2,129	462	1,338	268
Services	Cleaning and degreasing	0	0	0	0	0	29,490
Services	Hospital sterilization	0	0	0	0	0	20
Services	Hospital incineration	NE[b]	0	0	0	0	2
Services	Asphalt paving	NE	NE	0	0	0	100
Services	Sewage plants	0	0	0	0	0	42
Services	Airplane fuel provision	0	0	0	0	0	8
Services	Residential combustion	NE	186	1	846	5,880	221
Services	Commercial combustion	NE	166	3,586	332	1,952	73
Services	Graphic arts[a]	0	0	0	0	0	6,553
Services	Gasoline storage	0	0	0	0	0	97
Services	Industrial coating[a]	0	0	0	0	0	20,970
Services	Solvent use	0	0	0	0	0	46,856
Services	Sanitary landfills	0	0	0	0	0	119
Services	Open waste disposal	0	0	0	0	0	25
Services	Foreign railroad	NE	NE	40	43	381	15
Services	Local railroad	NE	NE	362	90	509	51
Services	Airports	NE	NE	3,215	2,284	859	NE
Services	Gasoline distribution	0	0	0	0	0	16,363
Services	LPG distribution[a]	0	0	0	0	0	76,414
Services	Dry cleaning	0	0	0	0	0	9,830
Services	Architectural surface coating[a]	0	0	0	0	0	22,281

(*Continued*)

Table 7.7. *Continued*

Category	Source	Pollutant					
		TSP	PM-10	SO2	CO	NO_x	HC
Services	Bakery	0	0	0	0	0	2,539
Services	Automotive painting[a]	0	0	0	0	0	2,243
Services	Road and street painting[a]	0	0	0	0	0	655
Soil	Erosion[a]	140,292	17,216	0	0	0	0
Transport	Private cars	NE	689	2,155	822,498	32,805	69,413[a]
Transport	Pickup	NE	18	73	33,068	1,221	2,806[a]
Transport	Combis and microbuses[c]	NE	71	282	208,492	6,489	19,052[a]
Transport	Taxis	NE	167	521	198,759	7,927	16,774[a]
Transport	Ruta-100 buses	NE	197	29	1,535	1,911	636
Transport	State of Mexico buses	NE	373	79	2,910	3,623	1,206
Transport	Light-duty trucks	NE	211	1,501	618,202	19,242	56,491[a]
Transport	Heavy-duty trucks[a]	NE	6,308	1,124	49,205	61,275	20,397
Vegetation	vegetation[a]	NE	NE	0	0	1,279	131,077
Total		146,985	31,303	28,600	1,947,767	174,020	569,099

[a] Methodological change between the 1994 and 1996 EIs.
[b] NE, not estimated.
[c] Taxis are cars with taxi license plates and generally with distinctive paint schemes. Combis are minivans, vans, or other vehicles licensed to carry fare-paying passengers (about nine). Microbuses have a capacity of about 20 passengers. The local authorities regulate the number of license plates for each of these vehicle types.

emission factors in the USEPA AP-42 procedure is based on the quality of the experimental methods and the adequacy of the sampling procedures used to determine the emission factors. Accordingly, the AP-42 emission factors are ranked in a relative quantitative order based on the precision of the methods used. However, the basis for such classification is not currently part of EI methodology for the MCMA. Even in the United States, ranking of the AP-42 emission factors is only qualitative, so estimating uncertainties within the EI is not a straightforward process. Much work is underway, both in the United States and Mexico, to improve the methodologies and measurements of emission factors used in creating the EIs.

An estimation of uncertainties in the EI may be determined by air-quality modeling. The Mexico City Air Quality Research Initiative study showed that HCs and NOx were underestimated by 400% and 150% in the PICCA Emission Inventories (IMP-LANL 1994a). The HC estimate in the 1998 EI indicates a slightly lower discrepancy. Thus, an uncertainty of 200% could be expected. The reconciliation among air-quality measurements, emission inventories, and modeled AQ data is a scientific issue at the present state of the art. Improvement of the MCMA EI is addressed by Molina et al. (2000), who recommended measurements of local emission factors for mobile sources, biogenic emissions, solvent use, and fuel storage to reduce EI uncertainty. Alternative methodologies such as tunnel measurements for mobile sources, remote-sensing techniques for stationary sources, and bottom-up calculations for the entire EI are also suggested.

Table 7.8. 1998 Mexico City Metropolitan Area emission inventory (tons/year)

Category	Source	Pollutant[a]					
			PM_{10}	SO_2	CO	NO_x	HC
Industry	Electric power generation	NE[a]	209	17	2,312	6,467	453
Industry	Several consumer products	NE	317	1,823	359	1,329	763
Industry	Printing products	NE	37	167	65	142	3,242
Industry	Metallic products	NE	116	816	624	2,554	2,172
Industry	Medium-life consumer products	NE	104	81	542	609	510
Industry	Long-life consumer products	NE	124	128	289	686	1,406
Industry	Others	NE	87	518	182	566	1,026
Industry	Refining/ petrochemical	NE	0	0	0	0	0
Industry	Chemical industry	NE	424	2,521	1,253	2,131	5,447
Industry	Metal processing	NE	688	716	1,074	621	232
Industry	Nonmetallic minerals	NE	378	1,620	629	4,460	644
Industry	Vegetables and animal products	NE	140	1,034	119	295	187
Industry	Wood and derivatives	NE	117	1,266	423	737	493
Industry	Food consumer industry	NE	294	1,227	845	1,153	427
Industry	Clothing industry	NE	138	666	235	345	593
Services	Cleaning and degreasing	NE	0	0	0	0	31,699
Services	Hospital sterilization	NE	0	0	0	0	23
Services	Hospital incineration	NE	9	22	20	75	3
Services	Asphalt paving	NE	0	0	0	0	122
Services	Sewage plants	NE	0	0	0	0	77
Services	Airplane fuel provision	NE	0	0	0	0	5
Services	Residential combustion	NE	118	6	535	4,056	151
Services	Commercial combustion	NE	208	4,385	365	1,617	51
Services	Graphic arts	NE	0	0	0	0	6,962
Services	Gasoline storage	NE	0	0	0	0	102
Services	Industrial coating	NE	0	0	0	0	22,702
Services	Solvent use	NE	0	0	0	0	78,714
Services	Sanitary landfills	NE	0	0	0	0	7,380
Services	Open waste disposal	NE	0	0	0	0	0
Services	Railroad (foreign and local)	NE	10	62	377	491	41
Services	Airports	NE	NE[b]	103	649	1,611	250
Services	Gasoline distribution	NE	0	0	0	0	516
Services	Liquid petroleum gas distribution	NE	0	0	0	0	77,147
Services	Dry cleaning	NE	0	0	0	0	11,038
Services	Architectural surface coating	NE	0	0	0	0	23,671
Services	Bakery	NE	0	0	0	0	2,698

(*Continued*)

Table 7.8. *Continued*

			Pollutant[a]				
Category	Source		PM_{10}	SO_2	CO	NO_x	HC
Services	Automotive painting	NE	0	0	0	0	2,338
Services	Road and street painting	NE	0	0	0	0	764
Soil	Erosion	NE	NE	0	0	0	0
Transport	Private cars	NE	751	2,379	873,301	35,870	75,683
Transport	Pickup	NE	31	63	33,961	1,252	2,942
Transport	Combis and microbuses[c]	NE	63	282	194,686	6,135	17,841
Transport	Taxis	NE	188	563	210,367	8,639	18,811
Transport	Buses (includes ruta-100)	NE	751	157	5,978	7,324	2,410
Transport	Light-duty trucks	NE	219	908	653,638	20,658	59,877
Transport	Heavy-duty trucks	NE	6,542	1,158	51,144	62,725	20,689
Vegetation	Vegetation	NE	0	0	0	11,802	72,670
Total		NE	12,064	22,687	2,033,973	184,351	554,972

[a] TSP was not estimated.
[b] NE, not estimated.
[c] Taxis are cars with taxi license plates and generally with distinctive paint schemes. Combis are minivans, vans, or other vehicles licensed to carry fare-paying passengers (about nine). Microbuses have a capacity of about 20 passengers. The local authorities regulate the number of license plates for each of these vehicle types.

Integration and Selection of Control Strategies

Different approaches have been implemented to translate general policy statements into specific control options or regulations to improve air quality. The general translation process is described in the following steps:

- Set goals, that is, the desired levels of reduction for each air pollutant.
- Assess uncertainties in the EI.
- Arrange the EI in decreasing order of the contributing emission source to identify the major emission sources and their relative contributions.
- Add the relative contributions until at least 90% of the total emissions are accounted for.
- Regroup by sectors.
- Assess which are the technologically feasible and/or politically feasible control options according to the goals. Modify or change control options that are not technologically or politically feasible.

Both a reduction in the maximum values of pollutant concentrations and a reduction in the number of ambient emergency days were established in 1995 as goals to be achieved by the year 2000 (DDF et al. 1996). This was conceived as a shift in the median concentration of the O_3 frequency distribution from 170 to 150 points in the Mexico City pollution indicator (IMECA). One hundred points of the IMECA correspond to the air quality standards for the regulated air pollutants: O_3, 0.11 ppm (1 hour); SO_2, 0.13 ppm (24 hours); NO_2, 0.21 ppm (1 hour); CO, 11 ppm (8 hours); TSP, 260 μg/m^3 (24 hours); PM_{10}, 150 μg/m^3 (24 hours). In terms of ambient emergency days, the goal was to decrease the number by 75% (DDF et al. 1996). Before 1995, the environmental authorities focused mainly on O_3, which exceeded the AQ standard at least 300 days per year. However, by 1995 it was recognized that PM_{10} is also of concern for AQ in the MCMA.

Control Strategies for Ozone

As a secondary pollutant, O_3 has no direct sources. It is formed in the atmosphere from its precursors: HCs and NOx. The maximum O_3 value in a given day is closely related to the HC/NOx ratio between 6 and 9 a.m. In particular, the HC/NOx ratios in the MCMA (Arriaga et al 1997) are two to three times higher than these same ratios for any city in the United States (McCauley 1999; National Research Council 1992). This suggests that NOx may be the controlling factor in O_3 formation, just as the less concentrated species controls the reaction in any other chemical process (Seinfeld 1986). Experiments with outdoor smog chambers indicate that a reduction in NOx would be the most effective option to lower the O_3 maximum in the MCMA (Sandoval et al. 2001). Air-quality modeling also supports this conclusion (IMP–LANL 1994a). The strategies for managing the O_3 pollution problem should be focused on the control of its precursors, NOx and HC, which are discussed in the following sections.

Sources of NOx and Hydrocarbons

Table 7.9 shows the top 15 sources of NOx in the 1998 EI in descending order of importance. All sources except soil are related to combustion processes. Control strategies for NOx emissions from soil are difficult to implement. The transportation and industrial sectors are the main combustion-related sources. Major contributors from the industrial sector are electric-power generation and nonmetallic mineral processes. The implementation of low NOx burners seem a good control strategy in these cases. The main contributor from the service sector is residential combustion. In the transportation sector, the most common sources include private cars, taxis, and heavy-duty and light-duty trucks. Therefore, some pollution-control options implemented by Federal District authorities seem appropriate. Such options are the inclusion of NOx emissions in the I/M and the replacement of catalytic converters in vehicles more than five years old. NOx certification became compulsory only after the year 2000. The replacement of catalytic converters in vehicles more than five years old stems from the fact that the converters in Mexico are guaranteed by the manufacturer for only one year or 20,000 km, but they are estimated to function properly for three years. There is a significant risk of angering consumers if they are held liable to a much greater degree than car manufacturers and fuel producers.

Sulfur in gasoline affects the performance of catalytic converters. However, this effect is reversible. Thus when the fuel is changed from high-sulfur to low-sulfur, the performance of catalytic converters improves (Benson et al. 1997), and

Table 7.9. Major sources of nitrogen oxides in the 1998 Mexico City Metropolitan Area emissions inventory[a]

Category	Emissions Ton/year	Emissions %	Accumulated emissions %
Heavy-duty trucks[b]	62,725	34.02	34.02
Private cars	35,870	19.46	53.48
Light-duty trucks	20,658	11.21	64.69
Soil	11,802	6.40	71.09
Taxis	8,639	4.69	75.78
Buses (includes ruta-100)	7,324	3.97	79.75
Electric power generation	6,467	3.51	83.26
Combis and micros[c]	6,135	3.33	86.58
Nonmetallic minerals processing	4,460	2.42	89.00
Residential combustion.	4,056	2.20	91.20
Metallic products processing	2,554	1.39	92.59
Chemical industry	2,131	1.16	93.75
Commercial sector combustion	1,617	0.88	94.62
Airport	1,611	0.87	95.50
Consumer products processing	1,329	0.72	96.22

[a] *Source*: Molina et al. (2000).
[b] Emissions from heavy-duty trucks may be overestimated by as much as 300%.
[c] Taxis are cars with taxi license plates and generally with distinctive paint schemes. Combis are minivans, vans, or other vehicles licensed to carry fare-paying passengers (about nine). Microbuses have a capacity of about 20 passengers. The local authorities regulate the number of license plates for each of these vehicle types.

emissions of NOx, HC and CO are expected to decrease. NOx emissions from soil, it should be remembered, are calculated for a greater area than combustion-related sources, as explained previously. Thus their apparent large contribution (6%) is largely a function of methodology.

Ninety percent of HC emissions arise from only 10 to 12 sources (10 main sources in 1994, 12 in 1996, and 10 in 1998). Table 7.10 presents 15 sources that contribute almost 96% of the total HC emissions. In contrast with NOx emissions, there are no major HC sources in the industry sector. Hydrocarbon emissions from the transportation sector originate from private cars, heavy-duty and light-duty trucks, combis, microbuses, and taxis. Both tailpipe and evaporative emissions are included in the transportation sector emissions estimates. The most important sources from the service sector are the distribution and use of LPG, solvent use, industrial and architectural coating, and car cleaning and degreasing. Distribution and use of LPG contribute as much as 76,414 tons per year to HC emissions. Domestic consumers generate 71% of these emissions, mainly as leaks from stoves and small water heaters (PEMEX Gas y Petroquímica Básica 1997). A program to eliminate domestic leaks will soon begin. It includes a public information campaign to raise public awareness and support for reducing escape of LPG from domestic LPG fittings and pipes and will also provide LPG equipment review information (INE 1999). Smog-chamber experiments in the MCMA indi-

Table 7.10. Major sources of hydrocarbons in the 1998 Mexico City Metropolitan Area emissions inventory[a]

Category	Emissions		Accumulated emissions %
	Ton/year	%	
Solvent use	78,714	14.18	14.18
Liquid petroleum gas distribution	77,147	13.90	28.08
Private cars	75,683	13.64	41.72
Vegetation	72,670	13.09	54.82
Light-duty trucks	59,877	10.79	65.61
Cleaning and degreasing	31,699	5.71	71.32
Architectural surface coating	23,671	4.27	75.58
Industrial coating	22,702	4.09	79.67
Heavy-duty trucks[b]	20,689	3.73	83.40
Taxi	18,811	3.39	86.79
Combis and microbuses[c]	17,841	3.21	90.01
Dry cleaning	11,038	1.99	91.99
Sanitary landfills	7,380	1.33	93.32
Graphic arts	6,962	1.25	94.58
Chemical industry	5,447	0.98	95.56

[a] Source: Molina et al. (2000).
[b] Emissions from heavy duty trucks may be overestimated by as much as 300%.
[c] Taxis are cars with taxi license plates and generally with distinctive paint schemes. Combis are minivans, vans, or other vehicles licensed to carry fare-paying passengers (about nine). Microbuses have a capacity of about 20 passengers. The local authorities regulate the number of license plates for each of these vehicle types.

cated that HCs associated with LPG contribute to the formation of only 15% of the O_3 formed in the Basin (PEMEX Gas y Petroquímica Básica 1997). As olefins are the most reactive precursors to O_3 formation, the olefin content of LPG used in the MCMA was reduced from 3% to 1% in 1997 (INE 1999).

Because the transportation sector is a major source for both HCs and NOx, control strategies for that sector should be a priority in the agenda for better air quality. In considering emissions of these two O_3 precursors from the transportation sector, it is important to take into account the size of the vehicular fleet, fuel consumption, and driving cycles under the conditions of the MCMA. Excluding TSP attributed to soil erosion in the 94–EI, the greatest uncertainties for emissions estimates are those for the transportation sector. These arise from uncertainties in the size and model distribution of the MCMA vehicular fleet (Table 7.5). The use of the MOBILE5-MCMA.a3 model, which is based on driving cycles in the United States, also contributes to uncertainty. These factors are likely to lead to underestimates of the emissions from this sector. The routes of microbuses and combis used for public transportation and emissions measurements at their terminal stations are other issues to consider for EI improvement. Some of the heavy-duty trucks drive across the MCMA en route to various destinations. There are also many vehicles which commute between the State of

Mexico and the Federal District, all of which contribute to emissions. Despite the great uncertainty in transportation emissions, it is clear that new car technology will continue to be the best option for reducing NOx emissions in the MCMA. Cars from 1996 on emit one fifth the NOx that 1990 and older models emit (Schifter et al. 2000a).

Ozone Pollution and Gasoline Specifications in the MCMA

The examination of gasoline specifications, fuel consumption trends, and comparisons with the trends for O_3 and other pollutants from 1986 to 2000 in the MCMA can provide useful insights into their relationships.

Gasoline specifications in the MCMA have undergone several changes since the mid 1980s. In 1986, only Nova Plus regular leaded gasoline was available. In June 1986, the Reid Vapor Pressure (RVP), a property related to gasoline evaporation, was mandated to be within the range of 7.0 and 9.5 PSIG (lb/in^2 gage), and an additive with detergent properties was also introduced. In January 1987, Extra Plus, a low-lead regular gasoline containing 0.05 g Pb/gal was introduced. In November 1989, MTBE was added to both types of gasoline, and a new dispersing additive also came into use. The maximum MTBE content was set at 10% by volume. In September 1990, Magna Sin, a new unleaded regular gasoline, was marketed as a substitute for Extra Plus. Magna Sin contained a maximum of 0.01 g Pb/gal, 1000 ppm of sulfur, and a RVP between 7.0 and 9.5 PSIG. A new regulation concerning gasoline specifications, put into place in December, 1992, established the maximum content of olefins at 15% by volume and aromatics at 30% by volume. In addition, the RVP was limited to a maximum of 8.5 PSIG. In August 1997, leaded gasoline was phased out in the MCMA, and a high-octane PREMIUM gasoline was introduced.

Thus, three critical periods can be identified in terms of gasoline specifications and the introduction of vehicular emission control devices: First, the period before MTBE was introduced (January 1986 to November 1989); then, the introduction of MTBE in the absence of a significant number of vehicles with catalytic converters (December 1989 to December 1991); and finally, the combined effect of a greater number of vehicles with emission-control devices and gasoline reformulation (January 1992 to December 1999).

From 1985 to 1987, the total gasoline consumption in the MCMA remained relatively constant. In 1987, EXTRA PLUS gasoline was introduced, reaching about 7% of total gasoline consumption by the end of the year. Between 1987 and 1988, gasoline consumption increased by less than 1%, but from 1989 to 1991, it increased at an annual average rate of 6.2%. In 1991 gasoline consumption increased by 7.5% with respect to 1990. This was the highest one-year increase in the MCMA over the entire 1986 to 1999 period (2.9% annual average). Because total gasoline consumption increased 22% from 1986 to 1991, and vehi-

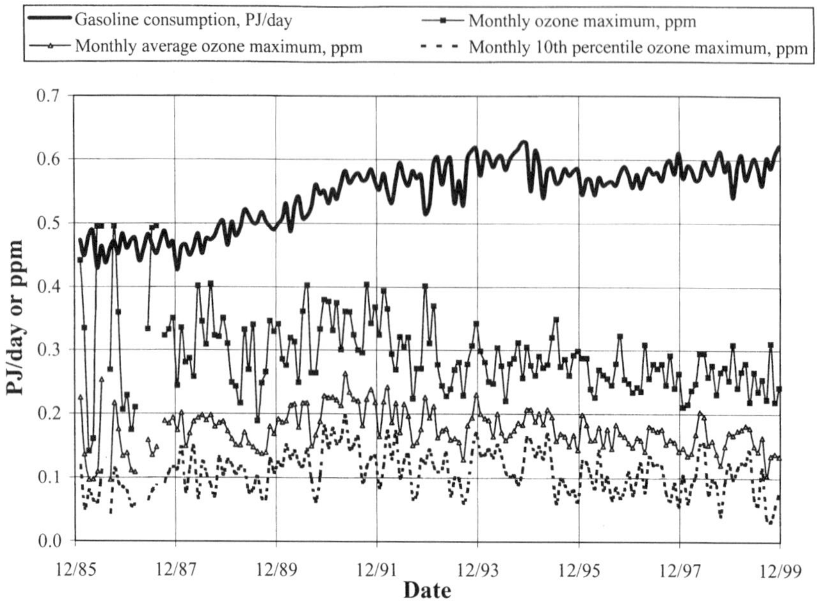

Figure 7.1. The relationship between gasoline consumption and ozone level indicators in the MCMA.

cles older than the 1991 model year were not equipped with catalytic converters, a continuous increase in vehicular emissions was expected.

Several compendia on AQ in the MCMA with statistical and historical reports of monitoring data have been published by environmental authorities in Mexico (GDF 2000b, INE 2000). They report one-hour AQ values since the first monitoring system started operation in February 1986.

In Figure 7.1, three indicators of O_3 trends between 1986 and 1999 are shown along with gasoline consumption. The indicators are: the highest hourly average value of O_3 in a month (monthly O_3 maximum), the monthly averages of daily O_3 maximums, and the tenth percentile value for O_3 (the upper threshold value for which only 10% of the values are lower, meaning that 90% of the values are higher). The AQ data correspond to the main monitoring stations in five different geographical areas in the MCMA (Tlalnepantla, NW; Xalostoc, NE; Merced, downtown; Pedregal, SW; and Cerro de la Estrella, SE). The monthly O_3 maximum value represents the highest hourly average record at any of the five monitoring stations during the day. The monthly O_3 maximums data show a decreasing trend from 1986 to 1999; the monthly averages of daily O_3 maximums and the tenth percentile of daily O_3 maximums increased in the 1986 to 1992 period, but decreased from 1992 to 1999.

For the June 1986 to November 1989 period, none of the three O_3 level indicators are significantly correlated with monthly gasoline consumption ($r = -0.29$,

–0.04, –0.05). Thus, it is difficult to say that the changes in gasoline properties during this period of relative constant gasoline consumption directly affected O_3 levels in the MCMA. As a matter of fact, high O_3 readings (near to 0.5 ppm) were observed in the MCMA before the beginning of the change gasoline composition in June 1986, although the O_3 data for 1986 and 1987 are scarce.

The introduction of MTBE into gasoline defines another important time period for analysis, from December 1989 to December 1991. MTBE was introduced because oxygenates were well known to reduce CO emissions from vehicles without sophisticated emission controls at high altitude and also because the addition of MTBE increased the gasoline octane number. In this period, the correlation coefficients between gasoline consumption and monthly O_3 maximums, monthly averages of daily O_3 maximums, and monthly tenth percentile values for daily O_3 maximums were 0.19, 0.43, and 0.27 respectively.

Emission tests in which ten pre-1991 MCMA vehicles were used showed that the addition of MTBE at 10% by volume to low-octane gasoline reduced HC, NOx, and CO emissions (Schifter et al. 2000c). Similar results were found during the Auto/Oil study for 1982 to 1985 cars in the United States (Auto/Oil Air Quality Research Improvement Program 1997). As the monthly average of daily O_3 maximums showed a moderate statistically significant correlation coefficient ($r = 0.43$), it is interesting to analyze the effect of MTBE on emissions, considering that MTBE has been postulated to increase the emission of formaldehyde and isobutene (highly reactive compounds involved in O_3 formation). In fact, the emissions of these compounds increased with MTBE in the pre-1991 vehicles, but the total O_3-forming potential of the HC emissions per kilometer decreased by 10% (Schifter et al. 2000c). In the December 1989 to December 1991 period the moderate correlation of monthly averages of daily O_3 maximums with gasoline consumption can be attributed more to the overall 13% increase in gasoline consumption than to the introduction of MTBE.

In 1991, some new model cars introduced into the MCMA were equipped with catalytic converters, but not until 1993 was the use of catalytic converters required for all types of new vehicles (vans and pickups included). In the emissions tests of ten post-1991 vehicles, the addition of 10% MTBE by volume to regular Magna gasoline decreased CO and HC emissions but increased NOx emissions marginally. As in the case of adding MTBE to leaded gasoline, formaldehyde and isobutene emissions increased; nevertheless the total O_3-forming potential of the HC emissions per kilometer decreased by 10.6% (Schifter et al. 2000c).

For the period of 1992 to 1999, the correlation coefficients between monthly gasoline consumption and monthly O_3 maximums, monthly averages of daily O_3 maximums, and monthly tenth percentile values of daily O_3 maximums were –0.14, –0.11 and –0.06 respectively, each of them without statistical significance. In the same period, gasoline consumption increased 4.3%. In summary, the control measures introduced between 1992 and 1999 resulted in a small reduction in O_3 levels during a period of slightly increasing gasoline consumption.

"A Day without a Car" and "Vehicular Emissions Verification" Programs

These two programs have been controversial control options for the transportation sector since their implementation in 1989. The original version of "a day without a car," "no car day," was also implemented in other cities with traffic congestion problems such as Athens, Greece. In Mexico City, however, it appears to have had the perverse effect of raising emissions, because a large number of car owners acquired extra vehicles to replace the driving-restricted car (Heil and Pargal 1998). This program was first implemented in 1989 as a voluntary program, but was made compulsory in 1990. In its first year, the "a day without a car" program reduced gasoline consumption by 400,000 L/day, but in 1994, the program reduced fuel use by only 132,000 L/day. In 1994, 27% of the car owners with more than one car acquired the extra one to avoid "a day without a car." Many families in Mexico City have one old extra car, so the suspension of this control option would now allow another family member to use the extra car, thus increasing the number of vehicles in circulation and increasing traffic congestion. Obviously, these extra cars (mostly older models) are higher emitters, so to encourage fleet renewal, CAM has combined the "a day without a car" and the I/M programs (Secretaría del Medio Ambiente 2000) by allowing cleaner and newer models to be driven every day. The emissions verification program, on the other hand, is costly to implement and manage (Heil and Pargal 1998). This program established different exhaust emission standards for CO and HC based on the vehicle model year.

The incorporation of NOx emissions verification into the program occurred in 2000. Vehicles employing new technology and cleaner vehicles are allowed to circulate every day if they meet stringent emission standards. However, as the beneficial AQ impacts (reduced O_3 levels) of the lower emissions values were not clearly communicated to the public, bribes and corruption frequently allow car owners to bypass fines and circulation restrictions. For this reason, the I/M option is undergoing a careful audit to assess its benefits.

Control Strategies for Particulate Matter

The approach to controlling particulate matter in the MCMA has been changing continuously over the years. Previously, only TSP were taken into account, and these were mainly attributed to erosion processes. PM_{10} emissions have been included in the EI since 1996 and contribute approximately 0.5% to the total emissions in the MCMA. There is considerable information on the impact of PM_{10} on human health [Edgerton et al. 1999; Ortiz et al. 1999; World Health Organization (WHO) 1998]. These emissions arise mainly from the industrial and transportation sectors (Table 7.11). The main industrial sources are mineral processing; chemical, food and textile industries; consumer goods; and electric power gen-

Table 7.11. Major sources of particulate matter less than 10 μm indiameter (PM_{10}) in the 1998 Mexico City Metropolitan Area emissions inventory[a]

Category	Emissions		Accumulated emissions %
	Ton/year	%	
Heavy-duty trucks[b]	6542	54.23	54.23
Private cars	751	6.23	60.45
Buses (includes ruta-100)	751	6.23	66.68
Metallic products processing	688	5.71	72.38
Chemical industry	424	3.51	75.90
Nonmetallic minerals processing	378	3.13	79.03
Consumer products processing	317	2.63	81.66
Food consumer products	294	2.44	84.10
Light-duty trucks	219	1.82	85.92
Electricity generation	209	1.73	87.65
Commercial sector combustion	208	1.72	89.37
Taxis	188	1.56	90.93
Vegetables and animal products	140	1.16	92.09
Clothing industry	138	1.14	93.23
Long-life consumer products	124	1.03	94.26

[a] *Source*: Molina et al. (2000).
[b] Emissions from heavy-duty trucks may be overestimated by as much as 300%.

eration. Heavy-duty trucks, buses, taxis, and private cars are the main emissions sources of the transportation sector. Combustion in commercial activities appears to be the most important source of the services sector.

The main strategy to control PM has been to institute the use of cleaner fuels, in particular, by reducing the sulfur content of diesel, gasoline, and gas oil for industry. The use of clean fuels has also led to reductions in SO_2 emissions. Top SO_2 sources are listed in Table 7.12. In the future, further reduction of sulfur in gasoline is expected to improve the efficiency of catalytic converters. However, as shown in Table 7.12, the industrial and services sectors are also major sources of SO_2. PEMEX decided to require substitution of low-sulfur diesel (0.05%) for industrial gas oil, based on the low sales volume of gas oil in the MCMA. The users of gas oil are low-income hotels, public baths, and small industries. The substitution is enforced by a gradual rise in prices for gas oil coupled with a modification in the standard of any fuel used by the industrial and service sectors to a maximum of 0.05% sulfur content starting in 2001. This new standard is currently under review.

The reduction of S in fuels will also reduce particle formation from transportation, services, and industry. Additional actions, directed toward the most polluting industrial facilities, are the installation of electrostatic precipitators and other control devices to reduce particle emissions. Unfortunately, particulate matter is also related to nonurban emissions from the surrounding areas of the MCMA, such as forest fires, controlled agricultural fire practices, and volcanic emanations, the latter arising as an issue since 1994. The contribution to PM_{10}

Table 7.12. Major Sources of SO_2 in the 1998 Mexico City Metropolitan Area emissions inventory

Category	Emissions Ton/year	Emissions %	Accumulated emissions %
Commercial combustion	4385	19.33	
Chemical industry	2521	11.11	30.44
Private cars	2379	10.49	40.92
Several consumer products	1823	8.04	48.96
Nonmetallic minerals	1620	7.14	56.10
Wood and derivatives	1266	5.58	61.68
Food consumer industry	1227	5.41	67.09
Heavy-duty trucks[a]	1158	5.10	72.19
Vegetables and animal products	1034	4.56	76.75
Light-duty trucks	908	4.00	80.75
Metallic products	816	3.60	84.35
Metal processing	716	3.16	87.50
Clothing industry	666	2.94	90.44
Taxis	563	2.48	92.92
Others	518	2.28	95.20

[a] May be overestimated by as much as 300%.

inventories of building and other construction activities and of resuspended dust from paved and unpaved roads is not taken into account in the 96–EI or in the 98–EI, although such PM_{10} sources are known to be important as well.

Other Control Strategies

Table 7.6 summarizes successful and unsuccessful control strategies proposed in different pollution control programs for the MCMA. These strategies were analyzed in the Mexico City Air Quality Research Initiative (MARI) project in 1994 (IMP–LANL 1994b). The main control strategies implemented to reduce O_3 levels have been the introduction of cleaner gasoline and fuels and the requirement of catalytic converters for 1991 vehicles and newer models, as discussed previously. The use of cleaner fuels conveniently reduced lead, CO and SO_2, in ambient air. Substituting fuel oil with NG in power plants and the lower content of sulfur in diesel (down to 0.05%) are directly related to the reduction in SO_2 levels. Most of the retrofit programs failed for technological reasons. For example, NOx emissions from LPG vehicles were 70% greater than emissions from gasoline vehicles equipped with catalytic converters. It is assumed that the carburation kits are calibrated to use a blend with at least 70% propane, but changes in LPG composition, which frequently occurs in Mexico, make them inefficient (Schifter et al. 2000b). Nevertheless, the vehicle fleets owned by large commercial and industrial companies have such devices because vehicles so equipped are allowed to run daily without restrictions. The main barrier to the removal of old private vehicles from circulation was financial, because the Treasury Secretariat refused to spend public money to support the acquisition of new models by private car owners.

In the early 1990s, there was a quite successful program, funded by the World Bank and other sources, to replace old taxis with new ones equipped with catalytic converters. On-road remote sensing measurements in the MCMA detected large reductions in the emissions from taxis as a result (Beaton et al. 1992; Bishop et al. 1997).

One of the major unsolved problems in the MCMA is the transportation system. In the past decade, transportation has become more dominated by private cars, minibuses, and combis, and the kilometer–passenger share of more efficient public transportation, such as the metro rail, trolleys, and buses, has declined. The local traffic management systems have failed to reduce congestion. A feebate system was also suggested, which would have included economic compensation for the owners of ecologically efficient cars, especially those used as taxis, or penalties for owners of high-polluting vehicles, but it was never implemented.

NOx Control Scenarios

For NOx emissions, it is not possible to correlate inventories with specific pollution control strategies in the AQ programs because of the many differences in methodologies and data collection among the 94, 96, and 98 EIs. For example, the apparent 43% reduction in NOx emissions from electric power plants from 1994 to 1996 is a result of the change in the emissions estimation methodology, rather than from a decrease in fuel consumption. In fact, fuel consumption in power plants increased about 7% in that period.

Future trends in NOx emissions, compared to the 1998 EI, can be projected by considering three incremental NOx emissions control scenarios for the year 2010. The first scenario, business as usual, continues the trends of the last decade. The vehicular fleet renewal rate remains the same. It also assumes that the I/M program would continue its current level of effectiveness. Consumption of LPG, gas oil, diesel, and gasoline are assumed to grow annually at rates of 1.4%, 1.85%, 2.5%, and 3% respectively, and if the market allows, natural gas is expected to increase from its historical 2.5% annual increase to 3%. Using this business-as-usual scenario, total NOx emissions are predicted to increase by 2.8% in 2010 compared to the 1998 emissions value.

The second scenario includes all the previous assumptions, plus 85% efficiency in controlling NOx emissions from natural gas electric utility boilers (USEPA 1995a). There are only three power plants in the MCMA, and they are the most important NOx point sources. In practice, it should be more successful to control these NOx sources than to focus on NOx control from mobile sources, which are more dispersed. Using this scenario, total NOx emissions are predicted to decrease by 1.3% from the 1998 value.

A third scenario adds to the assumptions of the second scenario an assumption of 85% efficiency in controlling NOx emissions from natural gas industrial utility boilers. The total NOx emissions for the year 2010, using this last scenario, would decrease by 9.3% from the 1998 value.

Summary

The 94–EI is assumed as the base case. This inventory considered particle emissions from sources besides erosion for the first time. So far, it is very difficult to correlate emissions figures with pollution control strategies in the MCMA because of the many differences in methodology and changes in activity data among the 1994, 1996, and 1998 emissions inventories. However, some control strategies implemented in the three MCMA pollution control programs may be considered successful, mainly those related to the use of cleaner reformulated fuels and the introduction of cars equipped with catalytic converters. Retrofit programs have failed, except when they were implemented by large commercial and industrial companies. The car-scrapping program to remove older vehicles from circulation failed for financial reasons. Efficiency and sustainability of the urban transportation system has decreased in recent years and is a major issue relevant to prospects for achieving further reductions in pollutant emissions. The Inspection and Maintenance program must be reinforced to avoid bribes and corruption, and the "no car day" (hoy no circula) program must be maintained, because canceling it could increase traffic congestion.

Results of smog-chamber experiments (Sandoval et al. 2001) and AQ modeling (IMP–LANL 1994a) suggest that a reduction in NOx emissions is the most effective option for reducing maximum O_3 concentrations. This strategy may also contribute to reduced levels of PM_{10} because NOx are precursors for the formation of secondary particulate matter. Except for emissions from soil, all of the important NOx sources are related to fuel combustion, mainly in the transportation sector.

In order to assure better AQ in the MCMA, it will be necessary to implement stringent regulatory emissions controls, including modernization of the mobile fleet, an improved inspection and maintenance program, and efficient NOx controls in both electric power plants and industrial boilers.

Acknowledgments. We acknowledge the support and information from many individuals and institutions. We especially want to thank the Massachusetts Institute of Technology. PEMEX support for several research projects at the Mexican Petroleum Institute is also acknowledged. In part, this work was done within the D.888 project of the IMP Environmental Research Program.

References

Arriaga, J.L., Escalona, S., Cervantes, A., Orduñez, R., and López, T. 1997. Seguimiento de COV en el aire urbano de la ZMCM. 1992–1996. In *Contaminación Atmosférica*, Vol. II, eds. L. García Colín, and J. Varela, pp. 67–96. México D.F.: El Colegio Nacional.

Auto/Oil Air Quality Research Improvement Program 1997. Final Report. Warrendale, Pennsylvania, U.S.A.: Society of Automotive Engineers.

Beaton, S.P., Bishop, G.A., and Stedman, D.H. 1992. Emission characteristics of Mexico City vehicles. *J. Air Waste Manage. Assoc.* 42:1424.

Belausteguigoitia, J.C. 1994. Algunas consideraciones sobre el uso de instrumentos económicos en la política ambiental. In *Medio Ambiente Problemas y Soluciones.* México D.F.: El Colegio de México.

Benson, J.D., Burns, V., Koehl, W.J., Gorse, R.A., Painter, L.J., Hochhauser, A.M., and Reuter, R.M. 1997. *Effects of Gasoline Sulfur Level on Mass Exhaust Emissions—Auto/Oil Air Quality Improvement Research Program*, Final Report. Warrendale, Pennsylvania, U.S.A.: Society of Automotive Engineering. Report No. 912323.

Bishop, G.A., Stedman, D.H., De la Garza, J., and Dávalos, F.J. 1997. On-road remote sensing of vehicle emissions in Mexico. *Environ. Sci. Technol.* 31:3505–3510.

CAM (Comisión Ambiental Metropolitana). 1999. *Inventario de Emisiones a la Atmósfera de la Zona Metropolitana del Valle de México 1996.* México D.F.: Comisión Ambiental Metropolitana.

CAM (Comisión Ambiental Metropolitana). 2000. *Informe sobre el Curso de Capacitación para el Manejo del Modelo Multiscale Climate Chemistry Model.* México D.F.: Comisión Ambiental Metropolitana.

Comisión Metropolitana de Transporte y Vialidad (COMETRAVI). 1999. *Estudio Integral del Transporte y Calidad del Aire en la ZMVM. Diagnóstico de las Condiciones del Transporte y Sus Implicaciones sobre la Calidad del Aire.* México D.F.: COMETRAVI.

DDF (Departamento del Distrito Federal), Gobierno del Estado de México (GEM), Secretaría de Medio Ambiente, Recursos Naturales y Pesca (SEMARNAP), and Secretaría de Salubridad y Asistencia (SSA). 1996. *Programa para Mejorar la Calidad del Aire en el Valle de México 1995–2000.* México D.F.: Instituto Nacional de Ecología.

Edgerton, S.A., Bian, X., Doran, J.C., Fast, J.D., Hubbe, J.M., Malone, E.L., et al. 1999. Particulate air pollution in Mexico City: A collaborative research project. *J. Air Waste Manage. Assoc.* 49:1221–1229.

Federal Aviation Administration. 1998. Aircraft engine emission user guide and database. FAEED v3.1, *http://www.epa.gov/otaq/regs/nonroad/aviation/faauser3.pdf.* The database is supported by the Office of Environment and Energy, Federal Aviation Administration, Washington D.C.

Finlayson–Pitts, B.J., and Pitts, J.N., Jr. 1986. *Atmospheric Chemistry: Fundamentals and Experimental Techniques.* New York: John Wiley and Sons.

GDF (Gobierno del Distrito Federal). 2000a. *Inventario 1998. http://sma.df.gob.mx/publicaciones/aire/inventario1998/inventario1998.htm.* Mexico D.F.: Secretaría del Medio Ambiente, Gobierno del Distrito Federal.

GDF (Gobierno del Distrito Federal) 2000b. *Compendio Estadístico de la Calidad del Aire 1986–1999.* México D.F: Secretaría del Medio Ambiente, Gobierno del Distrito Federal.

Gobierno del Distrito Federal–RADIAN. 1996. *Manual de Factores de Emisión Para Fuentes Móviles M5MCMA.a3.* México D.F.: Secretaría del Medio Ambiente, Gobierno del Distrito Federal.

Heil M., and Pargal, S. 1998. *Reducing Air Pollution from Urban Passenger Transport. A Framework for Policy Analysis.* Washington D.C.: World Bank.

IMP–LANL. 1994a. *Mexico City Air Quality Research Initiative. Modeling and Simulation*, Vol. III. Los Alamos, New Mexico: Instituto Mexicano del Petróleo and Los Alamos National Laboratory. LA—12699 Report.

IMP–LANL. 1994b. *Mexico City Air Quality Research Initiative. Strategic Evaluation*, Vol. V. Los Alamos, New Mexico: Instituto Mexicano del Petróleo and Los Alamos National Laboratory. LA—12699 Report.

INE. 1999. *Proyecto de Revisión y Eliminación de Fugas Domésticas de Gas Licuado de Petróleo en la Zona Metropolitana del Valle de México. http://www.ine.gob.mx/presidencia/asesores/pa_urbana/pag14.html.* México D.F.: Instituto Nacional de Ecología.

INE. 2000. *Almanaque de Datos y Tendencias de la Calidad del Aire en Ciudades Mexicanas.* México D.F.: Instituto Nacional de Ecología.

JICA (Japanese International Cooperation Agency). 1988. *Seminario sobre las Medidas contra la Contaminación Atmosférica de la Ciudad de México*, pp 89. México D.F.: JICA and Departamento del Distrito Federal.

Legorreta, J. 1989. *Transporte y Contaminación en la Ciudad de México.* México, D.F.: Centro de Ecodesarrollo.

McCauley, E. 1999. 1997 VOC to NOx ratios for three sites in the South Coast Air Basin. *http://www.arb.ca.gov/aqd/weekendeffect/vocnoxratio092299.doc.*

Molina, M.J., Molina, L.T., Sosa, G., Gasca, J., and West, J. 2000. *Analysis and Diagnostics of the Emissions Inventory of the Metropolitan Area of the Valley of Mexico. Integrated Program on Urban, Regional and Global Air Pollution*, August. Boston, MA: Massachusetts Institute of Technology. Report #5.

National Research Council. 1991. *Rethinking the Ozone Problem in Urban and Regional Air Pollution.* Washington D.C.: National Academy Press.

Ortiz, E., Alemón, E., Guzman, F., and Galván, J.L. 1999. Personal exposure to particulate material and volatile organic compound in microbuses at Mexico City. In *Urban Transport V. Urban Transport and the Environment for the 21st Century*, ed. L.J. Sucharov, pp.105–114. Boston, Massachusetts, USA: WIT Press.

PEMEX. 1999. *Informe de Ventas Mensuales de Gasolina.* Mexico City: PEMEX.

PEMEX Gas y Petroquímica Básica. 1997. *Efecto de los Componentes del Gas Licuado en la Acumulación de Ozono, en la Atmósfera de la ZMCM.* Mexico City: PEMEX Gas y Petroquímica Básica, Instituto Mexicano del Petróleo.

Sandoval J., Marroquín, O., Jaimes, J.L., Zúñiga, V., González, E., and Guzmán, F. 2001. Effect of hydrocarbons and nitrogen oxides on ozone formation in smog chambers exposed to solar irradiance of Mexico City. *Atmósfera* 14:17–28.

Schifter, I., Diaz, L., Vera, M., Castillo, M., Ramos, F., Avalos, S., et al. 2000a. Impact of engine technology on the vehicular emissions in Mexico City. *Environ. Sci. Technol.* 34:2663–2667.

Schifter, I., Diaz, L., Lopez-Salinas, E., Rodríguez, R., Avalos, S., and Guerrero, V. 2000b. An evaluation of the LPG vehicles program in the metropolitan area of Mexico City, *J. Air Waste Manage. Assoc.* 50:301–309.

Schifter, I., Diaz, L., Avalos, S., Vera, M., Barrera, A. and López-Salinas, E. 2000c. Effect of methyl tertiary butyl ether concentrations on exhaust emissions from gasoline used in the metropolitan area of Mexico City. *J. Air Waste Manage. Assoc.* 50:488–494.

Secretaría del Medio Ambiente. 2000. *Tercer Informe de Trabajo.* México D.F.: Secretaría del Medio Ambiente.

Secretariado Técnico Intergubernamental. 1990. *Programa Integral Contra la Contaminación Atmosférica en la Zona Metropolitana de la Ciudad de México.* México D.F.: Gobierno del Distrito Federal.

Seinfeld, J.H. 1986. *Atmospheric Chemistry and Physics of Air Pollution.* New York: John Wiley and Sons.

USEPA. 1990. *Background Documents for the Surface Impoundment Modeling Systems, SIMS. Version 2.0 PA-450/4-90019b.* Washington, D.C.: Environmental Protection Agency.

USEPA 1991. *The IM240 Transient I/M Dynamometer Driving Schedule and The Composite I/M Test Procedure*, *http://www.epa.gov/otaq/im240rpt.htm*, USA: Environmental Protection Agency. Report No. EPA-AA-TSS-91-1. See also *http://www.epa.gov/otaq/testproc.htm* for *Test Procedures and Laboratory Methods (Vehicles, Engines & Fuels).* Washington, D.C.: Environmental Protection Agency.

USEPA. 1994. *LANDFILL Model.* Washington, D.C.: Environmental Protection Agency.

USEPA. 1995a. *Compilation of Air Pollutants Emissions Factors*, Vol. I, II. Washington, D.C.: Environmental Protection Agency. AP-42.

USEPA. 1995b. *User's Guide to the Personal Computer Version of Biogenic Emission Inventory System (PC-BEIS2).* Washington, D.C., USA: Environmental Protection Agency. EPA-600/R-95-091.

USEPA. 1997. *TANKS 3.1 Installation Instructions. http://www.epa.gov/ttn/chief/tanks.html.* Washington, D.C.: Environmental Protection Agency.

World Bank. 1992. Transport air quality management in the Mexico City Metropolitan Area. Annex IV. *Estimated Motor Vehicle Emissions Inventory for the Mexico City Metropolitan Area and Emission Reductions due to Potential Control Measures.* Washington, D.C.: World Bank. Report No. 10045-ME.

WHO. 1998. *World Health Organization Air Quality Guidelines for Europe*, 2nd Second edn. Geneva: World Health Organization.

8. Heavy Metals in Forest Soils, Vegetation, and Drainage Waters in the Basin of Mexico

Mark E. Fenn, Juana María Castro-Servín,
Tomás Hernández-Tejeda, Natalie Krage, Chris Goodson,
and Thomas Meixner

Since the late 1970s, it has been recognized that air pollution is a major environmental problem in the Mexico City Air Basin (MCAB) (Bravo-Alvarez 1987; Vizcaíno-Murray 1975). Before the introduction of low-lead fuels in 1986, lead (Pb) was probably the most harmful pollutant in the MCAB (Bravo and Torres 2002). However since the mid 1980s, photochemical smog has been the primary concern of city officials and of the federal government because the frequency at which the air quality standard for ozone (O_3) is surpassed and because of the deleterious effects on people, plants, animals, and materials (Bauer and Krupa 1990; Krupa 1997; Quadri and Sánchez 1994). Forests to the south and southwest of Mexico City are located within a short distance of the urban zone and in the direction of the prevailing winds, which carry heavy atmospheric pollution loads into these forested areas (Bravo and Torres 2002; Jauregui 2002). The Desierto de los Leones (DL) National Park (Fig. 8.1) is the most impacted forested area in the entire (MCAB).

The Cultural and Recreational DL Park has great importance for the inhabitants of Mexico City as an important aquifer recharge area, a popular natural recreational area, an educational and cultural awareness site, a natural reserve providing ever-diminishing plant and wildlife habitat in the Basin, and a valuable forest site for ongoing scientific studies. The severe forest decline and dieback observed in sacred fir (*Abies religiosa*) stands in the park during last two decades is presumably a result of several biotic and abiotic factors impinging on these forests, including: air pollution from Mexico City, lack of forest

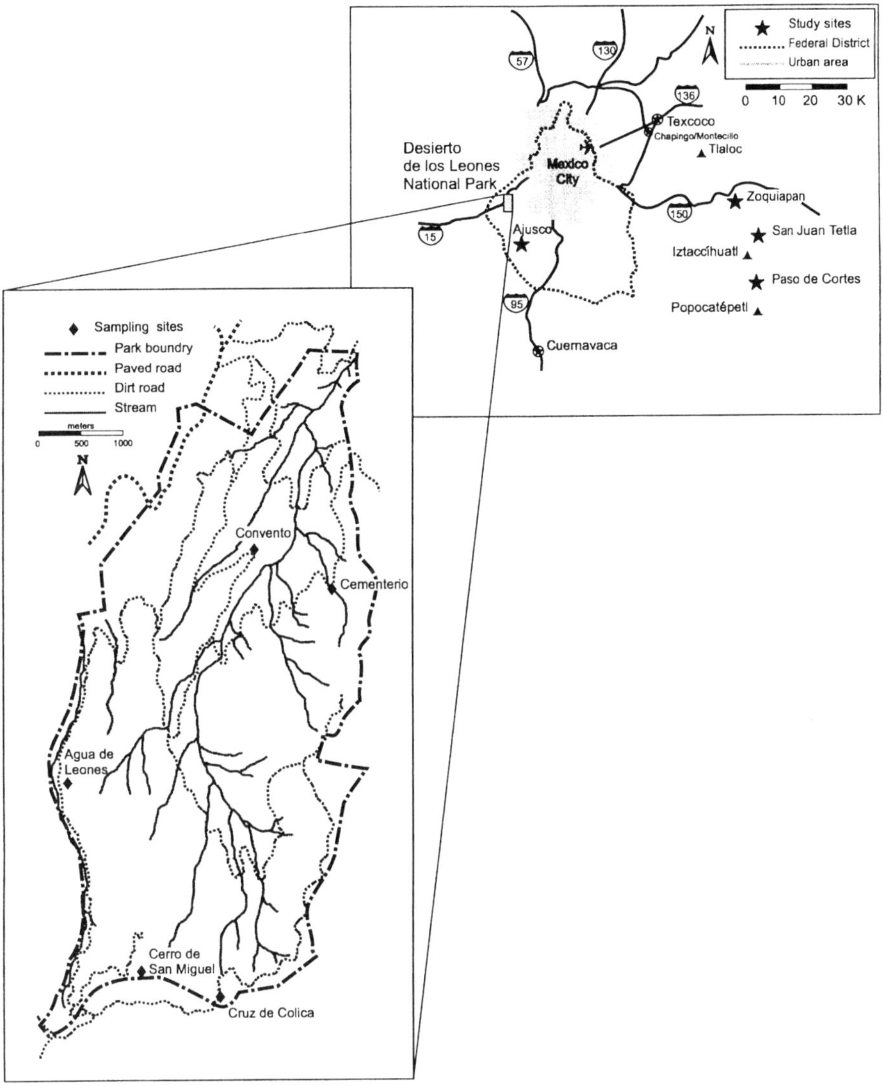

Figure 8.1. Location of the permanent soil and vegetation sampling sites in the Desierto de los Leones Forest Park (Modified from Vázquez-Soto, 1987.) and map showing the location of the study sites located along an air pollution gradient in the Mexico City Air Basin. DL and Ajusco (AJ) are high-pollution sites. Zoquiapan and Paso de Cortes are low-pollution sites.

management, overextraction of water from the forested watersheds, and forest pests and diseases (Alvarado-Rosales and Hernández-Tejeda 2002; Alvarado et al. 1991, 1993). Several studies have also addressed the effects of photochemical oxidants on sacred fir and *Pinus hartwegii* with emphasis on visual injury symptoms to foliage and crowns (Alvarado-Rosales and Hernández-Tejeda 2002; Alvarado et al. 1993; Miller et al. 2002).

Because of the severe dieback of sacred fir in certain areas of the park, and because of the difficulty in determining the primary causal factors of the dieback, a number of studies have addressed the problem from different points of view. Studies have been carried out related to surface geology, edaphology, and soil microbiology (Castro 1987; Castro et al. 1995b; Castro-Sevin 1999; Reyes et al. 1984; Romero 1986). Studies of heavy-metal content in soil and vegetation, and plant and soil nutritional studies of sacred fir have also been done (Alvarado et al. 1991; Castañeda-Gonzalez et al. 1995; Castro et al. 1993; Castro et al. 1995a; Castro et al. 1996; López and Rivera 1995). However, the cause of the fir decline is still not definitively known (Alvarado-Rosales and Hernández-Tejeda 2002).

The DL National Park is located directly downwind of the large urban pollution source area of greater Mexico City (Bravo-Alvarez and Torres-Jardon 2002; Jauregui 2002). Although atmospheric deposition of heavy metals has not been measured in forests of the MCAB, there is little doubt, based on trends of Pb emissions in the basin, that elevated levels of Pb deposition occurred in the decades prior to the widespread use of unleaded fuels. In the years prior to 1986, average atmospheric concentrations of Pb reached levels almost three times the Mexican air quality standard (Bravo-Alvarez and Torres-Jardon 2002). Because emissions have decreased dramatically since Pb levels in gasoline were reduced in January, 1987, because unleaded gasoline was introduced in September, 1990, and because leaded gasoline was phased out in August, 1997, it is also expected that deposition of Pb in forests downwind of Mexico City has decreased dramatically since 1987. Wet and dry deposition of atmospheric nitrogen (N) and sulfur (S) occurs at high levels in the park (Fenn et al. 1999). Although quantitative data on atmospheric inputs of heavy metals into the park are lacking, except for biomonitoring studies with lichens (Carrillo-Padrón 1993; Zambrano 2001), atmospheric inputs of heavy metals to forests downwind of Mexico City during the past several decades are undoubtedly quite large. Forest canopies are known to be effective receptors of air pollutants due to the high surface area and filtering capacity of tree canopies (Fenn and Bytnerowicz 1997). Ultimately, much of the atmospheric deposition reaches the forest floor and soil, either by direct deposition, as throughfall washed from the canopy by precipitation, or in litterfall. The organic litter layer is known to accumulate heavy metals (Friedland and Johnson 1985).

Although there are few studies of heavy-metal deposition, accumulation, or effects in forests of the MCAB, the purpose of this chapter is to review the available data on this subject. The primary studies of heavy metals in forests of the MCAB were carried out in the DL Park, and at control sites upwind of Mexico

City or just outside of the Basin. These data will be presented and discussed in light of related studies of heavy metals and phytotoxic effects.

Methods

Site Description of Desierto de los Leones National Park and Other Study Sites

The primary study site for all the studies reviewed in this chapter is the DL National Park. However, in a previously unpublished study, heavy-metal concentrations are reported for soil and vegetation samples collected in 2000 from four sites across an air pollution gradient in the Basin of Mexico. The DL site is the most polluted followed by the Ajusco (AJ) site. The two sites with relatively low pollution are in the Zoquiapan (ZOQ) National Park and the Paso de Cortes (PC) within the Izta-Popo national park (Fig. 8.1). Site characteristics of DL and ZOQ are described in Fenn et al. (1999), and all four sites are briefly described in Fenn et al. (2002) in this volume.

The cultural and recreational DL Park is located southwest of Mexico City in Cuajimalpa County, Federal District. It is located between the northern latitudes 19°16′18″ and 19°20′00″ and between the western longitudes 99°17′00″ and 99°20′00″. Elevation in the park ranges from 2750 to 3795 m, with an average elevation of 3250 m (Beltrán and Vázquez 1971). The park has an enlarged polygonal form with a total surface of 1529 hectares. In terms of land area, the DL is the thirtieth largest national park in Mexico and the fourth largest park in the Mexico City area (Vargas 1984).

The predominant climate, is temperate subhumid with rains in summer (García 1981). The mean annual precipitation is 1354 mm. Minimum and maximum temperatures are 8.4° C and 12.7° C, with dominant winds from the northwest and northeast. Frosts occur in the months from October to April. January, with an average of 22 frosts, is the month with the most frequent frosts, while April, on average, has one frost. The annual mean number of days with frost is 67, and the annual mean number of days with hail is 21. Hail can occur during all the months of the year except December, but mainly occurs during the rainy season. July is the month with the highest incidence of hail, with 4 days per year on average (DDF-COCODA 1986).

Soils

Soils in the Paso de Cortes and Zoquiapan sites are described in Marin et al. (2002) in this volume. Soils in the DL park are deep and rich in organic matter. The soils are classified as eutric cambisols mixed with andosols (FAO classification). Soil texture is predominantly sandy-clay to silty-clay. Soils can reach up to two meters in depth in the lower elevations of the park (Sosa 1952). In the location "La Venta," Aguilera (1962) described a soil profile two meters in depth of andesitic parent material in a sacred fir forest. The soil was described as a sandy-clay loam with a pH of 6.5, and the organic matter content of the soil ranged from 7% to 10%. The total N content of the soil (the top 15 cm) was reported to range from 0.4% to 0.6% in *Pinus hartwegii* stands in the park (Castro et al. 1997; Fenn et al. 1999). In an *Abies religiosa* stand, the N content was 0.5% (Castro et al. 1997). In the soil of a *P. hartwegii* stand in the park, Fenn et al. (1999) reported C:N ratios of 15 to 16 and a total S content of 0.04% to 0.06%.

The cation exchange capacities of soils in the DL park typically range from 40 to 60 meq/100 g (Castro et al. 1997; Marin et al. 2002). Soil pH values typically range from 5.3 to 6.7 (Aguilera 1962; Castro et al. 1997; Madrigal 1967; Marin et al. 2002), but Fenn et al. (1999) reported pH values of 4.5 to 4.6 for soils in *Pinus hartwegii* stands in the DL and in the ZOQ Experimental Forest east of Mexico City. Soil pH values under ten *Abies religiosa* canopies in the DL ranged from 4.4 to 6.4 compared with pH values of 6.6 to 6.9 under three trees at San Pablo, a site with relatively low pollution located approximately 45 km northeast of Mexico City (Watmough and Hutchinson 1999). The low pH values in soils in the DL area are attributed to pulses of acidic compounds intercepted by the tree canopies and deposited to the forest floor in throughfall and stemflow (Fenn et al. 1999).

Vegetation

The predominant forest vegetation at DL consists of *Abies religiosa* Schl. et. Cham., *Pinus patula* Schl. et. Cham., *P. hartwegii* Lindl., *Quercus laurina* HBK., *Q. mexicana* HBK and *Q. microphylla* Neé. Other species include *Prunus serotina* Ehrh sp. *capuli* (Cav) Mcbaugh, *Alnus firmifolia* Fern., *A. jorullensis* HBK, *Arbutus glandulosa* Mart. & Gal., *Buddleia cordata* HBK and *B. parviflora* HBK (Rzedowski and Calderón 1979). A complete description of the native flora in the park is given by Reiche (1977). Dominant overstory and understory species found at DL and in the other study sites reported in this chapter are listed in Table 8.1.

Sampling and Analysis of Soil, Vegetation, and Streams

Data from two sampling periods will be presented in this chapter, the first one in 1993 (Castro et al. 1997) and the second in 2000. Soil and plant samples were collected in April, 1993, from four permanent sampling sites at DL (Fig. 8.1) and from two permanent sampling sites in a low-pollution control site in San Juan Tetla in the adjacent state of Puebla (Table 8.1) (Castro et al. 1997). From each sampling site, three 1-kilogram soil samples were collected at 50 m intervals along a north–south transect. The soil samples were collected at two depths, 0 to

Table 8.1. Types of vegetation in the study sites described in this chapter

Study site	Overstory vegetation	Understory vegetation
DL Cementerio[a]	*P. ayacahuite, P. montezumae, P. patula, P. radiata*	(Understory vegetation is cut on a regular basis in this plantation)
DL Convento	*Abies religiosa*	*Senecio* sp., *Acaena* sp., *Fuchia* sp., *Vinca minor*, Grasses
DL Agua de Leones	*P. patula, P. radiata*	Grasses
DL Cruz de Colica	*Pinus* sp.	Grasses
San Juan Tetla, Puebla	*Pinus* sp. *Abies religiosa*	*Acaena* sp., *Senecio* sp. Grasses
DL, AJ, ZOQ and PC[b]	*Pinus hartwegii*	Tussock grasses

[a] The Cementerio site is a pine plantation that was planted in 1986.
[b] The following four pine forests were sampled in 2000: Desierto de los Leones, DL; Ajusco, AJ; Zoquiapan, ZOQ; and Paso de Cortes, PC.

15 and 16 to 30 cm, but in this chapter, the results are presented as the average of both depths because concentrations varied little between depths.

The following metals were extracted from the soil samples with diethylenetriaminepentaacetic acid (DTPA) and quantified with atomic absorption spectroscopy: Al, Pb, Fe, Zn, Mn (Castro et al. 1997). Current-year foliage and older foliage were sampled and analyzed separately from branches of pine and fir trees in order to compare the combined effect of foliar age and time of exposure to atmospheric pollutants on metal concentrations in foliage. Branches from trees and shrubs were sampled for analysis along with bark collected from the main stem. The above-ground tissues of herbaceous plants and grasses were also sampled for analysis. Samples were oven-dried and ground and then extracted with a nitric–perchloric acid mixture and analyzed for Ca, Mg, K, Fe, Zn, Pb, and Cu using an atomic absorption spectrophotometer (Castro et al. 1997).

In the more recent study, current-year foliage of *Pinus hartwegii* trees was sampled in the fall of the year 2000 at four sites along an air pollution gradient in the MCAB. DL and AJ were the two high-pollution sites, and ZOQ and PC were the two low-pollution sites in the eastern portions of the Basin of Mexico (Fig. 8.1). In the spring of 2000, soil was collected from the top 15 cm of soil under the canopies of *P. hartwegii* trees at all of the sites. Streamwater samples were also collected from springs and streams at the four sites from January through September, 2000. Soil, foliage, and water samples were analyzed for ten metals, Pb, Zn, Cd, Mn, Ni, Cu, Al, Fe, and Mg, and for Ca. Plant and water samples were also analyzed for Si, K, and Na.

For each soil sample, approximately 0.25 g of soil was microwave digested using concentrated HNO_3 and HF according to US EPA Method 3052 (U.S. EPA 1995). Samples of ground plant material (0.5 g) were digested for 20 min with a digest solution of 3 mL concentrated HCl, 9 mL concentrated HNO_3, and 1 mL double-deionized H_20 (modified from Millward and Kluckner 1989) in a CEM MARS 5 microwave digest system at full power. Digest solutions were then gravity filtered through Whatman 50 filter paper cones and diluted to 50 mL. Total concentrations of Al, Ca, Fe, Mg, and Na were determined using a Perkin-Elmer Optima 3000V ICP–OES. The total concentrations of Cd, Cu, Ni, and Pb were determined using a Perkin-Elmer AAnalyst 800 Graphite Furnace Atomic Absorption Spectrophotometer. Total concentrations of Mn and Zn were determined using a Perkin-Elmer AAnalyst 800 Flame Atomic Absorption Spectrophotometer. Streamwater samples were acidified with nitric acid and analyzed for metals using the same analytical methods as were used for plant and soil samples. Exchangeable bases in soil were determined from 2.5 g soil samples extracted with 25 ml of 1-*N*-ammonium acetate on a shaker for 15 min. Cation concentations in the extracts were analyzed with ICP as described above.

Results and Discussion

Atmospheric Deposition of Heavy Metals

Although atmospheric deposition of heavy metals in forests of the Basin of Mexico has not been studied, the premise that high levels of heavy metals have been deposited to forests to the south and southwest of Mexico City is supported by other available evidence: information on emissions of metals and other pollutants (Bravo-Alvarez and Torres-Jardon 2002); dominant wind patterns (Jauregui 2002); biomonitoring of lichens and mosses (Zambrano et al. 2002); and data on metal concentration in soil and plant tissue (Alvarado et al. 1993; Castro et al. 1997; López-López et al. 1998; Watmough and Hutchinson 1999). In 1990/1991, concentrations of six heavy metals in samples of epiphytic lichens (*Parmelia* sp. and *Parmotrema* sp.) were determined every three months in five forests surrounding Mexico City (Carrillo-Padrón 1993; Zambrano et al. 2002). There was a gradient of metal concentrations increasing from northeast to southwest, corresponding with the dominant regional wind circulation in the Basin of Mexico (Jauregui 2002). In most cases, the highest metal concentrations were from samples collected in the DL. This study adds further weight to the conclusion that high levels of heavy metals, along with O_3 and N and S pollutants, are transported to forests located downwind (southwest) of the urban pollution source area. However, it should be emphasized that current deposition levels for Pb, and possibly for other heavy metals, are undoubtedly much lower than in the 1970s and 1980s because the Pb content of gasoline was reduced in 1986, because unleaded gasoline was introduced in 1990, and as a result of other pollution control measures (Ruiz and Gasga 2002). However, increases in vehicle-miles in the Basin could result in greater emissions of some metals from wear of tires and other automotive parts.

Heavy Metals in Forest Soils

Concentrations of DTPA extractable Al, Pb, and Fe in soil sampled in 1993 were on average 8.6, 6.9 and 4.8 times greater at DL sites compared to the two control

Table 8.2. Characteristics of the sites, pH, and heavy-metal concentrations (μg/g, DTPA extractable) in the soils of the Desierto de los Leones Park (DL) and a control site

Site[a]	Vegetation	Aspect[b]	Elevation (m)	pH	Al	Pb	Fe	Zn	Mn
DL Cementerio	Pine Plantation[c]	N	2980	6.2	36	56	30	7	3
DL Convento	Fir	NW	2920	5.7	60	58	40	4	3
DL Agua de Leones	Fir–pine	NE	3320	5.0	141	61	6	5	3
DL Cruz de Colica	Pine	SU	3600	5.0	136	54	17	3	4
San Juan Tetla	Pine	SU	2980	6.0	13	10	8	5	2
San Juan Tetla	Fir	NW	3050	5.4	7	6	5	3	4

[a] Sites: DL, Desierto de los Leones, Distrito Federal, and S, San Juan Tetla, Puebla (control site).
[b] Aspect: N, North; NW, Northwest; NE, Northeast; and SU, Summit.
[c] Pine plantation was planted in 1986.
Data from Castro (1999) and Castro et al. (1997).
Soils were sampled in 1993 from a depth of 0 to 30 cm.

sites in San Juan Tetla, Puebla (Table 8.2). Average total Cd and Cu levels in soil under sacred fir trees in the park (0.8 and 25.4 ppm) were 2 to 4 times greater than at San Pablo (0.2 and 11.0 ppm), a control site upwind of Mexico City (Watmough and Hutchinson 1999). However, total Cr levels in soil were significantly greater at San Pablo (29.8 ppm) than in the DL (19.5). Levels of total Ni (14.9–33.1 ppm) in soil were not significantly different between DL and the control site (Watmough and Hutchinson 1999).

In soil samples collected in the spring of 2000 under pine trees at four sites across a pollution gradient, total Pb, Zn, and Cd concentrations in soil were highest at the Cerro de San Miguel site in the DL, moderately high at the DL Cementerio and AJ sites, and lowest at the ZOQ and PC sites (Fig. 8.2). Linear regressions of soil concentrations of these three metals are significant, as expected based on the assumption that fossil fuel combustion was the major source of each of these elements. This data also suggests that, of the two DL sites, the Cerro de San Miguel site is more exposed to atmospheric pollution than the Cementerio study site. Ajusco, the other high-pollution site, appears to be similar in air pollution exposure to the Cementerio site. Nickel and Cu concentrations in soil were also higher in the DL and AJ sites compared to ZOQ and PC, the two low-pollution sites. Manganese concentrations did not differ between the sites, except that concentrations were lower at the PC site (Fig. 8.2).

Lead

Lead concentrations in soil at DL are clearly higher than in forested sites upwind of Mexico City or outside of the Basin experiencing low deposition. Watmough and Hutchinson (1999) reported total Pb concentrations ranging from 24.7 to 205 ppm in ten soil samples (0–5 cm depth) collected under sacred fir canopies

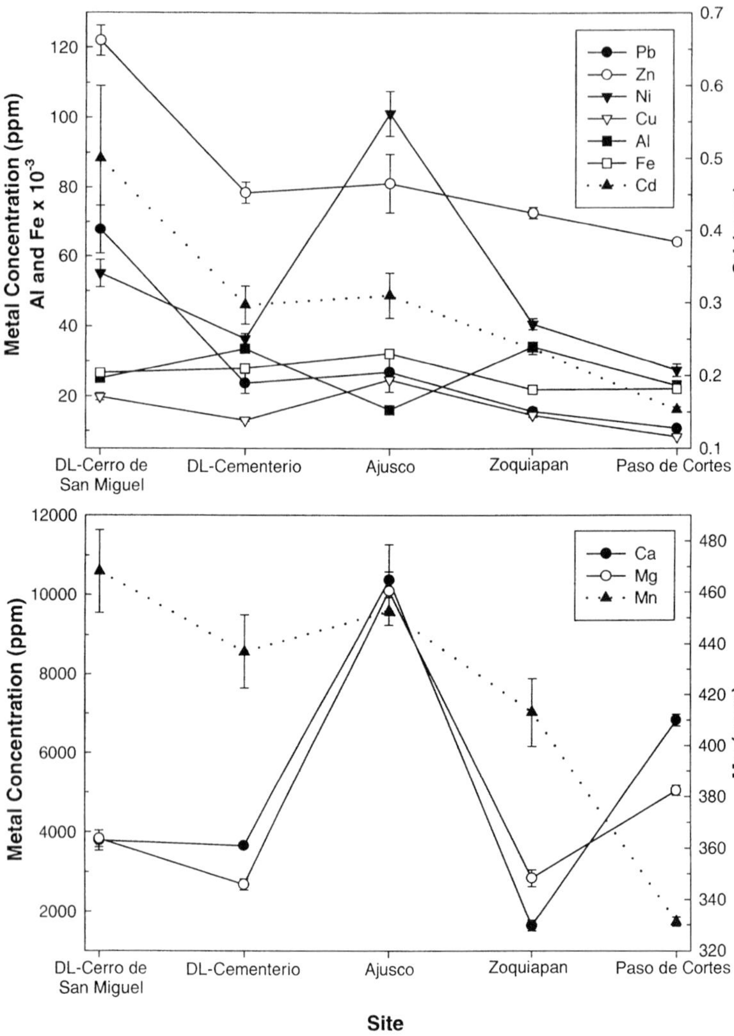

Figure 8.2. Average total concentrations of some heavy metals and other elements in soil collected in the spring of 2000 at sites across an air pollution gradient in the Basin of Mexico. DL, Desierto de los Leones.

at elevations between 2800 and 3200 m in the DL national park and total Pb concentrations from 9.6 to 11.2 ppm at the control site. Soil samples were collected within 1 m of the bases of fir trees between 98 and 135 years of age. Concentrations of DTPA-extractable Pb (52–61 ppm) (Table 8.2), from samples collected in the park in 1993 from the top 30 cm of soil along a linear transect within the forest are very high for extractable Pb levels in soil, suggesting that the values are overestimated. Furthermore, Coutiño-Alvarado (1989) reported an average value of 2.7 ppm extractable Pb (ranged from 0.02 to 30.1 ppm) in soil under three tree species in an urban site in southern Mexico City. Nonetheless, the data in Table 8.2 corroborate other data showing higher Pb concentrations at DL compared to low-pollution sites. Total Pb concentrations in the top 15 cm of soil collected under pine canopies at two locations in the DL park in 2000 were 68 and 24 ppm, compared with 11 to 15 ppm in soil under pine trees in low-pollution sites in the MCAB (Fig. 8.2). By comparison, 10 to 20 ppm of total Pb is considered as an average value for uncontaminated soils (Freedman and Hutchinson 1981; Nriagu 1978), but average values can vary depending on the type of soil and parent material.

Metal concentrations are expected to be higher directly under the canopies of trees as a result of stemflow and throughfall deposition of metals. The high levels of N and S deposition occurring in the park (Fenn et al. 1999) also contribute to high metal concentrations. Atmospheric deposition and nutrient cycling processes in soil under the canopy can be expected to increase soil acidity compared to open areas of the stand, thus increasing the solubility and availability of metals under tree canopies (Skeffington 1983). The data of Watmough and Hutchinson (1999) suggest that Pb concentrations in soil under sacred fir canopies are up to 20 times greater in the DL site than in forests outside of the Basin or upwind of Mexico City. Watmough and Hutchinson (1999) concluded that Pb concentrations in surface soils in the DL were similar to those occurring in the forest floor of high elevation spruce–fir forests in North America (Friedland et al. 1992) and in urban woodlands in Brussels (Pizl and Josens 1995), New York (Pouyat and McDonnell 1991), and Toronto (Watmough et al. 1998). Total Pb concentrations in soil under *Pinus radiata* trees in an urban area in southern Mexico City were 70.7 ppm (Coutiño-Alvarado 1989), similar to those reported in the DL Cerro de San Miguel site (Fig. 8.2) and within the range of values reported by Watmough and Hutchinson (1999). However, concentrations in foliage and roots of *Pinus radiata* were comparatively low (2.6 and 1.2 ppm) (Coutiño-Alvarado 1989).

Cadmium

Cadmium (Cd) concentrations in soil in the DL park (0.39–1.34) were significantly higher than in an unpolluted control site (0.18–0.21) (Watmough and Hutchinson 1999). We also found significantly higher concentrations in soil from the Cerro de San Miguel site in the park (0.50 ppm) than in other sites in the Basin. The lowest Cd concentrations (0.15 and 0.24 ppm) were at the low-pollution sites upwind of Mexico City (PC and ZOQ, respectively) (Fig. 8.2). Higher Cd concentrations at the sites downwind of Mexico City are possibly a result of a combination of industrial and vehicle emissions.

Iron, Zinc, and Manganese

Extractable Mn levels in soil were similar among all the sites sampled in 1993 (Table 8.2) as were total Mn levels in the sites sampled in 2000, except for lower concentrations at PC (Fig. 8.2). Extractable Zn levels in soil were also fairly similar among the sites. Watmough and Hutchinson (1999) reported that total Zn concentrations in the top 5 cm of soil under sacred fir canopies in the park ranged from 74 to 218 ppm compared to levels of 48 to 69 ppm in the control site. Average total Zn concentrations in the top 15 cm of soil under *P. hartwegii* canopies at the two DL sites were 122 and 78 ppm (Fig. 8.2), within the range of those reported by Watmough and Hutchinson (1999) for soil under sacred fir canopies in the DL. Concentrations of total Zn in soil were highest at the DL Cerro de San Miguel site and lowest at the low-pollution PC site. However, concentrations at ZOQ were similar to those at the DL Cementerio and AJ sites located downwind of Mexico City. Thus, there is some evidence of Zn accumulation in soils of the more exposed regions of the DL. Alvarado et al. (1993) reported DTPA-extractable concentrations in soil of Fe (58–88 ppm), Zn (8–20 ppm), and Mn (6–24 ppm) in the "Cemetery" area of the DL. Concentrations of the same metals in 1993 samples reported by Castro (1997, 1999) (Table 8.2) ranged from 6 to 40 ppm for Fe, 3 to 7 ppm for Zn, and 3 to 4 ppm for Mn. Extractable Fe concentrations were higher in soil at the Cementerio and Convento sites than at the other two sites in the DL or at the control sites (Table 8.2). Total Fe concentrations in soil were fairly constant (22–32 ppm) at sites across the air pollution gradient in the Basin (Fig. 8.2).

Aluminum and Base Cations

Aluminum (Al) concentrations in soil from the Agua de Leones and Cruz de Colica pine forest sites were 135.5 and 141.1 ppm. These are 2.3 to 3.9 times greater than concentrations at the other three sites within the park. Soil pH was lowest (5.0) at the Agua de Leones and Cruz de Colica sites, which are also the two highest elevation sites (Table 8.2). Castro et al. (1997) concluded that soil may be more acidic at these sites due to higher atmospheric deposition loads at these elevations. Soil acidification as a result of acidic deposition mobilizes Ca and Mg as well as Al. These conditions lead to nutrient cation losses and greater Al availability and plant uptake (Lawrence et al. 1996; Shortle et al. 1997). The soils and foliar data for the high elevation pine sites (Tables 8.2 and 8.3) suggest this may be occurring in the DL (Castro et al. 1997). Levels of Ca and Mg in the soil of pine stands was 3.4 and 6.2 times lower in the DL pine sites compared to the control site, while soil Al concentrations were 10.6 times higher in the DL sites (Castro et al. 1997).

No evidence was found for similar cation trends in soil and vegetation in sacred fir stands in the DL park, although Al was higher in DL soil than at the control site (but at least three times lower than at the higher elevation pine sites) (Castro et al. 1997). However, López-López et al. (1998) did find significantly lower levels of soil Ca, but not soil Mg, in a sacred fir plantation in the DL compared to levels found throughout the normal range of sacred fir. These authors also suggested the possibility that elevated atmospheric deposition may be causing excessive leaching of Ca from these soils. However, foliar levels of Ca and Mg in

Table 8.3. Average concentrations of some heavy metals and other elements in the vegetation of the Desierto de los Leones Park. D.F. and San Juan Tetla, Puebla

Site	Sample type[a]	Ca %	Mg %	K %	Fe[a] μg/g	Pb μg/g	Cu μg/g
Cementerio	*P. ayacahuite* young leaves	0.62	0.07	1.9	189.0	27.0	12.0
	P. ayacahuite old leaves	1.20	0.10	.73	211.3	23.5	2.5
	P. ayacahuite stem	0.66	0.10	.76	586.0	50.0	7.0
	P. montezumae young leaves	0.96	0.12	2.00	160.0	44.0	5.3
	P. montezumae old leaves	0.28	0.10	1.08	212.0	32.0	4.7
	P. montezumae stem	0.20	0.02	1.26	232.0	34.0	7.2
	Senecio sp. leaves	1.06	0.07	0.39	54.5	10.0	7.0
	Senecio sp. stem	3.00	0.07	0.92	544.0	40.0	20.0
	Vinca minor leaves	1.86	0.04	2.48	434.0	48.0	6.0
	Vinca minor stem	3.72	0.08	0.80	702.0	24.0	4.0
	Senecio sp. leaves	2.94	0.18	0.42	742	30	22
	Senecio sp. stem	3.0	3.0	2.56	442	56	2
Convento	*A. religiosa* young leaves	0.53	0.09	1.64	171.8	17.0	5.0
	A. religiosa old leaves	1.63	0.12	1.51	392.0	36.0	7.0
	A. religiosa stem	1.38	1.00	0.78	622.0	8.5	7.0
	Festuca sp. (grass foliage)	0.18	0.04	0.93	66.5	9.5	2.0
	Senecio sp. stem	3.16	0.66	1.18	422.0	20.0	34.0
	Acaena elongata leaves	1.99	0.74	0.58	626.0	36.0	14.0
	A. elongata stem	0.70	0.15	0.38	400.0	61.0	16.3
	Fuchsia microphylla leaves	1.86	0.54	4.19	528.0	46.0	20.0
	F. microphylla stem	2.48	0.38	3.49	538.0	24.0	10.0
	Vinca minor leaves	3.21	0.26	1.22	410.0	38.0	16.0
	V. minor stem	1.12	0.14	4.15	156.0	36.0	14.0
Agua de Leones	*P. hartwegii* young leaves	0.12	0.15	3.31	52.8	23.0	0.66
	P. hartwegii old leaves	0.09	0.05	1.31	299.0	11.3	1.20
	P. hartwegii stem	0.10	0.04	3.37	440.7	19.3	2.70
	Mühlenbergia sp.	0.10	0.01	3.64	173.5	17.7	3.00
Cruz de Colica	*P. hartwegii* young leaves	1.20	0.07	3.76	256.3	20.0	1.25
	P. hartwegii old leaves	0.27	0.01	1.34	354.0	29.3	2.70
	P. hartwegii stem	0.29	0.12	3.70	942.7	55.3	6.00
	Mühlenbergia sp.	0.47	0.02	1.26	175.0	22.0	3.00
San Juan Tetla, Puebla	*A. religiosa* young leaves	1.81	0.81	1.72	0.25	0.04	1.05
	A. religiosa old leaves	1.95	0.95	1.51	0.21	0.03	2.04
	A. religiosa stem	1.51	0.99	0.95	0.42	0.04	1.06
	Acaena elongata fruit	2.15	0.85	0.62	1.72	0.05	2.01
	Senecio sanguisorbae leaves & stem	1.74	0.64	0.91	0.03	0.03	0.93
	P. hartwegii young & old leaves &	1.49	0.72	1.70	0.08	0.01	1.04
	P. hartwegii stem	1.93	0.20	1.03	0.11	0.03	1.06

[a] Iron values at the San Juan Tetla, Puebla site are excessively low and considered to be in error. See text for discussion.

Data from Castro et al. (1997).

sacred fir growing in a young plantation were not deficient (López-López et al. 1998).

In a more recent study at five *P. hartwegii* sites (two in the DL) in the Basin of Mexico, the results at least partially supported earlier findings of a relationship between Al levels in soil and nutrient cation levels. At the Cerro de San Miguel DL site, exchangeable Al levels in soil were nearly seven times higher than in soil from the Cementerio DL site, while exchangeable Ca, Mg, and Sr levels were three-fold to 7-fold lower at the Cerro de San Miguel site (Fig. 8.3). No clear trends were observed in exchangeable cation levels in soil across the entire pollution gradient (Fig. 8.3), possibly because of inherent soil differences between the sites (e.g., PC had low available Al, Ca, and Mg), or it could be that the hypothesized effect of atmospheric deposition on Al availability and nutrient cation loss is only observable at the most highly exposed sites in the DL. In summary, several studies suggest the presence of greater available Al accompanied by decreases of Ca and Mg in soil in the DL. Further comparative and mechanistic studies of atmospheric deposition loading and soil nutrient cycling processes at these sites are needed before differences in soil pH, available Al, and nutrient cation losses can be more definitively attributed to air pollution.

Dendrochemical data from sacred fir trees in the park did not show the temporal patterns of decreasing Ca and Mg and increasing Al in tree rings in recent decades that has been observed in Europe and North America where base cation depletion and Al mobilization were found (Watmough and Hutchinson 1999). Aluminum concentrations in tree rings were nearly always higher in the DL compared with the control site, even in the early part of the 1900s.

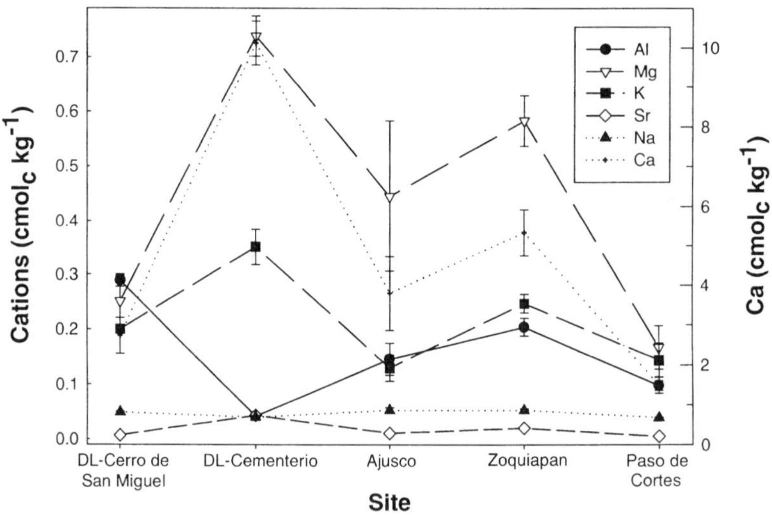

Figure 8.3. Concentrations of exchangeable cations in soil collected in the spring of 2000 at sites across an air pollution gradient in the Basin of Mexico. DL, Desierto de los Leones.

Spatial Variability in Metal Concentrations

Watmough and Hutchinson (1999) reported that spatial variability in trace metal concentrations in soil was high, and they concluded that the primary causal factor was variable canopy interception of pollutants and subsequent transfer to the forest floor. Altitudinal differences and variability in steepness of the slope, the amount and type of litter accumulation, and soil pH were also mentioned as possible factors. Soil pH was particularly important, and strong statistically significant negative correlations were reported between soil pH and concentrations of Pb, Cd, Zn, and Cu. Concentrations of Zn and Cu and, to a lesser degree, of Pb and Cd were also correlated with high organic matter content, suggesting that trace metals are associated with the acidic litter found beneath fir trees.

In a pine tree plantation located close to a paper mill in San Rafael east (upwind) of Mexico City, Riverón-Negrete (1994) reported very low levels of Pb, Zn, Cu, and Cr in the top 15 cm of soil, presumably because this site is upwind of Mexico City and there are no local emissions sources for these pollutants. However, levels of Al (122–200 ppm) and Fe (39–56 ppm) in soil were quite high at this site, presumably as a result of local emissions or as a result of soil acidification (pH 4.8–5.3; $CaCl_2$ method).

Heavy Metals in Forest Vegetation

There are few studies of heavy metals in forest vegetation in the Basin of Mexico. However, the data in Table 8.3 from Castro et al. (1997) indicate much higher heavy-metal concentrations (Fe, Pb, and Cu) in foliage and stems of trees and in understory species in the DL compared with a control site outside of the Basin. In a more recent study, Pb and Cu concentrations in pine and grass foliage were also higher at DL and AJ compared to most sites with lower air pollution exposure (Table 8.4). This was not true for Cu in grass at ZOQ.

Castro et al. reported a general trend of higher metal concentrations in stem tissue than in foliar tissue at DL (Castro et al. 1997). In pine and fir trees in the DL, Fe and, in some cases, Pb and Cu concentrations were usually higher in stem tissue than in foliage (Table 8.3). Levels of Fe, Pb, and Cu were higher in stem tissue of *Senecio* sp. than in foliage. Iron concentrations were also higher in stems of *Vinca minor* than in foliage.

Dendrochemical studies also confirm higher heavy metal concentrations in sacred fir trees at DL compared to similar stands at San Pablo, a control site upwind of Mexico City (Watmough and Hutchinson 1999). Lead and Cd was higher in tree rings formed since the 1960's. However, peaks in Cd and Pb were also found in rings formed between 1920 and 1940 corresponding to the heartwood–sapwood boundary, especially in trees growing in contaminated soils with low pH. The authors suggested that Pb and Cd taken up by roots of sacred fir accumulate in the heartwood, whereas Pb and Cd entering through the bark are transported radially to a much lesser extent and more accurately reflect

Table 8.4. Average concentrations of some heavy metals and other elements in current year foliage of *Pinus hartwegii* collected in fall 2000

Site		Pb ppm	Cd ppm	Ba ppm	Cu ppm	Sr ppm	Ti ppm	Zn ppm	Mn ppm	Fe ppm	Al ppm	Na ppm	Si ppm	Ca %	Mg %	K %
									Pine							
DL[a]	Average	6.09	0.083	2.85	8.70	4.45	4.87	44.3	72.2	87.7	301.1	51.3	175.5	0.194	0.141	0.536
	Std. Err.	0.63	0.021	0.60	1.16	0.95	1.42	5.3	15.6	15.7	39.5	9.5	65.8	0.020	0.015	0.013
AJ	Average	4.65	0.084	1.70	8.61	2.54	2.57	44.7	35.5	53.6	169.2	60.8	120.7	0.152	0.120	0.585
	Std. Err.	0.09	0.013	0.27	0.49	0.60	0.32	5.7	1.5	4.8	23.5	4.9	9.5	0.015	0.020	0.096
ZOQ	Average	3.39	0.039	1.23	6.01	3.37	1.37	36.9	89.7	36.1	204.1	54.8	141.2	0.146	0.131	0.622
	Std. Err.	0.11	0.009	0.23	0.92	0.85	0.33	2.1	17.1	2.3	29.1	12.5	74.5	0.024	0.005	0.048
PC	Average	3.64	0.044	1.62	7.33	4.70	2.27	39.2	46.9	55.1	188.0	71.0	241.4	0.180	0.138	0.580
	Std. Err.	0.89	0.022	0.33	1.06	0.73	0.62	4.4	6.2	5.8	30.8	9.8	96.1	0.013	0.013	0.022
									Grass							
DL[a]	Average	2.30	0.255	8.31	5.65	7.01	17.38	19.6	75.1	212.3	326.2	63.8	576.4	0.150	0.046	0.460
	Std. Err.	0.35	0.083	0.34	0.23	1.43	3.58	2.3	3.6	35.0	59.1	7.6	59.3	0.024	0.007	0.033
AJ	Average	2.25	0.277	9.85	6.40	10.29	22.45	36.8	38.7	268.9	362.2	84.1	1085.5	0.201	0.067	0.559
	Std. Err.	0.31	0.067	1.54	1.05	1.03	6.72	6.9	4.0	66.3	89.2	15.1	498.9	0.022	0.012	0.105
ZOQ	Average	0.55	0.122	12.33	7.34	13.09	14.36	19.4	84.7	178.2	301.0	74.4	393.5	0.202	0.075	0.456
	Std. Err.	0.16	0.021	0.74	2.60	0.38	2.86	2.1	17.1	29.2	62.1	10.5	77.9	0.002	0.004	0.025
PC	Average	0.97	0.029	10.39	4.40	9.95	25.65	32.1	43.9	323.7	451.1	93.4	2139.5	0.147	0.034	0.344
	Std. Err.	0.09	0.015	1.14	0.49	1.01	2.66	8.0	8.4	38.8	54.2	2.0	87.8	0.012	0.002	0.026

[a] This is the Cerro de San Miguel site within the Desierto de Leones National Park.
DL, Desierto de los Leones; AJ, Ajusco; ZOQ, Zoquiapan; PC, Paso de Cortes.

changes in trace metal deposition. The accumulation of Cd and Pb in the heartwood–sapwood boundary limits the usefulness of this technique, at least with *A. religiosa*. However, it is possible that this technique may be more appropriate in other species as a methodology for determining temporal patterns in pollutant uptake in forests of the MCAB. Bargagli (1998) suggested that whenever possible, ring-porous trees such as oaks, elms, and ashes should be used for these types of studies, because most of the water moves through only the current year's wood vessels in these species.

Lead

Prior to the introduction of unleaded fuel in 1990 and the phaseout of leaded fuel in 1997, Pb was a major air pollution problem in the MCAB (Bravo-Alvarez and Torres-Jardon 2002). It is believed that atmospheric deposition inputs of Pb to DL and other sites downwind of Mexico City during this time were very high. Lead concentrations in tree rings of *A. religiosa* in the DL have increased since the 1960s and were much greater than at a control site (Watmough and Hutchinson 1999). Lead concentrations in lichens averaged 12.8 ppm at a control site compared to 39.4 at the DL (Carrillo-Padrón 1993; Zambrano et al. 2002). Similarly, Pb concentrations in a moss (*Leptodontium viticulosoides*) which occurs naturally in *Abies religiosa* stands in the Basin ranged from 100 to 180 ppm in the region encompassing AJ and the DL (downwind of Mexico City). By comparison, in study areas upwind of Mexico with less air pollution exposure (the Izta-Popo National Park to the southeast and Monte Bajo to the northwest), Pb concentrations in moss samples were less than 40 ppm (Zambrano 1991).

Lead concentrations in pine foliage collected in 2000 were higher at DL (6.1 ppm) and AJ (4.7 ppm) compared to concentrations at sites within the MCAB with lower air pollution exposure (3.4–3.6 ppm) (Table 8.4). However, Pb concentrations in the pine foliage and grass samples (2.3 ppm) collected at DL in 2000 were approximately one third to one twelfth the values for pine foliage (20–27 ppm) and grass (*Mühlenbergia* sp.) (18–22 ppm) collected in 1993 (Table 8.3). In both cases, the values represent total plant tissue concentrations from digested samples, and concentrations at DL were significantly greater than at the low-pollution sites. However, the samples were analyzed in different laboratories, making the extent to which absolute differences in plant concentrations between the two dates are due to methodology unclear. Thus it is also unclear whether more recent samples are in fact lower in Pb content as a result of the introduction of unleaded gasoline in 1990 and the elimination of leaded fuel in 1997. By comparison, foliar Pb concentrations in *Pinus sylvestris* L. at low, intermediate, and high pollution sites in Poland were 1.29, 3.49, and 10.15 ppm (Dmuchowski and Bytnerowicz 1995).

Cadmium

The phytotoxicity of Cd is well documented (Påhlsson 1989). Temporal trends for Cd in tree rings in the DL park clearly demonstrate much greater plant uptake of Cd compared to the unpolluted control site. Cadmium concentrations in tree rings increased greatly after the late 1960s (Watmough and Hutchinson 1999), coinciding with the exponential increase in vehicle traffic. In foliar samples collected in 2000 across the pollution gradient in the Basin, there was a clear trend of higher Cd levels in pine (0.08 versus 0.04 ppm) and grass (0.26–0.28 versus 0.03–0.12 ppm) foliage at the DL and AJ sites compared to the low pollution sites ZOQ and PC (Table 8.4). Although Cd concentrations in pine foliage at DL and AJ were twice as high as in the low pollution sites, concentrations were at least an order of magnitude lower than levels reported to be phytotoxic to sensitive coniferous species (Påhlsson 1989). By comparison, foliar Cd concentrations in gymnosperms growing in uncontaminated sites are normally approximately 0.2 ppm (Freedman and Hutchinson 1981), and Cd concentrations in foliage of *Pinus sylvestris* L. at low, intermediate, and high pollution sites in Poland were 0.25, 0.61, and 1.53 ppm (Dmuchowski and Bytnerowicz 1995). Cadmium levels in exposed lichens were not different between a control site and the DL (0.5 ppm at both sites), although at a site near the industrial zone in northwest Mexico City, Cd levels in lichens were 0.9 ppm (Carrillo-Padrón 1993; Zambrano et al. 2002). From the available data, we conclude that plant uptake of Cd is elevated at sites with high air pollution exposure such as DL and AJ, but levels in plant tissue appear to be well below the threshold for observable phytotoxic effects.

Aluminum

Little data are available on Al concentrations in plants in the MCAB. Among the four sites sampled in the MCAB in 2000, Al concentrations appear to have been higher in pine foliage at the Cerro de San Miguel site in the DL. However in grass, concentrations tended to be higher at PC, a low-pollution site. On the other hand, Al concentrations in grass foliage were highly variable even within the same site (Table 8.4). In the dendrochemical study of Watmough and Hutchinson (1999) much higher concentrations of Al were found in in the rings of sacred fir trees from DL compared with the control site in most years. Significantly, they did not find temporal patterns of decreasing Ca and Mg and increasing Al concentrations in later years as reported for forest decline areas in Europe and North America. The relatively high base cation content of forest soils in the Basin has presumably ameliorated the excessive acidification and base cation depletion of some forest soils in the Basin, although as discussed in the previous section, preliminary data suggest that soil acidification, greater Al availability, and cation depletion may have occurred to some degree at higher elevation pine sites in the DL.

Zinc

Results of a biomonitoring study of lichens demonstrated high Zn inputs and bioavailability in the park. Zinc concentrations in lichens were 83.9 μg/mg dry weight at the DL compared to 35.2 μg/mg at el Chico National Park, outside of the Basin (Carrillo-Padrón 1993; Zambrano et al. 2002). The Zn concentrations in plant tissues reported by Castro et al. (1997) seem to be in error and are therefore not included in Table 8.3. Concentrations reported for foliage and stems from the DL were unusually high, ranging from 327 to 1827 ppm. On the other hand, concentrations reported in plant tissue (Castro et al. 1997) in the control site in San Juan Tetla (0.1–0.9 ppm) were extremely low and certainly in error, considering that the normal range for plant tissue is 15 to 100 ppm (Raven et al. 1981) and that for pine species, approximately 35 ppm, is considered normal (Bargagli 1998).

Zinc concentrations in pine foliage from the four sites along the air pollution gradient ranged from 37 to 45 ppm, showing only a slight trend of increasing values at DL and AJ. On the other hand, there was no clear relationship between Zn levels in grass tissue (19–37 ppm) and air pollution exposure at these same four sites (Table 8.4). López-López et al. (1998) reported a median Zn concentration of 41 ppm in a seven-year-old fir plantation in the DL, and Alvarado et al. (1993) reported an average Zn concentration of 29 ppm in fir foliage at DL. These values were within the expected range. By comparison, foliar Zn concentrations in *Pinus sylvestris* L. at low, intermediate, and high pollution sites in Poland were 40.9, 49.9, and 173.7 ppm (Dmuchowski and Bytnerowicz 1995). Critical levels, the minimum concentrations at which toxic effects occur, vary from 60 to 900 ppm, but most species are sensitive to the 200 to 300 ppm range (Påhlsson 1989). The critical value for *Picea sitchensis* was 226 ppm in foliage (Jordan 1975), and, the critical value was 250 ppm for *Quercus rubra* (Burton et al. 1983). Both levels are several times greater than those in leaves in unpolluted areas. In conclusion, Zn concentrations in plant tissue at DL do not appear to be within the range expected to cause toxic effects.

Copper

At high soil concentrations, Cu can be extremely phytotoxic. Copper concentrations in young pine foliage from the 1993 samplings in the DL ranged from 0.7 to 12.0 ppm compared with 1.0 ppm in pine foliage at the control site (Table 8.3). Concentrations as high as 34 ppm were reported in stem tissue. Average Cu concentrations in pine foliage collected across the pollution gradient in 2000 were 8.6 and 8.7 ppm at DL and AJ compared with 6.0 and 7.3 ppm at two low-pollution sites (Table 8.4). By comparison, Cu concentrations in *Pinus sylvestris* L. foliage at low, intermediate, and high pollution sites in Poland were 2.92, 3.49, and 3.24 ppm (Dmuchowski and Bytnerowicz 1995). Copper concentrations in foliage typically range between 2 and 20 ppm (Breckle 1991). Most plant species are affected by foliar Cu concentrations between 15 and 25 ppm (Påhlsson 1989), but others (e.g., *Picea sitchensis*) tolerate needle concentrations as high as 88 ppm (Burton et al. 1983). Tissue concentratrions of Cu in epiphytic lichens were 6.5 ppm at a control site compared with 15.3 at the DL site (Carrillo-Pardón 1993; Zambrano et al. 2002), further suggesting pollution inputs of Cu at some forested sites within the Basin. However, the available data, suggest that Cu levels in forest vegetation in the Basin of Mexico are not at phytotoxic levels.

Iron

Iron levels in current-year pine and fir foliage at DL (53–256 ppm) were also much higher than at San Juan Tetla, the control site, (0.08–0.25 ppm) (Table 8.3). However, the Fe values for San Juan Tetla were extremely low compared to the normal ranges in plants (described later) and are almost certainly in error. Alvarado et al. (1993) measured Fe levels in sacred fir foliage in the DL and reported concentrations (175–363 ppm) that were similar to those reported for DL (172–392 ppm) by Castro et al. (1997) for 1993 samples (Table 8.3). Iron concentrations in foliage of *P. hartwegii* sampled in 2000 were significantly higher at DL (88 ppm) than at the other three sites studied (36–55 ppm) (Table 8.4). In this same study, Fe concentrations in grass varied from 178 to 324 ppm at the four sites, but concentrations were not significantly different between the sites. For a variety of plant species, Fe concentrations range from 60 to 1500 ppm (Breckle et al. 1991), although 100 ppm is considered adequate as a general rule (Salisbury and Ross 1985) and for pine species (Bargagli 1998).

That Fe concentrations may be somewhat elevated in tissue of some plant species at DL is also supported by biomonitoring results with epiphytic lichens. Concentrations in lichens in the DL were 1788 ppm compared with 1388 ppm at the control site (Carrillo-Padrón 1993; Zambrano 2001). In summary, Fe levels, at least for some species, may be elevated in plant tissue in the DL, but these Fe concentrations are not believed to be phytotoxic. Iron toxicity is mainly a problem in flooded soils such as in rice paddies (Marschner 1995).

Chromium

The importance of Cr as a pollutant in the DL is not clear, largely because the available data is insufficient. In the lichen study of Carrillo-Pardón (1993) and reported in Zambrano et al. (2002), Cr concentrations in lichen tissue were 2.5 and 4.8 ppm at the control site (El Chico National Park) and in the DL respectively. However, Watmough and Hutchinson (1999) found significantly higher Cr concentrations in soil at their control site (San Pablo, 45 km northeast of Mexico City) than in the DL.

Biological Effects of Metals

We are not aware of any studies investigating the effects of heavy metals on plant growth or other biological processes in forests of the Basin of Mexico. Thus we cannot definitively identify the ecological effects of heavy metals in the Basin. Based on plant tissue concentrations of Pb shown in Table 8.3 (8.5–61.0 ppm), it may be that some plant species have been significantly impacted. However, Pb concentrations from samples collected in 2000 were much lower (4.7–6.1 ppm at DL) (Table 8.4). A concentration of 19 ppm Pb (by dry weight) in needles of *Picea sitchensis* caused a biomass decrease (Burton et al. 1983), whereas a concentration of 35 ppm was needed to alter growth in barley (Davis et al. 1978). Impeded root development and altered root architecture are commonly the first, or at least the most conspicuous, symptoms of heavy-metal injury in plants (Breckle 1991; Davies 1991; Tyler et al. 1989). The biomass of seedlings of *Fagus sylvatica* was relatively unaffected by Pb concentrations similar to those in soil in the DL, although root development was changed by Pb from a loose to a more compact and more branched root system (Breckle 1991).

The phytotoxicity of Pb is lowered by the limited availability and uptake of Pb from soil (Påhlsson 1989). High Ca concentrations, such as those in soils of the Basin of Mexico, reduce Pb toxicity, and apparently the toxicity of other metals as well (Davies 1991). Lead uptake by plants is favored by low soil pH, low cation exchange capacity, low organic matter, and low phosphate levels (Zimdahl and Koeppe 1979). Lead taken up from soil commonly accumulates in roots, but translocation to aerial shoots is often limited due to binding at root surfaces and cell walls. Lead phosphate deposits may partially explain the limited toxicity of Pb in some plants (Påhlsson 1989; Zimdahl and Hassett 1979).

As a general rule, vascular plants seem to be less sensitive to heavy-metal pollution than other biological groups such as soil arthropods and some microbial processes (Tyler et al. 1989). In the forest floor, N transformation processes can be affected at levels that are only three times background heavy-metal concentrations (Bååth 1989; Tyler et al. 1989). Many studies have shown that litter decomposition is retarded by elevated levels of heavy metals (Laskowski et al. 1994). Thus, in forests of the Basin of Mexico, and more especially in highly exposed sites such as the DL, the literature suggests that sensitive organisms and processes, and possibly some sensitive plant species as well, may have been affected to some degree. However, these speculative conclusions must be confirmed with studies in the Basin, and it must be kept in mind that emissions of Pb, and possibly emissions of other metals, are believed to have decreased in recent years. On the other hand, while Pb emissions from fossil fuel combustion has undoubtedly decreased, the increase in vehicle miles in the region may have resulted in sustained or greater emissions of metals such as Cd, Zn, Ni, V, Cr from the wear of tires and metallic automotive parts (Freedman and Hutchinson 1981) in addtion to industrial emissions in the Basin.

Decreased Foliar Calcium and Magnesium

Calcium concentrations in pine foliage (old and current-year foliage concentrations combined) were twice as high at the control site in San Juan Tetla as at the DL sites. In young sacred fir needles, Ca concentrations were 3.4 times higher at the control site, while Ca concentrations in older foliage levels were only 1.2 times higher than in the foliage of trees in the DL (Table 8.3). Nonetheless, Ca levels in fir needles were within the range considered normal for plants in general (Salisbury and Ross 1985), and López-López et al. (1998) concluded that Ca levels in fir foliage in the Cementerio plantation site within the DL were sufficient. Calcium levels in *Pinus hartwegii* foliage in the DL sites were also within the normal range, except for a value reported for the Agua de Leones site (0.12%) which was approximately half typical values for pine foliage (Huettl et al. 1990;

Salisbury and Ross 1985) and 12.4 times lower than Ca concentrations in pine foliage in San Juan Tetla (Table 8.3).

Concentrations of Mg in pine foliage (old and current-year foliage concentrations combined) were 7 to 18 times higher at the control site in San Juan Tetla compared to the DL sites. Young and old fir needles from the control site had magnesium concentrations 9.0 and 7.9 times higher respectively than the foliage from the DL (Table 8.3). Magnesium levels in pine and fir foliage at the DL sites (Table 8.3) were at the low end of the typical range in some cases (Salisbury and Ross 1985), although López-López et al. (1998) reported that Mg levels were normal in young fir trees in the Cementerio site. Potassium levels were generally within the normal range for all species at all sites.

The data from pine and grass samples collected at four sites across an air pollution gradient in 2000 do not support the hypothesis that Ca and Mg levels in foliage in the high pollution forest sites has decreased. Foliar cation levels were similar at all four sites despite their different pollution levels (Table 8.4). Levels of Ca, Mg, and K in pine foliage were within the range of values expected for pine foliage (Huettl et al. 1990; Salisbury and Ross 1985). Thus, the evidence is not consistent in showing reductions in foliar cation concentrations in the high-pollution sites. The evidence, to date, from several studies does, however, generally support a possible relationship between higher available Al in soil and reduced exchangeable Ca and Mg. Further studies will be necessary to verify if greater soil acidification, Al mobilization, and loss of cations are occurring in the high-elevation pine sites at DL and to test the hypothesis that atmospheric deposition is a causal factor in this phenomenon.

Lead in Drainage Waters

In a hydrogeological study in the AJ mountains located south of Mexico City and to the east of the DL (Morales 1998), average Pb concentrations in drainage water samples collected in 1996 ranged from 0.168 to 0.259 ppm, an order of magnitude greater than the drinking water standard for Mexico (0.025 ppm). By comparison, streamwater concentrations of Pb averaged 0.6 ppm in a high-elevation forest in northern Vermont (United States) with a high level of Pb deposition (700 g/ha/yr) (Friedland and Johnson 1985). Average Fe and Mn concentrations in water in the AJ mountains ranged from 0.073 to 0.117 ppm (Fe), and 0.008 to 0.032 ppm (Mn), values which are lower than the drinking water standard for Mexico (0.30 for Fe and 0.15 ppm for Mn). Similarly, we found low concentrations of Fe and Mn in streamwater in the Basin. Average Fe concentrations were 0.033 to 0.036 ppm at AJ and DL, and Mn concentrations were 0.003 and 0.007 ppm. The thin litter layer in some forested areas in the Basin of Mexico may contribute to problems with Pb in drainage waters that occur because of decreased Pb retention in soil. However, this has not been studied in the Basin of Mexico.

The data from the AJ mountains (Morales 1998) suggest that Pb may be leached from forests downwind (SSW) of Mexico City at levels unhealthful for human consumption. Lead concentrations were often near background levels (ca. 1 ppb) in 21 streamwater and springwater samples collected from DL, AJ, ZOQ, or PC at various times from January, 2000, to September, 2000. However, in four of the samples, two each from DL and AJ, values were greater than 2 ppb (Fig. 8.4). The highest value measured in this study (13.0 ppb) was only a little more than one twentieth of the highest value (259 ppb) reported by Morales (1998) but was 52% as high as the Mexican drinking water standard for Pb (25 ppb). These same water samples from DL and AJ also had elevated nitrate concentrations as a result of high atmospheric inputs of nitrogen from air pollution originating in greater Mexico City (Fenn et al. 1999; Fenn et al. 2002). In summary, Pb levels do appear to be higher in streams draining watersheds with high air pollution exposure, but further studies are needed to determine the geographic extent and risk potential for Pb contamination of runoff from these forested watersheds. Although Pb emissions have decreased greatly since 1990, Pb accumulation in litter and soil could possibly be a source of Pb in runoff for many years after atmospheric inputs have decreased. In forests of the northeast United States, it has been reported that, following decreasing trends in atmospheric deposition of Pb as a result of the introduction of unleaded fuels in 1974, Pb has been slowly migrating from the forest floor to the mineral soil fraction, with minimal export into streamwater (Friedland et al. 1992). However, this accumulated Pb, which is being redistributed in the soil profile, could possibly begin

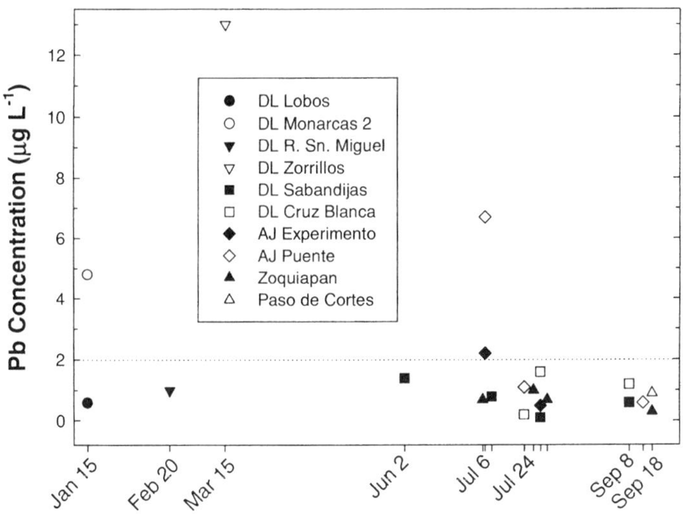

Figure 8.4. Concentrations of Pb in drainage waters from forested watersheds at four sites across an air pollution gradient in the Mexico City Air Basin. Samples were collected from January, 2000 to September, 2000. The first four Desierto de los Leones Forest Part (DL) sites listed in the figure legend are springs. AJ, Ajusco.

to be released to streams in greater amounts within the next 50 years (Miller and Friedland 1994).

Summary

The results of previous studies of heavy-metal concentrations in soil, plant tissue, and lichens in the highly-polluted DL National Park and at control sites are reviewed in this chapter. New data on heavy-metal concentrations at four sites spanning a range of air pollution exposure in the MCAB are also reviewed. The DL is the most exposed site, followed by the AJ mountains. ZOQ and PC are the two relatively low-deposition sites sampled in this study. Biomonitoring of lichens and mosses, Pb emissions data, dominant wind patterns in the Basin of Mexico, and data on heavy-metal concentrations in soil and vegetation indicate that large amounts of heavy metals have been atmospherically deposited to forested areas to the SSW of Mexico City. Elevated concentrations of the following metals have been documented in soil and/or plant samples in forested sites downwind of greater Mexico City: Pb, Fe, Zn, Al, Cd, and Cu. Data on tree rings and soil under sacred fir trees in the DL indicated that Ni levels were not elevated compared to a control site, but we found evidence of elevated Ni in forest soil at DL and AJ. Soil concentrations of Cr were not elevated in the DL, but Cr levels in lichen tissue from the DL park were higher than at a control site.

Preliminary evidence suggests a relationship between greater soil acidity, Al mobilization, and loss of Ca and Mg at high-elevation pine sites in the DL. More research is needed to test the validity of these findings, to determine the ecological impacts of this phenomenon, and to determine if atmospheric deposition is a causal factor. A hydrogeological study in the AJ a south of Mexico City indicated that Pb concentrations in springwater from these forested watersheds exceed the Mexican drinking water standard (25 ppb). However in a subsequent study investigating of Pb in springwater and streamwater samples from DL, AJ, ZOQ, and PC, only 4 of 21 samples were greater than 2 ppb. The highest value (13.0 ppb) was 52% as high as the Mexican drinking water standard for Pb. Further studies are warranted to more fully evaluate the severity of Pb export into drainage waters of these forest watersheds, because these watersheds are an important drinking water source for Mexico City. Atmospheric emissions of Pb, and possibly of some other metals, have decreased dramatically as a result of the introduction of unleaded gasoline in 1990, but little is known of the potential ecological and biological effects of Pb and other metals which have accumulated in these systems. Furthermore, current deposition rates of heavy metals from sources such as the wear of tires and metallic automotive parts, from fossil fuel combustion, and from industry in the Basin have not been measured. The available data suggests that current levels of heavy metals in forest vegetation are not sufficient to cause major direct phytotoxicity effects. However, the effects on more sensitive species, organisms, microbial processes, and water quality are largely unknown.

References

Aguilera, N. 1962. *Suelos: Problema Básico de Silvicultura. Seminario y Viaje de Estudio de Coníferas Latinoamericanas.* México, D.F.: SAG (Secretaría de Agricultura y Ganadería). Publicación especial INIF (Instituto Nacional de Investigaciones Forestales) #1.

Alvarado, D., de Bauer, L.I., and Galindo, J. 1991. Declinación y muerte del bosque de oyamel (*Abies religiosa*) en el sur del Valle de México. *Agrociencia* 1:130–143.

Alvarado, D., de Bauer, L.I., and Galindo, J. 1993. Decline of sacred fir (*Abies religiosa*) in a forest park south of Mexico City. *Environ. Pollut.* 80:115–121.

Alvarado-Rosales, D., and Hernández-Tejeda, T. 2002. Decline of sacred fir in the Desierto de los Leones National Park. In *Urban Air Pollution and Forests: Resources at Risk in the Mexico City Air Basin*, Ecol. Stud., eds. M.E. Fenn, L.I. de Bauer, and T. Hernández-Tejeda, New York: Springer-Verlag.

Bååth, E. 1989. Effects of heavy metals in soil on microbial processes and populations (A review). *Water Air Soil Pollut.* 47:335–379.

Bargagli, R. 1998. *Trace Elements in Terrestrial Plants: An Ecophysiological Approach to Biomonitoring and Biorecovery.* Heidelberg, Germany: Springer-Verlag and R.G. Landes Company.

Beltrán, E., and Vázquez, R. 1971. *En Defensa del Parque Nacional Desierto de los Leones*, Folleto 36. México, D.F.: Instituto Mexicano de Recursos Naturales Renovables (IMRNR).

Bravo-Alvarez, H. 1987. *La Contaminación del Aire en México.* Mexico, D.F.: Ed. Fundación Universo Veintiuno.

Bravo-Alvarez, H., and Torres-Jardón, R. 2002. Air pollution levels and trends in the Mexico City metropolitan area. In *Urban Air Pollution and Forests: Resources at Risk in the Mexico City Air Basin*, Ecol. Stud., eds. M.E. Fenn, L.I. de Bauer, and T. Hernández-Tejeda, New York: Springer-Verlag.

Breckle, S.-W. 1991. Growth under stress. In *Plant Roots: The Hidden Half*, eds. Y. Waisel, A. Eshel, and U. Kafkafi, pp. 351–373. New York: Marcel Dekker, Inc.

Burton, K.W., Morgan, E., and Roig, A. 1983. The influence of heavy metals upon the growth of sitka-spruce in South Wales forests. I. Upper critical and foliar concentrations. *Plant Soil* 73:327–336.

Carrillo-Padrón, V.H. 1993. *Líquenes Foliosos como Biomonitores de Contaminación Atmosférica por Metales Pesados en los Alrededores de la Zona Metropolitana de la Ciudad de México (ZMCM).* B.Sc. Thesis. México, D.F.: Facultad de Ciencias, Universidad Nacional Autónoma de México.

Castañeda-González, M.I., López-López, M.A., and Velásquez-Martínez, A. 1995. Efecto de tratamientos de fertilización sobre algunos síntomas de declinación en una plantación de oyamel. In *Resúmenes del II Congreso Mexicano sobre Recursos Naturales*, pp. 52. Montecillo, Estado de México, México: Colegio de Postgraduados.

Castro, J.M. 1987. Estudio microbiológico de los suelos del parque cultural y recreativo Desierto de los Leones, D.F. In *Primera Reunión Científica Forestal y Agropecuaria*, pp. 10. México, D.F.: Centro de Investigación Forestal, Agrícola y Pecuaria del Distrito Federal (CIFAP-D.F.) and Instituto Nacional de Investigaciones Forestales y Agropecuarias-Secretaría de Agricultura y Recursos Hidráulicos (INIFAP–SARH).

Castro, J.M., González, V., and Hernández-Tejeda, T. 1993. Determinación preliminar de algunos metales pesados en los suelos del Desierto de los Leones. In *I Congreso Mexicano Sobre Recursos Forestales*, p. 59. Universidad Autónoma Agraria Antonio Narro, Saltillo, Coahuila, México: Sociedad Mexicana de Recursos Forestales, A.C.

Castro, J.M., González, V., and Hernández-Tejeda, T. 1995a. Metales pesados en los suelos del Desierto de los Leones, Distrito Federal. Revista Ciencia Forestal en México 20: 101–111.

Castro, J.M., González, V., and Hernández-Tejeda, T. 1996. Metales pesados en los suelos del Desierto de los Leones. In *Primera Reunión Científica Forestal*, pp. 10. México, D.F.: Centro Nacional de Investigación Disciplinaria en Conservación y Mejoramiento de los Ecosistemas Forestales (CENID-COMEF) and Instituto Nacional de Investigaciones Forestales y Agropecuarias–Secretaría de Agricultura, Ganadería y Desarrollo Rural (INIFAP–SAGAR).

Castro, J.M., González, V., and Hernández-Tejeda, T. 1997. Contenido de metales pesados en algunos suelos y vegetación del Desierto de los Leones, Distrito Federal. *Rev. Cienc. Forestal México* 22:109–127.

Castro-Servín, J.M. 1999. Estudio de los suelos forestales del Desierto de los Leones Distrito Federal. In *North American Science Symposium: Toward a Unified Framework for Inventorying and Monitoring Forest Ecosystem Resources*, compilers C. Aguirre-Bravo and C. Rodriguez-Franco, pp. 98–102. November 2–6, 1998, Guadalajara, México. Proceedings RMRS-P-12. Fort Collins, Colorado: U.S. Department of Agriculture, Forest Service, Rocky Mountain Research Station.

Castro-Servin, J. Ma., Riverón-Negrete, L., González-Kladiano, V., and Hernández-Tejeda, T. 1995b. Contenido de metales pesados en suelos forestales. In *Partnerships for Sustainable Forest Ecosystem Management: Fifth Mexico/U.S. Biennial Symposium*, Tech. Eds. C. Aguirre-Bravo, L. Eskew, A.B. Villa-Salas, and C.E. González-Vicente, pp. 97. October 17–24, 1994, Guadalajara, Jalisco, Mexico. General Technical Report RM-GTR-266. Fort Collins, Colorado: USDA Forest Service.

Coutiño-Alvarado, M. 1989. *Acumulación y Translocación de Plomo en Vegetales, en una Zona Urbana del Distrito Federal.* M.Sc. Thesis. México, D.F.: Facultad de Ciencias, Universidad Nacional Autónoma de México.

Davies, M.S. 1991. Effects of toxic concentrations of metals on root growth and development. In *Plant Root Growth: An Ecological Perspective*, ed. D. Atkinson, pp. 211–227. Oxford, U.K.: Blackwell Scientific Publications. Special Publication No. 10 of the British Ecological Society.

Davis, R.D., Beckett, P.H.T., and Wollan, E. 1978. Critical levels of twenty potentially toxic elements in young spring barley. *Plant Soil* 49:395–408.

de Bauer, L.I., and Krupa, S. 1990. The Valley of Mexico: Summary of observational studies on its air quality and effects on vegetation. *Environ. Pollut.* 65:109–118.

DDF–COCODA (Departamento del Distrito Federal–Comisión Coordinadora para el Desarrollo Agropecuario). 1986. *Parque Cultural y Recreativo Desierto de los Leones.* México, D.F.: Informe Técnico Inédito.

Dmuchowski, W., and Bytnerowicz, A. 1995. Monitoring environmental pollution in Poland by chemical analysis of Scots pine (*Pinus sylvestris* L.) needles. *Environ. Pollut.* 87:87–104.

Elliott, H.A., Liberati, M.R., and Huang, C.P. 1986. Competitive absorption of heavy metals in soils. *J. Environ. Qual.* 15:214–219.

Enríquez, E.G. 1975. *Integración Sistemática del Medio Natural en el Parque Nacional Desierto de los Leones.* Tesis profesional. México, D.F.: Facultad Filosofía y Letras, Colegio de Geografía, Universidad Nacional Autónoma de México.

Fenn, M.E., and Bytnerowicz, A. 1997. Summer throughfall and winter deposition in the San Bernardino Mountains in southern California. *Atmos. Environ.* 31:673–683.

Fenn, M.E., de Bauer, L.I., Quevedo-Nolasco, A., and Rodriguez-Frausto, C. 1999. Nitrogen and sulfur deposition and forest nutrient status in the Valley of Mexico. *Water Air Soil Pollut.* 113:155–174.

Fenn, M.E., de Bauer, L.I., Zeller, K., Quevedo, A., Perea, V., and Hernández, T. 2002. Nitrogen and sulfur deposition in the Mexico City Air Basin: Impacts on forest nutrient status and nitrate levels in drainage waters. In *Urban Air Pollution and Forests: Resources at Risk in the Mexico City Air Basin*, Ecol. Stud., eds. M.E. Fenn, L.I. de Bauer, and T. Hernández-Tejeda, New York: Springer-Verlag.

Freedman, B., and Hutchinson, T.C. 1981. Sources of metal and elemental contamination of terrestrial environments. In *Effect of Heavy Metal Pollution on Plants. Metals in the Environment*, Vol. II, ed. N.W. Lepp, pp. 35–94. London: Applied Science Publishers Ltd.

Friedland, A.J., and Johnson, A.H. 1985. Lead distribution and fluxes in a high-elevation forest in Northern Vermont. *J. Environ. Qual.* 14:332–336.

Friedland, A.J., Craig, B.W., Miller, E.K., Herrick, G.T., Siccama, T.G., and Johnson, A.H. 1992. Decreasing lead levels in the forest floor of the northeastern USA. Ambio 21:400–403.

García, E. 1981. *Modificaciones al Sistema de Clasificación Climática de Köppen (para Adaptarlo a las Condiciones Climáticas de la Republica Mexicana).* México, D.F.: Colegio de Geografía, Universidad Nacional Autónoma de México.

Huettl, R.F., Fink, S., Lutz, H.-J., Poth, M., and Wisniewski, J. 1990. Forest decline, nutrient supply, and diagnostic fertilization in southwestern Germany and in southern California. *Forest Ecol. Manage.* 30:341–350.

Jáuregui, E. 2002. The climate of the Mexico City Air Basin: Its effects on the formation and transport of pollutants. In *Urban Air Pollution and Forests: Resources at Risk in the Mexico City Air Basin*, Ecol. Stud., eds. M.E. Fenn, L.I. de Bauer, and T. Hernández-Tejeda, New York: Springer-Verlag.

Jordan, M.J. 1975. Effects of zinc smelter emissions and fire on a chestnut-oak woodland. *Ecology* 56:78–91.

Krupa, S. 1997. *Air Pollution, People and Plants.* St. Paul, MN, U.S.A.: APS Press.

Laskowski, R., Niklińska, M., and Maryański, M. 1994. Heavy metals in forest litter: A chemical time bomb. In *Conference Papers 19: Climate and Atmospheric Deposition Studies in Forests*, ed. J. Solon, E. Roo-Zielińska, and A. Bytnerowicz, pp. 211–217. Warsaw, Poland: Institute of Geography and Spatial Organization, Polish Academy of Sciences.

Lawrence, G.B., David, M.B., and Shortle, W.C. 1996. Aluminum mobilization and calcium depletion in the forest floor of red spruce forests in the northeastern United States. In *Proceedings of the 1995 Meeting of the Northern Global Change Program*, eds. J. Hom, R. Birdsey, and K. Obrian. General Technical Report 214. Radnor, PA. U.S.A.: U.S. Department of Agriculture, Forest Service, NE Experiment Station.

López, M.A., and Rivera, A.R. 1995. Caracterización nutrimental de follaje de oyamel en proceso de declinación. In *Proceedings of the II Congreso Mexicano sobre Recursos Forestales. Sociedad Mexicana de Recursos Forestales*, pp. 50. Montecillo, Mexico: Colegio de Postgraduados.

López-López, M.A., Velázquez-Mendoza, J., Velázquez-Martínez, A., González-Romero, V., and Cetina-Alcalá, V.M. 1998. Estado nutrimental de *Abies religiosa* en una área con problemas de contaminación ambiental. *Agrociencia* 32:53–59.

Madrigal, X. 1967. *Contribución al Conocimiento de la Ecología de los Bosques de Oyamel (Abies religiosa (H.B.K.) Schl et Cham.) en el Valle de México.* Boletín Técnico INIF (Instituto Nacional de Investigaciones Forestales) #18. México, D.F.: Secretaría de Agricultura y Ganadería (SAG).

Marin, L.E., Escolero-Fuentes, O., and Trinidad-Santos, A. 2002. Physical geography, hydrogeology and forest soils of the Basin of Mexico. In *Urban Air Pollution and Forests: Resources at Risk in the Mexico City Air Basin*, Ecol. Stud., eds. M.E. Fenn, L.I. de Bauer, and T. Hernández-Tejeda, New York: Springer-Verlag.

Marschner, H. 1995. *Mineral Nutrition of Higher Plants*, 2nd edn. Cambridge, U.K.: Academic Press.

Miller, E.K., and Friedland, A.J. 1994. Lead migration in forest soils: Response to changing atmospheric inputs. *Environ. Sci. Technol.* 28:662–669.

Miller, P.R., de Bauer, Ma. L.I., Hernández-Tejeda, T. 2002. Oxidant exposure and effects on pines in forests in the Mexico City and Los Angeles, California air basins. In *Urban Air Pollution and Forests: Resources at Risk in the Mexico City Air Basin*, Ecol.

Stud., eds. M.E. Fenn, L.I. de Bauer, and T. Hernández-Tejeda, New York: Springer-Verlag.

Millward, C.G., and Kluckner, P.D. 1989. Microwave digestion technique for the extraction of minerals from environmental marine sediments for analysis by inductively coupled plasma atomic emission spectrometry and atomic absorption spectrometry. *J. Analy. Atomic Spectro.* 4:709–713.

Morales, R. 1998. *Estudio Geohidrológico del Valle de Monte Alegre, Ajusco, México.* Tesis de Maestría en Ciencias. Montecillo, México: Colegio de Postgraduados.

Nriagu, J.O. 1978. Lead in soils, sediments and major rock types. In *The Biogeochemistry of Lead in the Environment, Part A: Ecological Cycles*, ed. J.O. Nriagu, pp. 15–72. Amsterdam, The Netherlands: Elsevier/North–Holland Biomedical Press.

Påhlsson, A.-M.B. 1989. Toxicity of heavy metals (Zn, Cu, Cd, Pb) to vascular plants. *Water Air Soil Pollut.* 47:287–319.

Pizl, V., and Josens, G. 1995. Earthworm communities along a gradient of urbanization. *Environ. Pollut.* 90:7–14.

Pouyat, R.V., and McDonnell, M.J. 1991. Heavy metal accumulation in forest soils along an urban-rural gradient in southeastern New York, USA. *Water Air Soil Pollut.* 57–58:797–807.

Quadri, G., and Sánchez, L.R. 1994. *La Ciudad de México y la Contaminación Atmosférica.* México, D.F.: Editorial Limusa.

Raven, P.H., Evert, R.F., and Curtis, H. 1981. *Biology of Plants*, 3rd edn. New York: Worth Publishers, Inc.

Reiche, C. 1977. *Flora Excursoria en el Valle Central de México.* México, D.F.: Editorial Porrúa.

Reyes, R., Castro, J.M., and Camargo, T. 1984. *Estudio de los Suelos Forestales en el Parque Cultural y Recreativo Desierto de los Leones.* México, D.F.: Informe Técnico Inédito.

Riverón-Negrete, L. 1994. *Efecto de los Metales Pesados Tóxicos en una Plantación de Coníferas en San Rafael, México.* Tesis de licenciatura. Facultad de Ciencias. México, D.F.: Universidad Nacional Autónoma de México.

Romero, J. 1986. *Estudio de Reconocimiento de los Suelos Forestales del Parque Cultural y Recreativo Desierto de los Leones.* Tesis profesional. Escuela Nacional de Estudios Profesionales-Zaragoza, México, D.F.: Universidad Nacional Autónoma de México.

Ruiz, M.E., and Gasca, J. 2002. Emissions trends and pollution control strategies in the Mexico City Metropolitan area. In *Urban Air Pollution and Forests: Resources at Risk in the Mexico City Air Basin*, Ecol. Stud., eds. M.E. Fenn, L.I. de Bauer, and T. Hernández-Tejeda, New York: Springer-Verlag.

Rzedowski, J., and Calderón, G. 1979. *La Flora Fanerogámica del Valle de México.* México, D.F.: Editorial Limusa.

Salisbury, F.B., and Ross, C.W. 1985. Plant Physiology, 3rd edn. Belmont, California, U.S.A.: Wadsworth Publishing Company.

Shortle, W.C., Smith, K.T., Minoch, R., Lawrence, G.B., and David, M.B. 1997. Acidic deposition, cation mobilization, and biochemical indicators of stress in healthy red spruce. *J. Environ. Qual.* 26:871–876.

Skeffington, R.A. 1983. Soil properties under three species of tree in southern England in relation to acid deposition in throughfall. In *Effects of Accumulation of Air Pollutants in Forest Ecosystems*, eds. B. Ulrich and J. Pankrath, pp. 219–231. Dordrecht, Holland: D. Reidel Publishing Company.

Sosa, H. 1952. *Parque Nacional Desierto de los Leones*, Serie: Los Parques Nacionales de México. México, D.F.: Secretaría de Agricultura y Ganadería.

Tyler, G., Påhlsson, A.-M.B., Bengtsson, G., Bååth, E., and Tranvik, L. 1989. Heavy-metal ecology of terrestrial plants, microorganisms and invertebrates. *Water Air Soil Pollut.* 47:189–215.

U.S. EPA (Environmental Protection Agency). 1995. Test Methods for Evaluating Solid Waste. Laboratory manual Physical/Chemical Methods, Vol. 1A, 3rd edn. Washington, D.C.: U.S. Gov. Print. Office. SW 846.

Vargas, F. 1984. *Los Parques Nacionales de México y Reservas Equivalentes. Colección: Grandes Problemas Nacionales*, Serie: Los Bosques de México. IIE. México, D.F.: Universidad Nacional Autónoma de México.

Vásquez-Soto, J. 1987. *El Saneamiento y la Limpia Forestales en el Desierto de los Leones*. México, D.F.: Comisión Coordinadora para el Desarrollo Rural (COCODER), Departamento del Distrito Federal.

Vizcaíno-Murray, F. 1975. *La Contaminación en México*. México, D.F.: Fondo de Cultura Económica.

Watmough, S.A. 1997. *Heavy Metals in Soils and Bark of Sacred Fir in Desierto de los Leones Park*. Internal Report. Peterborough, Ontario, Canada: Trent University.

Watmough, S.A., and Hutchinson, T.C. 1999. Change in the dendrochemistry of sacred fir close to Mexico City over the past 100 years. *Environ. Pollut.* 104:79–88.

Watmough, S.A., Hutchinson, T.C., and Sager, E.P.S. 1998. Changes in tree ring chemistry in sugar maple (*Acer saccharum* Marsh.) along an urban–rural gradient in southern Ontario. *Environ. Pollut.* 101:1–11.

Zambrano, A., Nash III, T.H., and Herrera-Campos, A. 2002. Lichens and air pollution in the forests surrounding Mexico City. In *Urban Air Pollution and Forests: Resources at Risk in the Mexico City Air Basin*, Ecol. Stud., eds. M.E. Fenn, L.I. de Bauer, and T. Hernández-Tejeda, New York: Springer-Verlag.

Zambrano-García, A. 1991. *Evaluación del Contenido de Plomo en el Musgo Leptodontium viticulosoides de los Bosques de Oyamel de la Cuenca de México*. Tesis de Maestría en Ciencias, Facultad de Ciencias. México, D.F.: Universidad Nacional Autónoma de México.

Zimdahl, R.L., and Hassett, J.J. 1979. Lead in soil, In *Lead in the Environment*, eds. W.R. Boggess and B.G. Wixson, pp. 93–98. National Science Foundation, Castle House Publications, Ltd.

Zimdahl, R.L., and Koeppe, D.E. 1979. Uptake by plants. In *Lead in the Environment*, eds. W.R. Boggess and B.G. Wixson, pp. 99–104. National Science Foundation, Castle House Publications, Ltd.

3. Ecological Effects of Air Pollutants in the Mexico City Air Basin: Case Studies

[illegible] ant Exposure and Effects on Pines [illegible] in the Mexico City and Los Angeles, California, Air Basins

[illegible], L.I. de Bauer, and Tomás Hernández-Tejeda

Fore[illegible] hwest of Mexico City and east of Los Angeles, California, share several similarities, including: the location of mixed conifer forests downwind of large urban "smog" source areas, some of the highest seasonal and year-round tropospheric ozone exposures in the world, foliar and canopy injury symptoms typical of severe ozone exposure, and a prominent pine component afflicted with decline and tree death. These similarities eventually led to joint studies using similar research methods, in which differences in climate and seasonal occurrence of elevated ozone concentrations became apparent. The smaller contrast between the summer and winter temperatures of the Mexico City region may eventually help researchers predict how forests in California might respond to ozone under suggested trends of a warming climate. In this chapter, the early studies of air pollution symptoms exhibited by vegetation indigenous to the forests surrounding the Valley of Mexico will be reviewed along with more recent joint Mexican–US studies which further define the problem.

Description of the Locations and Climate

The wet and dry tropical climate of Mexico is under control of the Intertropical Convergence Zone (ITCZ) as discussed elsewhere in this volume (Jáuregui 2001). The ITCZ moves north in summer, and a deep northeasterly flow brings major rains to the Basin of Mexico. In winter, when the ITCZ moves south again, the region comes under the influence of the subtropical high pressure belt. During this season, the air is very dry and stable, and air pollution is at its worst in the region of Mexico City.

Mexico City is located in a broad valley at the southern end of the central Mexican plateau. To the south rises the Chichinautzin range, to the west the Sierra de las Cruces, and to the east are the prominent volcanoes Popocatépetl (5465 m) and Iztaccíhuatl (5230 m). The mountain ranges circumscribing much of the Basin of Mexico (See Figs. 3.1 and 3.2 in this volume) limit horizontal advection of polluted air from Mexico City (2240 m).

The most similar circumstance of these two forested regions (Fig. 9.1), is that they are both situated downwind from a large metropolitan complex which is a source of precursor pollutants for ozone formation (i.e., forests situated as little as 20 km SSW of Mexico City, and forests in the San Bernardino Mountains 80 km east of Los Angeles, California). The characteristics of the terrain of the two regions are also similar. In California, this includes mountains on three sides which limit advection and guide polluted air flow into the San Bernardino mountains (Edinger et al. 1972).

A major difference between the two regions is the presence of a marine inversion influence in the Los Angeles Air Basin, where the western side of the Air Basin is open to the Pacific Ocean. The marine inversion persists and penetrates inland because the cooler air slides in below warmer air (created by subsidence) inland. The inversion is maintained until it is disrupted by the warming of air near the surface, and the accumulated air pollution is transported by thermal advection along with the afternoon sea breeze in an easterly flow into the San Bernardino Mountains.

The seasonal patterns of temperature and precipitation of the two regions are also very different (Donley et al. 1979; U.S. Department of State 1988). The comparisons of monthly precipitation (Fig. 9.2a) and of temperature (Fig. 9.2b) show higher summer and fall temperatures and a greater differential between summer and winter temperatures in the Los Angeles region. Storms and precipitation are confined to the winter, that is, November through April, in the Los Angeles region, but in the Mexico City region, precipitation occurs almost entirely in summer, that is, June through September. The summer rains in the Mexico City area contribute to the cloudiness and cooler summer temperatures compared to Los Angeles.

The climate also determines the season during which poor ventilation, low rainfall, and sunlight can result in polluted air as measured by the accumulation of ozone. Differences in these factors between forest sites in the Los Angeles, California, and in the Mexico City Air Basin result in different hourly ozone exposure profiles during both the May through October period (Fig. 9.3a) and in the November through April period (Fig. 9.3b). In the Mexico City region, the daily peak in ozone during the winter is more defined (November–April) than at the Crestline station in California. But, in summer (May–October), data from 1990–1991 (Miller et al. 1994) indicates that Crestline experiences higher ozone peaks. Seasonal patterns of ozone uptake by tree foliage, timing of symptom

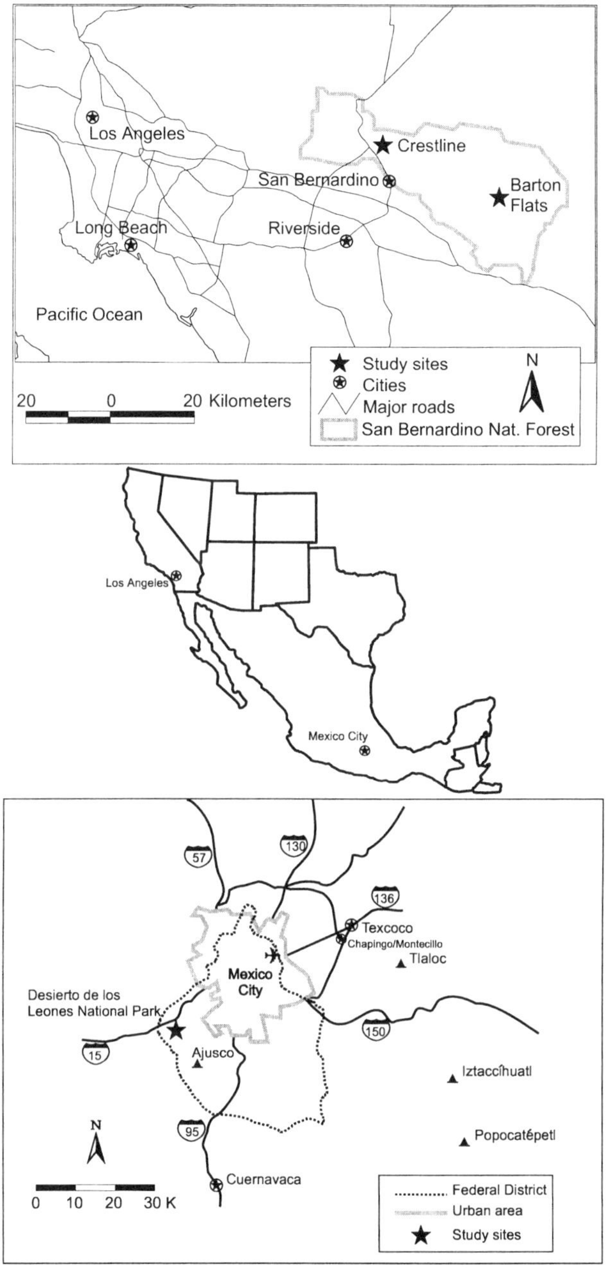

Figure 9.1. Map of the Los Angeles and Mexico City regions showing the location of monitoring and ozone injury study sites.

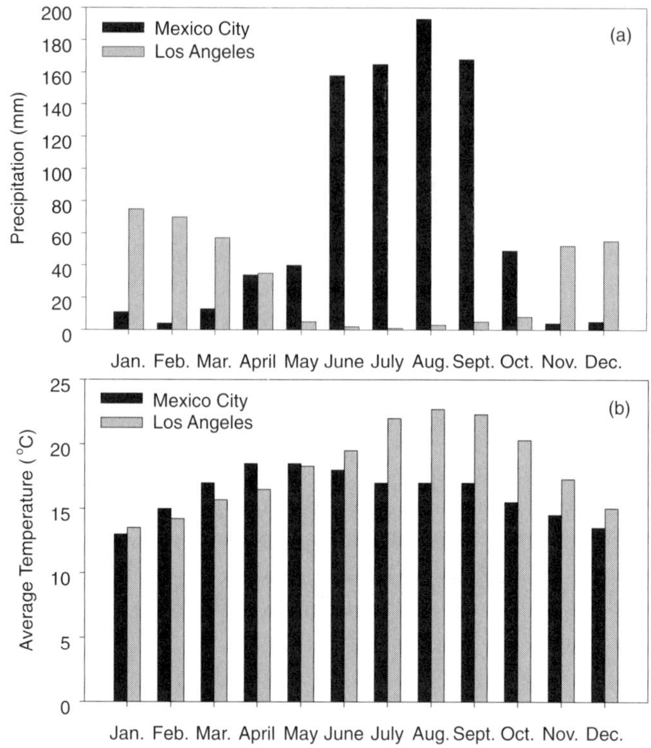

Figure 9.2. Average monthly precipitation (a) and temperature (b) in the Los Angeles and Mexico City regions.

development, and the possible co-occurence of other stresses, particularly soil water stress, is likely to be different between the two areas as well. However, the difference between the two regions in soil water availability during the summer is considered to be a very important difference.

It would be desirable to have a record of ozone representing long-term trends in the rural forested areas, but there is none available. There is an internet site maintained by the Instituto Nacional de Ecologia (*http://www.ine.gob.mx/dggia/cal_aire/libros/segu_info/mexico.html*) which has air pollution data from about 1988 to 1997. This web site is in Spanish and includes data from six ozone-measuring stations in the southeastern part of Mexico City and slightly upwind from forested areas. Information is also available from the air pollution monitoring networks in Mexico City. Ozone is the most important factor in the pollution index used. In its worst year the ozone standard was exceeded 88% of the time (or 322 of 365 days). Ozone was highest in 1991, but the gradual decrease, if it is indeed real, has been slight in subsequent years. Although this web site represents urban locations, it provides information on trends of upwind air quality for ozone in the Mexico City Air Basin.

Discovery of Ozone Injury to Bioindicator Species in the Basin of Mexico and to Native Pine Species Southwest of Mexico City

Bioindicator plants were first used to demonstrate the occurrence of phytotoxic ozone exposures in Mexico City and environs in 1971 (de Bauer 1972; de Bauer and Hernández-Tejeda 1986). Indicator plants first exposed in the metropolitan area of Mexico City were of the following species: *Lactuca sativa* var. Parris Island, *Spinacea oleracea* var. Viroflay and H-7, *Phaseolus vulgaris* var. Pinto 133, *Medicago sativa* var. African, *Nicotiana tabacum* var. Bel W-3 and *Capsicum annum* var. 3-79-1 and 2-5-805. Later, other photochemical oxidant bioindicators were used, including: *N. tabacum* var. Bel C and Bel B, *Raphanus*

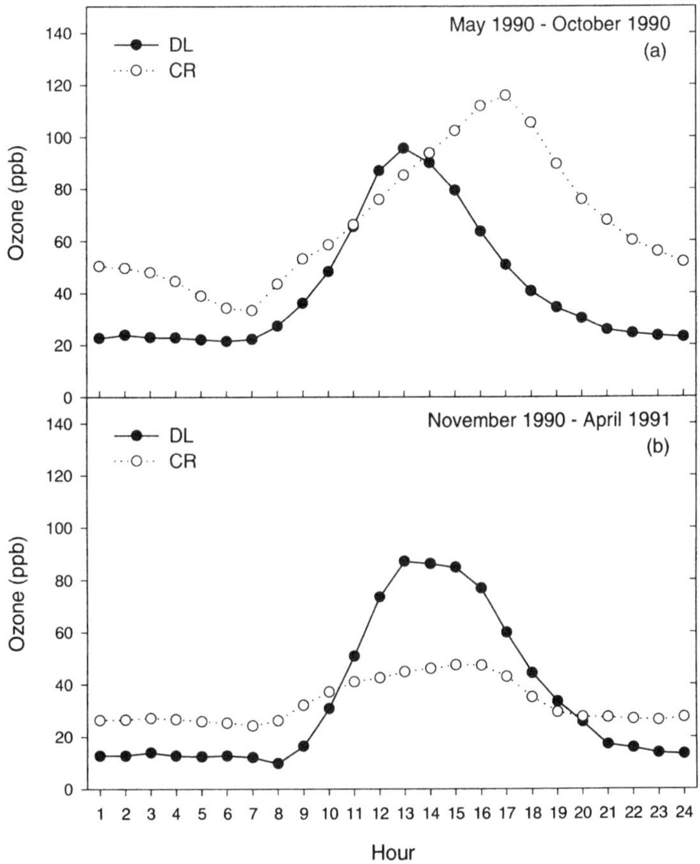

Figure 9.3. Hourly ozone averages at Crestline, California, (CR) and the Desierto de Los Leones (DL) in (a) May through October 1990 and (b) November 1990 through April 1991.

sativus var. Cherry Belle, *P. vulgaris* var. Pinto 111, *Poa annua*, *Urtica uhrens* (burning nettle or dog nettle), and *Petunia axillaris*. The indicator plants were exposed at six different sites within the urban area and nearby zones of Mexico City for three weeks. At the end of the exposure period, the plants were evaluated for height, foliar injury, abnormal defoliation, pigment production, and general observations of the plants. In five of the six sites within the urban area, the tobacco plants exhibited injury characteristically induced by ozone, including white stippling, and later on, necrotic lesions on the upper leaf surface of the mature leaves. At one urban site, bean, spinach, pepper, and especially lettuce plants, developed a metallic bronzing on the lower leaf surface, a symptom characteristic of peroxyacetyl nitrate injury. Bean and tobacco plants exhibited both ozone and sulfur dioxide injury, the latter evidenced as a leaf interveinal and marginal necrosis (de Bauer and Hernández-Tejeda 1986).

Once the occurrence of phytotoxic levels of oxidants was confirmed in the Mexico City area, an attempt was made to trace the pathway or corridor of ozone damage. With the help of Dr. Sagar V. Krupa of the University of Minnesota in the United States, symptoms of oxidant injury in agricultural crops and in natural vegetation were surveyed in areas downwind (to the south and southwest) of the urban zone, particularly towards Xochimilco and Ajusco. Krupa and Bauer (1976) detected injury to vegetables such as spinach, lettuce, and the native plant known as "chayotillo" or as star or bur cucumber, (*Sicyos* sp.) in Xochimilco. Symptoms were similar to those known and described for related plant species (Jacobson and Hill 1970). Ozone injury symptoms were also commonly observed on a few other native species such as *Sambucus mexicana* (known locally as *sauco* and as blue elderberry) (Fig. 9.4, see color plate). Ozone injury symptoms are also seen

Figure 9.4. Incipient symptoms of ozone injury in *Sambucus mexicana* (*Sauco*). (See color plate.)

Figure 9.5. Severe chlorotic mottle injury symptoms on foliage of *Pinus hartwegii*. Symptoms are highly similar to visible foliar ozone injury symptoms on *Pinus ponderosa* in California. (See color plate.)

in this species in the San Bernardino Mountains (Temple 1999). However, the most important finding was the discovery of chlorotic mottle and banding of the needles of *Pinus hartwegii* (Fig. 9.5, see color plate) and *Pinus leiophylla*. Such symptoms are diagnostic of ozone injury to the foliage of *P. ponderosa* in California.

Initial observations of injury to *P. hartwegii* and *P. leiophylla* downwind from Mexico City in the direction of Cuernavaca indicated that the symptoms detected were similar to those induced by ozone in California. *P. leiophylla* was included in these early surveys because its natural presence at lower elevations near Cuernavaca served to extend observations over a greater distance. In subsequent studies, along this transect between Mexico City and Cuernavaca and at Ajusco closer to Mexico City, *P. hartwegii* and *P. montezumae* var *lindleyi* were evaluated.

In the late 1970s, leaf injury symptoms like those caused by ozone were seen on *Eucalyptus globulus* seedlings, especially in newly reforested areas (Fig. 9.6, see color plate). In short-term fumigation studies in 1980, exposures to 0.40 ppm ozone for only 2 or 4 h resulted in *Eucalyptus* seedlings showing symptoms similar to those observed on seedling growing in Mexico City (de Bauer and Hernández-Tejeda 1986; Hernández-Tejeda et al. 1981). These results support the hypothesis that ozone is the cause of foliar injury observed in trees growing in forested sites more distant from the urban center.

The effects of transported ozone on forests of the San Bernardino mountains east of Los Angeles were first reported just prior to the early observations in Mexico (Miller et al. 1963; Richards et al. 1968). Because of similarities in the symptoms, the results in California provided clues to the cause of pine injury

Figure 9.6. Foliar ozone injury symptoms on two-year-old seedlings of *Eucalyptus globulus* Labill. in a reforestation area south of Mexico City. Similar symptoms were observed in artificially fumigated seedlings. (Reproduced by permission of Colegio de Postgraduados, Montecillo, Mexico.) (See color plate.)

observed in Mexico. The work carried out in California, and in particular in the San Bernardino Mountains, is reviewed and supplemented in Miller and McBride (1999). Similarly, a recent assessment of the air pollution problems in the Valley of Mexico in comparison to air pollution problems in Central American locations is presented by de Bauer et al. (2000).

In 1981, a study of eight plots in the Ajusco mountains, including 95 *P. hartwegii* and 28 *P. montezumae* var. *lindleyi* trees, was established (de Bauer and Hernández-Tejeda 1986). The symptoms of injury were evaluated during the months of January, March, May, and August. These observations established that needle injury increased most rapidly from January to May and remained at its highest through the summer and at the August evaluation. The assessment of the amount of injury was based on the oxidant injury score utilized in California (Miller 1973) because the major symptoms were similar, namely, yellow mottle and chlorotic banding of remaining needles and abscission of needles. Ozone was invoked as the cause with greater confidence after the appearance of thin needle sections of *P. hartwegii* was observed. These thin needle sections contained random mesophyll cells with chloroplasts affected. The injury observed in locally grown oats also closely resembled the symptoms expressed by the ozone bioindicator oat variety, Clintland. The observation of typical ozone symptoms in other bioindicators, such as Bel W3 tobacco, also supported the diagnosis that ozone was the cause of injury to pine. However, Bel W3 is highly sensitive to ozone and could show symptoms at ozone exposures that may not cause visible injury in pine species. *P. hartwegii* proved to be more sensitive than *P. montezumae* var. *lindleyi*. Throughout these studies, *P. hartwegii* showed more severe symptoms than any of the other pine species (Figs. 9.5, 9.7, see color plate, and 9.8, see color plate) present in the Ajusco mountain range, namely *P. patula* and *P. teocote. P. ponderosa* and *P. hartwegii*, which are both highly ozone-sensitive, have similar taxonomic characteristics according to Martínez (1963).

Since chlorophyll pigment destruction is one of the consequences of air pollutant injury, photosynthetic pigments were extracted in June 1983 from *P. hartwegii* needles that had been protected by either 0%, 50%, or 100% activated carbon filtration in plastic branch chambers that were ventilated by natural convection at Ajusco. The branch chambers were of the design described by Hernández-Tejeda (1984). The branches under treatment were protected for seven months, and the fully-developed current-year needles were used for chlorophyll extraction. The leaves were protected from air pollution injury from the time they started to grow. Pigment loss was nonexistent in the completely protected needles but ranged from 20 to 31% in the 50% protected or unprotected needles. The needles protected by 100% charcoal filtration developed no visible symptoms of ozone injury; however, the unprotected needles developed the typical ozone symptoms, which consisted of chlorotic mottling and banding on the older needles. Foliage protected by 50% charcoal filtration also developed ozone symptoms, but the symptoms were lower in intensity. These results helped narrow the range of possible causes for the pine injury to some atmospheric contaminant, probably ozone (de Bauer and Hernández-Tejeda 1986).

To further test the hypothesis that exposure to ozone was the cause for pine injury, observations were made along the presumed gradient of decreasing ozone concentration southwest of Mexico City along the highway to Cuernavaca. Ten plots were established, each 0.10 ha, at average interval distances of 2.55 km. The 17 natural *P. hartwegii* and 21 *P. montezumae* in the plots (on average, 4 study

Figure 9.7. Comparison of branches from ozone-sensitive (*left*) and ozone-tolerant (*right*) *Pinus hartwegii* trees. Both trees are from el Ajusco, a montane area to the south of Mexico City. (Reproduced by permission of Colegio de Postgraduados, Montecillo, Mexico.) (See color plate.)

Figure 9.8. A *Pinus hartwegii* tree from el Ajusco with dieback symptoms as a result of chronic ozone injury. (See color plate.)

trees per plot) were evaluated in January, 1984 (de Bauer and Hernández-Tejeda 1986). Of the trees in the very severe injury class, 35% were *P. hartwegii* and none were *P. montezumae*, but no trend for injury to *P. hartwegii* was found along the transect, because this species was found only on the first five plots (38 to 47 km from Mexico City), where the ozone exposures were sufficient to cause severe injury to this sensitive species. *P. montezumae* was more widely distributed along the transect, occurring on eight of the plots. Based on ozone injury scores, it was concluded that *P. montezumae* at greater distances from Mexico City suffered less ozone injury. Observations in this study and at El Ajusco suggest that *P. montezumae* is slightly more ozone sensitive than *P. montezumae* var. *lindleyi* (de Bauer and Hernández-Tejeda 1986). Numerous field observations of pine species exposed to ambient ozone have provided information leading to a general classification of ozone sensitivity (de Bauer and Hernández-Tejeda, personal communication) (Table 9.1).

Table 9.1. Sensitivity of pine species to photochemical oxidants in the Mexico City Air Basin

Pine tree species	Sensitivity
Pinus hartwegii	Very high
P. radiata[a]	High
P. montezumae	High
P. montezumae var. lindleyi	High
P. leiophylla	Moderately high
P. patula	Moderately high
P. teocote	Moderate
P. rudis	Moderate
P. pseudostrobus	Moderate
P. maximartinezii[a]	Moderately low
P. cembroides[a]	Low
P. ayacahuite	Very low

[a] Species introduced to the Mexico City area (Hernández-Tejeda and Nieto de Pascual-Pola 1996).

More recently, pollen grains of *Pinus hartwegii* were artificially fumigated with ozone at facilities of the Canadian Forest Service in Fredericton, New Brunswick. The pollen grains were fumigated under moist conditions for five days for five hours per day with 0.02 ppm (control), 0.10, 0.20 and 0.30 ppm concentrations of ozone. Pollen sensitivity to ozone was unusually high in *P. hartwegii*. Pollen grain germination and tube length of the pollen grains was greatly reduced by ozone exposure (Tomás Hernández and Roger Cox, unpublished data). These studies suggest that pine reproduction, in addition to the health and vigor of established *P. hartwegii* trees, may be impacted under specific conditions by ozone exposure in forests downwind of Mexico City. Benoit et al. (1983) showed a significant reduction ($P = < 0.01$) in pollen germination when pollen collected from *P. strobus* was exposed under moist conditions to 0.15 ppm ozone for four hours. However, when sensitive, intermediate, and tolerant *P. strobus* L. trees were exposed to 0.10 ppm for 4 or 8 hours a day, there was no reduction in pollen production or germination under dry conditions.

Studies at the Desierto de los Leones

Within the Sierra de las Cruces mountain range, to the west of Ajusco, is the Desierto de los Leones (DL) national park, containing 1529 ha between an elevation of 2800 and 3200 m (Fig. 9.1). It is a popular recreational area because of its close proximity to Mexico City. Desierto de los Leones has been a primary air pollution research site because of its location directly downwind of Mexico City. A study of *P. hartwegii* in DL showed that 96% of the trees were in the very severe injury class in June of 1985 (de Bauer and Hernández-Tejeda 1986). Visible injury and decline of *Abies religiosa* (sacred fir) in the park are described in a separate chapter of this volume. The evidence of ozone injury to *P. hartwegii*, combined with the location of the park immediately downwind of Mexico City, implicated ozone as the causal or predispositional agent responsible for the massive mortality of *Abies religiosa* at the site. Because of the abundance of ozone injury symptoms in trees within the DL park, and because a secured site with electric power was available within the park, an ozone monitor and a weather station were established there to confirm that atmospheric contamination, namely ozone, was present in amounts sufficient to account for the observed injury to trees.

A comparison trial of four planted pine species was carried out in the so-called "cemetery" areas of the DL for several years ending in 1995 (Tomás Hernández, personal communication). This trial, designed to detect tolerance to ozone exposure, included *P. ayacahuite*, *P. montezuma*, *P. patula* and *P. radiata*. The least amount of foliar symptoms and the best growth occurred in *P. ayacahuite*, and the worst symptoms and growth occurred in *P. radiata*. The other species were intermediate in their response (see Table 9.1). *P. ayacahuite* may be a suitable species for reforestation in high ozone areas where mortality of *P. hartwegii* and sacred fir has been heavy.

In mid-June 1995, the upper and lower crowns of black cherry trees (*Prunus serotina* var. *capuli*) in the DL National Park were evaluated. Of these trees, 72% showed upper leaf surface stipple, leaf reddening, and premature senescence (Fig. 9.9, see color plate). Forty-one percent of a more extensive sample, 169 trees spread over a broader area, had similar symptoms. The asymptomatic trees were located at lower elevations and situated in well-shaded concave slope positions (Skelly et al. 1997).

Specific Comparisons of Seasonal Ozone Exposure and Climate at the Desierto de los Leones in Mexico and in the San Bernardino Mountains in Southern California

The ozone monitoring equipment has been maintained since 1990 at the DL to gather an ozone record that, when compared to exposures in California, would provide the data needed to evaluate the role of ozone as a principal cause of forest damage in the Mexico City Air Basin. Ozone exposure profiles in the Desierto are compared with Crestline, California, in Figure 9.3. In California, we observed that conditions which encourage ozone uptake, for example, adequate soil moisture, also increase needle injury to *P. ponderosa* and *P. jeffreyi* (Temple and Miller 1998). Summers are dry in California, but the water stored in the soil by winter precipitation is important to transpiration and hence to ozone flux to foliage. In California, this pertains only to the summer season, because ozone concentrations are low in winter. In Mexico, high ozone concentrations occur in between rain events during the wet summer months and follow the rain during the fall and winter months (Figs. 9.3). The simultaneous combination of sufficient soil moisture and high ozone concentrations in the DL suggests that there

Figure 9.9. Ozone injury symptoms in *Prunus serotina* var. *capuli*. Symptoms include upper leaf stipple, leaf reddening, and premature senescence. This species is highly sensitive to ozone and can be used as a bioindicator of ozone injury, although individual trees can show considerable variation in ozone sensitivity. (See color plate.)

may be a greater potential for ozone uptake and injury than in California forests where high ozone concentrations are more prevalent during the long summer drought, during which stomatal conductance and ozone uptake is low. The dry summer conditions and wet cool winters are characteristic of the Mediterranean climate in California.

According to historical records, the air pollution problems affecting Mexico City have worsened during the last 30 years, mainly due to the rapid population growth and the number of cars circulating in the Basin. The results of dendrological studies of *P. hartwegii* (Alarcón et al. 1993) also reflect the increasing severity of air pollution within the Basin during this time.

In 1995, we compared temperature, relative humidity, and ozone in July through December at the DL in Mexico and Barton Flats in the San Bernardino Mountains of California (Figs. 9.10a–f). The results were consistent with our first

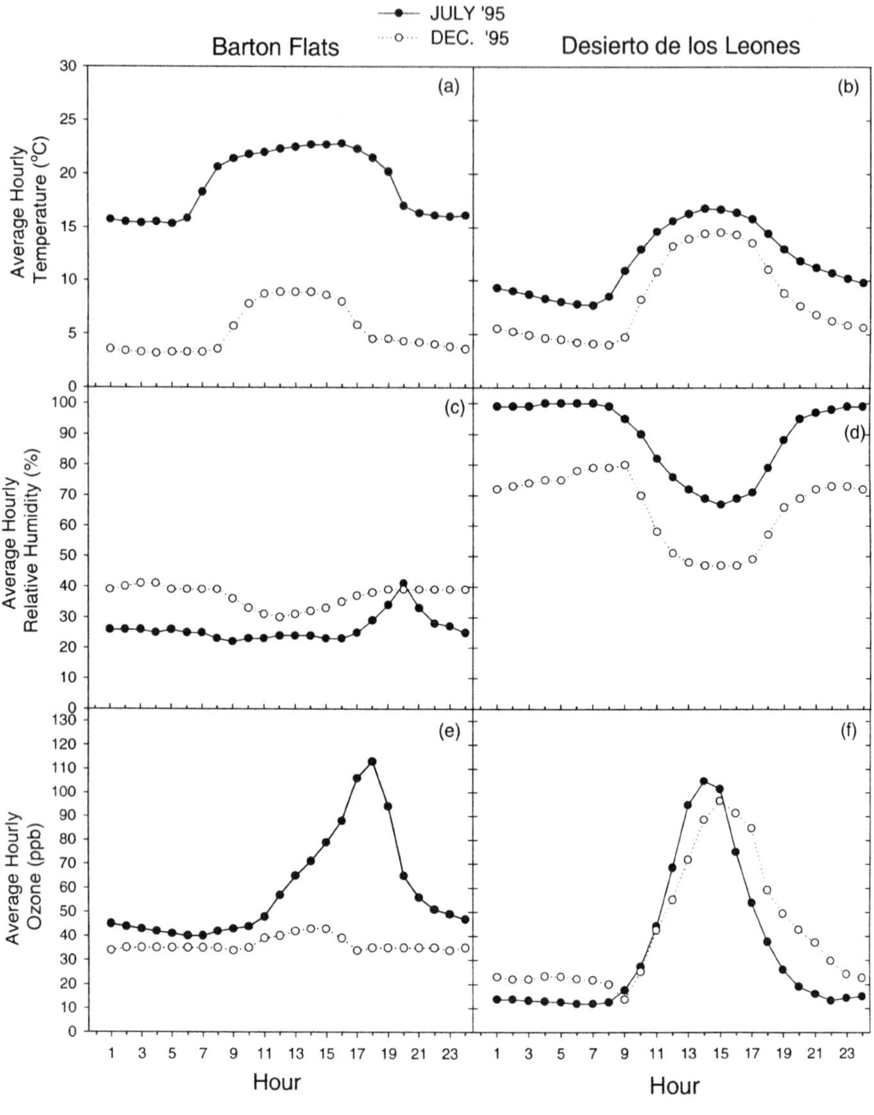

Figure 9.10. Hourly diurnal temperature (a, b), relative humidity (c, d) and ozone concentration (e, f) at Barton Flats and Desierto de los Leones for July and December 1995.

comparison (Miller et al. 1994). The average summer (July) temperature was higher at Barton Flats than at DL, and the winter temperature (December) was higher at DL (Fig. 9.10a and 9.10b). The relative humidity at Barton Flats did not average higher than 40% in either July or December, while relative humidity remained between 40% and 100% at the Desierto in both July and December

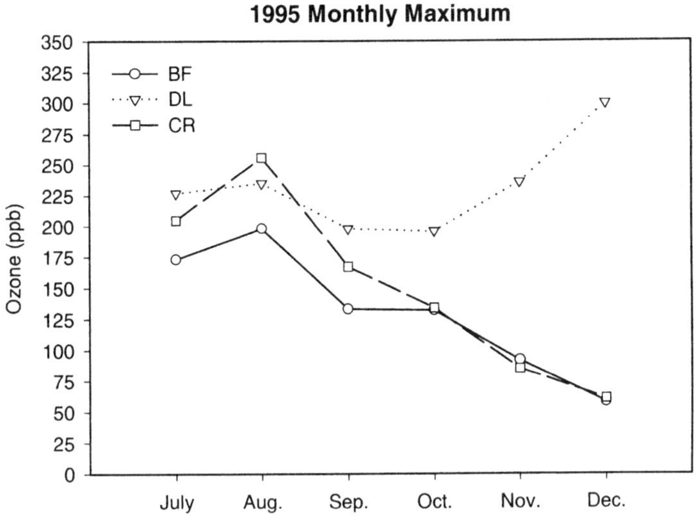

Figure 9.11. Monthly maximum hourly ozone concentrations at Barton Flats (BF) and Crestline (CR) in the San Bernardino Mountains, California, and at the Desierto de los Leones (DL), Mexico, from July to December 1995.

(Fig. 9.10c and 9.10d). Average hourly ozone concentrations were high at the Desierto in summer and winter (Figs. 9.10f and 9.11), especially during November and December. At Barton Flats, California, however, ozone concentrations were only elevated in summer (Figs. 9.10e and 9.11). These results demonstrate that, at the Desierto, the climate combines with the seasonal distribution of ozone in such a way that high ozone exposure coincides with the period of lower drought stress. This greater availability of water makes possible greater ozone flux and tree injury at the Desierto than at Barton Flats, California, where the greatest exposure to ozone coincides with high drought stress. We can speculate that a warming of California's winter climate would also lengthen the seasonal exposure to ozone and result in greater forest damage, especially if soil water availability is sufficient to promote high ozone uptake rates.

Comparison of Injury to Pines at the Desierto de los Leones, Mexico, with Pines at Barton Flats and Camp Paivika, California

A standard procedure (Miller et al. 1996) for evaluating ozone injury was used in evaluating injury to *P. hartwegii* at the DL and to pines at two sites in the San Bernardino Mountains. Barton Flats is a climate and air monitoring site in the eastern San Bernardino mountains with a mixture of *P. ponderosa* and *P. jeffreyi*. Ozone injury was also evaluated in *P. ponderosa* at Camp Paivika (close to

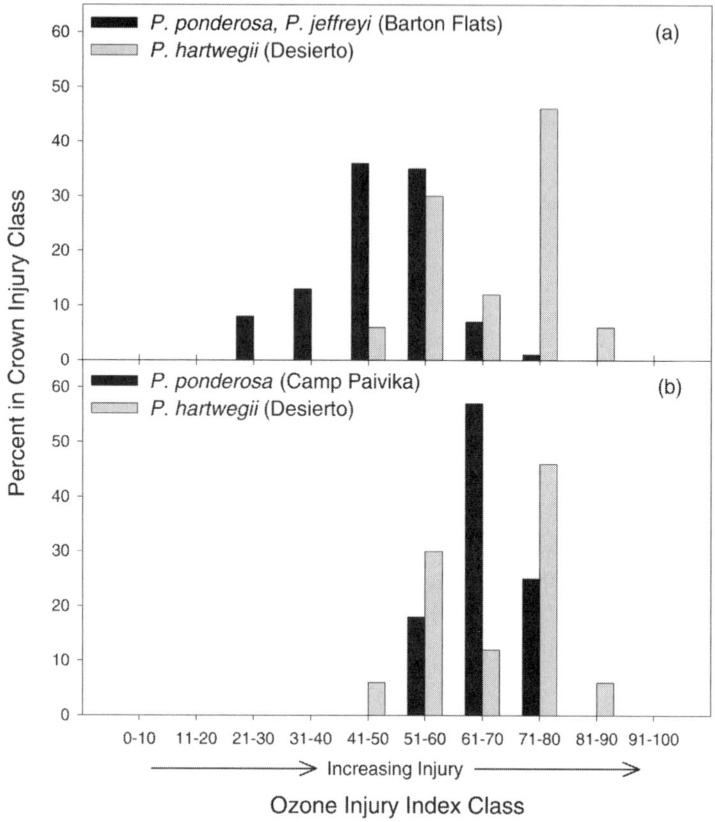

Figure 9.12. Crown injury (ozone injury index) (Miller et al. 1996) distribution of a population of 50 *Pinus hartwegii* at the Desierto de los Leones, Mexico, compared with similar populations of *P. jeffreyi* and *P. ponderosa* at Barton Flats, California (a), and *P. ponderosa* at Camp Paivika, near Crestline, California (b).

Crestline) (Fig. 9.1) in the western part of the San Bernardino Mountains where there is greater exposure to ozone. A population of 130 trees at Barton Flats was not as severely injured as the 50 trees at the DL (Fig. 9.12a), but the 50 trees at Camp Paivika had a level of injury nearly identical to the trees at DL (Fig. 9.12b) (Miller et al. 1998).

Summary

Various field experiments and ozone monitoring have established ozone as the cause of injury to pines downwind from Mexico City. There are many similarities in the injury to pines in the Mexico City Air Basin with the injury observed earlier in forests downwind from Los Angeles. The regions differ in seasonal patterns of precipitation, and much greater ozone exposure is experienced during winter months in Mexico. A comparison of the amount of ozone injury to *P. hartwegii* in Mexico with *P. ponderosa* in California reveals that the injury to *P. hartwegii* is highly similar to that of *P. ponderosa* in high-exposure sites in California. However, ozone concentrations have been declining over the past 10 to 15 years in the Los Angeles Basin, but have remained higher in the Mexico City Air Basin. Thus ozone injury is less severe in the San Bernardino Mountains than in past decades, but it is not decreasing as rapidly in highly-exposed forests in the Mexico City Air Basin.

References

Alarcón-C., M.A., de Bauer, Ma. L.I., Jasso-M., J., Segura-W.G., and Zepeda-B., E.M. 1993. Patrón de crecimiento radial en árboles de *Pinus hartwegii* afectados por contaminación atmosférica en el suroeste del Valle de México. *Agrocienc. Serie Recursos Naturales Renovables* 3:67–80.

Benoit, L.F., Skelly, J.M., Moore, L.D., and Dochinger, L.S. 1983. The influence of ozone on *P. strobus* L. pollen germination. *Can. J. For. Res.* 13:184–187.

de Bauer, L.I. 1972. Uso de plantas indicadoras de aeropolutos en la Ciudad de México. *Agrociencia* 9:139–141.

de Bauer, L.I., Hernandez-Tejeda, T., and Skelly, J.M. 2000. Air pollution problems in the forested areas of Mexico and Central America. In *Air Pollution and the Forests of Developing and Rapidly Industrializing Regions*, IUFRO Research Series 5, eds. J.L. Innes, and A.H. Haron, pp. 35–61. Wallingford, U.K.: CABI Publishing.

de Bauer, L.I., and Hernández-Tejeda, T. 1986. *Contaminación: Una Amenaza para la Vegetación en México*. Chapingo, State of Mexico, Mexico: Colegio de Postgraduados.

Donley, M.W., Allan, S., Caro, P., and Patton, C.P. 1979. *Atlas of California*, pp. 132–135. Portland, OR: Academic Book Center.

Edinger, J.G., McCutchan, M.H., Miller, P.R., Ryan, B.C., Shroeder, M.J., and Behar, J.V. 1979. Penetration and duration of oxidant air pollution in the South Coast Air Basin of California. *J. Air Pollut. Control Assoc.* 22:882–886.

Hernández-Tejeda, T. 1984. *Efecto de los Gases Oxidantes sobre Algunas Especies del Genero Pinus Nativas del Valle de México*. M. Sc. Thesis. Chapingo, State of Mexico, Mexico: Colegio de Postgraduados.

Hernández-Tejeda, T., and Nieto de Pascual-Pola, C. 1996. Effects of oxidant air pollution on *Pinus maximartinezii* Rzedowski in the Mexico City region. *Environ. Pollut.* 92:79–83.

Hernández-Tejeda, T., Krupa, S.V., Platt, G.C., and de Bauer, L.I. 1981. Sensibilidad de plántulas de eucalipto (*Eucalyptus globulus* Labill.) al ozono. *Agrociencia* 43:89–95.

Jacobson, J.S., and Hill, A.C. 1970. *Recognition of Air Pollution Injury to Vegetation: A Pictorial Atlas*, Informative report 1 of the TR-7 Agricultural Committee. Pittsburg, Pennsylvania, U.S.A.: Air Pollution Control Association.

Jáuregui, E. 2001. The climatc of the Mexico City Air Basin—Its effects on the formation and transport of pollutants. In *Urban Air Pollution and Forests: Resources at Risk in the Mexico City Air Basin*. Ecological Studies series, eds. M.E. Fenn, L.I. de Bauer, and T. Hernández-Tejeda, (this volume). New York: Springer-Verlag.

Krupa, S.V., and de Bauer, L.I. 1976. La ciudad daña los pinos del Ajusco. *Panagfa* 4:5–7.

Martínez, M. 1963. *Las Pináceas Mexicanas*, 3rd edn. Mexico City, Mexico: Universidad Autónoma de México.

Miller, P.R. 1973. Oxidant-induced community change in a mixed conifer forest. *Adv. Chem. Ser.* 122:101–117.
Miller, P.R., and McBride, J.R. 1999. *Oxidant Air Pollution Impacts in the Montane Forests of Southern California*. Ecol. Stud. 134. New York: Springer-Verlag.
Miller, P.R., Parmeter, J.R., Taylor, O.C., and Cardiff, E.A. 1963. Ozone injury to the foliage of *Pinus ponderosa. Phytopathology* 53:1072–1076.
Miller, P.R., Stolte, K.W., Duriscoe, D., and Pronos, J. (Tech. Coordinators). 1996. *Monitoring Ozone Air Pollution Effects on Western Pine Forests*, Gen. Tech. Report 155. Albany, California: Pacific Southwest Research Station, Forest Service, U. S. Department of Agriculture.
Miller, P.R., de Bauer, L.I., Nolasco-Quevedo, A., and Hernández-Tejeda, T. 1994. Comparison of ozone exposure characteristics in forested regions near Mexico City and Los Angeles. *Atmos. Environ.* 28:141–148.
Miller, P.R., Fenn, M., Skelly, J., Savage, J., Bauer, L.I., Alvarado, A., and Hernández-Tejeda, T. 1998. *Assessment of Forest Climate and Air Pollution Impacts in the Western US and the Valley of Mexico*, Final Letter of Intent Report. Riverside, California: USDA Forest Service.
Richards Sr., B.L., Taylor, O.C., and Edmunds Jr., S.G. 1968. Ozone needle mottle of pines in southern Califorrnia. *J. Air Pollut. Control Assoc.* 18:73–77.
Skelly, J.M., Savage, J.E., de Bauer, L.I., and Alvarado, D. 1997. Observations of ozone-induced foliar injury on black cherry (*Prunus serotina*, var. *capuli*) within the Desierto de los Leones National Park, Mexico. *Environ. Pollut.* 95:155–158.
Temple, P.J. 1999. Effects of ozone on understory vegetation in the mixed conifer forest. In *Oxidant Air Pollution Impacts in the Montane Forests of Southern California*. Ecol. Stud. 134, eds. P.R. Miller and J.R. McBride, pp. 208–222. New York: Springer-Verlag.
Temple, P.J., and Miller, P.R. 1998. Seasonal influences on ozone uptake and foliar injury to ponderosa and Jeffrey pines at a southern California site. In *Proceedings of the International Symposium on Air Pollution and Climate Effects on Forest Ecosystems*, technical coordinators A. Bytnerowicz, M.J. Arbaugh, and S.L. Schilling, pp. 221–228. February 5–9, 1996, Riverside, California, U.S.A. Albany, California: USDA Forest Service, Pacific Southwest Research Station. Gen. Tech. Rep. GTR-166. Electronic version is available at *http://www.rfl.psw.fs.fed.us/pubs/psw-gtr-164/index.html.*
U.S. Department of State. 1988. *Background Notes*, Series Publication 7865. Washington, D.C.: Department of State.

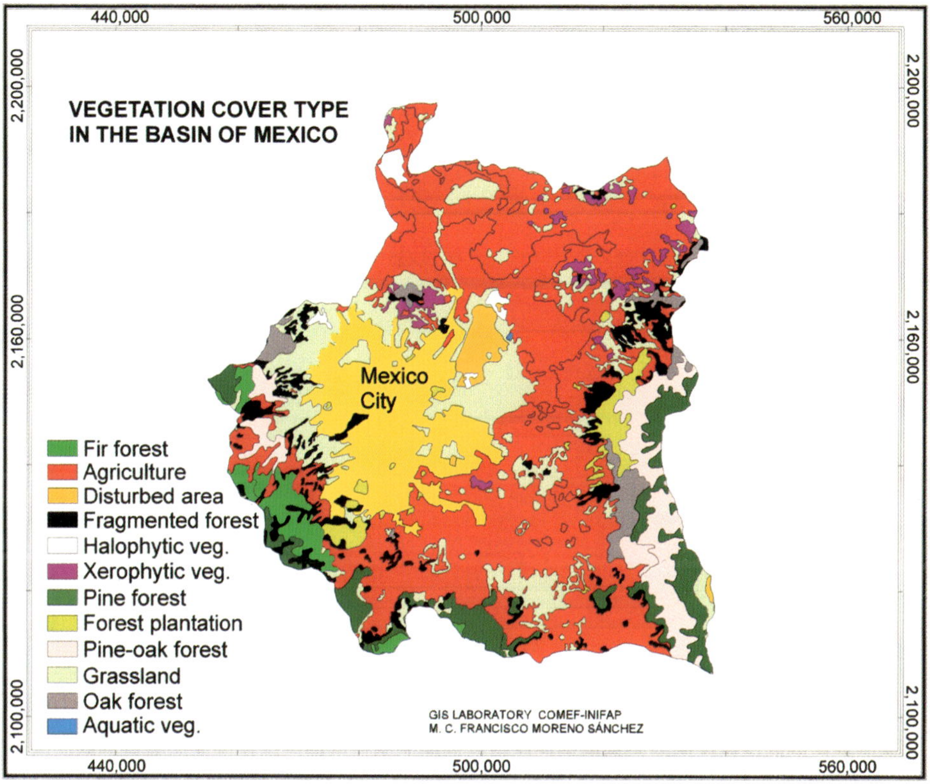

Figure 4.1. Location of major vegetation types within the Basin of Mexico.

Figure 9.4. Incipient symptoms of ozone injury in *Sambucus mexicana* (*Sauco*).

Figure 9.5. Severe chlorotic mottle injury symptoms on foliage of *Pinus hartwegii*. Symptoms are highly similar to visible foliar ozone injury symptoms on *Pinus ponderosa* in California.

Figure 9.6. Foliar ozone injury symptoms on two-year-old seedlings of *Eucalyptus globulus* Labill. in a reforestation area south of Mexico City. Similar symptoms were observed in artificially fumigated seedlings. (Reproduced by permission of Colegio de Postgraduados, Montecillo, Mexico.)

Figure 9.7. Comparison of branches from ozone-sensitive (*left*) and ozone-tolerant (*right*) *Pinus hartwegii* trees. Both trees are from el Ajusco, a montane area to the south of Mexico City. (Reproduced by permission of Colegio de Postgraduados, Montecillo, Mexico.)

Figure 9.8. A *Pinus hartwegii* tree from el Ajusco with dieback symptoms as a result of chronic ozone injury.

Figure 9.9. Ozone injury symptoms in *Prunus serotina* var. *capuli*. Symptoms include upper leaf stipple, leaf reddening, and premature senescence. This species is highly sensitive to ozone and can be used as a bioindicator of ozone injury, although individual trees can show considerable variation in ozone sensitivity.

Figure 10.1. Sacred fir forest decline in a "cemetery" area of Desierto de los Leones National Park.

Figure 10.2. Close-up of the red-brownish coloration on the upper leaf surface of the older leaves of sacred fir.

Figure 14.2. Range of ozone-induced injury in Tempo trifoliate bean (*Phaseolus vulgaris* L.) leaves based on upper surface purple stipple/ bronzing and bifacial necrosis.

Figure 14.2. (*Continued*)

10. Decline of Sacred Fir in the Desierto de los Leones National Park

Dionicio Alvarado-Rosales and Tomás Hernández-Tejeda

The sacred fir (*Abies religiosa* (H.B.K.) Schl. et Cham.) forest in the Mexico City Air Basin (MCAB) is a highly valued and appreciated resource for the more than 20 million inhabitants of the Basin. This forest provides several important benefits, among which are cultural, economic, ecological, and aesthetic values and enhanced soil retention, water infiltration, aquifer recharge, and oxygen production. A prime location for *A. religiosa* in the Basin is the Desierto de los Leones National Park, located southwest of Mexico City. The park covers a land area of 1529 ha with an altitude from 2800 to 3800 m. The most important tree species are *Abies religiosa* and *Pinus* sp., which represent 96% of the total timber volume (Vázquez-Soto 1987).

In the early 1980s, the fir forest in the Desierto de los Leones Park began to show decline symptoms never seen before in Mexican forests (Fig. 10.1, see color plate). During the 1980s, the area impacted by fir decline increased. In some areas, later called cemeteries, tree mortality was widespread and severe, and the visual effects were dramatic. "Dead and dying trees by tens of thousands brown the once-lush slopes of this park above smog-beset Mexico City, and the accusing finger points to air pollution as the cause", wrote Ciesla and Macías-Sámano in 1987. These authors also mentioned that insects had killed many of the dead trees, but that the underlying cause was suspected to be something more insidious—air pollution. The public became alarmed at the loss of these highly valued stands. In recent years, mortality rates have declined.

Figure 10.1. Sacred fir forest decline in a "cemetery" area of Desierto de los Leones National Park. (See color plate.)

This chapter provides an overview of previous studies of fir decline. The analysis of possible causal factors begins with the abiotic factors: air pollution, drought, and nutritional deficiencies; it continues with the biotic agents: insects and pathogens; and concludes with the role of lack of forest management. Relevant results are examined, and a conceptual model to explain the origin of the problem is suggested based on the presented information. Finally, research needs are discussed.

Mexican Fir Forests

Martínez (1963) reported eight species and five varieties of the genus *Abies* in Mexico: *A. concolor*, *A. durangensis*, *A. durangensis* var. *coahuilensis*, *Abies guatemalensis*, *A. guatemalensis* var. *tacanensis*, *A. guatemalensis* var. *jaliscana*,

A. hickeli, *A. mexicana*, *A. religiosa*, *A. religiosa* var. *emarginata*, *A. oaxacana*, *A. vejari*, and *A. vejari* var. *macrocarpa*. Most of them grow naturally in the cold and humid mountains of Mexico at altitudes from 2800 to 3500 masl.

In general, the *Abies* forests of Mexico grow on the slopes of the mountains, often in areas protected from strong winds and intense insolation. These forests are not present in flat areas or on gentle slopes. They also require an annual precipitation greater than 1000 mm, and an annual mean temperature from 7° to 15° C. The geologic substrates of *Abies* forests are diverse, however those of volcanic origin predominate. The Mexican *Abies* forests are physiologically active year-round, although perhaps there is a reduction in physiological activity during the cold and dry periods of the year (Rzedowski 1978).

Abies religiosa forests are located in the central part of the country, in the States of Michoacán, Jalisco, Guerrero, Mexico, Puebla, Hidalgo, Veracruz, and Tlaxcala, and in the Federal District. This species is one of the most elegant trees in the Mexican forests. Its height is from 35 to 45 m and its stem diameter from 1.0 to 1.8 m.

The wood is white-colored, lightweight, and of poor quality. The timber is used primarily in paper production but also for making long thin strips of wood called *tejamaniles* used to cover the roofs of houses. Sacred fir wood is also used to make beams (Martínez 1963).

Sacred fir forests very often occur as pure stands, but sometimes sacred firs are mixed with other species (Rzedowski and Rzedowski 1979). The normal lifespan of sacred fir trees is 119 years in the Desierto de los Leones park (Vázquez-Soto 1987). The soil where *A. religiosa* stands grow is typically deep, well drained, and moist during the entire year. These soils usually have developed A, B, and C horizons in the soil profile and have a dark brown color and a sandy loam texture. Soil pH values are from 5 to 7, and organic matter is abundant, as much as 35% in the A_1 horizon. The soil C:N ratio is typically approximately 20, and the cation exchange capacity ranges from 10 to 30 meq/100 g of soil (Rzedowski 1978).

The name of this species originates from the religious use of this tree. The indigenous people used the branches in certain religious ceremonies to decorate their churches because of the cruciform structure of the younger branches. Occasionally the tips of the trees were used as Christmas trees, but this practice is now prohibited (Martínez 1963). For more detailed descriptions of *Abies religiosa* the reader is referred to Martínez (1963), Rzedowski and Rzedowski (1979), and Sánchez-Sánchez (1974).

Progressive Development of the Sacred Fir Decline Symptoms

Sacred fir decline in the Desierto de los Leones Park was first observed in the early 1980s. Since then, there have been many studies of cause-effect relationships, and much is yet to be learned about the causes of the decline. There are relatively few cases of well-documented regional declines in forest health that can be directly attributed to the effects of air pollution, yet Innes (1993) stated that one of the best examples is the decline of trees in the vicinity of Mexico City, particularly in the Desierto de los Leones Park, where the most affected species is sacred fir. However, we are of the opinion that there is not yet enough evidence to make a definitive conclusion concerning the most important causal factors in the decline.

Tree Decline Development

During sacred fir decline, the trees exhibit branch mortality and reduced foliar retention; however, the most impressive symptom appears on the leaves, which develop a whitish stippling. Later, these lesions coalesce and turn reddish-brown in color, resulting in a uniform discoloration on the upper leaf surface (Fig. 10.2, see color plate). The lower leaf surface does not show any visible symptoms (Alvarado et al. 1993). Branch mortality usually begins in the lower crown and develops upward. However, branch mortality can also move from the upper crown to the lower crown. Leaf retention is poor in declining trees, and occasionally the affected trees only retain current-year needles plus foliage of the previous two years (Alvarado et al. 1987). The normal age range of foliage retained on branches of healthy trees is 4 to 6 years. There have been no detailed studies of foliar length or of the root systems of impacted trees.

Stand Decline Development

Forest inventory results, completed in October, 1983, indicated that trees were damaged in about 28% of the park area covered by sacred fir stands (Ciesla and

Figure 10.2. Close-up of the red-brownish coloration on the upper leaf surface of the older leaves of the sacred fir. (See color plate.)

Macías-Sámano 1987). Five "cemetery" areas within the Park were detected, encompassing a surface of 274 ha (Vázquez-Soto 1987). Four years later in 1987, aerial infrared photography showed that a total of 873 ha of forest was impacted to some degree (Macías-Sámano and Cibrián-Tovar 1989). Cibrián-Tovar (1989) reported that approximately 36% of the total volume of sacred fir wood in the park was deadwood.

More recent data show that tree mortality is decreasing, but decline symptoms are still present in the sacred fir stands. The remaining trees appear to be less susceptible to the syndrome. Possible causes of decreasing mortality rates include: age (the remaining trees are younger), location (many remaining trees are located on exposures other than north, northeast, or northwest and show less damage), and genetic factors (the remaining trees are resistant trees) (Alvarado-Rosales and Hernández-Tejeda 1998).

The spatial and temporal patterns of the sacred fir decline are still largely unknown. In a continuous 12-year study in the first "cemetery", located in a ravine of 20 ha facing the metropolitan area of Mexico City, a total of 123 sacred firs were evaluated. In the first two years, it was observed that the decline spread 100 meters per year, but only towards the southern end of the study plot where the ravine ends (Alvarado-Rosales 1989; Alvarado et al. 1991; Alvarado-Rosales and Hernández-Tejeda 1995; Alvarado-Rosales and Hernández-Tejeda 1998). The general distribution of the cemetery areas is shown in Figure 10.3.

There are other sacred fir forests in the central part of the country, which do not show symptoms of decline. Why the decline has ocurred only in the MCAB may depend on stress factors such as: photochemical air pollution, excessive water extraction, pest and diseases, drought, nutritional deficiencies, heavy metals, and solar radiation, among others.

Reforestation of "Cemetery" Areas

The forest areas called "cemeteries" in the Desierto de los Leones Park are those areas where sacred fir trees have died in great number. The local people started to call these areas "cemeteries" because the trees, although they were dead, remained upright. Although they had no live branches, the main stems remained reaching toward the sky. These areas were completely harvested and subsequently reforested with native and exotic pine species and with sacred fir. However, most of the pine species used for replanting were ozone sensitive and the trees have been impacted by ozone injury. The only pine species growing very well at lower elevations in the Desierto de los Leones Park is *Pinus ayacahuite* Ehren. Sacred fir seedlings did not grow well in such conditions. This may be because the seedlings require moist soil and high relative humidity, shade from canopy cover, a different seedling density, or some other unknown environmental conditions that a natural sacred fir forest provides.

There are now four different pine species (*Pinus ayacahuite*, *P. montezumae* Lamb., *P. patula* Schl. & Cham. and *P. teocote* Schl. & Cham.) growing in most of the "cemetery" areas. Exclusion of sacred fir from these areas may reduce

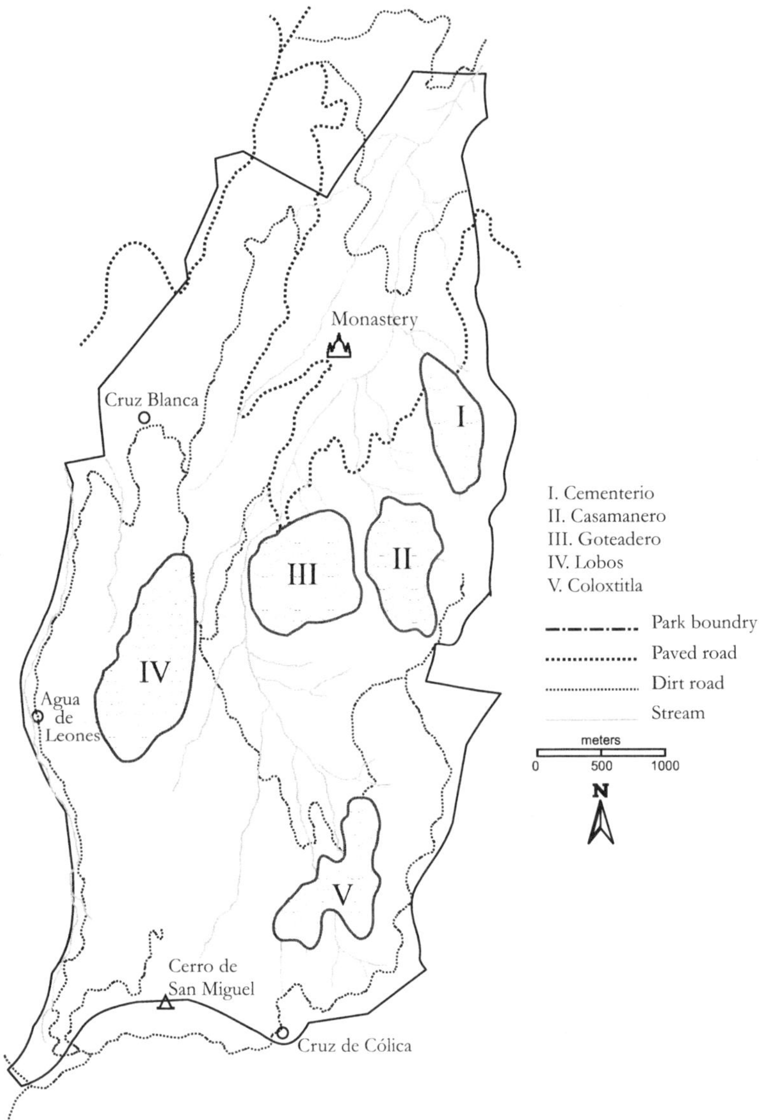

Figure 10.3. Distribution of the cemetery areas in Desierto de los Leones National Park.

the genetic biodiversity of *Abies religiosa* in the Desierto de los Leones Park. However, it was critical to maintain tree cover in order to reduce soil erosion and improve water retention, thus preventing further problems for the people of the Mexico City Basin.

Air Pollution and Its Potential Role in the Decline

Air pollution, as a health factor for many organisms, is a particular threat to vegetative species because they are generally more sensitive than humans and animals (Hernández-Tejeda and de Bauer 1989). The Mexico City Air Basin has been recognized as one of the most polluted geographic areas in the world (de Bauer and Krupa 1990), and photochemical oxidant exposures are particularly severe (Bravo and Torres 2002; Miller et al. 1994). Deposition of nitrogen and sulfur is also highly elevated in forests downwind of Mexico City including the Desierto de los Leones Park (Fenn et al. 1999). Ozone causes needle injury and tree decline of *Pinus hartwegii* Lindl. at high altitudes in the mountains around Mexico City, especially in the Desierto de los Leones Park (de Bauer et al. 1985; Miller et al. 1994, 2002).

Alvarado-Rosales (1989) hypothesized the presence of high levels of ozone in the Park based on the symptoms of exposure in tobacco plants sensitive to ozone at several elevations. Fortunately, in 1990, an ozone monitoring station was established in the park. Data from this monitoring station demonstrated that ozone concentrations were high throughout the year as a result of the daily upslope flow of polluted air from Mexico City. Winter ozone concentrations remain nearly as high as summer concentrations in the MCAB, because winter skies are clearer, and incident sunlight remains high (Miller et al. 1994).

At the beginning of the decline phenomenon, several hypotheses were proposed to explain it. However, based on the available information, air pollution was considered the main cause of sacred fir decline. Unfortunately, at that time only indirect evidence was available to support this hypothesis. It was known that the prevailing winds came from the north and northeast (Jáuregui 1958), pushing the cloud of smog southward into the forested areas of the Desierto de los Leones.

Alvarez-Camarena (1996) studied the progression of injury development in *Abies religiosa* at a site with high ozone exposure. She reported that the symptoms in the leaves of sacred fir were mainly manifest on the upper leaf surface and consisted of whitish stippling that turned into larger and reddish lesions. Alvarado and coworkers had previously described these symptoms in the sacred fir forest of the Desierto de los Leones Park (Alvarado-Rosales 1989; Alvarado et al. 1987; Alvarado et al. 1993). Branches protected by plastic chambers did not develop the symptoms described above. Likewise, needles protected by an antitranspirant retained their typical dark green color, and needle retention time on protected branches was longer. The fact that foliar injury symptoms observed in declining fir trees did not develop in branches protected by plastic branch chambers or treated with antitranspirant suggests that air pollution was a primary stress factor in the decline, and that other factors such as insect and disease attacks, and possibly nutritional disorders, were secondary factors affecting predisposed trees.

Another study with sacred fir showed a reduction in tracheid size of the affected trees. Tracheid wall thickness, tracheid length, and ring width had all decreased in size since the late 1970s. This trend was observed in most of the sampled trees.

These types of responses have been reported for other species exposed to high ozone levels (Bernal-Salazar 1999). Reduced ring widths have been observed in *P. hartwegii* in the Desierto de los Leones Park and in el Ajusco, another high pollution site south of Mexico City since the 1970s (Alarcón 1993).

Native *Pinus hartwegii* growing on the top of the mountains within the park are clearly ozone sensitive and show typical ozone symptoms. Nevertheless, this species does not show the typical forest decline observed with sacred fir. *P. hartwegii* grows in open stands, while sacred firs grow in much denser stands. *P. hartwegii* trees also show a less dramatic type of decline characterized by reduced vigor, poor needle retention and condition, reduced needle length, and branch mortality.

Ciesla and Macías-Sámano (1987) attributed the sacred fir decline to air pollution, specifically to ozone, based on the effects of ozone on *Pinus hartwegii* to the south and southwest of Mexico City. These authors pointed out that symptoms in *P. hartwegii* were most severe on the slopes and ravines directly facing Mexico City. This pattern bolstered the hypothesis that some form of air pollution might be at least partially to blame for the problem; however, no ozone monitoring data were available within the Desierto de los Leones Park in 1987.

Typical symptoms caused by ozone on bioindicator species like tobacco, *Pinus hartwegii*, and native herbaceous plants, the direction of dominant winds, the location of the stands with the highest mortality (particularly at the end of ravines), the absence of foliar symptoms when sacred fir branches were protected by charcoal-filtered branch chambers or by antitranspirant sprayed on the upper surface of the needles (Alvarado et al. 1993), and data on ozone concentrations in the park, all support the hypothesis that ozone is an important factor in the sacred fir decline. Air pollution levels in the neighboring healthy stands are not as severe as in the cemeteries, mainly because of different exposure to the dominant winds from Mexico City. However, a glaring deficiency in the evidence is the lack of controlled ozone exposure studies with sacred fir. Only with such studies can it be conclusively determined that ozone causes the observed visual foliar injury.

Drought

Vázquez-Soto (1987) described two droughts in Mexican forests, the first one in 1976, and the second one at the end of 1982 and the beginning of 1983. The second drought killed a large number of firs, white cedars, and pines in the central part of Mexico. He pointed out that another disturbance factor in the Desierto de los Leones Park is excessive water extraction. Since 1786, the Mexico City government has owned the legal rights to capture and use the spring waters coming from the Desierto de los Leones Park. In 1899, the park and the Santa Fe forest provided the City of Mexico with 150L water/sec (Vázquez-Soto 1987). The steep, heavily forested slopes of the Desierto are an important source of water for the many small villages on the lower slopes (Ciesla and Macías-Sámano

1987). There is concern about the over collection of water in these catchments, because overcollection greatly reduces water infiltration and may impact ecosystem sustainability as a consequence of chronic moisture deficits (CORENA 1998). The "cemetery" areas, where mortality reached 100%, are located in some of the places where water is trapped and conveyed to the city. In these areas of water capture, it is expected that soil water available for plant uptake is greatly reduced, and plant water stress is increased. Although data on the amount of water extracted from the Desierto is limited, the few existing reports suggest that the lack of moisture may have contributed to the decline phenomenon since its inception.

Nutritional Deficiencies

In Mexico, there are few studies of nutrient relations in sacred fir forests. Browning of the upper leaf surface of older foliage was associated with N, Zn, K, Mg, and Mn deficiencies in the Desierto (López-López 2002; Rivera 1998). In other surveys in the park, chlorosis of the older foliage was attributed to low N, P, K, Mg, and Mn concentrations (Castañeda-Gonzáles et al. 1995; López-López and Rivera-Amaro 1995; López-López et al. 1995). In a Desierto de los Leones *Abies religiosa* plantation that was approximately six years old, there was a generalized K and Mn deficiency, and S was the element with the highest foliar concentration, possibly as a result of elevated S deposition (Fenn et al. 1999). Concentrations of P, Zn, and N were almost normal. Exchangeable K levels in the soils of the Park were statistically identical with those of other nonpolluted areas with healthy sacred fir stands (López-López et al. 1998).

According to the studies cited above, it can be hypothesized that nutritional deficiency plays a role in the decline by reducing the capacity of trees to produce photosynthates. Thus the trees are more susceptible to the effects of other stressors, including air pollutants. However, another posibility is that nutritional problems are secondary factors in the decline. For management purposes, it is important to diagnose nutrient excesses or deficiencies in order to correct nutritional disorders in the Desierto de los Leones Park.

Insects

In woody plants, physiological changes that are induced by environmental stresses are commonly prerequisites for attack by certain insects. Survival and growth of insects in stressed trees have been attributed to lowered chemical defenses, improved nutrition for the insect, and a better environment for insect growth. In conifers, successful attack by bark beetles for example, depends to a large degree on lowered defensive capacity in trees. Decreased defensive capacity is manifested as a change in the amount of oleoresin available at the time of attack and the capacity of trees to mobilize it near the point of attack. Production of defensive chemicals requires large amounts of carbohydrates (Koslowzky et al. 1991). Christiansen et al. (1987) speculated that the reduced capacity of stressed trees to withstand bark beetle attacks and the associated fungal colonization is correlated with low availability of carbohydrates for use in defensive wound reactions. Hence, a number of environmental stresses that reduce crown size or photosynthetic efficiency might lower the resistance of trees to bark beetle attack (Koslowzky et al. 1991).

In the Desierto de los Leones forest, trees have been attacked on multiple occasions by several insects and diseases (Vázquez-Soto 1987). Researchers have found several infestations of different species of bark beetles in dead and dying trees. More significanly, they found many of the living trees to be visibly stressed. The species of bark beetles infesting these trees were somewhat unusual in that they normally do not cause extensive damage. *Scolytus mundus* Wood, a bark beetle, was occasionally found invading the upper crowns and larger branches. Large numbers of a second bark beetle, *Pseudohylesinus variegatus* Blandford, infested the main boles. This last species had not previously been observed to be this aggressive, usually killing only isolated groups of trees during periods of drought (Ciesla and Macías-Sámano 1987).

Muñiz (1989) reported a general lack of correlation between the degree of tree damage in the park and the presence of some orders of insects. However, he documented the presence of Trichoptera, Plecoptera, Neuroptera, Collembola, and Thysanoptera in healthy trees in some instances. He also found the presence of some insect pests at very low levels. Gatica (1989) reported the presence of 10 insect orders in association with the foliage of declining fir trees. However, the author concluded that their action was not decisive in the decline syndrome. In this study, several Lepidoptera families were identified, including the genus *Cinara*, and *Pithyopthorus*. A field study in the Desierto de los Leones Park showed that *Pseudohylesinus variegatus* attacked only trees with transparent crowns, less than 25% live crown, and yellowish green foliage; otherwise, fir trees survived attacks by this insect (Paz and Aguirre 1988).

Based on the above information and direct observations, the role attributed to insects in fir decline and mortality is secondary. Insect infestations occur only when trees are already stressed by other factors.

Pathogens

Abiotic stresses often predispose woody plants to biotic stress. Outbreaks of stem cankers, diebacks, declines, and some root rots follow loss of tree vigor as a result of drought, flooding, mineral deficiency, or air pollution, as well as combinations of these (Kozlowski 1985). Such stresses render the host plant more susceptible to facultative parasites, including weak and nonaggressive parasites that would not be a threat to vigorous plants. In many instances, woody plants are invaded regardless of prevailing stresses, but pathogens are usually more aggressive when the host has become physiologically predisposed under stress (Schoeneweiss 1978).

There are few reports on microorganisms associated with sacred fir. In 1989, *Nectria* sp. was reported as a new fungus attacking fir saplings in the Desierto de los Leones Park (Cibrián and Macías-Sámano 1989). In a survey carried out by Reséndiz and Salinas (1989), several fungi were found associated with buds and foliage of declining firs. Among these fungi were the following: *Alternaria*, *Torula*, *Hormodendrum*, *Pestalotia*, *Pullularia*, *Chaetomium* and *Cladosporium*. From stems and branches they isolated: *Penicillium*, *Alternaria*, *Hormodendrum*, *Mucor*, *Zygodesmus*, *Stemphyllium*, *Aspergillus*, *Trichoderma*, *Rhizopus*, *Sporotrichum*, *Zygodesmus* and *Pullularia*. From the rhizosphere, they isolated species of *Phoma*, *Phomitopsis*, and *Pythiopsis* as possible pathogenic fungi. Finally, among the macroscopic fungi they found *Boletus*, *Lactarius*, and *Russula*. The pathogenic nematode *Tylenchus* was also isolated.

Lachnellula sp., a canker fungus, and Armillaria root rot are two fungi recently associated with declining fir trees in the Desierto de los Leones. Incidence of the canker fungus has been increasing in recent years, which may be an indicator of the stressful condition prevailing in the park. A few trees have also been infected by Armillaria (Alvarado, unpublished data). Finally, in a survey of mycorrhizal fungi associated with *Abies religiosa*, García (1997) found that the highest percentage of species and families of such fungi were associated with the healthiest fir stand. The greater mycorrhizal diversity was attributed to the wider range of tree age classes in the healthy stand. In a stand with greater symptoms of fir decline, fungal diversity was slightly lower; however, most of the fungi were saprophytes. No parasitic fungi were found in any of the sampled plots (García 1997).

The papers reviewed above show the lack of research in this area. At this point, no conclusions can be made about the role of pathogens in the sacred fir decline. We can only say that there is no evidence that a plant pathogen is the primary cause of the observed symptoms of sacred fir decline.

Lack of Forest Management

Quantity and quality of plant growth depend on the hereditary potential of the plant and the environment in which a plant is growing, operating through its physiological processes. The full potential for growth of woody plants is seldom attained because various environmental stresses continually inhibit essential physiological processes. These include abiotic stresses such as drought, extreme temperatures, shading, wind, air pollution, fire, infertility, poor soil structure (e.g., soil compaction), and inadequate soil aeration. Biotic stresses include plant competition, human activities, attacks by insects and plant pathogens, and occasionally, feeding by higher animals such as deer and beavers.

Most of the increases in forest yield have resulted from improvements in management. Foresters, arborists, farmers, and gardeners have quite different objectives in terms of products, but they have one common objective: to manage their crops in ways that will minimize the effects of the omnipresent stresses on growth and yield. Important cultural practices include site preparation for planting, fertilization, irrigation, stand thinning, pruning, and application of chemicals.

Naturally regenerated stands are often greatly overstocked, resulting in severe competition for water and minerals and in slow growth early in the life of the stand. Natural thinning is inefficient, because it results from severe competition that reduces the rate of growth. On the other hand, if stand density is too low, growing space is wasted, and self-pruning is delayed, resulting in trees with many knots, strongly tapered stems, and a reduced economic return over time per unit of land. When a stand of trees is thinned, the amount of growing space for the crowns and roots of residual trees is increased. Usually the rates of vegetative and reproductive growth of the trees left standing are accelerated because of greater availability of light, water, and mineral nutrients (Koslowzky et al. 1991). Thinning of stands often alters the susceptibility of the residual trees to insects and diseases and reduces the incidence of pathogenic fungi.

Based on the above information and observations at the Desierto Park and in adjacent healthy stands, it seems that thinning or lack of thinning as a part of forest management, plays an important role in the decline phenomenon in the park. In this forest, no management is being performed. Sanitation cuts, in particular, are not allowed, so the risk of pest incidence and disease spread increases. Aging of the forest enhances this situation. Forest managers in the Desierto Park recognize the importance of stand thinning. However, Mexican law does not allow forest management in national parks. Public opinion and communication media have also argued against the use of sanitation cuts. The main effect of the lack of forest management is overmaturation of the trees, lack of natural regeneration, increased physiological stress due to competition for resources, and increased susceptibility to pollution, insects and diseases.

In the 1983 inventory in the Desierto de los Leones Park, the oldest fir trees found were between 107 and 119 years old (Vázquez-Soto 1987). In the 1987 inventory, the number of trees per hectare was 200 or more, the wood volume was more than 600 m^3 per hectare, and the crown cover was 3000 m^2/ha (Vázquez-Soto 1988). In contrast, in neighboring fir forests which have been harvested for timber for many years (La Venta in Cuajimalpa Municipality and San Lorenzo Acopilco), wood volumes were 116.3 and 290.0 m^3/ha and most of the trees were in the 15 and 30 cm dbh size class, meaning that natural regeneration is occurring in these forests (Vázquez-Soto 1988). The San Lorenzo Acopilco and La Venta forests remain healthy, and Vázquez-Soto (1987) hypothesized that the main reason was the sanitation cuts carried out for several years in order to eliminate older, dead, and dying trees. Currently, forest recovery is occurring in areas of the Desierto de los Leones where sanitation cuts were carried out in 1984. The remaining previously declining trees show some signs of recovery in that many epicormic branches are growing along the main bole. The rate of fir decline is now lower than it was in the late 1980s.

The Fir Decline Model in the Desierto de los Leones Park

Sinclair (1964) was the first to suggest that decline problems may develop from the interaction of three or more sets of factors: the predisposing, inciting, and contributing factors. Predisposing factors are long-term, slowly changing factors such as soil, site, and climate. Predisposing factors alter the tree's ability to withstand or respond to injury-inducing agents. Inciting factors are short in duration. They may be physiological or biological in nature and generally produce dieback of small branches. Among these are insect defoliators, late spring frost, drought, and salt spray. Contributing factors have been suggested to include a collection of environmental conditions as well as biotic factors such as canker and decay fungi and bark and wood boring insects. Later, Manion (1991) expanded, modified, and developed the model into a teaching tool based on additional information from similar decline problems. The outcome of this was the Manion decline spiral. Based on Sinclair´s concepts, Manion´s decline spiral, and the best available evidence, we propose a decline spiral as a working conceptual model of the process of sacred fir decline (Fig. 10.4).

Long-term predisposing factors in the fir decline model include overmaturation of the stand (lack of forest management), soil moisture deficit, the urban heat-island effect (Jáuregui 1997), and air pollution. Vázquez-Soto (1987) states that commercial and sanitation cuts on neighboring forests are beneficial and prevent overmaturation of the stand. Insufficient soil moisture is also a likely long-term predisposing factor. Currently there is an extensive water capture system in the park. Based on field observations, a correlation exists between water capture sites and the locations of most of the "cemeteries." These ceme-

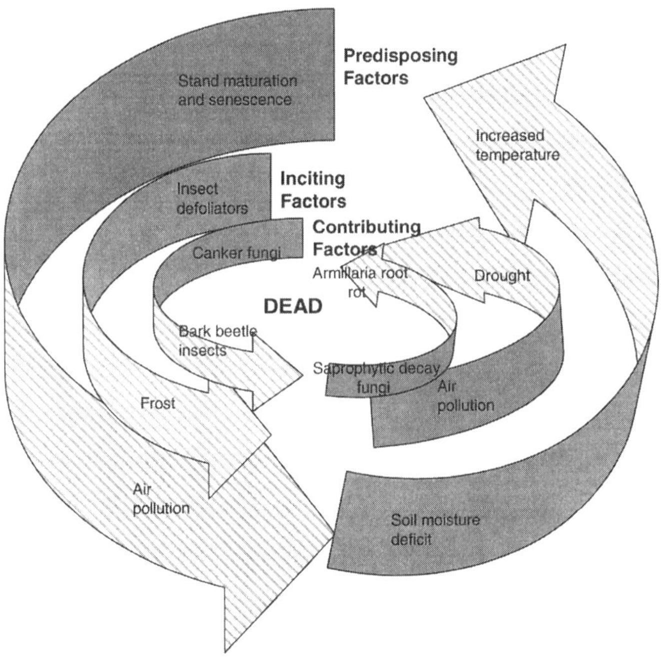

Figure 10.4. Decline spiral proposed to explain sacred fir mortality in the Desierto de los Leones National Park.

tery areas are located at the ends of the ravines where the normal water supply has been greatly reduced as a result of water capture for human consumption. Over the last century, the atmospheric temperature in the MCAB has increased 0.5° C, which likely exacerbates drought stress effects (Ernesto Jáuregui, personal communication).

Among the inciting factors, drought, frost, air pollution, and insect defoliators appear to be the most important. These factors, over time, have produced dieback, and they have also exerted stress effects for short periods of time. As mentioned before, two droughts have occurred during the onset and development of the fir decline problem. The first one occurred in 1976, and the second one in 1982/1983. Damaging frosts occur in winter periodically. High air pollution levels occur throughout the year, but physiological insults are based on plant uptake of pollutants. High pest incidence occurs when environmental conditions are conducive. All of the above are considered inciting factors.

Finally, contributing factors include bark beetles, canker fungi, and decay fungi. Bark beetles, *Pseudohylesinus variegatus* and *Scolytus mundus*, have been reported as the final cause of death of declining fir trees. The canker fungi, *Nectria* and *Lachnellula*, are currently killing branches and small stems. The presence of these opportunistic canker fungi is an indicator of the stressful conditions prevailing in some areas of the park. Decay fungi present in the trees are considered saprophytes, and the incidence of pathogenic root fungi is still at low levels.

Based on the presented information, we hypothesize the following scenario leading to fir decline. Stand densification and the stress caused thereby, in combination with physiological injury induced by exposure to high pollution levels, predispose sacred fir to decline. Pollution levels high enough to cause this predisposition occur predominantly in ravines directly downwind of the urban pollution source, where pollutants can accumulate to abnormally high levels. Predisposed trees are further impacted by drought stress, which occurs as a result of periodic droughts and because of excessive water capture from these watersheds for urban use. Increasing temperature, occurring as a result of the large heat-island effect from Mexico City (Jáuregui 1997), may further exacerbate the problem. Overstressed sacred fir trees ultimately succumb to insect infestations and the combination of multiple stresses.

Causal Factors of the Decline and Management Options: Research Needs

The cause or etiology of forest or tree declines is notoriously difficult to conclusively diagnose. Sacred fir decline is no exception. Various factors such as air pollution, stand density and maturation, drought stress, nutritional problems, and insect pests, possibly involved in the development of fir decline, have been studied and are reviewed in this chapter. A conceptual model of the factors leading to fir decline is proposed, although the most important factors leading to the decline are not definitively known. This leaves forest managers in a dilemma, because without knowing the causal factors, the managers cannot know the best management options. There is evidence that thinning of the affected stands may allow sacred fir regeneration or reestablishment; however, field trials are needed to test this hypothesis. Some of the priority research questions, which need to be addressed to increase our understanding of the decline and to evaluate the best management options, are listed below:

- To determine by controlled ozone exposure whether the observed foliar symptoms on sacred fir are caused by ozone.
- To determine the ozone exposure at which physiological impacts and visual injury occur.
- To determine the effects of chronic deposition of N and S and heavy metals on nutrient cycling and plant/soil nutrient relations in sacred fir stands.
- To study root and mycorrhizal development, growth dynamics, and turnover in healthy and declining stands, including root and mycorrhizal community responses to air pollution exposure.
- To characterize air pollution exposure in the park, including spatial variability across the landscape, using passive monitors.
- To study, in depth, the role in the decline of other stress factors (e.g., drought, pests, nutritional deficiencies or toxicity, stand densification), including multiple stress interactions.
- To determine by physiological studies the major stress factors and their effects on plant function.
- To carry out applied research on silvicultural management options such as thinning to determine if sacred fir can be reestablished in sites of decline as opposed to replanting with other species.
- To select appropiate genetic lines for replanting in the cemeteries, while still maintaining genetic heterogeneity in the population, through research on genetic variation in sacred fir.
- To determine the spatial and temporal patterns of development of fir decline and the environmental and site factors associated with decline.

Summary

Sacred fir decline (*Abies religiosa* Schl. et Cham.) in the Desierto de los Leones Park southwest of Mexico City has attracted public and scientific attention since it was first reported in 1987 (Alvarado et al. 1987; Ciesla and Macías-Sámano 1987). Numerous studies have been carried out to elucidate the causes and to help foresters manage the problem. Unfortunately, the available information is fragmented, and studies often focused on only one or two possible factors. More integrative studies, controlled pollution exposures, and field tests of silvicultural options to manage for forest health are badly needed in order to better understand the causal factors and to develop effective control measures. In consideration of the lack of publications that integrate the available information on the phenomenon, this chapter was written to fill this void. A conceptual model of the most likely factors in the decline is proposed as a hypothesis to guide future research. Further research needs are also suggested.

References

Alarcón, M.A., de Bauer, L.I., Jasso, J., Segura, G., and Zepeda, E.M. 1993. Patrón de crecimiento radial en árboles de *Pinus hartwegii* afectados por contaminación atmosférica en el suroeste del Valle de México. *Agrociencia. Serie Recursos Naturales Renovables* 3:67–80.

Alvarado-Rosales, D. 1989. *Declinación y Muerte del Bosque de Oyamel Abies religiosa en el Sur del Valle de México.* M.Sc. Thesis. Montecillo, State of México: Colegio de Postgraduados.

Alvarado-Rosales, D., and Hernández-Tejeda, T. 1995. Condición fitosanitaria actual de un bosque de oyamel al suroeste de la Ciudad de México. In *Resúmenes del II Congreso Mexicano sobre Recursos Naturales*, pp. 49. Montecillo, State of México: Colegio de Postgraduados and Sociedad Mexicana de Recursos Naturales.

Alvarado-Rosales, D., and Hernández-Tejeda, T. 1998. Twelve years evaluation of sacred fir decline (*Abies religiosa*) evolution at the Desierto de los Leones Park, southern Mexico City. In *Abstracts of the Annual Meeting of the American Phytopathological Society in Las Vegas, Nevada*, pp. 88:S3. St. Paul, Minnesota: American Phytopathological Society.

Alvarado-Rosales, D., de Bauer, L.I., and Galindo-Alonso, J. 1991. Declinación y muerte del bosque de oyamel (*Abies religiosa*) en el sur del Valle de México. *Agrociencia. Serie Recursos Naturales Renovables* 1:123–143.

Alvarado-Rosales, D., de Bauer, L.I., and Galindo-Alonso, J. 1993. Decline of sacred fir (*Abies religiosa*) in a forest park south of Mexico City. Environmental Pollution 80:115–121.

Alvarado-Rosales, D., Hernández-Tejeda, T., de Bauer, L.I., and Galindo-Alonso, J. 1987. Declinación y muerte del bosque de oyamel en el sur del Valle de México: Síntomas, desarrollo y posibles causas. In *Resúmenes del XIV Congreso Nacional de Fitopatología*, pp. 72. Morelia, Michoacán, México: Sociedad Mexicana de Fitopatología.

Alvarado-Rosales, D., Hernández-Tejeda, T., de Bauer, L.I., and Galindo-Alonso, J. 1989. Decaimiento del oyamel (*Abies religiosa*) en el Desierto de los Leones, D.F. In *Memoria del IV Simposio Nacional sobre Parasitología Forestal*, pp. 790–801. Publicación Especial 60, II. México, D.F.: (INIFAP-SARH) Instituto Nacional de Investigaciones Forestales, Agrícolas y Pecuarias-Secretaria de Agricultura y Recursos Hidraulicos.

Alvarez-Camarena, D.E. 1996. *Determinación de Alteraciones a Nivel Citológico e Histoquímico en Abies religiosa (H.B.K.) Schl et Cham del Desierto de los Leones del D.F. y Su Relación con el Ozono.* M.Sc. Thesis. México, D.F.: Universidad Nacional Autónoma de México.

Bernal-Salazar, S.I. 1999. *Efecto del Estrés Ambiental sobre el Tamaño Celular de la Madera en Abies religiosa Schl. et Cham.* M.Sc. Thesis. Montecillo, México: Colegio de Postgraduados in Ciencias Agrícolas.

Bravo-Alvarez, H., and Torres-Jardón, R. 2002. Air pollution levels and trends in the Mexico City metropolitan area. In *Urban Air Pollution and Forests: Resources at Risk in the Mexico City Air Basin.* Ecological Studies series, eds. M.E. Fenn, L.I. de Bauer, and T. Hernández-Tejeda, (this volume). New York: Springer-Verlag.

Castañeda-González, M.I., López-López, M.A., and Velásquez-Martínez, A. 1995. Efecto de tratamientos de fertilización sobre algunos síntomas de declinación en una plantación de oyamel. In *Memoria del II Congreso Mexicano sobre Recursos Naturales*, pp. 52. Montecillo, State of Mexico: Colegio de Postgraduados and Sociedad Mexicana de Recursos Naturales.

Christiansen, E., Waring, R.H., and Berryman, A.A. 1987. Resistance of conifers to bark beetle attack: Searching for general relationships. Forest Ecology Management 22: 89–106.

Cibrían-Tovar, D. 1989. Air pollution and forest decline near Mexico City. Environmental Monitoring and Assessment 12:49–58.

Cibrían-Tovar, J., and Macías-Sámano, J.E. 1989. Presencia de un nuevo patógeno sobre renuevo de oyamel (*Abies religiosa*) en el Parque Recreativo y Cultural Desierto de los Leones. In *Memoria del IV Simposio Nacional sobre Parasitología Forestal*, pp. 653–666. Publicación Especial 60, II. México, D.F.: (INIFAP–SARH) Instituto Nacional de Investigaciones Forestales, Agrícolas y Pecuarias-Secretaria de Agricultura y Recursos Hidráulicos.

Ciesla, W.M., and Macías-Sámano, J.E. 1987. Desierto de los Leones: A forest in crisis. *Am. For.* 93:29–31; 72–74.

CORENA (Comisión de Recursos Naturales). 1998. *Memoria del Taller Regional de Cuajimalpa de Morelos: Programa Regional de Desarrollo Sustentable de la Zona Rural del Distrito Federal.* México, D.F.: Gobierno del Distrito Federal.

de Bauer, L.I. 1972. Uso de plantas indicadoras de aeropolutos en la Ciudad de México. *Agrocienc.* 9:139–141.

de Bauer, L.I., and Hernández-Tejeda, T. 1986. *Contaminación: Una Amenaza Para la Vegetación en México.* Chapingo, México: Colegio de Postgraduados.

de Bauer, L.I., and Krupa, S.V. 1990. The Valley of Mexico: Summary of observational studies on its air quality and effects on vegetation. *Environ. Pollut.* 65:109–118.

de Bauer, L.I., Hernández-Tejeda, T., and Manning, W.J. 1985. Ozone causes needle injury and tree decline in *Pinus hartwegii* at high altitudes in the mountains around Mexico City. *J. Air Pollut. Control Assoc.* 35:838.

Fenn, M.E., de Bauer, L.I., Quevedo-Nolasco, A., and Rodríguez-Frausto, C. 1999. Nitrogen and sulfur deposition and forest nutrient status in the Valley of Mexico. *Water Air Soil Pollut.* 113:155–174.

García, M. 1997. *Distribución de Hongos Ectomicorrízicos en Rodales de Abies religiosa (H.B.K.) Schl. et Cham. y la Relación con su Declinación.* Tesis de Licenciatura. México, D.F.: Universidad Nacional Autónoma de México. Departamento de Biología.

Gatica, E. 1989. Incidencia de insectos en suelo y follaje de oyamel en el Desierto de los Leones, D.F. In *Memoria del IV Simposio Nacional sobre Parasitología Forestal*, pp. 775–789. Publicación Especial 60, II. México, D.F.: (INIFAP–SARH) Instituto Nacional de Investigaciones Forestales, Agrícolas y Pecuarias-Secretaria de Agricultura y Recursos Hidraulicos.

Hernández-Tejeda, T. 1981. *Reconocimiento y Evaluación del Daño por Gases Oxidantes en Pinos y Avena del Ajusco, D.F.* Tesis profesional. Chapingo, México: Universidad Autónoma Chapingo.

Hernández-Tejeda, T., and de Bauer, L.I. 1989. *La Supervivencia Vegetal ante la Contaminación Atmosférica.* Montecillo, Mexico: Colegio de Postgraduados.

Innes, J.L. 1993. *Forest Health: Its Assessment and Status.* Wallingford, Oxon, UK: CAB International.

Jáuregui, E. 1958. El aumento de la turbiedad del aire en la Ciudad de México. *Ing. Hid. México* 4:A77–A81.

Jáuregui, E. 1997. Heat island development in México City. *Atmos. Environ.* 31: 3821–3831.

Kozlowsky, T.T. 1985. Tree growth in response to environmental stresses. *J. Arboric.* 11:97–111.

Kozlowsky, T.T., Kramer, P.J., and Pallardy, S.G. 1991. *The Physiological Ecology of Woody Plants.* San Diego, CA: Academic Press, Inc.

López-López, M.A. 2002. Efectos del ambiente aéreo y del suelo sobre el desarrollo de síntomas de declinación de oyamel. Revista Terra (Mexico) (in press).

López-López, M.A., and Rivera-Amaro, R. 1995. Caracterización nutrimental de follaje de oyamel en proceso de declinación. In *Memoria del II Congreso Mexicano sobre Recursos Naturales*, pp. 50. Montecillo, State of Mexico: Colegio de Postgraduados.

López-López, M.A., Castañeda-González, M.I., and Velásquez-Martínez, A. 1995. Sistema de evaluación del grado de daño de árboles de oyamel en proceso de declinación. In *Memoria del II Congreso Mexicano sobre Recursos Naturales*, pp. 51. Montecillo, State of Mexico: Colegio de Postgraduados.

López-López, M.A., Velásquez-Mendoza, J., Velásquez-Martínez, A., González-Romero, V., and Cetina-Alcalá, V.M. 1998. Estado nutrimental de *Abies religiosa* en un área con problemas de contaminación ambiental. *Agrociencia* 32:53–59.

Macías-Sámano, J., and Cibrían-Tovar, J. 1989. Evaluación mediante fotografía aérea infrarroja de la mortalidad de *Abies religiosa* en el parque Desierto de los Leones. In *Memoria del IV Simposio Nacional sobre Parasitología Forestal*, pp. 889–901. Publicación Especial 60, II. Mexico, D.F.: (INIFAP–SARH) Instituto Nacional de Investigaciones Forestales, Agrícolas y Pecuarias-Secretaria de Agricultura y Recursos Hidraulicos.

Manion, P.D. 1991. *Tree Disease Concepts*, 2nd edn. Englewood Cliffs, N.J.: Prentice Hall.

Martínez, M. 1963. *Las Pináceas Mexicanas*, 3rd edn. México, D.F.: Instituto de Biología, Universidad Nacional Autónoma de México.

Miller, P.R., de Bauer, L.I., Quevedo-Nolasco, A., and Hernández-Tejeda, T. 1994. Comparison of ozone exposure characteristics in forested regions near Mexico City and Los Angeles. *Atmos. Environ.* 28:141–148.

Miller, P.R., de Bauer, L.I., and Hernández-Tejeda, T. 2002. Oxidant exposure and effects on pines in forests in the Mexico City and Los Angeles California Air Basins. In *Urban Air Pollution and Forests: Resources at Risk in the Mexico City Air Basin*. Ecological Studies series, eds. M.E. Fenn, L.I. de Bauer, and T. Hernández-Tejeda, (this volume). New York: Springer-Verlag.

Muñiz, R. 1989. Los insectos en el bosque de oyamel del Desierto de los Leones, D.F. In *Memoria del IV Simposio Nacional sobre Parasitología Forestal*, pp. 765–774. Publicación Especial 60, II. Mexico, D.F.: (INIFAP–SARH) Instituto Nacional de Investigaciones Forestales, Agrícolas y Pecuarias-Secretaria de Agricultura y Recursos Hidraulicos.

Reséndiz, M.J.F., and Salinas, Q.R. 1989. Observaciones preliminares sobre la micoflora asociada a *Abies religiosa* (H.B.K.) Schl. et Cham. In *Memoria del IV Simposio Nacional sobre Parasitología Forestal*, pp. 695–699. Publicación Especial 60 II. Mexico, D.F.: (INIFAP–SARH) Instituto Nacional de Investigaciones Forestales, Agrícolas y Pecuarias-Secretaria de Agricultura y Recursos Hidraulicos.

Rivera, R. 1998. *Nutrición y Síntomas de Declinación de Abies religiosa* (H.B.K.) *Schl. et Cham. en el Desierto de los Leones, D.F.* Tesis de Licenciatura. Chapingo, México: Universidad Autónoma Chapingo.

Rzedowski, J. 1978. *Vegetación de México.* Mexico, D.F.: Editorial Limusa, S.A.

Rzedowski, J., and de Rzedowski, G.C. 1979. *Flora Fanerogámica del Valle de México. Generalidades, Gymnospermae, Dicotiledoneae*, Vol. I., Mexico City: Editorial CEC S.A.

Sánchez-Sánchez, O. 1974. *La Flora del Valle de México.* Mexico, D.F.: Editorial Herrero, S.A.

Schoeneweiss, D.F. 1987. Water stress as a predisposing factor in plant disease. In *Water Deficits and Plant Growth, Water and Plant Disease,* Vol. V., ed. T.T. Kozlowski, pp. 61–99. New York: Academic Press.

Sinclair, W.A. 1964. Comparisons of recent declines of white ash, oak, and sugar maple in Northeastern woodlands. *Cornell Plantations* 20:62–67.

Vázquez-Soto, J. 1987. *El Saneamiento y la Limpia Forestales en el Desierto de los Leones.* México, D.F.: Comisión Coordinadora para el Desarrollo Rural (COCODER), Departamento del Distrito Federal.

Vázquez-Soto, J. 1988. *Los Tratamientos Silvícolas del Desierto de los Leones. Sus Fundamentos.* México, D.F.: Comisión Coordinadora para el Desarrollo Rural (COCODER), Departamento del Distrito Federal.

11. Histological Symptoms of Air Pollution Injury in Foliage, Bark, and Xylem of *Abies religiosa* in the Basin of Mexico

Teresa Terrazas and Sergio Bernal-Salazar

Histological research has provided valuable information contributing to our understanding of the impacts of air pollution on plant species (Thomson 1975; Weber and Grulke 1995). Tissue modifications may appear before visible symptoms in leaves and stems. Tissue modifications can also be used as indicators of air pollution effects (Bernal et al. 2002; Fincher and Alscher 1992). Anatomical studies help elucidate biochemical and physiological changes that may possibly occur in plants exposed to pollutants, especially those changes related to net photosynthesis, allocation of resources (carbohydrates), and mechanisms of injury response. Xylem studies using dendrochronological techniques make it possible to reconstruct past events in forest communities and to determine tree growth rates in species exposed to air pollutants based on changes in ring width, density, and element composition (Weber and Grulke 1995).

Forest decline in the Mexico City Air Basin (MCAB) has been observed since the late 1970s (Vázquez 1987, 1988). Evaluation of the contribution of air pollutants to this decline was not initiated until the late 1980s when a rapid decline was observed (Alvarado-R et al. 1993; de Bauer and Krupa 1990; Miller et al. 1994). In this chapter, we review studies on the histological symptoms of air pollution injury to foliage and symptoms in bark and xylem observed in species growing in the MCAB. We compare these symptoms with those reported for other species in America and Europe, with emphasis on foliar and xylem modifications. In particular, we present results of studies with the special case of *Abies religiosa*, a species which has suffered severe decline in forests southwest of Mexico City.

Anatomical and Physiological Responses of Foliage to Air Pollution

Foliar injury symptoms and reduced needle retention, symptom associated with photochemical air pollutants, have been observed in conifers like *Pinus hartwegii* Lindl., *P. leiophylla* Schlecht et Cham., *P. montezumae* Loudon. (Montezuma pine), and *Abies religiosa* (H.B.K.) Schlecht. et Cham. (sacred fir) (Alvarado-R et al. 1993; de Bauer et al. 1985; Hernández-Tejeda and de Bauer 1984, 1986). These symptoms were also reported in broadleaf species (e.g., *Prunus serotina* and *Eucalyptus globulus* Labill.) that have been influenced by air pollutants (Hernández-Tejeda et al. 1981; Skelly et al. 1997). Symptoms related to nutrient content in healthy and unhealthy needles of *Abies religiosa* were studied by Rivera-Amaro (1998). More severe reddish mottling was associated with low concentrations of N, P, and K. Other authors (Alvarado-R et al. 1993; López-López et al. 1998) have also reported deficiencies of K, Mg, and Mn in mature foliage *A. religiosa*. It was suggested that chlorosis in mature foliage has its origin in nutrient retranslocation to new foliage. Furthermore, López-López et al. (1998) proposed that pollutant deposition to soil could cause nutrient disorders.

Anatomical responses of needles exposed to air pollution have been investigated in only a few studies in the MCAB (Alvarado-R 1993; Alvarez et al. 1998). Cross-sections of symptomatic *P. hartwegii* leaves show that the endodermis, transfusion, and phloem cells are collapsed. In more severely injured needles, the cells were necrotic (Hernández-Tejeda 1981). In *A. religiosa* the cells of palisade tissue showed an accumulation of substances that appeared to be phenolic compounds (Alvarado-R et al. 1993). A summary of tissue modifications occurring in different coniferous species in association with air pollutants is presented in Table 11.1. However, caution should be exercised with species which did not develop any visible symptoms attributed to air pollutant such as ozone (O_3) (Fincher and Alscher 1992). Chlorosis is associated with chloroplast deterioration and reduction of chlorophyll content of affected cells. These lesions give rise to chlorotic flecks or mottle (Treshow 1984) and are common in pine in which the mesophyll is not differentiated (Davis and Wood 1972). In leaves with differentiated mesophyll cells, injury symptoms are conspicuous in the adaxial side (Lucas et al. 1993).

After an extensive investigation, Alvarez et al. (1998) described fully the etiology of foliar injury in *A. religiosa* growing in the MCAB. They selected three stands at Desierto de los Leones (DL) where mountainsides and ravines face Mexico City, the source of air pollution. A control site, on the mountainsides of the Popocatepetl volcano, at which air pollutants occur at much lower concentrations was also selected. The leaves collected in the DL stands showed visual evidence of injury that varied in severity among the stands (Alvarado-R et al. 1993). The needles collected at the control site appeared healthy. Foliar injury at the DL was characterized initially as whitish stippling on the adaxial surface, confirming the observations of Alvarado-R et al. (1993). Injury may start early in the development of one-year-old needles (Alvarez 1996). The white stippling

Table 11.1. Summary of affected foliar tissues in different coniferous species[a]

Species	Country	Mesophyll modification	Cell wall modification	Chloroplast modification	Production of inducible phenolic compounds
Abies religiosa	Mexico	X	X	X	X
Pinus hartwegii	Mexico	X			X
Abies alba	Europe	X	X		
Pinus nigra	Europe	X			X
Pinus resinosa	USA	X			X
Pinus sylvestris	Europe	X	X		X
Pinus ponderosa	USA	X	X	X	
Picea rubens	Europe	X			X
Picea abies	Europe	X	X		
Sequoiadendron giganteum	USA	X	X		

[a] Data from Alvarez et al. 1998; Evans and Leonard 1991; Fincher and Alscher 1992; Hernández-Tejeda 1981; Kärenlampi 1986; Maier-Muercker and Koch 1992; Schiffgens-Gruber and Lutz 1992; Sutinen et al. 1998.

is commonly located at the base of the leaves and progresses towards the tip. As injury progresses, there is a color change to reddish mottling, which begins to coalesce laterally to the midrib. When injury becomes severe, the reddish mottling occurs not only in the margins of the leaf but also on the abaxial surface. The entire needle eventually turns yellow, and injury culminates with premature abscission after only three years. These symptoms were observed in the three stands located in DL, but only moderate leaf symptoms were seen in the control stand in the southeast part of the Basin.

Cross-sections of conspicuously injured needles show that palisade parenchyma is the first tissue to be altered (Alvarez et al. 1998; Alvarado-R et al. 1993). The injury may involve plasmolysis and necrosis. The first anatomical symptoms are observed as a cellular alteration with separation of the plasmalemma from the cell wall, degradation of the protoplast, and modifications of the vacuole and chloroplast. The separation of the plasmalemma has been described in vegetation with chronic exposure to ozone and other air pollutants, and is frequently found in the event of acute exposure (Thomson et al. 1996). Other signs of injury in *Abies religiosa* include a drastic reduction in cytoplasm together with an accumulation of phenolic compounds in the vacuole. These symptoms distinguish the various degradation stages in *A. religiosa* (Fig. 11.1) as in other coniferous species (Evans and Leonard 1991; Kärenlampi 1986; Maier-Maercker and Koch 1992; Zobel and Nighswander 1991).

Alvarez (1996) observed many cellular alterations in *A. religiosa* leaves exposed to air pollution in the MCAB. These ultrastructural alterations were compaction, greatly increased cytoplasmic density, cytoplasmic degradation, separation of the plasmalemma from the cell wall, cellular collapse, and degradation of

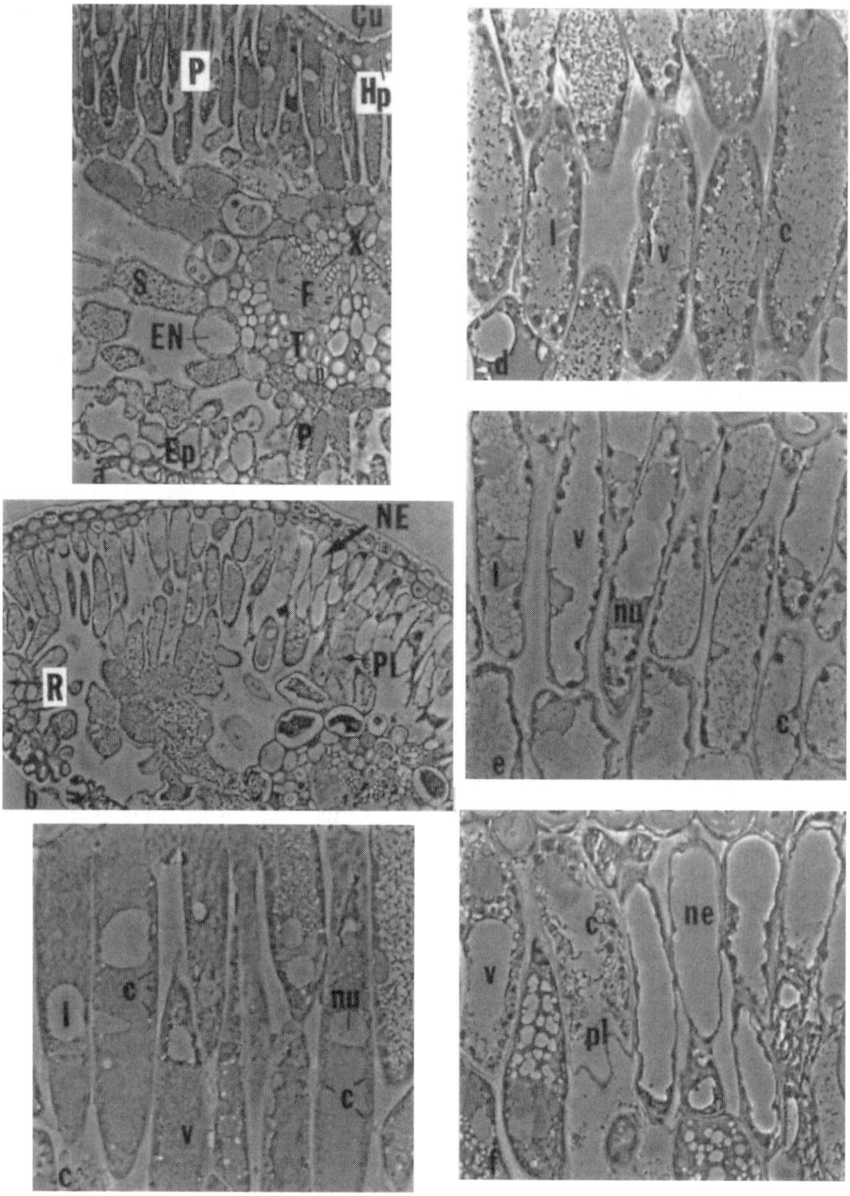

Figure 11.1. Leaf anatomy of healthy and unhealthy *Abies religiosa* trees. (a) Healthy leaf with cuticle, *Cu*; single epidermis, *Ep*; one-layered hypodermis, *Hp*; palisade parenchyma, *P*; spongy parenchyma, *S*; endodermis, *EN*; transfusion tissue, *T*; and two vascular bundles with xylem, *X*, and phloem, *F*. (b) Unhealthy leaf: plasmolysis, Pl; and necrosis *NE*; (c) Three-year-old leaf in Popocatepetl control stand (healthy leaf). (d) Three-year-old leaf in Desierto de los Leones (DL) (moderate unhealthy leaf). Cells with granular vacuoles. (e) Two-year-old leaf in DL (unhealthy leaf). Cells with fine and clear vacuoles. (f) Three-year-old leaf in DL (unhealthy leaf). Plasmolysed cells with protoplasm degradation and empty vacuoles. The symbols are c, chloroplast; v, vacuole; nu, nucleus; l, lipid; pl, plasmolysis; ne, necrosis; and R, resin canal. (Reprinted from *Environmental Pollution*, Vol. 103, D. Alvarez, G. Laguna, and I. Rosas, Macroscopic and microscopic symptoms in *Abies religiosa* exposed to ozone in a forest near Mexico City, pp. 251–259, Copyright 1998, with permission from Elsevier Science.)

all organelles. However, thylakoids were unaltered. These modifications constitute various manifestations of membrane alteration. Compaction and large increases in the cytoplasmic density of *A. religiosa* cells correspond to the chronic-type injury observed in other species (Miyake et al. 1989; Ojampera et al. 1992) with exposure to relatively low concentrations of O_3. Cytoplasmic degradation, separation of plasmalemma, and cellular collapse correspond to the type of severe injury typically observed with high concentrations of ozone (Pell and Weissberger 1976).

Alvarez's results (1996) showed that the internal membrane system of chloroplasts is altered in injured leaves; the amount of plastoglobuli increases as thylakoids decrease. In addition, plastoglobuli electrodensity tends to decrease as injury increases. It has been reported that the increase in electrodensity of the stroma, accompanied by subsequent weakening of chloroplast structure in severely injured cells, is a characteristic associated with high O_3 exposure (Sutinen et al. 1998; Thomson 1975). Similar ultrastructural results were detected in *Pinus sylvestris* L. (Scotch pine) seedlings exposed to O_3 (Utriainen et al. 1998) and in mature trees of this species exposed to other air pollutants (excess sulfur, heavy metals, and acidic deposition) (Kukkola et al. 1997). The increase in the number and size of plastoglobuli found in unhealthy leaves of *A. religiosa* is characteristic of chloroplast senility, as is foliar bleaching. The effect of O_3 on chloroplast structure is very similar to the onset of senility in terms of the changes seen in the lipid behavior of healthy leaves (Miyake et al. 1989). Plastoglobuli probably also disappear due to the effect of lipase activity. The mechanisms that act on chloroplast membranes may also affect other cellular membranes, particularly plasmalemma and tonoplast of *A. religiosa*. These results are supported by Le Thiec et al. (1994), who found that during the needle aging process, element concentrations in guard cells reflect the foliar tissue variation associated with senescence.

Alvarez (1996) noted a depletion in starch content in injured leaves from DL compared with the control stand. In both areas, the younger leaves had a higher carbohydrate content than the older ones. Tingey et al. (1976) mentioned that chronic exposure to O_3 reduces the photosynthetic rate in the oldest leaves, but it stimulates photosynthesis in younger leaves. A similar interpretation was given for several nutrient deficiencies with reduced nutrient concentrations in older foliage (Alvarado-R et al. 1993; López-López et al. 1998). These impacts on photosynthesis and starch are observed before visual injury is observable. Under field conditions, where O_3 exposure is continuous, tissue starch levels initially decrease, chlorosis becomes evident for an extended period of time, and ultimately foliar necrosis occurs (Kärenlampi 1986). In spite of clear evidence of tissue injury, leaves from one site in DL maintained normal starch levels. This behavior has also been reported for other cases of severe exposure (Wellburn and Wellburn 1994). Numerous studies have been carried out to determine the effect of O_3 on carbohydrate content with varying results. This makes it difficult to establish a consistent relationship between O_3 exposure and carbohydrate content. Several authors have reported an increase or decrease. However, both phenomena are a manifestation of dysfunction in the metabolism of polysaccharides caused by O_3 (Kainulainen et al. 1994; Kozlowski and Constantinidou 1986) as has been pointed out for *A. religiosa* in forest stands in DL (Alvarez 1996). The distribution of carbohydrates in leaf tissues from two DL stands showed that starch is translated from palisade parenchyma towards tissues of the midrib. The carbohydrate levels in phloem indicate that there is starch movement to the rest of the plant (Alvarez 1996). However, starch transport is low because there are few carbohydrates stored in the leaves. This result suggests that the problem is a decrement of starch synthesis in the photosynthetic parenchyma cells. In the third stand in DL, the high accumulation of carbohydrates in palisade parenchyma suggests an altered transport. Most likely, the transport path is altered in the chloroplast (Schiffgens-Gruber and Lutz 1992), since the source-sink is interrupted.

In *A. religiosa*, inducible phenolics detected in unhealthy or stressed tissue are observed as spherical and reticular bodies and as deposits in walls of necrosed cells (Alvarez 1996). The spherical bodies have also been observed in foliage of *Pinus nigra* Arnold. (Austrian pine) and *P. resinosa* Ait. (American red pine) after spraying with sulfuric acid (H_2SO_4) (Zobel and Nighswander 1991). Phenolic compound accumulation induced by O_3 exposure is reported in many species. The accumulation of these compounds is associated with necrosis. Phenols are present in chloroplasts and vacuoles of plant cells. Several enzymes of the phenolase complex are bound to thylakoids or stroma or in the cytoplasm. These enzymes do not react until a physical alteration occurs, or until the process of natural senility alters the permeability or integrity of the membranes, exposing the phenols to enzymatic oxidizers. This process can be induced directly by O_3 due to its capacity to alter the structure of the cellular membranes (Howell and Kremer 1973). In *A. religiosa*, O_3 may be responsible for the pigmentation of necrotic cells because of its effect on enzymes and substrata associated with the cytoplasm and vacuoles (Alvarez 1996). Polyphenols might function to protect cells against oxidative stress (Fincher and Alscher 1992).

In some samples of *A. religiosa*, calcium oxalate crystals were observed in the walls of the palisade parenchyma cells. Fink (1991) also found that *Picea abies* leaves exposed to O_3 in fumigation studies show an abnormal accumulation of calcium oxalate crystals. The crystals may be indicative of a strong flow of Ca toward the cytoplasm that alters membrane permeability.

Alvarez (1996) and Alvarez et al. (1998) reported that leaves of *A. religiosa* in DL are subjected to numerous air pollutants. Ozone is the most phytotoxic pollutant that occurs at high levels in the southwest part of the MCAB. The symptoms observed correspond to injury caused by O_3. The participation of other photochemical oxidants is not distinguished histologically in the environmental concentrations present. Moreover, O_3 exposures in DL can be chronic and severe (Miller et al. 1994). Injury begins in the cellular membrane, affecting structural aspects of the cell and physiological processes that alter its metabolic content and ultimately causes necrosis. Chronic injury similar to the symptoms that develop in the process of natural aging was observed even in one-year-old needles of

A. religiosa. Premature foliar senescence is a primary consequence of ozone injury. Injury symptoms such as necrosis, carbohydrate loss, and phenolic compound production, which are manifested in mature trees, are much more severe in leaves of seedlings (Alvarez 1996), suggesting that *A. religiosa* seedlings are more sensitive to ozone.

Bark Response to Air Pollutants

The pollutant content of tree bark has been used as a bioindicator of air pollution for many years. It is well known that tree bark is capable of accumulating and retaining diverse pollutants due to its biochemical and physical properties. When bark is exposed to air pollutants, either directly from the atmosphere or from rainwater running down the stem, the chemical composition of the surface layers of the bark change. The pH, electrical conductivity, and ash content of the bark have been used as bioindicators of environmental pollution. Some researchers prefer to use electrical conductivity of bark extract as bioindicators of sulfur dioxide (SO_2) in the air, because Härtel and Grill (1972) found that conductivity is a more sensitive indicator than the pH of the bark. This is because even low emissions of SO_2 produce changes in conductivity, whereas pH remains unchanged in certain species. The bark of broadleaf species provides a better indication of air pollution than that of conifers, because of its lower acidity (Härtel and Grill 1972). The accumulation of Ca and S in bark reflects a general deposition of air pollutants (Santamaria and Martin 1997). Sulfur, in the form of SO_2, sulfurous acid (H_2SO_3), and H_2SO_4, is one of the main components of acid deposition. Calcium is a major part of alkaline deposition coming from fly ash, erosion, and industrial dust (Huhn et al. 1995). Bark also absorbs certain metals and has been used to survey heavy-metal emissions from traffic and industry (Lippo et al. 1995; Poikolainen 1997). Pollutants extracted from bark may be used to generate spatial patterns of metal deposition in relation to emission sources.

Bark is a highly complex tissue that varies in structure and thickness among species (Mauseth 1988). Conifer bark is generally considered to be rather simple compared to broadleaf species. In addition to protecting the secondary plant body from external threats, other functions of bark are gas exchange, long distance transport of organic nutrients, storage of material during dormant periods, and various roles in plant metabolism (Zimmermann and Brown 1974). While certain coniferous taxa, such as *Pinus*, develop a thick rhytidome composed of completely inert cells, others have a thinner rhytidome or none at all (Fig. 11.2). Every new layer developed from cork cambium and vascular cambium is well protected from the direct effects of weather, because only a few layers of cork are exposed to the atmosphere. Thus, the accumulation of pollutants in these dynamic tissues is useful as a tool for studying spatial and temporal deposition patterns.

To use the content of bark as a bioindicator, the researcher must prepare extracts so that the pH and conductivity measures can be correlated with the health of sampled trees and the degree of pollution within the ecosystem. In

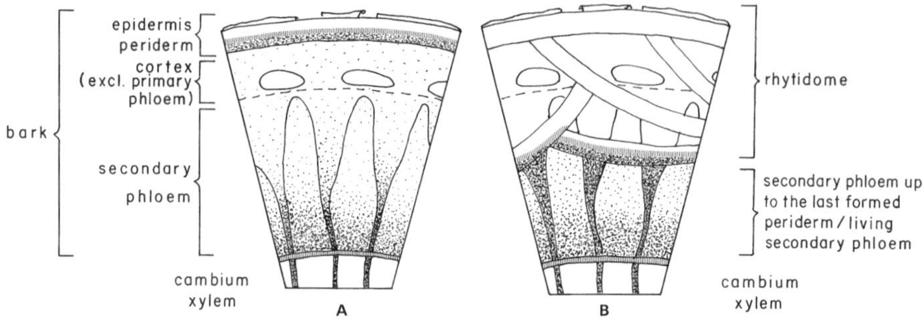

Figure 11.2. Diagram of bark tissue. (A) Only one periderm. (B) Rhytidome. Meristematic tissue (vascular cambium and cork cambium) is shown with vertical lines. Living tissue is represented by stipples. Dense stippling represents areas where all cells are living. Light stippling represents areas where few cells are living, and dead cells are represented by white areas. (Modified from Trockenbrodt 1990.)

Mexico, several studies have measured pH and conductivity in extracts of bark from various tree species found in Mexico City and surrounding forests (Calva-Vázquez 1999; Saavedra-Romero 1996; Sánchez-Reyna 1996; Soto-Tellez 1996). The pH values reported for Mexican species growing in the MCAB are similar to those reported elsewhere (Table 11.2). Differences in acidity between species may be related to distinctive differences in bark anatomy among species (Yáñez-Espinosa and Terrazas 1998).

Tracheid size and ring width decrease in trees impacted by air pollution (Bernal-Salazar et al. 2002; Mitchell and Denne 1997). A similar pattern may

Table 11.2. Summary of pH values of bark extracts in different tree species[a]

Species	Country	pH Range
Abies religiosa	Mexico	2.5–6.1
Cupressus lusitanica	Mexico	3.7–6.2
Pinus montezumae	Mexico	2.2–4.2
Pinus patula	Mexico	2.8–3.3
Pinus nigra	Spain	3.8–4.2
Pinus sylvestris	Spain	2.6–3.4
Pinus strobus	USA	2.6–3.4
Fraxinus uhdei	Mexico	3.7–4.6
Populus canadiensis	Mexico	4.2–4.9
Quercus robur	Spain	4.5–4.8
Quercus ilex	Spain	4.5–5.2
Fagus sylvatica	Spain	5.0–5.5

[a] Data from Calva-Vázquez 1999; Saavedra-Romero 1996; Sánchez-Reyna 1996; Santamaria and Martin 1997.

occur in phloem cells, since both develop from vascular cambium. In O_3-fumigated *Picea abies* saplings, it has been demonstrated that the number of sieve cells and the diameter of latewood tracheids decreased with increasing O_3 concentrations (Kurczynska et al. 1998). However, the number of mucilage cells in living bark in *Abies alba* trees could not be correlated with tree health (Torelli et al. 1989). There is little evidence to propose a general trend for the responses of sieve cells or parenchyma cells of bark to air pollution, except that, in most species, the inner bark remains rather constant in thickness. This is a rough equilibrium, because harsh conditions that are severe enough to inhibit vascular cambium activity and thus stop the production of new secondary phloem may encourage the production of new deeper cork cambium. This deeper cork cambium may cause functional bark to become unusually thin. No evidence that cork cambium is affected by air pollution is available; however, permeability may be altered.

Xylem Response to Air Pollutants

Tree diameter growth is highly correlated with secondary xylem cell accumulations commonly referred to as wood. In other words, woody biomass increases with tree age. Secondary xylem structures of conifer and broadleaf species are fundamentally different. Conifers have tracheid cells, which are responsible for both water transport and mechanical support. Broadleaf species have vessels for water transport and fibers for mechanical support. It is well known that environmental stresses such as fire, disease, drought, and pollutants affect wood production (Elling 1993; Shortle et al. 1995; Weber 1997; Weber and Grulke 1995). With wood tissue, it is possible to reconstruct biological events at the individual or ecosystem level through the interpretation of changes in radial growth of xylem expressed by ring width and cellular modifications (de Kort et al. 1991; Watmough 1997; Wimmer et al. 1996).

Biomass Accumulation

To evaluate the effect of air pollutants on biomass accumulation, 24 six-year-old *A. religiosa* saplings were planted in 1992 and grown for five years under the same natural environmental conditions at DL (Acosta-Montoya 1998). The saplings were harvested at the end of the growing season in 1996. Three sapling populations were delineated based on the degree of needle injury: weak (0), moderate (1), and severe (2). Foliage, wood, and root biomass were evaluated as well as total above-ground biomass. Table 11.3 shows the probability values for the effects of injury class on biomass of foliage of different ages and on total foliar biomass. The biomass of foliage formed in the period from 1992 to 1994 (when the trees were 3 to 5 years old) was not significantly different for the three needle-injury classes. However, there were differences for needle injury classes 0 and 1 between one-year-old and current year foliage (1995, 1996). Foliar biomass

Table 11.3. Summary of variance analysis for annual foliar biomass and total foliar biomass of 6-year-old *Abies religiosa* seedlings by injury class (0, slight; 1, moderate; 2, severe)

Foliage year	1992	1993	1994	1995	1996	All years combined
Pr > F	0.166	0.245	0.166	0.0001	0.0017	0.0001
Injury class			Foliar biomass (g)			
0	0.000^{a}	0.018^{a}	0.91^{a}	10.62^{a}	28.23^{a}	39.83^{a}
1	0.029^{a}	0.052^{a}	0.43^{a}	4.30^{b}	16.75^{b}	21.57^{b}
2	0.000^{a}	0.000^{a}	0.03^{a}	1.27^{b}	14.39^{b}	12.23^{b}

Plants were harvested at the end of the 1996 growing season. Different letters indicate statistically significant differences (Tukey at 0.05) in foliar biomass (g) from Acosta-Montoya 1998.

values for the period of 1992 to 1995 represent only those needles present on the tree at the time of harvesting and not the ones produced during each respective year, because large amounts of needles had already abscised. Only in 1996 did the harvested foliage represent all the needles produced during that year. Foliar biomass differed significantly in relation to foliar injury (Table 11.3) for the needle injury classes. The highest values for foliar biomass corresponded to the trees with little injury. Reduced foliar biomass is due to two factors, reduced leaf size and premature needle abscission.

Woody biomass accumulation was also significantly different for the needle injury classes (Table 11.4). The results indicate that as injury increased, woody biomass accumulation decreased. This suggests that the allocation of carbohydrates to the growth of these organs was reduced by foliar injury. Moreover, reduction in biomass accumulation in both foliage and wood is reflected as a decrement in total above-ground biomass accumulation. Variance analysis shows

Table 11.4. Summary of variance analysis for annual wood biomass and total wood biomass accumulation in the stems of 6-year-old *Abies religiosa* seedlings by injury class (0, slight; 1, moderate; 2, severe)

Year wood produced	1992	1993	1994	1995	1996	All years combined
Pr > F	0.004	0.046	0.10	0.01	0.01	0.006
Injury class			Wood biomass (g)			
0	26.11^{a}	12.98^{a}	10.22^{a}	13.19^{a}	10.92^{a}	73.42^{a}
1	16.19^{b}	7.78^{b}	9.03^{a}	8.63^{ab}	7.57^{ab}	49.20^{b}
2	13.34^{b}	8.30^{b}	6.37^{a}	6.32^{b}	5.29^{b}	39.62^{b}

Plants were harvested at the end of the 1996 growing season. Different letters indicate statistically significant differences (Tukey at 0.05) in woody biomass (g) among the three injury classes within a given year. From Acosta-Montoya 1998.

that variation in total above-ground biomass accumulation is related to severity of foliar injury in trees ($F = 10.77$, $P > 0.0006$, $df = 2$, $N = 23$). Significant differences were found between trees with slight injury (mean above-ground biomass = 113.25 g) and those with moderate (mean above-ground biomass = 70.77 g) or severe (mean above-ground biomass = 55.27 g) injury. Foliar biomass generally decreased in response to O_3 injury at a greater rate than that of stem biomass. Thus the differential effect on stem biomass compared with the effect on foliage and roots probably increases gradually with time. Waring and Schlesinger (1985) pointed out that as trees grow, the amount of tissue respiration also increases. If photosynthetic area does not increase sufficiently, respiration requirements will not be satisfied, and eventually the tree may die (Oliver and Larson 1990). A reduction in photosynthetic area could accelerate the process of forest decline, because trees used for reforestation in this experiment showed visual injury symptoms beginning in the first year (Acosta-Montoya 1998). Results of biomass accumulation in *A. religiosa* growing in the MCAB are consistent with those presented for other species studied elsewhere (Frost et al. 1991; Reich and Lassoie 1985). Concentrations of air pollutants such as O_3 increase foliar senescence and reduce tree biomass.

Radial Xylem Variation

Dendrochronological techniques have been used to assess temporal growth responses to stress in *P. hartwegii* and *A. religiosa* growing in the MCAB (Alarcón et al. 1993; Alvarado-R et al. 1993; Bernal-Salazar et al. 2002). Alarcón et al. (1993) studied *P. hartwegii* trees growing in two areas with high air-pollution exposure. They sampled unhealthy trees in Ajusco and DL stands (Fig. 11.3) and evaluated their ring-width patterns. The health status of the sampled trees was determined by the typical foliar air pollutant injury symptoms. Raw ring-width and standardized chronologies for both stands showed a ring-width reduction in two periods (Fig. 11.3), one near the 1930s and a second from the 1970s onward. The decrement during the 1930s cannot be attributed to pollution stress from Mexico City; instead, it is associated with decreased rainfall. Tree ring-width reduction in the 1970s is greater and more conspicuous in trees from DL than from the Ajusco stands. However, because the variability among trees was high, trees were grouped by age. Two age groups, A1 (mean age 47 years) and DL3 (mean age 80 years), clearly showed radial growth decrements (Alarcón et al. 1993). Growth rates for the periods 1951 to 1960, 1961 to 1970, and 1971 to 1980 were higher than for 1981 to 1990 in the group of trees from Ajusco stand. This growth trend supports the hypothesis that ring width is affected by air pollutants because concentrations of pollutants were higher in 1981 to 1990 than in previous decades (Bravo et al. 1990, Miller et al. 1994), and no association with rainfall was detected. A similar behavior was observed in the DL3 group. However, in the period from 1971 to 1980, the growth rate was higher than in the period from 1961 to 1970. This may be related to sanitary thinning practices in this forest.

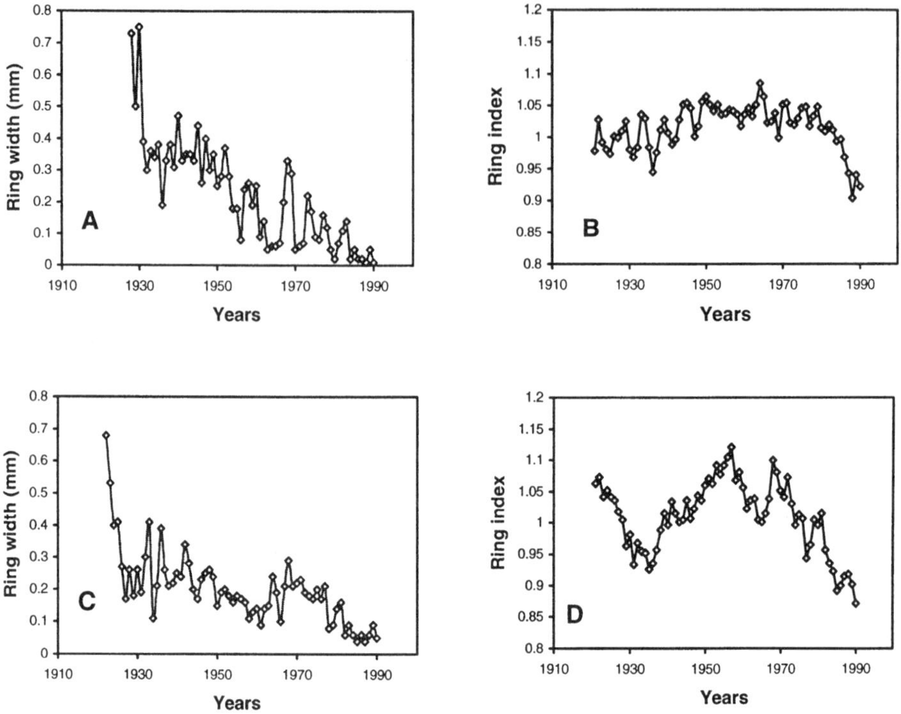

Figure 11.3. Raw and standardized (ring index) chronology of growth ring for 50 unhealthy trees of *Pinus hartwegii* at El Ajusco (A, B) and Desierto de los Leones (C, D). (Modified from Alarcón et al. 1993.)

Alvarado-R et al. (1993) detected a drastic ring-width decrement in nearly 100 unhealthy *A. religiosa* trees during the last three decades. Average widths of the last 30 rings showed similar values for most tree-age categories (Table 11.5). This decrement was similar for trees of all size classes and began in the early 1970s. The results imply that because growth rates have been slower since the 1970s, it takes longer for trees to move to the next size class. The low xylem accumulation associated with higher needle loss was one of the pieces of evidence the authors used to support their assertions that polluted air may be an important causal factor in the decline of *A. religiosa* (Alvarado-R et al. 1993). A decrement of ring width was also detected in two periods for *Abies religiosa* trees growing in DL (Bernal-Salazar et al. 2001). One period corresponded to the 1930s and a second to the 1970s onwards (Fig. 11.4). Although no standardized ring index has been calculated for *A. religiosa*, a similar annual ring-width reduction was observed in *P. hartwegii* at DL stands (Alarcón et al. 1993).

More recently, Bernal-Salazar et al. (2001) were the first to evaluate radial variation of several tracheid features in unhealthy trees at the Desierto stands. Their sampled trees had between 50% and 80% leaf loss. Stands were selected

Table 11.5. Average width of the last 30 rings and number of rings in the last 2 to 5 cm observed in 100 *Abies religiosa* stumps in Desierto stands

Average age	Diameter class (cm)	Average width of the last 30 rings (cm)	Number of trees
57	25	0.86	2
71	30	1.30	3
83	35	1.26	7
88	40	1.60	4
73	45	1.56	5
99	50	1.66	10
100	55	1.20	8
95	60	1.30	18
108	65	1.13	12
103	70	1.00	6
104	75	1.43	2
103	80	1.26	7
131	85	1.50	4
133	90	1.70	4
110	95	1.26	5
121	100	1.06	3
148	110	0.66	1

Modified from Alvarado-R et al. (1993).

to represent both aspects of a ravine highly exposed to air pollutants from Mexico City. Two cores were taken per tree in four stands. Tracheid length and tracheid wall thickness in latewood show a reduction beginning in the 1970s (Figs. 11.4 and 11.5). However, there is a wider range of variation among trees in tracheid

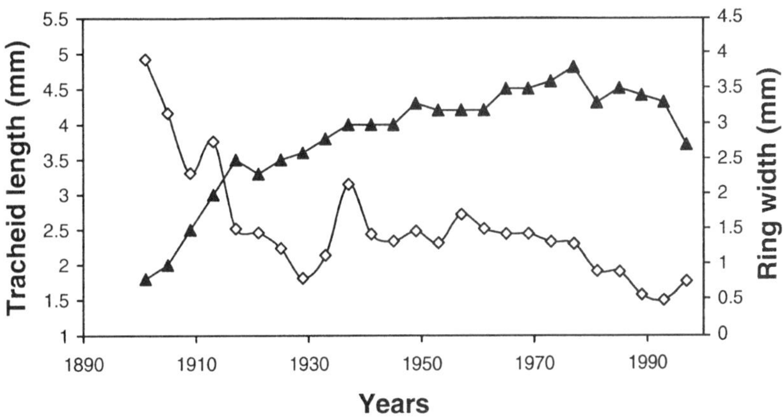

Figure 11.4. Raw ring width and tracheid length for unhealthy *Abies religiosa* trees at Desierto de los Leones. Tracheid length (*triangles*) and ring width (*diamonds*). (Redrawn from Bernal-Salazar et al. 2002).

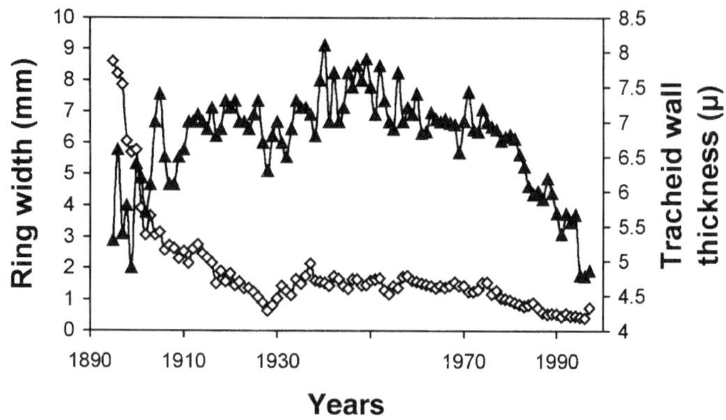

Figure 11.5. Raw ring width and tracheid wall thickness for unhealthy *Abies religiosa* trees at Desierto de los Leones. Tracheid wall thickness (*triangles*) and ring width (*diamonds*). (Redrawn from Bernal-Salazar et al. 2002).

diameter of earlywood and latewood than in tracheid wall thickness and length. Most of these tracheid features are similar to those reported for other species (Bauch 1986; Baucker et al. 1996; de Kort 1986; Kawakami and Furukawa 1992; Liu and Liu 1996; Mitchell and Denne 1997; Nieweglowska 1995; Schweinruker and Voronin 1996), where O_3, SO_2, and heavy metals are involved in forest decline.

Other tracheid modifications observed in unhealthy *A. religiosa* trees include slight differences between earlywood and latewood tracheids viewed in cross sections and the occurrence of dark-staining deposits, mostly composed of phenolic compounds (Fig. 11.6). These deposits occluded ray cell lumina and occasionally tracheid lumina (Bernal-Salazar and Terrazas 2002). When these authors compared tracheid features of unhealthy trees with tracheid features of healthy trees growing in a control stand outside the Basin at El Campanario, differences were dramatic (Fig. 11.6). Similar tracheid modifications have been observed in species growing in forests affected by air pollution in other areas (Schweingruber and Voronin 1996). In a few highly unhealthy trees, Bernal-Salazar and Terrazas (2002) observed cell wall separation. This behavior was also reported in latewood tracheids of severely injuried *Picea abies* trees (Wagenfuhr 1996). Baucker et al. (1996) characterized a decrease in cell-wall thickness and larger microfibril angles within the S_2 cell-wall layer in latewood of *Picea abies* trees that had been severely injuried by SO_2 pollution. Other authors have not detected differences between healthy and unhealthy conifer trees in tracheid wall thicknesses and chemical compositions of cell walls (Dunisch et al. 1996; Kucera 1989). Abundant traumatic resin canals were observed in unhealthy and healthy *A. religiosa* trees, but no correlation between this feature and tree health was found. A similar interpretation was given for *Abies alba* trees growing in a declining forest in Slovenia (Torelli et al. 1989).

Anova between trees growing in the Basin and outside it, for the period of 1975 to 1997, show statistical differences for all anatomical tracheid features analyzed (Bernal-Salazar et al. 2002). While trees growing in the DL had greater tracheid length and diameter in latewood and earlywood, trees from El Campanario had thicker latewood tracheid walls confirming the radial growth pattern

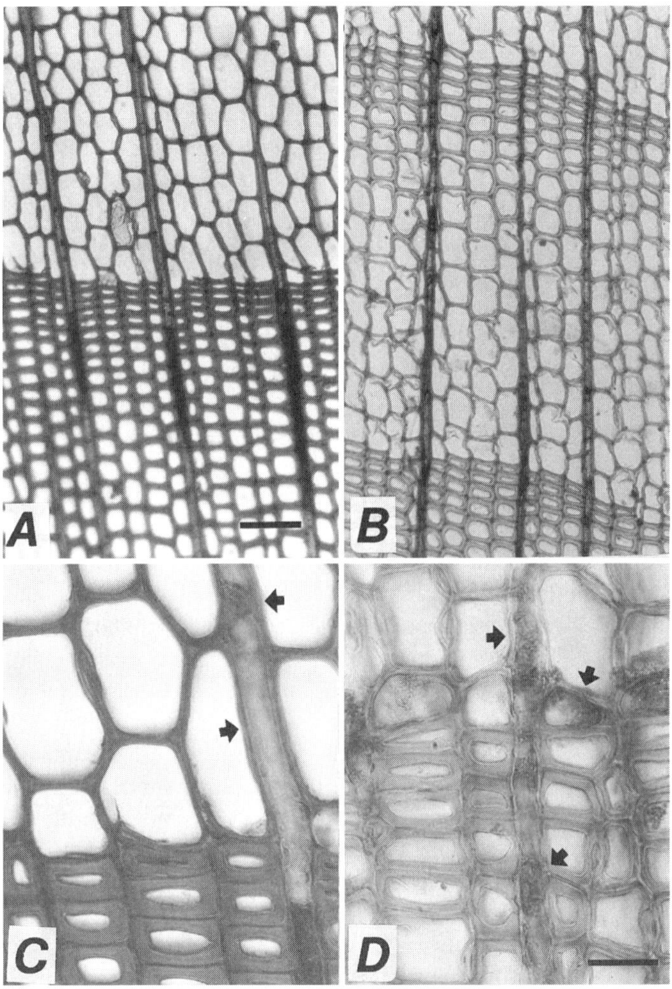

Figure 11.6. Cross-sections showing earlywood and latewood tracheid modifications in *Abies religiosa* trees in Desierto de los Leones and El Campanario forest. (A) Healthy trees at El Campanario. (B) Unhealthy trees at Desierto. (C) Ray cells without dark-staining deposits (*arrow*) in the healthy forest El Campanario. (D) Ray cells with dark-staining deposits (phenolic compounds, *arrows*) in the unhealthy forest Desierto. Scale bar: A, B = 100 µm; C, D = 20 µm.

for both stands (Bernal-Salazar and Terrazas 2000; Bernal-Salazar et al. 2002). Trees at El Campanario were younger, and their cambial maturation curve was in the steepest region, where tracheid features and ring-width variation is considerable (Bernal-Salazar and Terrazas 2000). The DL stand had only mature trees in which tracheids had reached their optimal size with lower variation. Nonetheless, a drastic reduction in tracheid-wall thickness was found in wood produced in the years when air pollution concentrations greatly increased (Bernal-Salazar et al. 2002).

Modifications in tracheid features for unhealthy trees in DL stands were confirmed when the 1961 to 1974 and 1975 to 1997 growth periods were compared for tracheid size. Both periods corresponded to the asymptotic region of the growth curve (Figs. 11.4 and 11.5). However, the results do not support the asymptotic growth curves, and statistical differences were detected between periods. There was a reduction in tracheid wall thickness ($F = 247$, $df = 1$, $p < 0.001$, $N = 17{,}700$) and tracheid length ($F = 887$, $df = 1$, $p < 0.001$, $N = 17{,}700$). Growth decrement may be associated with the greatly reduced foliar area of most trees growing in these DL stands. A reduction in photosynthesis would affect secondary wall accumulation and modify cambial growth in terms of cell divisions. It would also affect the period of activity. In chronically polluted *Picea abies* forests in Germany, Dunisch et al. (1996) found that reduced annual growth of trees subject to heavy pollution compared with uninjured trees was due to a reduced period of cambial cell divisions and a lower rate of periclinal cell divisions. A similar behavior may be affecting unhealthy *A. religiosa* trees growing in the MCAB. Studies on cambial activity are currently underway to test this hypothesis.

When cambial electrical resistance was used to measure the vitality of *Abies alba* (silver fir) in the forest of Slovenia (Torelli et al. 1999), the authors found that cambial electrical resistance and the crown status index were inversely correlated with annual ring-width increment. From these results it was inferred that ring growth decreased prior to the development of canopy symptoms. Variation in cambial electrical resistance was due both to the vascular cambium and to the thickness of the nonconducting phloem. The use of this technique in trees growing in forests in the MCAB may help to detect nonsusceptible and susceptible trees and species before foliar injury is detected.

Dendrochemical changes in *A. religiosa* were recently studied in DL (Watmough and Hutchinson 1999). High metal concentrations were associated with low soil pH. Lead and Cd increased in tree rings formed since the 1960s, reflecting the increase in urbanization and vehicle numbers in Mexico City. However, peaks in Cd and, most obviously, in Pb were found in rings formed in the period of 1920 to 1940, corresponding to the heartwood–sapwood boundary, suggesting that Pb and Cd taken up by roots of *A. religiosa* accumulate in the heartwood. Such radial translocation severely restricts the potential use of *A. religiosa* in dendrochemical studies. However, other species growing in the Basin should be studied to determine if this methodology can help to elucidate temporal patterns in pollutant uptake in forests of the MCAB.

Summary

In air pollution studies, only selected components of trees are commonly studied, because of their complexity; however, modifications to one part, leaves, for example, may affect other parts, for example, bark or wood. Our understanding of the effects of air pollution on trees can benefit from more integrative studies of the responses of various tissue types to air pollution stress. Dendrochronological studies will allow us to reconstruct the impacts of historical stresses (e.g., air pollution) and climate variation on tree growth and vitality. Changes in ring width in *P. hartwegii*, and in radial tracheid variation in *A. religiosa*, corroborated other evidence that tree growth reduction started in these species in the 1970s. Similar studies should be performed in other conifer and broadleaf species in order to identify tolerant species as well as resistant individuals of susceptible species. Tolerant species and resistant individuals may be useful for reforestation programs.

In the Mexico City Air Basin, there have been few studies on the use of foliar, bark, and wood anatomy to evaluate the role of air pollutants as as a factor contributing to *A. religiosa* decline or to the injury symptoms observed in *P. hartwegii*. Anatomical studies of foliage and wood of *A. religiosa* and of *P. hartwegii* wood support other studies which indicate that air pollution may be the main causal factor for the symptoms observed in these species in the Basin of Mexico. However, data for other species are not available. Further detailed studies on other conifer or broadleaf species such as those by Alvarez (1996) and Alvarez et al. (1998) for *A. religiosa* leaves are needed. Studies on epicuticular waxes together with anatomical data may help to detect air pollutant injury, even in the absence of visible symptoms.

Bark anatomy studies for most species sensitive to air pollution in the Basin are nonexistent. To date, only descriptive anatomical studies have been published. Studies of modification of bark tissue in pollution-stressed trees may help in understanding the relationship between crown vitality and transport of photosynthates in the stem. In other countries, where forest decline has been associated with air pollutants, bark has been used as a bioindicator of heavy metals and S deposition, but only limited studies of this type in the Basin of Mexico have been reported.

References

Acosta-Montoya, J. 1998. *Comportamiento de la Distribución de Biomasa en Brizales de Abies religiosa (H.B.K.) Schl. et Cham. en Declinación.* Tesis de Licenciatura. Chapingo, Estado de México, México: División de Ciencias Forestales. Universidad Autónoma de Chapingo.

Alarcón-C., M.A., de Bauer, Ma. L.I., Jasso-M., J., Segura-W., G., and Zepeda-B., E.M. 1993. Patrón de crecimiento radial en los árboles de *Pinus hartwegii* afectados por contaminación atmosférica en el suroeste del Valle de México. *Agrociencia Serie Recursos Naturales Renovables* 3:67–80.

Alvarado-R., D., de Bauer, L.I., and Galindo-A., J. 1993. Decline of sacred fir (*Abies religiosa*) in a forest park south of Mexico city. *Environ. Pollut.* 80:115–121.

Alvarez-C., D.E. 1996. *Determinación de Alteraciones a Nivel Citológico e Histoquímico en Abies religiosa (H.B.K.) Schl. et Cham. del Desierto de los Leones del D.F. y su Relación con el Ozono.* Tesis de Maestría. México, D.F.: Division de Estudios de Postgrado, Facultad de Ciencias, Universidad Nacional Autónoma de México (UNAM).

Alvarez, D., Laguna, G., and Rosas, I. 1998. Macroscopic and microscopic symptoms in *Abies religiosa* exposed to ozone in a forest near Mexico City. *Environ. Pollut.* 103:251–259.

Bauch, J. 1986. Characteristics and response of wood in declining trees from forests affected by pollution. *IAWA Bull. n.s.* 7:269–276.

Baucker, E., Bemmann, A., Bues, C.T., and Nuys, G.J. 1996. Wood properties of heavily pollution-damaged spruce (*Picea abies* (L.) Karst.) at the high-altitude zones of the eastern parts of Erzgebirge. *Holz Roh-Werkst.* 54:215–258.

Bernal-Salazar, S., and Terrazas, T. 2000. Influencia climática sobre la variación radial de los caracteres anatómicos de la madera en *Abies religiosa. Madera y Bosques* 6:73–86.

Bernal-Salazar, S., and Terrazas, T. 2002. Pollution effects on wood anatomical features in *Abies religiosa* (H.B.K.) Schl. et Cham. *Environ. Exper. Bot.* (in press).

Bernal-Salazar, S., Terrazas, T., and Alvarado, D. 2002. Wood anatomical characters and annual ring width of *Abies religiosa* (H.B.K.) Schl. et Cham. trees comparing two forests with contrasting health characteristics. *Environ. Pollut.* (in press).

Bravo, H., Roy-Ocotla, G., Sosa, R., and Torres, R. 1990. Weekdays and weekend days variation in ozone and nitrogen oxides concentrations in the atmosphere of Mexico City. In *Proceedings of the 83rd Air and Waste Management Annual Meeting and Exhibition*, June 24–29. Paper 90–95,6. Pittsburgh, PA: Air and Waste Management Association.

Calva-Vázquez, G. 1999. *Monitoreo de la Depositación Ácida en Zonas Boscosas de la Cuenca de México, Usando como Superficie de Exposición la Corteza Arbórea de Tres Pináceas.* Tesis de Maestría. Montecillo, Texcoco, Estado de México, México: Colegio de Posgraduados.

Davis, D.D., and Wood, F.A. 1972. The relative susceptibility of eighteen coniferous species to ozone. *Phytopathology* 62:14–19.

de Bauer, M.L.I., and Hernández-Tejeda, T. 1986. *Contaminación: Una Amenaza para la Vegetación en México.* Chapingo, State of Mexico, Mexico: Colegio de Postgraduados.

de Bauer, L.I., and Krupa, S.V. 1990. The valley of Mexico: a summary of observational studies on its air quality and effects on vegetation. *Environ. Pollut.* 65:109–115.

de Bauer, M.L.I., Hernández, T., and Manning, W.J. 1985. Ozone causes needle injury and tree decline in *Pinus hartwegii* at high altitudes in the mountains in Mexico City. *J. Air Pollut. Control Assoc.* 35:838.

de Kort, I. 1986. Wood structure and growth ring width of vital and nonvital Douglas fir (*Pseudotsuga menziesii*) from a single stand in the Netherlands. *IAWA Bull. n.s.* 7:309–318.

de Kort, I., Loeffen, V., and Baas, P. 1991. Ring width, density and wood anatomy of douglas fir with different crown vitality. *IAWA Bull. n.s.* 12:453–465.

Dunisch, O.J., Bauch, J., Puls, J., and Muller, M. 1996. Biological and chemical wood properties of long-term polluted spruce (*Picea abies* (L.) Karst.) at high altitude stands of the Erzgebirge. *Holzforschung* 50:497–506.

Elling, W. 1993. Immissions in the complex of causes involved in damage and dying of silver fir. *Allg. Forst–Zeitschr.* 48:87–95.

Evans, L.S., and Leonard, M.R. 1991. Histological determination of ozone injury symptoms of primary needles of giant sequoia (*Sequiadendron giganteum* Bucch.) *New Phytol.* 117:557–564.

Fincher, J., and Alscher, R.G. 1992. The effect of long-term ozone exposure on injury in seedlings of red spruce (*Picea rubens* Sarg). *New Phytol.* 120:49–59.

Fink, S. 1991. Unusual patterns in the distribution of calcium oxalate in Spruce needles and their possible relationship to the impact of pollutants. *New Phytol.* 119:41–51.

Frost, D.L., Taylor, G., and Davies, W.J. 1991. Biophysics of leaf growth of hybrid poplar: impact of ozone. *New Phytol.* 118:407–415.

Härtel, O., and Grill, D. 1972. Die Leitfähigkeit von Fichten-brorken-Extrakten als empfindlicher indikator lif Luftverun-reinigungen. *Eur. J. For. Pathol.* 2:205.

Hernández-Tejeda, T. 1981. *Reconocimiento y Evaluación del Daño por Gases Oxidantes en Pinos y Avena del Ajusco*, D.F. Tesis profesional, Departamento de Fitotecnia. Chapingo, Estado de México, México: Universidad Autónoma de Chapingo.

Hernández-Tejeda, T., and de Bauer, M.L.I. 1984. Evaluación del daño por gases oxidantes en *Pinus hartwegii* and *P. montezumae* var *lindleyi* en El Ajusco, Mexico, D.F. *Agrociencia* 56:183–194.

Hernández-Tejeda, T., and de Bauer, L.I. 1986. Photochemical oxidant damage on *Pinus hartwegii* at the Desierto de los Leones, D.F. *Phytopathology* 76:337.

Hernández-Tejeda, T., de Bauer, L.I., and Ortega-Delgado, M.L. 1983. Identificación y determinación de los principales pigmentos fotosintéticos de hojas de *Pinus hartwegii* afectadas por gases oxidantes. *Agrociencia* 55:71–82.

Hernández-Tejeda, T., Krupa, S.V., Platt, G.C., and de Bauer, L.I. 1981. Sensibilidad de plántulas de eucalipto (*Eucalyptus globulus* Labill.) al ozono. *Agrociencia* 43:89–95.

Hill, A.C., Pack, M.R., Treshow, M., Downs, R.J., and Transtrum, L.G. 1961. Plant injury induced by ozone. *Phytopathology* 51:356–363.

Howell, R.K., and Kremer, D.K. 1973. The chemistry and physiology of pigmentation in leaves injured by air pollution. *J. Environ. Qual.* 2:434–438.

Huhn, G., Horst, S., Hans-Joachim, S., Rainer, T., and Gerrit, S. 1995. Evaluation of regional heavy metal deposition by multivariate analysis of element contents in pine tree barks. *Water Air Soil Pollut.* 84:367–383.

Kainulainen, P., Holopainen, J.K., Hyttinen, H., and Oksanen, J. 1994. Effect of ozone on the biochemistry and aphid infestation of Scots pine. *Phytochemistry* 35:39–42.

Kärenlampi, L. 1986. Relationships between macroscopic symptoms of injury and cell structural changes in needles of ponderosa pine exposed to air pollution in California. *Ann. Bot. Fennici* 23:255–264.

Kawakami, K., and Furukawa, I. 1992. Horizontal variation of the annual-ring width and the tracheid length in the declined pine trees. *Res. Bull. Tottori Univ. For.* 21:153–159.

Knudson, L.L., Tibbitts, T.W., and Edwards, G.E. 1997. Measurement of ozone injury by determination of leaf chlorophyll concentration. *Plant Physiol.* 60:606–608.

Kozlowski, T.T., and Constantinidou, H. 1986. Responses of woody plants to environmental pollution. *For. Abstr.* 47:5–51.

Kucera, L.J. 1989. Biological and technological properties of the wood of damaged Norway spruce. *Schweiz.-Z.-Forstwes.* 140:203–215.

Kukkola, E., Huttunen, S., Back, J., and Rautio, P. 1997. Scots pine needle injuries at subarctic industrial sites. *Trees* 11:378–387.

Kurczynska, E.U., Bastrup-Birk, A., and Mortensen, L. 1998. Influence of ozone and soil nitrogen content on the stem anatomy of Norway spruce saplings grown in open-top chambers. *Environ. Exper. Bot.* 40:113–121.

Le Thiec, D., Rose, C., Garrec, J.P., Laffray, D., Louguet, P., Galaup, S., and Loosvaldt, P. 1994. Alteration of element contents in guard cells of Norway spruce (*Picea abies*) subjected to ozone fumigation and (or) water stress: X-ray microanalysis study. *Can. J. Bot.* 72:86–92.

Lippo, H., Poikolainen, J., and Kubin, E. 1995. The use of moss, lichen, and pine bark in the nationwide monitoring of atmospheric heavy metal deposition in Finland. *Water Air Soil Pollut.* 85:2241–2246.

López-López, M.G., Velázquez, M.J., Velázquez, M.A., González, R.V., and Cetina, V.M. 1998. Estado nutricional de *Abies religiosa* en un área con problemas de contaminación ambiental. *Agrociencia* 32:53–59.

Liu, P., and Liu, P. 1996. Properties of wood from Masson pine stand affected by air pollution and acid rain. II: Wood structure. *Sci.–Silvae–Sinica* 32:69–77.

Lucas, P.W., Rantanen, L., and Mehlhorn, H. 1993. Needle chlorosis in Stika spruce following a three-year exposure to low concentrations of ozone: changes in mineral content, pigmentation and ascorbic acid. *New Phytol.* 124:264–275.

Maier-Maercker, U., and Koch, W. 1992. Histological examination of spruce needles from a long-term gas exchange experiment in pure and polluted air in the field. *Trees* 6:186–194.

Mauseth, J.D. 1988. *Plant Anatomy.* Menlo Park, Califormia U.S.A.: Benjamin/Cummings Publ. Co.

Miller, P.R., de Bauer, M.L., Quevedo-Nolasco, A., and Hernández-Tejeda, T. 1994. Comparison of ozone exposure characteristics in forested regions near Mexico City and Los Angeles. *Atmos. Environ.* 28:141–148.

Mitchell, M.D., and Denne, M.P. 1997. Variation in density of *Picea sitchensis* in relation to within-tree trends in tracheid diameter and wall thickness. *Forestry (Oxford)* 70:47–60.

Miyake, H., Matsumara, H., Fujinuma, Y., and Totsuka, T. 1989. Effects of low concentrations of ozone on the fine structure of radish leaves. *New Phytol.* 111:187–195.

Nieweglowska, G.B. 1995. Cytological and anatomical changes in pine tree xylem caused by sulfur compounds. *Acta Soc. Bot. Pol.* 64:233–238.

Ojanpera, K., Sutinen, S., Pleijel, H., and Selldén, G. 1992. Exposure of spring wheat, *Triticum aestivum* L. Cv. Drabant, to different concentrations of ozone in open-top chambers. Effect on the ultrastructure of flag leaf cells. *New Phytol.* 120:39–48.

Oliver, C.D., and Larson, B.C. 1990. *Forest Stand Dynamics.* New York: MacGraw-Hill.

Pell, E.J., and Weissberger, W.C. 1976. Histopathological characterization of ozone injury to soybean foliage. *Phytopathology* 66:856–861.

Reich, P.B., and Lassoie, J.P. 1985. Influence of low concentrations of ozone on growth, biomass partitioning and leaf senescence in young hybrid poplar plants. *Environ. Pollut.* 39:39–51.

Rivera-Amaro, J. 1998. *Nutrición y Síntomas de Declinación de Abies religiosa (H.B.K) Schl. et Cham. en el Desierto de los Leones, D.F.* Tesis de Licenciatura. Chapingo, Estado de México, México: División de Ciencias Forestales, Universidad Autónoma de Chapingo.

Saavedra-Romero, L.L. 1996. *Análisis Espacio-Temporal de la Depositación Ácida, Utilizando como Biomonitor de Exposición la Corteza Arbórea de Especies Perennes del Parque Nacional Desierto de los Leones.* Tesis de Licenciatura en Biología. México, D.F.: Facultad de Estudios Superiores Zaragoza (FES–Z), Universidad Nacional Autónoma de México (UNAM).

Sánchez-Reyna, M.G. 1996. *Estudio de la Depositación Ácida en la Ciudad de México, Mediante Bioindicadores (Corteza Arbórea).* Tesis de Licenciatura en Biología. México, D.F.: Facultad de Estudios Superiores Zaragoza (FES–Z), Universidad Nacional Autónoma de México (UNAM).

Santamaria, J.M., and Martin, A. 1997. Tree bark as a bioindicator of air pollution in Navarra, Spain. *Water Air Soil Pollut.* 98:381–387.

Schiffgens-Gruber, A., and Lutz, C. 1992. Ultrastructure of mesophyll cell chloroplasts of spruce needles exposed to ozone, sulfur dioxide and nitrogen dioxide alone and in combination. *Environ. Exp. Bot.* 32:243–254.

Schweingruber, F.H., and Voronin, V. 1996. A study of the dendrochronology and soil chemistry of the forest damage area at Norilsk, Siberia, and consequences for the interpretation of large scale crown assessment inventories. *Allg. Forst u. Jagdzeit.* 167:53–67.

Shortle, W.C., Smith, K.T., Minocha, R., Alexeyev, V.A., and Stott, P. 1995. Similar patterns of change in stem wood calcium concentration in red spruce and Siberian fir. *J. Biogeogr.* 22:467–473.

Skelly, J.M., Savage, J.E., de Bauer, M.L., and Alvarado, D. 1997. Observations of ozone-induced foliar injury on black cherry (*Prunus serotina* var. *capuli*) within the Desierto de los Leones National Park. Mexico City. *Environ. Pollut.* 95:155–158.

Soto-Tellez, J.A. 1996. *Estimación de la Depositación Ácida en Tres Parques Urbanos del Distrito Federal.* Tesis de Licenciatura en Biología. México, D.F.: Facultad de Estudios Superiores Zaragoza (FES–Z), Universidad Nacional Autónoma de México (UNAM).

Sutinen, S., Wallin, G., Karlsson, P.E., Skarby, L., Sellden, G., and Paoletti, E. 1998. Cell ultrastructure of needles from saplings of Norway spruce, *Picea abies* (L.) Karst., exposed to ozone and low phosphorus supply in open-top chambers. *Chemosphere* 36:691–696.

Thomson, W.W. 1975. Effects of air pollutants on plant ultrastructure. In *Response of Plants to Air Pollution*, eds. J.B. Mudd, and T.T. Kozlowski, pp. 179–194. New York: Academic Press.

Thomson, W.W., Dugger, W.M., and Palmer, R.L. 1996. Effects of ozone on the fine structure of the palisade parenchyma cell of bean leaves. *Can. J. Bot.* 44:1677–1682.

Tingey, D.T., Wilhour, R.G., and Tandley, C. 1976. The effect of chronic ozone exposures on the metabolite content of ponderosa pine seedlings. *For. Sci.* 22:234–241.

Torelli, N., Cufar, K., and Robic, D. 1989. Mucilage cells in bark and traumatic resin canals in wood as possible symptoms of silver fir dieback. *Gozdarski-Vestn.* 47:163–167.

Torelli, N., Shortle, W.C., Cufar, K., Ferlin, F., and Smith, K.T. 1999. Detecting changes in tree health and productivity of silver fir in Slovenia. *Eur. J. For. Pathol.* 29:189–197.

Treshow, M. 1984. Diagnosis of air pollution effects and mimicking symptoms. In *Air Pollution and Plant Life*, ed. M. Treshow, pp. 97–115. Chichester: John Wiley and Sons, Ltd.

Trockenbrodt, M. 1990. Survey and discussion of the terminology used in bark anatomy. *IAWA Bulletin n.s.* 11:141–166.

Utriainen, J., Holopainen, T., and Paoletti, E. 1998. Ultrastructural and growth responses of young Scots pine seedlings (*Pinus sylvestris*) to increasing carbon dioxide and ozone concentrations. *Chemosphere* 36:795–800.

Vázquez, S.J. 1987. *El Saneamiento y la Limpia Forestales en el Desierto de Los Leones.* México, D.F.: Comisión Coordinadora de Desarrollo Rural, Departamento del Distrito Federal.

Vázquez, S.J. 1988. *Los Tratamientos Silvícolas del Desierto de Los Leones: Sus Fundamentos.* México, D.F.: Comisión Coordinadora de Desarrollo Rural, Departamento del Distrito Federal.

Wagenfuhr, R. 1996. Studies on the wood anatomy of emission-damaged Norway spruce from the higher parts of the eastern Erzgebirge. *Holz-Zentralblatt* 122:140–147.

Waring, R.H., and Schlesinger, W.H. 1985. *Forest Ecosystems: Concepts and Management.* New York: Academic Press.

Watmough, S.A. 1997. An evaluation of the use of dendrochemical analyses in environmental monitoring. *Environ. Rev.* 5:180–201.

Watmough, S.A., and Hutchinson, T.C. 1999. Change in the dendrochemistry of sacred fir close to Mexico City over the past 100 years. *Environ. Pollut.* 104:79–88.

Weber, J.A., and Grulke, N.E. 1995. Response of stem growth and function to air pollution. In *Plant Stems: Physiology and Functional Morphology*, ed. B.L. Gartner, pp. 343–363. San Diego California: Academic Press.

Weber, U.M. 1997. Dendroecological reconstruction and interpretation of larch budmoth (*Zeiraphera diniana*) outbreaks in two central alpine valleys of Switzerland from 1470–1990. *Trees* 11:277–290.

Wellburn, F., and Wellburn, A.R. 1994. Atmospheric ozone affects carbohydrate allocation and winter hardiness of *Pinus halapensis*. *J. Exp. Bot.* 45:607–614.

Wimmer, R., Grabner, M., Schäy, E., and Halbwachs, G. 1996. How useful is xylem anatomy as an indicator for SO_2 pollution? *IAWA J.* 17:266–267.
Yañez-Espinosa, L., and Terrazas, T. 1998. Anatomía de la corteza de algunas gimnospermas. *Bol. Soc. Bot. México* 62:15–28.
Zimmermann, M.H., and Brown C.L. 1974. *Trees, Structure and Function.* New York: Springer-Verlag.
Zobel, A., and Nighswander, J.E. 1991. Accumulation of phenolic compounds in the necrotic areas of Austrian and red pine needles after spraying with sulphuric acid: a possible bioindicador of air pollution. *New Phytol.* 117:575–585.

12. Lichens and Air Pollution in the Forests Surrounding Mexico City

Angel Zambrano, T.H. Nash III, and M.A. Herrera-Campos

Very little is known about the lichens of the Basin of Mexico. Lichenologists are just starting to systematically analyze the local lichen flora and communities. Anyone familiar with lichens would, however, recognize in these communities signs of decline resembling the lichen impoverishment reported for many polluted areas throughout the world. This is especially apparent at Sierra de las Cruces, the nearest forested area SSW of Mexico City, where lichens are very scarce compared to similar areas farther away from the city.

In this chapter, following a brief overview concerning the use of lichens as biomonitors of air pollution, we will summarize the results of recent studies that document lichen decline and provide evidence that Mexico City's heavily polluted atmosphere is a major cause of the decline.

Lichens and Air Pollution

It is well established that lichens and cities are rarely compatible. Early in the nineteenth century, a reduction in lichen abundance was noticed in some European cities (Richardson 1992). Today, this same impoverishment involving the local extinction of sensitive lichen species occurs in many urban areas all over the world (Gries 1996; Hawksworth and Hill 1984). Among the various environmental factors potentially affecting lichen survival in urban areas, air pollution is the one more consistently associated with lichen decline (Nash 1988).

Reestablishment of lichens in some previously polluted areas after successful abatement of air pollution has confirmed that air pollution was a main factor responsible for the original decline (e.g. Gilbert 1992; Henderson-Sellers and Seaward 1979).

Lichens are specialized fungi symbiotically associated with photosynthetic partners (green algae, cyanobacteria, or both). Ecologically speaking, such an association is defined as a unique microecosystem (Tehler 1996). Lichens lack roots, cuticles for protection against dessication, and the gas exchange controls (stomata) found in vascular plants. In contrast with most vascular plants, lichens do not obtain their nutrients by roots from the soil but depend primarily on the atmosphere as a source of water and mineral nutrients. Thus, lichens tend to accumulate material deposited from the atmosphere onto their surfaces. In heavily polluted areas, some airborne chemicals often reach critical concentrations in lichen cells disrupting the physiological coordination between the symbiotic partners (Nash 1996).

Air pollution studies involving lichens include a variety of field and laboratory techniques. For instance, in field studies, sites are compared by means of species lists to identify missing species (floristic approach, Wetmore 1988). Adding species abundance information allows comparisons that take into account structural characteristics of the lichen communities (e.g. Nash 1975). The transplantation of lichens from relatively clean (control) sites to polluted sites is an example of a field experimental approach for evaluating responses to polluted conditions across time (e.g. Boonpragob and Nash 1991). Controlled laboratory experiments quantify uptake of pollutants by lichens and physiological responses to known doses of pollutants (e.g. Gries et al. 1997; Sanz et al. 1992). There is good agreement between field and laboratory studies at the species level (Nash 1988).

Lichens are used in studies of air pollutants such as sulfur dioxide (SO_2), ozone (O_3), oxides of nitrogen (NO_x), fluorides, and a variety of toxic metals (Gries 1996). Lichen distribution maps document that lichen diversity is poorest in and near industrial regions and city centers. These patterns often match the typical shape of the pollutant dispersion plume toward the outskirts, suburban, and rural areas (Hawksworth and Hill 1984; Richardson 1975). In Europe, lichen maps and the decline and recovery in lichen species richness across time, correlated best with SO_2 levels (Richardson 1992; van Dobben 1996). In areas where SO_2 has not been a major pollutant, such as southern California, lichen distribution patterns correlated better with photochemical oxidant gradients (Nash and Sigal 1998). In sites polluted by both SO_2 and O_3, such as Indianapolis, lichen parameters correlated better with SO_2 than O_3 (McCune 1988). We are utilizing lichen community parameters, lichen transplant responses, and fumigation studies to assess the ecological importance of several primary and secondary pollutants that reach chronically high levels in the environs of Mexico City.

Lichen Diversity Near Mexico City

Although there are over fifty reports dealing with lichens of the Basin of Mexico (Godínez and Ortega 1989), only one provides a floristic inventory. An unpublished master thesis by Gibert (1935) reported 220 lichen species. A detailed review of this work is, however, needed to determine that his identifications were correct and to update nomenclature. Unfortunately, Gibert's collections at the National Herbarium (MEXU, Instituto de Biología, UNAM) are incomplete, making it impossible to confirm all species names. His work would have provided an excellent baseline for evaluating the status of the current lichen flora, if his species distribution data had been more complete. Unfortunately, he described each species as if it occurred at only one of 33 sampling sites. Thus, it impossible to compile accurate species distribution patterns for the Basin from Gibert's data or to compile a species list for each site. Table 12.1 summarizes the number of species assigned by Gibert to his sampling sites. It might be inferred from these data that Sierra de Las Cruces, represented by montane sites such as Desierto de los Leones (DL), Contreras, and Salazar, and foothill sites, such as Tlalpan and Pedregal, were at that time the richest parts of the Basin in terms of lichen species diversity. That is not surprising, because Sierra de Las Cruces is the most humid area within the Basin, and it is generally known that the number of lichen species usually increases with precipitation (Adams and Risser 1971).

The lack of sufficient historical data on the local lichen flora presents a serious problem for making across-time evaluations of the effects of the diminished air quality of Mexico City on the lichen comunities. However, substituting space for time, by comparing the current lichen diversity of stressed sites against that of environmentally similar sites exposed to much lower stress levels, can be a useful alternative approach (Wetmore 1988).

We applied this research approach to the epiphytic lichen community of sacred fir trees (*Abies religiosa*) in the DL National Park (Fig. 12.1) (Zambrano et al. 2000). This site is located approximately 8 km southwest of Mexico City's urban limit. Severe symptoms of photochemical injury on the foliage of *A. religiosa*, *Pinus hartwegii* and *Prunus serotina* have been reported for this site (Alvarado et al. 1993; Skelly et al. 1997). The DL lichen community was compared with a community growing under fairly similar environmental conditions in El Chico National Park (CH), located approximately 100 km northeast of Mexico City. El Chico is a good reference site because it lies outside the Basin of Mexico and is located upwind of Mexico City (Fig. 12.1). Consequently, it is much less exposed to urban emissions than DL. Both sites belong in the same floristic and geological region, the Neovolcanic Axis (Rzedowski 1978), and share a similar macroclimate, an important factor in shaping lichen communities (Adams and Risser 1971). These two sites also have comparable mountain topography, altitude, and vegetation. Because the sites were environmentally similar, and because lichens were sampled exclusively on sacred firs, the two lichen communities were expected to have similar diversity. In part, this was true, because nearly 80% of the DL lichen species were common to CH. However, the two communities

Table 12.1. Lichen collection sites visited by Gibert (1935) and number of species per site[a]

Collection site	Number of species[a]
Desierto de los Leones	38
Contreras	25
Talpan	14
Pedregal	13
Salazar	12
Tenayuca	10
Río Frío	9
Monte de Guadalupe	8
San Juanico	7
Cerro de la Estrella	6
El Chico	6
Jalalpa	6
Santa Fe	6
Acopilco	5
Mixcoac	5
La Venta	5
Tacubaya	4
Peñón	4
Colonia del Valle	3
San Jerónimo	3
Texcoco	3
Zoquiapan	3
Acueducto de Guadalupe	2
Carretera a Puebla	1
Chapultepec	1
Contadero	1
Culhuacán	1
Lerma	1
Los Morales	1
Montes de Pachuca	1
San Rafael	1
Santa Rosa	1
Tepoztlán	1

[a] Since Gibert assigned each different species exclusively to one of his collection sites, total species numbers for the sites are inaccurate.

were very different in lichen species richness, defined as the number of species, and abundance. We found only 28 lichen species at DL, compared with 53 species at CH. Trees at DL had, on average, half the species richness of trees at CH (9 versus 18 species) (Fig. 12.2a), and had 62% lower lichen abundance, measured as the percentage of the first two meters of the tree bole covered with lichen (Fig. 12.2b).

The structural differences between the DL and CH lichen communities are so strong that they are not likely to be due to chance or natural causes, such as

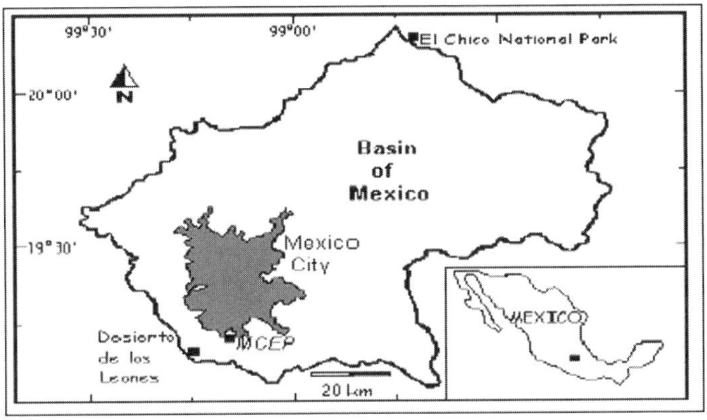

Figure 12.1. Sampling sites for lichen transplant and community studies (Desierto de los Leones, El Chico National Park, and Mexico City Ecological Park [MCEP]).

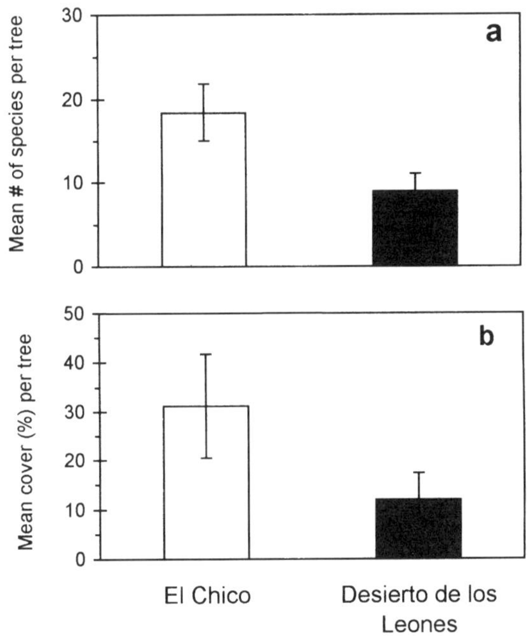

Figure 12.2. Mean lichen species richness (a) and cover (b) per tree. Percent cover refers to the average percentage of the first two meter of the tree bole covered with lichen. Error bars represent one standard deviation. Number of sampled trees: El Chico, 46; Desierto de los Leones, 58.

general climate or forest structure differences; instead, they may be related to anthropogenic disturbance. The abundance of lichens per tree was positively correlated with forest stand density and cover in the DL, but it varied independently from these factors in the more homogeneous forest at CH (Zambrano et al. 2000). It is reasonable to assume, although no measurements were made, that the more patchy and open DL forest canopy has greater light penetration than the CH forest. Such a difference in light intensity reaching the forest interior is likely to be the cause of differences in other correlated variables such as temperature and evaporation rate. Lichens are, in general, sensitive to changes in these factors (Adams and Risser 1971), but this is a point requiring further study in the DL. In contrast with lichen abundance, variations in species richness per tree were independent of forest density and cover at both sites. This suggests that the loss of lichen cover at DL is occurring faster than the loss of species. This is consistent with the 62% lower cover and 47% fewer species at DL compared with CH.

The species-abundance pattern of the DL lichen community reveals that it has been subjected to severe disturbance (Zambrano et al. 2000). The CH community is numerically dominated by species having moderate abundance (lognormal pattern, as defined by Whitaker 1975), whereas at DL, this group of species is strongly depleted (geometric pattern, according to Whitaker 1975). A simplified version of this difference is illustrated in Figure 12.3. The DL community pattern closely resembles those frequently described for plant and animal communities of stressed environments (e.g., Gray 1986). Nash (1975) reported a similar example for lichen communities exposed to emissions from a zinc factory. Although there is no direct way to prove whether the DL and CH patterns were originally more similar, indirect evidence indicates that a much richer community existed at DL in past decades. Table 12.2, which is based on historical

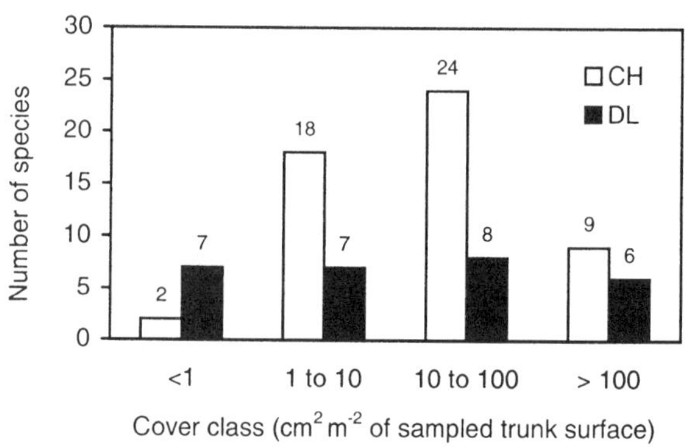

Figure 12.3. Lichen species frequency per abundance class on sacred fir trunks at El Chico National Park, CH, and Desierto de los Leones, DL. Lichen cover classes are subdivided to represent four logarithmic cycles.

Table 12.2. Preliminary list of lichens collected in Desierto de los Leones prior to 1977

Species	Collector	Year of collection	Observation[a]
Cetrelia cf. *cetrarioides* (Duby) Culb. & Culb.	Gibert	1935	1
Everniastrum pseudonepalense (Hale & Wirth) Hale	Ruíz Oronoz	1955, 1957	2
Heterodermia diademata (Taylor) D. D. Awasthi	Ruíz Oronoz	1941	1
H. leucomela (L.) Poetl	Ruíz Oronoz	1945	1
H. squamulosa (Degel) Culb.	Brizuela	1984	1
Hypotrachina livida (Taylor) Hale	Ruíz Oronoz	1941	1
Leptogium azureum (Swartz) Mont.	Gibert	1935	1
L. corticola (Taylor) Tück.	Lowy, Herrera and Ruíz Oronoz	1962	1
L. furfuraceum (Harm.) Sierk	Gibert	1935	1
L. hildebrandii (Garov.) Nyl.	Gibert	1935	1
L. phyllocarpon (Pers.) Mont.	Gibert, Herrera and Duboboy	1955, 1956, and 1962	1
Peltigera spp.	Several collectors	1908–1958	2
Pseudociphellaria aurata (Ach.) Vainio	Gibert	1935	1
Pseudoevernia consocians (Vainio) Hale & Culb.	Several collectors	1941–1976	2
P. intensa (Nyl.) Hale & Culb.	Several collectors	1941–1970	2
Punctelia hypoleucites (Nyl.) Krog	Gibert	1935	2
P. rudecta (Ach.) Krog	Gibert	1935	2
Usnea arizonica s. l. Motyka	Ruíz Oronoz	1953	2
U. ceratina Ach.	Gibert, Herrera, and Ruíz Oronoz	1924–1958	2
U. cirrosa Motyka	Ruíz Oronoz	1940, 1953	1
U. cornuta s. l. Koerb.	Ruíz Oronoz	1941	1
U. filipendula Stirton	Ruíz Oronoz	1940	1
U. cf. florida (L.) Wigg	Ruíz Oronoz	1953	2
U. subscabrosa Nyl. ex Motyka	Bautista	1976	1

[a] Observations: 1, species not found recently and suspected to be locally extinct; 2, species with low or extremely low current abundance compared with El Chico National Park.

collections held at the National Herbarium, lists some species collected at DL between the early 1900s and 1970s but not found in our recent sampling of the same general locality. Thus, we hypothesize that the current lichen species-abundance pattern of DL may have resulted from the depletion of a lognormal pattern via extinction of an unknown number of species, perhaps most of the species with intermediate abundance.

Such lichen decline is not restricted to DL. Apparently, it is a generalized process across the entire Sierra de las Cruces. We have preliminary unpublished data from an ongoing survey on the epiphytic lichens of oaks (*Quercus rugosa*) in the Mexico City Ecological Park (MCEP) (Fig. 12.1), a site within the same mountain range, also showing impoverishment of the lichen community.

Is this general pattern of lichen decline caused by unsuitable local environmental conditions? To what extent do factors such as air pollution and deforestation affect patterns of change in lichen communities? It is still not possible to answer such complex questions because of the lack of data on the ecology and ecophysiology of the native lichens and because quantitative information on the air pollutants reaching those forests is *insuffient*. Regardless, air pollution appears to be a major influence, as suggested by the species and genera that have disappeared or are disappearing from the area. For example, *Usnea ceratina*, and the genus *Usnea* in general, *Teloschistes*, and *Peltigera* are known to be highly sensitive to air pollution, and especially to SO_2 (Hawksworth and Rose 1976). Furthermore, as shown by trace metal concentrations, in lichen tissues (such as Pb, see next section), there is no doubt that lichen communities to the south of the city have been exposed to urban emissions for many years.

Heavy Metals and Lichens

Lichens accumulate metal ions abundantly because of the high binding capacity of their cell walls (Brown and Beckett 1984) and by particulate entrapment (Garty et al. 1979). The rationale for using lichens as natural monitors of airborne metals is based upon a demonstrated positive correlation between the bulk deposition of many metals and their content in lichen tissues (e.g., Sloof 1995a). Thus, the measurement of site-to-site variability in lichen metal concentrations allows the detection of relative differences in metal deposition within a region (Puckett 1988) and the identification of sources of metallic air pollution (e.g., Sloof 1995b).

In the early 1990s, as part of an international project on air quality in Mexico City, relative patterns of atmospheric metal deposition were studied by assaying pollutant concentrations in lichens from five forests surrounding the city (Carrillo-Padrón 1993; IMP-LANL 1994; Streit and Guzmán 1996). Concentrations of Ca, Cd, Cr, Cu, Fe, K, Pb, and Zn were determined in samples of epiphytic lichens (*Parmelia* sp. and *Parmotrema* sp.) collected every three months between May, 1990, and February, 1991. Figure 12.4 shows the grand mean (the mean of the 4 sampling dates) concentrations of six metals at each site. There were higher values of most metals at sites near the city (DL, San Francisco Chimalpa, and Sierra de Guadalupe) than at sites farther from the city (CH and Zoquiapan (ZOQ)). This trend is best observed for concentrations of Cu, Pb, Zn, and to a lesser degree, Cr. Lichens showed a metal concentration gradient increasing from northeast to southwest, a gradient corresponding with the dominant regional wind circulation in the Basin of Mexico as described by Jáuregui and Luyando (1992). The highest thallus concentrations for most metals occurred at

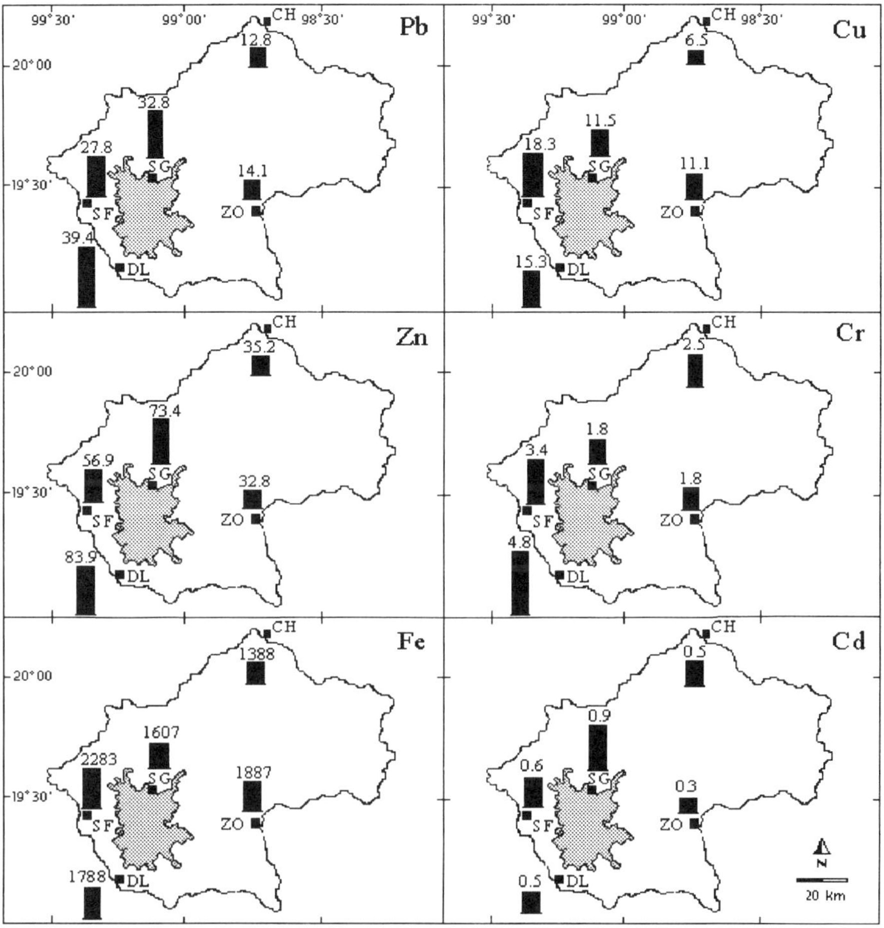

Figure 12.4. Mean concentration of six metals in lichens from four sacred fir forests around Mexico City and Sierra de Guadalupe (SG) an oak forest. Samples were collected in May, August, and November, 1990, and February, 1991. Values above bars are the grand mean (in ppm) for that site calculated from mean values for each sampling period (after Carrillo-Padrón 1993). Sample size for each site = 5 samples/date × 4 dates = 20. DL, Desierto de los Leones; SF, San Francisco Chimalpa; ZO, Zoquiapan National Park; and CH, El Chico National Park.

DL, a site downwind of the city, whereas samples from sites upwind or crosswind (CH and ZOQ) had the lowest values. Thus, concentration of metals at each site clearly depended on the distance and direction from the city. The contribution of Mexico City to metal concentrations in lichens was clearly demonstrated by the geographical distribution of Pb, a typical indicator of gasoline combustion emissions (Carrillo-Padrón 1993). It must be added that only unleaded gasolines are currently sold in the local market (PEMEX 1999). Thus, the previous results, obtained at a time when leaded gasoline was still in use, may not correspond to current levels of lead in lichen tissues.

Previously, Cantoral-Herrera (1986) reported Pb concentrations for the lichens *Ramalina* cf. *farinacea* (28 ppm) and *Usnea strigosa* (54 ppm) collected at DL from slopes not directly exposed to emissions from Mexico City. Concentrations of 139 and 91 ppm, respectively, were reported for samples from slopes facing the city. The latter values are considerably higher than those reported by Carrillo-Padrón (1993) for *Parmelia* and *Parmotrema* (39.4 ppm) from the same DL. Such difference may be related to interspecific differences in the ability of lichens to accumulate Pb. For example, both of Cantoral-Herrera's species were fruticose (shrubby) lichens characterized by a greater surface/volume ratio. Consequently, these lichens must have a greater metal exchange surface than foliose (laminar) species such as those used by Carrillo-Padrón (1993).

A preliminary lichen transplant experiment was conducted in the urban area of Mexico City to assess the usefulness of these organisms as bioaccumulators of metal pollution (Jiménez-Pardo 1993). Branches of *Juniperus* sp. with *Parmotrema* sp. were transplanted from CH to an urban site in the northern part of Mexico City, the most industrialized sector of the city. Thallus concentrations of Cd, Cr, Cu, Fe, Pb, and Zn were measured at the beginning of the transplantat study and every three months afterwards from May, 1990, to February, 1991. Figure 12.5 shows the mean concentrations of metals in lichen tissues for each sampling period. Elements with the highest concentrations included Cu, Pb, and Zn, which all increased by an order of magnitude from initial values. Unfortunately, no other transplant sites were established; therefore, comparison of relative metal deposition rates in Mexico City was not possible.

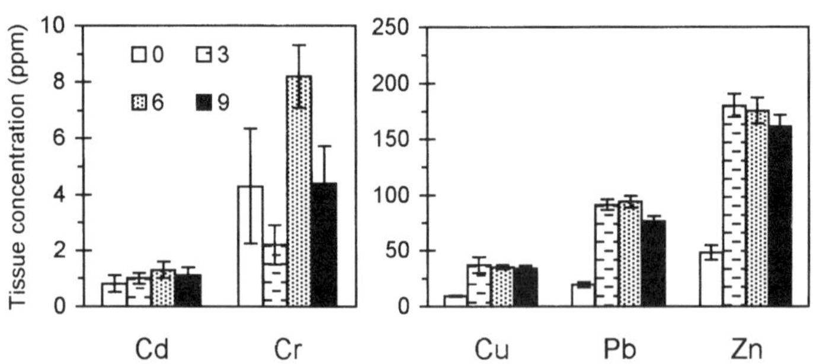

Figure 12.5. Mean metal concentrations (± standard deviation) in *Parmotrema* sp. transplants (Jimenez-Pardo 1993). Samples were transplanted from El Chico National Park to a site in northern Mexico City. Metal contents were measured at the beginning of the experiment (May 1991, month 0) and 3, 6, and 9 months after transplantation. Sample size, 10, except for the last period (9). Concentrations are expressed in ppm (lichen dry weight).

Lichen Photosynthetic Responses Near Mexico City

Carbon fixation is one of the best documented physiological variables for pollution-sensitive lichens (e.g., Fields 1988). Among the air pollutants currently monitored in Mexico City, O_3 very often reaches levels known to negatively affect the physiology of some sensitive lichens. The SO_2 concentrations in the city were also high enough to harm lichens in previous decades, but they have decreased greatly in recent years as a result of improved fuel quality (Jáuregui 1995; PEMEX 1999). The section of the city most polluted by O_3 is the southwest. Over the last two decades, the hourly standard for O_3 (0.11 ppm) has been violated an average of 250 to 300 days per year in that part of the city (Jáuregui 1995; WHO-UNEP 1993). The frequency of such violations at DL is also high, and O_3 events as high as 0.27 ppm have been reported (Miller et al. 1994). Together with symptoms of photochemical injury in conifers at several sites in the Sierra de las Cruces (de Bauer and Krupa 1990), high levels of O_3 in DL suggest that the entire mountain range is exposed to elevated concentrations of ozone.

Lichen sensitivity to O_3 is not well understood. However, some experimental reports show that lower concentrations than those occurring in Mexico City and DL are sufficient to impair lichen photosynthetic performance (e.g., Ross and Nash 1983) and to partially destroy their photosystems (Scheidegger and Schroeter 1995). One of the declining lichen species in DL, *Flavoparmelia caperata* (Zambrano et al. 2000), is among those that tested positive for O_3 sensitivity under controlled fumigation with concentrations comparable to the Mexican standard (Ross and Nash 1983).

Short-term lichen transplant experiments (January 24 to March 20, 1997) showed that environmental conditions in Mexico City impaired photosynthetic performance of two epiphytic species, *Everniastrum neocirrhatum* and *U. ceratina* (Zambrano and Nash 2000; Zambrano et al. 1999). Healthy samples were transplanted from CH into an oak forest (*Quercus rugosa*) at the MCEP (Fig. 12.1) and into a sacred fir (*A. religiosa*) stand in the DL. Two different sets of control samples per lichen species and vegetation type were used. One included samples transplanted within the control site (CH) for comparison with lichens transplanted to the urban site, while the second *in situ* lichen samples at CH were used to assess possible transplant effects.

The transplanted lichen at CH did not show significant physiological changes. In contrast, the lichens transplanted near the city showed an overall 30% decrease in maximum net photosynthesis. Chlorophyll *b* decreased approximately 25% in *E. neocirrhatum* at the site closest to the city (MCEP) and 18% at DL. No interspecific difference was detected in these responses. Chlorophyll *a* was slightly degraded (ca. 15%) only at MCEP, with no difference between species. Carotene content varied between species but was indistinguishable between control and experimental sites; therefore, this effect cannot be attributed to urban air pollution.

These transplantation experiments were conducted during the local dry season when air pollution exposures are highest. During the transplant period (54 days), episodes in which O_3 was over 0.11 ppm occurred on 44 days, and O_3 was above this level, for a cumulative total of 171 hours as recorded at Pedregal, the nearest air-monitoring station (ca. 5 km distant) in southwest Mexico City (Zambrano and Nash 2000; Zambrano et al. 1999). In contrast, average SO_2 levels during the same period were well below the local 24-hour standard (0.13 ppm). The decline in photosynthesis described earlier, along with previous reports based on O_3 fumigation experiments (Ross and Nash 1983; Scheidegger and Schroeter 1995), suggests that pollutant levels were sufficiently high to cause a response in the transplant material. The concurrent degradation of lichen chlorophyll *b*, the photosynthetic pigment most often reported as sensitive to O_3 according to Runnecles and Chevonne (1992), also supports this view.

Conclusion and Research Needs

The evidence from recent lichenological studies near Mexico City shows a severe decline of lichen communities in forests around the city. The decline appears to coincide with the period of accelarated industrial and population growth experienced by Mexico City since the 1930s and 1940s. Compared to many European cities, where industrialization occurred many decades earlier, this is a relatively short period of exposure to pollution to cause local extinction of native lichen species. The decline now documented for DL suggests that the local lichen communities may have been exposed to much higher air pollution levels than most European cities during the last 50 years.

Current levels of O_3 in the study area are high enough to threaten lichen communities. We also have unpublished data showing that some lichen species now absent or disappearing from DL are sensitive to SO_2. Levels of SO_2, one of the most deleterious pollutants for lichens (Fields 1988; Gries et al. 1995; Sanz et al. 1992), are now declining in urban Mexico City (Jáuregui 1995). However, SO_2 was a major local contaminant for many years (Bravo-Alvarez 1987). Recent measurements of atmospheric deposition at Desierto de los Leones revealed that atmospheric inputs of sulphate were more than twice as high as at ZOQ, a similar forest approximately 50 km east of Mexico City (Fenn et al. 1999). Thus, it is difficult to prove but not unreasonable to infer that elevated exposures to SO_2 at DL in previous decades was an important factor causing lichen impoverishment. Average rain pH values reported in the late 1980s for DL (4.7) and Tlamacas (5.1), a site near ZO (Padilla 1989), reinforce such possibility.

Considering the role of lichens as indicators of environmental stressors which threaten the sustainability of forest ecosystems, further lichenological studies could provide valuable insights on the geographic extent of ecological risks from air pollution in local forested areas. For instance, because none of these areas is covered by Mexico City's air pollution monitoring network, we may greatly benefit from using lichens as relatively inexpensive biomonitors to alleviate the lack of information on pollution exposure levels in the forests of the Basin of Mexico. Thus, more work on the native lichens is needed, including taxonomic, ecological and ecophysiological studies, as well as controlled experimentation on the effects of single and mixed air pollutant exposures.

References

Adams, D.B., and Risser, P.G. 1971. Some factors influencing the frequency of bark lichens in north central Oklahoma. *Amer. J. Bot.* 58:752–757.

Alvarado, D.R., de Bauer, L.I., and Galindo, J.A. 1993. Decline of sacred fir (*Abies religiosa*) in a forest park south of Mexico City. *Environ. Pollut.* 80:115–121.

Boonpragob, K., and Nash III, T.H. 1991. Physiological responses of the lichen *Ramalina menziesii* Tayl. to the Los Angeles urban environment. *Environ. Exp. Bot.* 31:229–238.

Bravo-Álvarez, H. 1987. *La Contaminación del Aire en México.* Mexico City: Universo Veintiuno.

Brown, D.H., and Beckett, R.P. 1984. Uptake and effect of cations on lichen metabolism. *Lichenologist* 16:173–188.

Cantoral-Herrera, M.T. 1986. *Comunidades Liquénicas Epífitas en Abies religiosa (H.B.K.) Cham. & Schl. como Indicadoras de Contaminación Atmosférica en el Parque Cultural y Recreativo Desierto de los Leones.* B.Sc. Thesis. Mexico City: Escuela Nacional de Estudios Profesionales Iztacala, Universidad Nacional Autónoma de México.

Carrrillo-Padrón, V.H. 1993. *Líquenes Foliosos como Biomonitores de Contaminación Atmosférica por Metales Pesados en los Alrededores de la Zona Metropolitana de la Ciudad de México (ZMCM).* B.Sc. Thesis. Mexico City: Facultad de Ciencias, Universidad Nacional Autónoma de México.

de Bauer, L.I., and Krupa, S.V. 1990. The Valley of Mexico: Summary of observational studies on its air quality and effects on vegetation. *Environ. Pollut.* 65:109–118.

Fenn, M.E., de Bauer, L.I., Quevedo-Nolasco, A., and Rodríguez-Frausto, C. 1999. Nitrogen and sulfur deposition and forest nutrient status in the Valley of Mexico. *Water Air Soil Pollut.* 113:155–174.

Fields, R.F. 1988. Physiological responses of lichens to air pollution fumigations. In *Lichens, Bryophytes and Air Quality*, Bibliotheca Lichenologica 30, eds. T.H. Nash III, and V. Wirth, pp. 175–200. Berlin–Stuttgart: J. Cramer.

Garty, J., Galun, M., and Kessel, N. 1979. Localization of heavy metals and other elements accumulated in the lichen thallus. *New Phytol.* 82:159–168.

Gibert, A. 1935. *Líquenes del Valle de México.* M.Sc. Thesis. Mexico City: Facultad de Ciencias, Universidad Nacional Autónoma de México.

Gilbert, O.L. 1992. Lichen reinvasion with declining air pollution. In *Bryophytes and Lichens in a Changing Environment*, eds. J.W. Bates, and A.M. Farmer, pp. 159–177. Oxford: Clarendon Press.

Godínez, J.L., and Ortega, M.M. 1989. *Liquenología de Mexico: Historia y Bibliografía*, Cuadernos 3. Mexico City: Instituto de Biología, Universidad Nacional Autónoma de México.

Gray, J.S. 1986. Species-abundance patterns. In *Organization of Communities: Past and Present, 27th Symposium of the British Ecological Society*, Special Publication in Series of the British Ecological Society 27, eds. J.H.R. Gee, and P.S. Giller, pp. 53–67. London: Blackwell Sci. Pub.

Gries, C. 1996. Lichens as indicators of air pollution. In *Lichen Biology*, ed. T.H. Nash III, pp. 240–254. New York: Cambridge Univ. Press.

Gries, C., Sanz, M.J., and Nash III, T.H. 1995. The effect of SO_2 fumigation on CO_2 gas exchange, chlorophyll fluorescence and chlorophyll degradation in different lichen species from western North America. *Crypt. Bot.* 5:239–246.

Gries, C., Sanz, M.J., Romagni, J.G., Goldsmith, S., Kuhn, U., Kesselmeier, J., and Nash III, T.H. 1997. The uptake of gaseous sulphur dioxide by non-gelatinous lichens. *New Phytol.* 135:595–602.

Hawksworth, D.H., and Hill, D.J. 1984. *The Lichen-Forming Fungi.* New York: Chapman and Hall.

Hawksworth, D.L., and Rose, F. 1976. Lichens as Pollution Monitors. London: E. Arnold.

Henderson-Sellers, A., and Seaward, R.D. 1979. Monitoring lichen reinvasion of ameliorating environments. *Environ. Pollut.* 19:207–213.

IMP-LANL. 1994. *Mexico City Air Quality Research Initiative, Characterization and Measurement*, Final Report, Volume IV, Los Alamos, NM: Instituto Mexicano del Petróleo-Los Alamos National Laboratory. LA-12699.

Jáuregui, E. 1995. Algunas alteraciones de largo período del clima de la Ciudad de México debidas a la urbanizacion—revisión y perspectivas. *Bol. Inst. Geog. UNAM* 31:9–44.

Jáuregui, E., and Luyando, E. 1992. Patrones de flujo de aire superficial y su relación con el transporte de contaminantes en el Valle de México. *Bol. Inst. Geog. UNAM* 24:51–78.

Jiménez-Pardo, A.H. 1993. *Acumulación de Metales en Líquenes Transplantados a un Área Industrial en la Zona Metropolitana de la Ciudad de México.* B.Sc. Thesis. Mexico City: Facultad de Ciencias, Universidad Nacional Autónoma de México.

McCune, B. 1988. Lichen communities along O_3 and SO_2 gradients in Indianapolis. *Bryologist* 91:223–228.

Miller, P.R., de Bauer, M.L., Quevedo-Nolasco, A., and Hernández-Tejeda, T. 1994. Comparison of ozone exposure characteristics in forested regions near Mexico City and Los Angeles. *Atmos. Environ.* 28:141–148.

Nash III, T.H. 1975. Influence of effluents from a zinc factory on lichens. *Ecol. Monogr.* 45:183–198.

Nash III, T.H. 1988. Correlating fumigation studies with field effects. In *Lichens, Bryophytes and Air Quality*, Bibliotheca Lichenologica 30, eds. T.H. Nash, III, and V. Wirth, pp. 201–216. Berlin-Stuttgart: J. Cramer.

Nash III, T.H. 1996. Nutrients, elemental accumulation and mineral cycling. In *Lichen Biology*, ed. T.H. Nash III, pp. 136–153. New York: Cambridge Univ. Press.

Nash III, T.H., and Sigal, L.L. 1998. Epiphytic lichens in the San Bernardino Mountains in relation to oxidant gradients. In *Oxidant Air Pollution Impacts in the Montane Forests of Southern California: A Case Study of the San Bernardino Mountains.* Ecol. Stud. 134, eds. P.R. Miller, and J.R. McBride, pp. 223–234. New York: Springer-Verlag.

Padilla, H. 1989. *Estudio Analítico de la Composición Química de la Precipitación Pluvial en el Valle de México.* M.Eng. Thesis. Mexico City: Facultad de Ingeniería, Universidad Nacional Autónoma de México.

PEMEX. 1999. *Seguridad, Salud y Medio Ambiente: Informe 1999.* Mexico City: Dirección Corporativa de Seguridad Industrial y Protección Ambiental de Petróleos Mexicanos.

Pucket, K.J. 1988. Bryophytes and lichens as monitors of metal deposition. In *Lichens, Bryophytes and Air Quality*, Bibliotheca Lichenologica 30, eds. T.H. Nash III, and V. Wirth, pp. 231–267. Berlin–Stuttgart: J. Cramer.

Richardson, D.H.S. 1975. *The Vanishing Lichens.* Vancouver, British Columbia: David and Charles.

Richardson, D.H.S. 1992. *Pollution Monitoring with Lichens*, Naturalist's Handbooks 19. Slough, England: Richmond Publishing Co.

Ross, L.J., and Nash III, T.H. 1983. Effects of ozone on gross photosynthesis of lichens. *Environ. Exp. Bot.* 23:71–77.

Runeckles, V.C., and Chevone, B.I. 1992. Crop responses to ozone. In *Surface Level Ozone Exposures and their Effects on Vegetation*, ed. A.S. Lefohn, pp. 189–270. Chelsea, Michigan, U.S.A.: Lewis Publishers.

Rzedowski, J. 1978. *La Vegetación de México.* México, D.F.: Limusa.

Sanz, M.J., Gries, C., and Nash III, T.H. 1992. Dose-response relationships for SO2 fumigation in the lichens *Evernia prunastri* (L.) Ach. and *Ramalina fraxinea* (L.) Ach. *New Phytol.* 122:313–319.

Scheidegger, C., and Schroeter, B. 1995. Effects of ozone fumigation on epiphytic macrolichens: ultrastructure, CO_2 gas exchange and chlorophyll fluorescence. *Environ. Pollut.* 88:345–354.

Skelly, J.M., Savage, J.E., de Bauer, M.L., and Alvarado, D. 1997. Observations of ozone induced foliar injury on black cherry (*Prunus serotina*, var. *capuli*) within the Desierto de los Leones National Park, Mexico City. *Environ. Pollut.* 95:155–158.

Sloof, J.E. 1995a. Lichens as quantitative biomonitors for atmospheric trace-element deposition, using transplants. *Atmos. Environ.* 29:11–20.

Sloof, J.E. 1995b. Pattern recognition in lichens for source apportionment. *Atmos. Environ.* 29:333–343.

Streit, G.E., and Guzmán, F. 1996. Mexico City air quality: progress of an international collaborative project to define air quality management options. *Atmos. Environ.* 30:723–733.

Tehler, A. 1996. Systematics, phylogeny and classification. In *Lichen Biology*, ed. T.H. Nash III, pp. 217–239. New York: Cambridge Univ. Press.

Van Dobben, H.F. 1996. Decline and recovery of epiphytic lichens in an agricultural area in the Netherlands (1900–1988). *Nova Hedwigia* 62:477–485.

Wetmore, C.M. 1988. Lichen floristics and air quality. In *Bryophytes and Lichens in a Changing Environment*, eds. T.H. Nash III, and V. Wirth, pp. 55–65. Berlin-Stuttgart: J. Cramer.

Whitaker, R.H. 1975. *Communities and Ecosystems*. New York: MacMillan.

WHO-UNEP. 1993. *Urban Air Pollution in Megacities of the World, Global Environment Monitoring System*. Oxford: World Health Organization and United Nations Environment Program, Blackwell Publishers.

Zambrano, A., and Nash, T.H. 2000. Lichen responses to short-term transplantation in Desierto de los Leones, Mexico City. *Environ. Pollut.* 107:407–412.

Zambrano, A., Nash, T.H., and Gries, C. 1999. Physiological effects of the Mexico City atmosphere on lichen transplants on oaks. *J. Environ. Qual.* 28:1548–1554.

Zambrano, A., Nash, T.H., and Herrera-Campos, M.A. 2000. Lichen decline in Desierto de los Leones, Mexico City. *Bryologist* 103:428–441.

13. Nitrogen and Sulfur Deposition in the Mexico City Air Basin: Impacts on Forest Nutrient Status and Nitrate Levels in Drainage Waters

Mark E. Fenn, L.I. de Bauer, Karl Zeller, Abel Quevedo, Claudio Rodríguez, and Tomás Hernández-Tejeda

Based on published emission inventories for metropolitan Mexico City (Bravo 1987; INEGI 1999; Ruiz and Gasca 2002), elevated deposition of nitrogen (N) and sulfur (S) pollutants is expected to occur in forests downwind of Mexico City. Air pollution concentrations are elevated in montane areas to the south-southwest (SSW) of Mexico City because of their proximity to the large urban pollution sources and because of the prevailing northeasterly winds (de Bauer 1991). Fossil fuel consumption, mainly for transportation, is the primary source of atmospheric N pollutants in Mexico City. Sulfurous air pollutants come primarily from industrial activities and burning of fossil fuels (Ruiz and Gasca 2002). There were 128,646 tons of nitrogen oxides (NOx) emissions in 1994 compared with 45,468 tons of SO_2 (INEGI 1999). Data from the acid precipitation wet deposition network in Mexico City (INEGI 1999) and from a recent throughfall study in an adjacent forested area southwest (SW) of the metropolitan area (Fenn et al. 1999) suggest that sulfate deposition is still greater than nitrate deposition, but similar to the combined deposition of inorganic N forms (nitrate and ammonium).

Forest production in northern temperate forests is normally N limited. However, a large body of evidence now indicates that in forested areas exposed to chronic atmospheric deposition, the N status of the forest shifts from N limitation to a condition of N excess. The syndrome of effects or forest responses that occur in forests with elevated atmospheric N deposition is referred to as N saturation (Aber et al. 1989). Nitrogen saturation can be concisely defined as the long-term removal of N limitations on biotic activity accompanied by a decrease in N retention capacity. Nitrogen saturated forests, while not widespread, are predominantly found in forests downwind of strong source areas of atmospheric N pollutants. Such sources include large urban or industrial centers or livestock or poultry operations. Nitrogen-saturated forests are also found in highly populated regions with numerous dispersed sources of N emissions.

Nitrogen saturation, according to the conceptual model (Aber et al. 1989), is hypothesized to occur in stages. Stage 1 occurs when N concentrations in foliage increase, and brief periods of N losses from the system are temporarily amplified. Tree production may also increase. In stage 2, N losses from the forest are sustained, and nitrification increases. In most instances, reports of N-saturated forests refer to stage 2 of the Aber model. In stage 3, chronic N inputs lead to forest decline or decreased productivity. Stoddard (1994) defined a set of progressive stages for watershed N saturation analogous to the terrestrial N saturation model. The cardinal sign of a N-saturated forest or watershed is excessive loss of N, usually as leached nitrate, and to a lesser degree, as trace nitrogenous gas emissions from soil (Fenn et al. 1998).

The phenomenon of N excess or N saturation poses a threat to the natural resources and ecosystem services provided by wildland watersheds and forests (Fenn et al. 1998). Protection of water quality is a primary concern in N-saturated watersheds. High nitrate levels in streamwater and groundwater of N-saturated watersheds is a health concern because of reported links between high nitrate levels in drinking water and blue-baby syndrome and stomach cancer (Stevenson and Cole 1999). Furthermore, water supplies from montane watersheds are often relied upon to improve water supplies tainted by contaminants from agriculture, industry, or urban runoff. Thus, even though water from N-saturated forest watersheds rarely exceeds established drinking water standards for nitrate, its effective dilution power for blending with other water supplies is reduced. The ecological impacts of excess N, especially over the long term, is a major research question being investigated by many research groups in North America and Europe. Many key ecological processes and organisms are impacted by high levels of available N. Excess N constitutes one more ecological perturbation or environmental stress with potential long-term consequences for sensitive plants and microorganisms in forested areas with chronic atmospheric N deposition. Nitrogen saturation of terrestrial ecosystems is analogous to the well-know phenomenon of eutrophication of aquatic systems and over-fertilization of agricultural crops. Furthermore, in forested areas where N deposition is elevated, concentrations of ozone (O_3) and other air pollutants may also be causing important ecological effects.

Many studies have demonstrated a close relationship between levels of N deposition and effects on chemical processes in soil such as soil acidification, stimulation of nitrification, and nitrate leaching, leaching of counterbalancing cations, and alleviation of N deficiency (Fenn et al. 1998). These types of responses have been documented in forest sites located across air pollution gradients within the Los Angeles California Air Basin (Fenn and Poth 1999; Fenn et al. 1996). Because of similarities in oxidant air pollution exposures in forests downwind of

Los Angeles and Mexico City, we hypothesized that forests with high air pollution levels in the Mexico City Air Basin (MCAB) would also express the symptoms of N saturation or excess. To this end, atmospheric deposition of N and S pollutants was measured and soil, plant, and hydrologic indicators of forest nutrient status were evaluated at two low-deposition and two high-deposition sites near Mexico City. Results from this research are summarized in this chapter.

Methods

Description of Study Sites

Zoquiapan (ZOQ) Experimental Forest and Paso de Cortes (PC) are the two low-deposition sites selected for this study, and the Desierto de los Leones (DL) National Park and Recreation Area and the Ajusco (AJ) National Park were the two high-deposition sites (Fig. 13.1). The DL National Park is located in the southwest end of the MCAB in the direction of the prevailing winds from Mexico City. The AJ site is SSW of Mexico City and is also highly exposed to air pollutants from the city, although it is not located as directly in the path of the predominant winds from Mexico City as the DL site. The ZOQ Experimental Forest is located in the southeast corner of ZOQ National Park (Fig. 13.1), to the east and upwind of Mexico City. Paso de Cortes, the other low-pollution site, is southeast of Mexico City within the Iztaccíhuatl–Popocatépetl National Park (Fig. 13.1).

Precipitation in the Valley of Mexico occurs predominantly from May through September. Precipitation at the study sites occurs as rainfall or hail. Snowfall rarely occurs in the MCAB except at sites above 4000 m in elevation. The dominant tree species at the study sites is *Pinus hartwegii* Lindl (Rzedowski 1978). There are frequent local fires in the forest in the Zoquiapan area, but fires are suppressed in the DL and PC, and at the AJ site. Annual precipitation is 1455 mm at the DL and 778 mm at ZOQ. The average annual precipitation at AJ is approximately 1000 mm/yr (Jáuregui 2002). The dominant soil types are humic andosol, ochric at DL and mollic andosol, humic at ZOQ. Further soils descriptions are given in Marin et al. (2002). The elevation of all the study sites is between 3210 and 3570 m.

Throughfall Collection

Throughfall consists of solutes collected in wet deposition under forest canopies. Throughfall deposition provides a useful estimate of atmospheric deposition inputs to forests, because it includes wet deposition and also dry deposition, which is washed from canopy surfaces by precipitation. Because of the high cost and difficulty of measuring dry deposition fluxes in forest stands, quantifying N and S deposition in throughfall provides a useful alternative measure of annual atmospheric inputs at the stand level. Throughfall deposition provides a reliable estimate of total S deposition to a forest (Butler and Likens 1995; Lovett 1994).

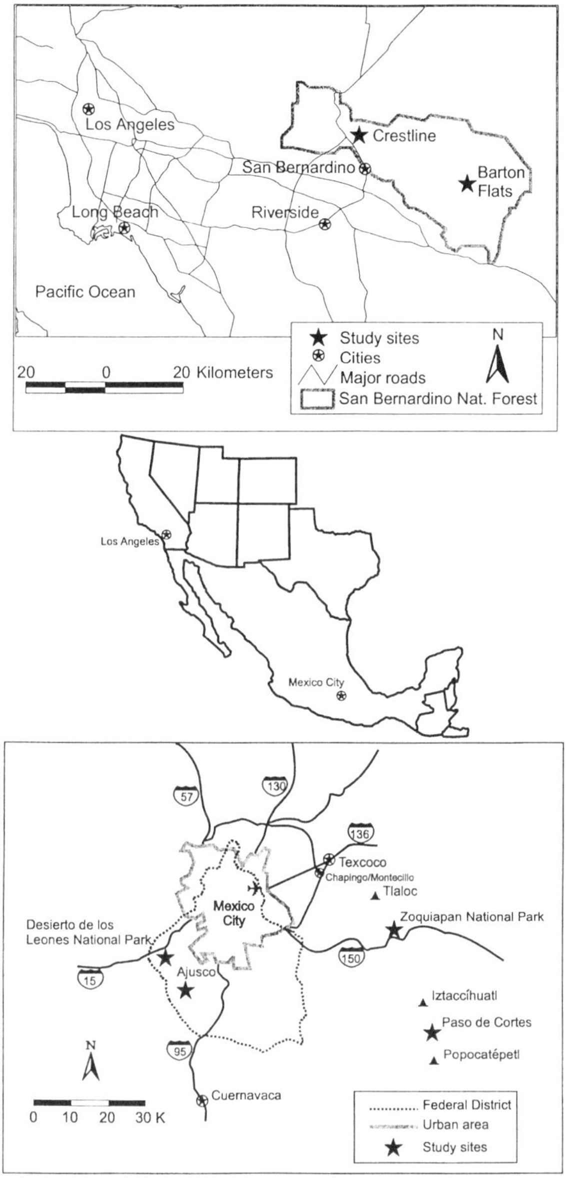

Figure 13.1. Map showing the location of the four study areas in the Mexico City Air Basin, the two study sites in the San Bernardino Mountains in California, and the location of the associated urban centers (Mexico City and Los Angeles).

However, in areas of chronic N deposition, throughfall frequently underestimates total (wet plus dry) N deposition (Fenn and Bytnerowicz 1997; Lovett and Lindberg 1993). Underestimates of deposition by the throughfall method are attributed to canopy uptake and retention of atmospheric N compounds, especially from dry deposition (Garten et al. 1998). Notwithstanding this limitation, throughfall estimates of N deposition are useful in helping to set the bounds of what the minimum levels of N deposition are. Because of the large worldwide database of throughfall deposition to forests (Lovett and Lindberg 1993; Parker 1983; Thimonier 1998), throughfall N inputs to any given area can be compared to throughfall N inputs in other areas, thus providing an indication of the relative severity of N deposition.

Bulk throughfall was collected in *P. hartwegii* stands for one year (March, 1996 to March, 1997) at DL and ZOQ in standard funnel collectors located in four transects with eight collectors per transect (Fenn et al. 1999). Samples were preserved in the field with phenyl mercury acetate and analyzed for nitrate, ammonium, and sulfate using liquid ion chromatography at the United States Department of Agriculture (USDA) USDA Forest Fire Laboratory in Riverside, California. Most of the throughfall was collected from May through October, which is the rainy season in the Basin of Mexico (Fenn et al. 1999).

Dry Deposition Methods

To date, only limited measurements of dry deposition of N and S have been made in forested areas in the MCAB. Dry deposition fluxes of nitrate, ammonium, and sulfate to surrogate surfaces (Whatman No. 41 paper filters) were measured during the 1994 winter dry season to compare relative dry deposition fluxes at a high-pollution (DL) site and at a relatively-low pollution forest site (ZOQ) located upwind of Mexico City (Fenn et al. 1999). The filters were exposed for several consecutive two-week periods.

In addition, weekly average atmospheric concentrations of gaseous and particulate N and S compounds at DL were determined during 39 exposure periods, mainly in winter and spring, between September, 1992, and May, 1995, using battery-powered portable filter-pack collectors (Zeller et al. 1997a). The filter pack inlet is 0.95 cm in diameter, which allows for the collection of aerosols smaller than approximately 10 μm diameter when the inlet is oriented downward. During exposure, the filter pack was mounted under a 20.3 cm diameter polyvinyl chloride (PVC) rain shield. Air flow is volumetrically controlled with a diaphragm pump, pelton-wheel flow meter. The filter-pack preparation chemistry employed at the Desierto de los Leones site is the same used for the US Environmental Protection Agency National Dry Deposition Network (Edgerton et al. 1992). Three filters are placed in series within an inert teflon holder: first a teflon "Zeflour" 2 μm filter which traps suspended particulate aerosols; next a nylon "Nylasorb" 1 μm filter which absorbs acidic gases such as nitric acid and some sulfur dioxide; and last, two Whatman 41 ashless cellulose filters impregnated with a 25% K_2CO_3 solution with glycerol. SO_2 is oxidized to sulfate when passed over

K_2CO_3 and retained as sulfate on the Whatman filters. The filter pack collectors generally ran for one-week intervals, although exposure times were extended to two weeks on two occasions. A detailed discussion of this technique is given by Zeller et al. (1997a).

Plant, Soil, and Water Analyses

Various soil, plant, and hydrologic nutrient analyses were carried out on samples from both sites to determine whether elevated atmospheric deposition of N and S was reflected in nutrient concentrations of plants, soil, and streamwater. Soil variables measured were aqueous extracts of nitrate, ammonium, and sulfate, and total C, N, and S of the mineral soil. Plant analyses included total C, N, P and S of current and one-year-old foliage of *P. hartwegii*. Streams and springs were also sampled for nitrate analysis to evaluate possible watershed-level influences of N deposition. Water samples were collected from streams at the four sites, with more intensive sampling of streams and springs at DL.

Results

Dry Deposition

Average concentrations ($\mu g/m^3$) of the N pollutants were 1.7 ± 1.4 for nitrate, 1.3 ± 0.8 for ammonium, and 1.5 ± 0.8 for nitric acid (HNO_3). Average S concentrations were 2.8 ± 1.7 and 3.8 ± 2.4 $\mu g/m^3$ for sulfate and SO_2 (Fig. 13.2). Sulfate concentrations were similar to those found in many parts of Europe (Ruijgrok et al. 1995) but below those of more polluted regions (Zeller et al. 1997a, 1997b). Sulfur dioxide concentrations were high, indicating significant emissions from Mexico City. Nitrate concentrations were greater than most parts of Europe (Ruijgrok et al. 1995) and similar to those measured at a forested Czech site (Zeller et al. 1997b). The ammonium concentrations were also similar to the forested Czech site. The concentrations reported in Figure 13.2 are based almost exclusively on winter to early spring measurement periods (the dry period) and should be compared to other studies with caution because measurements were not taken in summer and rarely in autumn.

The data from DL suggest that atmospheric concentrations of sulfate and SO_2 were greater than for the analogous oxidized N pollutants nitrate and HNO_3. However, if we include atmospheric concentrations of ammonium and ammonia (NH_3) (the latter not measured), the total atmospheric load of inorganic N compounds is likely greater than the total atmospheric load of inorganic S pollutants. Official reports indicate that emissions of SO_2 in the MCAB have decreased by approximately 45% compared to pre-1993 dates as a result of the use of cleaner diesel fuels and lower polluting energy sources for industry, while NOx emissions have remained relatively constant since 1989 (INEGI 1999; Ruiz and Gasca 2001). Thus, atmospheric concentrations and deposition of S at DL and in other forested sites downwind of Mexico City have probably decreased since 1993. The

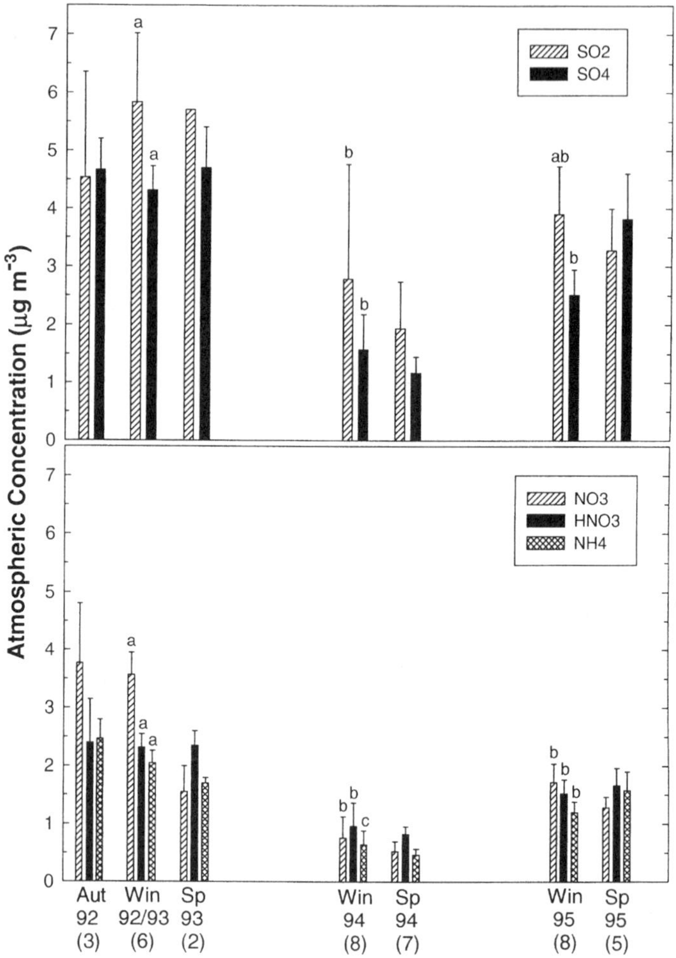

Figure 13.2. Atmospheric concentrations of N and S pollutants at the Desierto de los Leones National Park as measured by portable filter-pack collectors (Zeller et al. 1997a). Numbers in parentheses are the number of sampling periods for that season. Letters above bars indicate significant differences between winter exposures in 1992/1993, 1994 and 1995. The 1992/1993 sampling period was in winter and spanned parts of two calendar years. The other seasonal samplings occurred within a single calendar year.

filter-pack data support this trend, as evidenced by generally lower atmospheric concentrations of SO_2 and sulfate in 1994 and 1995 compared to 1993 (Fig. 13.2). Notwithstanding the reduced SO_2 emissions in Mexico City, atmospheric sulfate concentrations (5–6 μg/m^3) in fine particulate aerosols were still twice as high as concentrations of nitrate or ammonium (2–3 μg/m^3) across several urban sites in 1997 (Edgerton et al. 1999).

Filter packs provide a convenient, low-cost method for measuring acidic air pollutants (Anlauf et al. 1985). However, under certain environmental conditions and pollution exposures, filter-pack systems are not as reliable as annular denuder techniques, especially in differentiating gaseous and particulate pollutants (Andersen and Hovmand 1994; Febo et al. 1993). For example, filter packs have a propensity to underestimate nitrate and overestimate HNO_3 (Zeller et al. 1997b). However, in a side-by-side comparison near Bondville, Illinois, filter-pack results were similar to annular denuder values, except that SO_2 was underestimated by 12% to 20% (Sickles et al. 1999). Thus, the filter-pak data presented in Figure 13.2 are not considered as reliable for determining absolute concentrations as measurements made with annular denuder systems, especially in differentiating between gaseous and particulate forms. The data are, however, believed to effectively indicate temporal trends in N and S pollutants at DL.

Dry deposition fluxes of nitrate and sulfate to surrogate surfaces were significantly greater at DL compared to the Papayo site in ZOQ (Fig. 13.3) according to repeated measures ANOVA. However, sulfate fluxes were similar at both sites on the last two collection dates. Fluxes of ammonium to the paper filter disks were not significantly different between the two sites, presumably because fossil fuel emissions of N pollutants, the main source of pollution in Mexico City, are

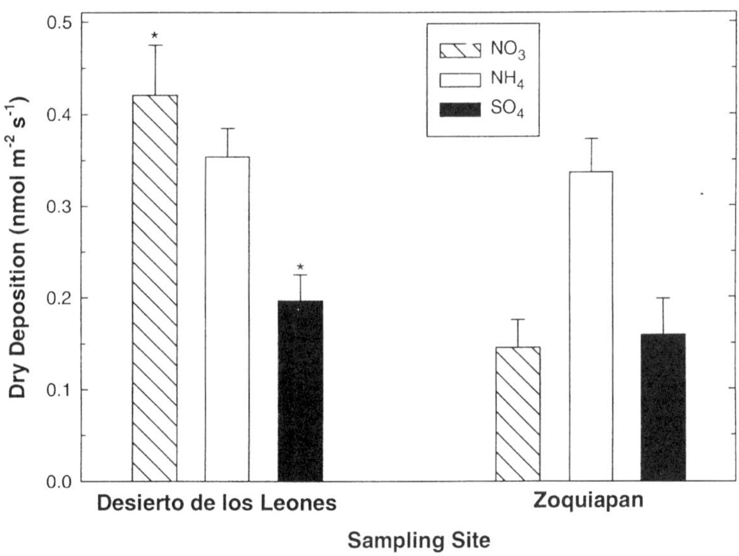

Figure 13.3. Average dry deposition fluxes to paper filter disks at Desierto de los Leones and Zoquiapan National Parks in the winter of 1994. Error bars are standard errors of the mean from five collection periods of two weeks each. Modified from Fenn et al. 1999. Asterisks above the bars indicate significant differences between DL and ZOQ for each ion ($p = 0.1$).

dominated by oxidized N compounds. Deposition fluxes of nitrate to the paper filter collectors were on average twice as high as sulfate fluxes (Fig. 13.3). The relatively low deposition fluxes of sulfate to the paper filters may be due to their lower efficiency as a collection surface for sulfate compared to nitrate and ammonium (Fenn and Bytnerowicz 1993), but is probably also a result of lower SO_2 emissions after 1993, as mentioned above.

Wet Deposition

Wet deposition is composed of dry deposition which is scavenged or washed out of the atmosphere. Wet-deposition data is of limited value as a measure of total atmospheric deposition to a forest, because it includes only a small fraction of dry-deposited pollutants, which are frequently the largest deposition input. Forest canopies have large surface areas that serve as effective receptors for atmospheric pollutants, especially dry-deposited and fog-deposited pollutants (Lovett 1994; Fenn et al. 2000). However wet deposition is measured in some monitoring programs, because it is relatively simple to collect and measure, results can be compared to other monitoring sites and networks, and because wet-deposition data has been used as an index to total deposition when other data is lacking. However, this should be done only as a last resort, when more comprehensive data including dry deposition is not available. In this chapter, we report wet-deposition data because of the limited amount of dry-deposition available data for the MCAB. Data for the forested regions is especially lacking. Wet deposition data is also reported because it provides an indication of the relative importance of the various N and S ions in the MCAB.

Bulk wet deposition has been measured within a network of urban sites beginning in 1987. Independent data from the early 1980s for six rural monitoring sites, including DL, and for nine urban monitoring sites demonstrated the much greater importance of S deposition compared to N deposition throughout the MCAB. At the urban sites sulfate concentrations were, on average, 3.9 times greater than nitrate concentrations. At the four rural sites which were located downwind (south or southwest) of the metropolitan area, the sulfate : nitrate ratio was 2.7. At DL, in the period from 1982 to 1984, average annual concentrations of sulfate ranged from 6.0 to 7.4 mg/L and nitrate concentrations were 2.0 to 2.2 mg/L (Aguilar-A. et al. 1985). However, as noted previously and in the throughfall section later, since 1993, S emissions have decreased, and as a result, N and S deposition in the MCAB are now more closely balanced.

In 1996/1997 the sulfate : nitrate concentration (as mg/L) ratio in throughfall was 1.2 at DL and 1.8 at ZOQ, a site east (upwind) of Mexico City receiving relatively low deposition. Sulfate : nitrate ratios of annual deposition inputs to the forest stands ($kg\,ha^{-1}$) at DL and ZOQ were 1.8 and 2.7, respectively. These more recent data from DL suggest that the sulfate : nitrate deposition ratio has decreased since the 1980s, as suggested by downward trends in atmospheric SO_2 concentrations in Mexico City from 1992 to 1997 (INEGI 1999). A downward trend in the sulfate : nitrate ratio of wet deposition is also confirmed by comparing 1997 data from the urban monitoring network with data from the early 1980s. In the early 1980s, the average ratio for the urban sites was 3.9 (Aguilar-A. et al. 1985) compared with a ratio of 1.8 for the urban sites in 1997 (INEGI 1999). During this time span, nitrate concentrations in precipitation increased by 39% and sulfate concentrations decreased by 33%. Increased N deposition is due to increasing numbers of motor vehicles in the MCAB, and reduced S deposition is attributed to regulations requiring the use of cleaner industrial fuel sources and diesel with a maximum S content of 0.05% (INEGI 1999). Nonetheless, S deposition remains high relative to nitrate deposition and compared with many other forested regions. For example, in pollution-impacted forests of the Los Angeles California Air Basin, N deposition is typically 6 to 10 times greater than S deposition (Fenn and Bytnerowicz 1993). It should be noted that an alternative and intermittent source of atmospheric S in the southeastern side of the MCAB is the active volcano Popocatépetl (Fenn et al. 1999).

Throughfall: Deposition to Forest Canopies

Annual inorganic N (as ammonium and nitrate) and S (as sulfate) deposition at DL were 18.5 and 20.4 kg per hectare. Analogous values at ZOQ were 5.5 and 8.8 kg/ha/yr (Fenn et al. 1999). Bulk throughfall deposition of N and S at DL was relatively high compared to many forests throughout the world. This was expected considering the high emissions of N and S oxides from Mexico City (INEGI 1999; Ruiz and Gasca 2001). Annual throughfall inputs were 1455 mm at DL and 778 mm at ZOQ. Patterns of cumulative deposition inputs of nitrate and ammonium paralleled cumulative precipitation volume at DL, but at ZOQ, deposition of nitrate and ammonium was less responsive, lagging behind the trend of cumulative precipitation. Cumulative sulfate deposition trends generally tracked cumulative precipitation at both sites. Volume-weighted annual mean concentrations of ammonium and nitrate in throughfall were 1.6 and 1.8 times higher at DL than at ZOQ, while sulfate concentrations in throughfall were only 1.2 times higher at DL (Fenn et al. 1999).

Sulfur deposition in throughfall is generally considered to be a reasonable estimate of total S deposition to a forest. However, many studies demonstrate that 20% to 40% of atmospheric N deposited to a forest is retained within the canopy and thus not measured in throughfall analysis (Cadle et al. 1991; Friedland et al. 1991; Hanson and Garten 1992; Lovett 1992; Lovett and Lindberg 1993; Marshall and Cadle 1989; Schulze 1989). Assuming that significant canopy retention of N also occurred in this study, total N deposition to the forest expressed on a mass basis (kg ha^{-1}) was slightly higher than total S deposition. For example, if the forest canopy retained from 20% to 40% of atmospherically-deposited N, total N deposition would be 22.2 to 25.9 kg/ha/yr at DL and 6.6 to 7.7 kg/ha/yr at ZOQ. In any case, deposition of N and S to the forest at DL is 10 to 20 times greater than levels measured in forests with minimal anthropogenic atmospheric deposition, assuming that background levels of N and S deposition in relatively pristine forests are approximately 1 to 2 kg/ha/yr (Young et al. 1988). Deposition of N and S at ZOQ was approximately three to nine times greater than background levels.

Concentrations of nitrate and ammonium in throughfall at DL were 1.8 and 1.6 times greater than at ZOQ, but sulfate concentrations were only 1.2 times greater at DL. The lower DL : ZOQ concentration ratio for S compared to N, suggests that there are sources of atmospheric S other than air pollution from Mexico City which contribute to S loading at ZOQ. Dry-deposition fluxes of sulfate to the surrogate surfaces at DL and ZOQ were highly similar on some dates, which also supports the hypothesis that there are significant atmospheric sources of S pollution in forests at ZOQ. Possible additional sources of S include local industrial sources and volcanic S emissions. The volcano Popocatépetl, located 28 km south of ZOQ, has increased in fumarolic activity since 1991 (Del Pozzo et al. 1996) and is a likely source of SO_2 at ZOQ. In recent years, Popocatépetl has been degassing and generating SO_2 (2.6×10^6 tons in 617 days) at sustained high rates (Kress 1997). Volcanic activity in March/April, 1996, coincided with the beginning of throughfall sampling (March 1996) at ZOQ and DL. In late March to mid-June, 1996, S deposition in throughfall at ZOQ increased at a much greater rate than did throughfall volume, possibly because of high S inputs from volcanic emissions. Meanwhile, at DL, which is not located downwind of Popocatépetl, S deposition and throughfall volumes increased at similar rates. Increased wet deposition of sulfate, nitrate, and other ions has been observed downwind of Popocatépetl (Baez et al. 1997a).

Effects of Atmospheric Deposition on Forest Nutrient Status

Chronic atmospheric deposition of N and S pollutants can have significant effects on nutrient cycling and ecosystem function. One possible effect of chronic atmospheric inputs of N is that it can lead to the maintenance of high N levels in soil and soil solution, which in turn can sustain active populations of N cycling microorganisms such as nitrifiers, and as a result, nitrification rates are also stimulated (Johnson 1992). Chronic N-deposition inputs and soils with high nitrification activity, along with low soil and litter C : N ratios, are major factors in the the development of N saturation of forests (Emmett et al. 1998; Fenn et al. 1998; Nohrstedt et al. 1996). These forests or watersheds are highly prone to lose excess N as streamwater nitrate. The leached nitrate can originate mainly from organic N which has been mineralized (often the major source), but recent studies suggest that atmospheric N can also leach directly through the ecosystem without biological assimilation (Durka et al. 1994).

Edaphic Indicators of Forest N Status

Nitrification is a key process controlling N retention or mobility within forest ecosystems. The end product of nitrification is, of course, nitrate, a highly mobile inorganic form of N. Soils can vary widely in the degree at which mineralized N is nitrified. A four-category classification system has been proposed to describe soil nitrification activity, ranging from non-nitrifying soils to total nitrifying soils

(Gundersen and Rasmussen 1990; Zöttl 1960). Soils in the San Bernardino Mountains in southern California can be considered total nitrifying soils. This characteristic is believed to predispose these forests to N saturation under conditions of chronic N deposition (Fenn et al. 1998). Recent data in forested sites in the MCAB suggest that the soils there are also total nitrifying soils. Soil from ZOQ, PC, AJ, and DL were collected from the top 15 cm in early December, 1999, and incubated in the laboratory after adding water to field capacity. Subsamples were removed weekly, and levels of ammonium and nitrate were determined in potassium chloride (KCl) extracts (Fenn et al. 1993). Ammonium levels in soils from all the sites were always low and near a constant baseline level. In contrast nitrate levels increased throughout the eight-week incubation, suggesting that ammonium is rapidly and nearly completely converted to nitrate (Fig. 13.4). Mineralization rates were low at PC, although nitrification seems to be complete in soils from PC as well. The high nitrification capacity of these soils suggests that if chronic atmospheric N deposition enriches these forests with N, the risk is high that as this N cycles through the plant/microbial complex, most of the excess N will be nitrified and exported as leached nitrate (Fenn et al. 1998). In fact, watershed nitrate export data described in the next section suggest that this is what is happening in forested areas southwest of Mexico City.

The soils used for the incubation studies presented in Figure 13.4 were collected beneath *P. hartwegii* trees with and without N fertilization (300 kg N/ha) in the form of ammonium nitrate. Half of the fertilizer was applied in late May, 1999, and half in late June, 1999. Nitrate production was significantly higher in the fertilized soil from ZOQ and AJ compared to the unfertilized soils, although the fertilization response was much greater at ZOQ than at AJ (Fig. 13.4). At DL (Cerro de San Miguel site) and PC, nitrate levels were similar in the fertilized and unfertilized soils. Similar results were found in a repeat experiment, except that at DL, nitrate production was significantly greater in the unfertilized soils. These assays gave mixed results, but suggest that at the most polluted site (DL), nitrification may not be substrate limited, at least under laboratory conditions. *In situ* field incubations to determine annual net N mineralization and nitrification rates at sites across the pollution gradient would be useful for comparing N cycling rates under field conditions.

The soil C:N ratio is a standard indicator of site fertility in agriculture and in forestry. Soil C:N ratios at five sites across a N deposition gradient in the MCAB do not suggest a clear trend. C:N ratios at three high-pollution sites and at PC (a low pollution site) ranged from 14.8 to 15.5 and were not significantly different. The soil C:N ratio at ZOQ was 16.5, higher than any of the other sites (Table 13.1). Total N levels in soil were highest at ZOQ and lowest at PC, while DL and AJ had intermediate levels. Extractable nitrate levels were highest at the DL Cementerio site and AJ (Table 13.1). In summary, these findings suggest that differences in N fertility or N enrichment across the N deposition gradient in the MCAB are not as great as differences across the N deposition gradient in the San Bernardino Mountains (SBM) in southern California. This may be because the volcanic forest soils in the MCAB have inherently greater N fertility. Soil C:N

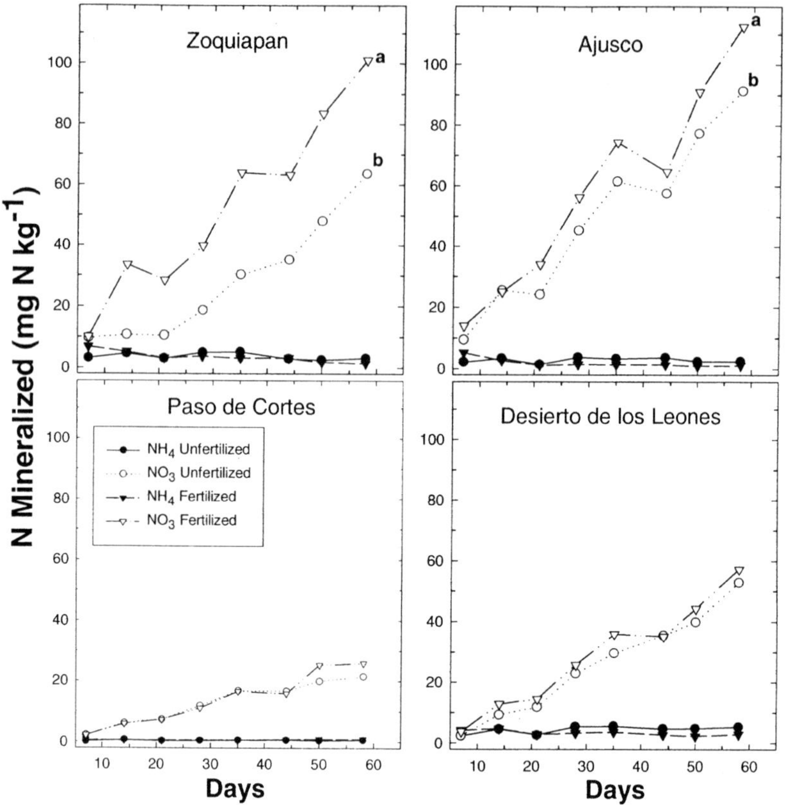

Figure 13.4. Net production of ammonium and nitrate in soil in laboratory incubations. Letters at the end of the line plots indicate significant differences between fertilized and unfertilized soils. Cerro de San Miguel is the location of the fertilized and unfertilized soils within the Desierto de los Leones National Park.

Table 13.1. Nutrient concentrations and C : N ratio of soil collected under *Pinus hartwegii* canopies from five forest sites with varying air pollution exposure[a]

Site	Extractable NO_3 (mg/kg)	Extractable NH_4 (mg/kg)	Total N (g/kg)	Total S (g/kg)	C : N
Paso de Cortes	25.8 (9.4)c	4.6 (1.1)b	1.7 (0.1)c	0.22 (0.02)c	14.8 (0.5)b
Zoquiapan	29.4 (4.4)c	9.7 (1.6)b	5.6 (0.5)a	0.79 (0.09)a	16.5 (0.3)a
Ajusco	49.2 (5.7)a	6.8 (5.2)b	3.6 (0.8)b	0.34 (0.09)bc	15.0 (0.3)b
DL—Cerro de San Miguel	18.0 (3.7)c	7.0 (0.4)b	3.9 (0.5)b	0.55 (0.04)ab	15.5 (0.4)ab
DL—Cementerio	36.3 (6.7)b	0.4 (0.1)a	3.1 (0.3)b	0.38 (0.03)bc	14.8 (0.2)b

[a] Numbers in parentheses are standard errors of the mean. Letters following the parentheses indicate significant differences between sites according to one-way ANOVA and Tukey's means comparison test.

ratios ranged from 19 to 27 at sites across the pollution gradient in the SBM (Fenn et al. 1996; Kiefer and Fenn 1997). The lower value was found at Camp Paivika, a severely N-saturated site with high N deposition. As noted above, the C : N ratios in the Mexican forests were nearly always close to 15, even in sites with low N deposition. Differences in total N content between the California and Mexican soils also suggest greater N fertility in the Mexican soils. Total N in soil from four sites across the deposition gradient in the SBM ranged from 0.56 to $1.06\,g\,kg^{-1}$ (Kiefer and Fenn 1997) compared to a range of 1.7 to 5.6 g/kg in the Mexican soils. However, in four of the five Mexican forest sites, total soil N values were greater than 3.1 g/kg (Table 13.1).

Foliar Indicators of Forest N Status

In a previous study, it was concluded, based on foliar N content and on N:P, N:S, and C:N ratios, that N fertility was greater at DL than at ZOQ (Fenn et al. 1999). Although, differences were statistically significant, differences between DL and ZOQ were not as dramatic as comparisons between high–N-deposition and low–N-deposition sites in the SBM (Fenn 1991; Fenn et al. 1996). In four subsequent analyses of current-year foliage, collected in April, 1999, November, 1999, January, 2000, and April, 2000, N concentrations in foliage were not significantly different at the high-deposition and low-deposition sites except for the November, 1999, sampling (Fig. 13.5), for which N concentrations were approximately 55% lower than on previous samplings. In the November, 1999, sampling, N and S concentrations in current-year foliage were significantly higher, and C:N ratios were lower for the two high-pollution plots than for the two low-pollution plots (Fig. 13.5). The most plausible explanation for the lower foliar N

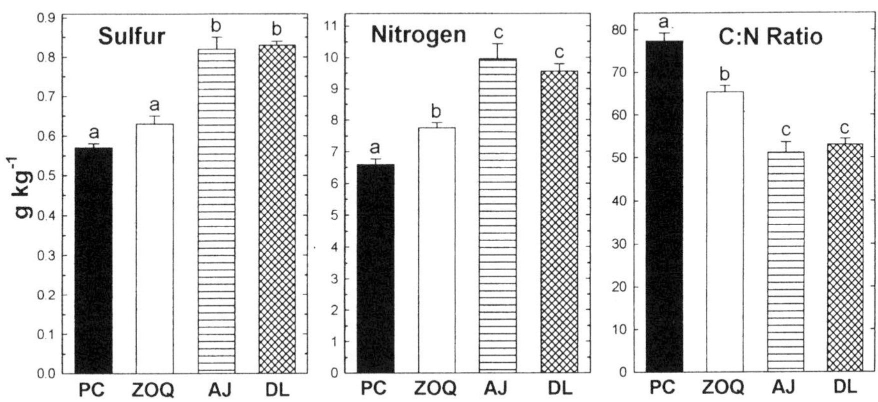

Figure 13.5. Total S and N concentrations and C:N ratios of current-year foliage harvested in November, 1999, from *P. hartwegii* trees at two low-deposition (Paso de Cortes and Zoquiapan) and two high-deposition sites (Ajusco and Desierto de los Leones). Letters above the bars indicate significant differences between sites ($p \leq 0.05$).

levels in November 1999 is that the plants were still physiologically active, with a high demand for nutrients. In the other samplings, which occurred in the dry winter season or prior to the summer rains, N concentrations were higher in current-year foliage as a result of retranslocation of N from older foliage in preparation for dormancy. It may be that as a result of retranslocation of mobile nutrients to the current-year foliage prior to dormancy, there were no differences in current-year foliar N concentrations among the four sites, except during the active growing period. The active growing season in the MCAB is during the warmer months of the wet season, typically from May to October (Miller et al. 1994). In any case, these foliar nutrient studies confirm the soil studies, suggesting that forests in the MCAB are less N deficient, and the N status of *P. hartwegii* is less affected by N deposition than *P. ponderosa* in the SBM near Los Angeles, California, even with similar N deposition inputs (Fenn et al. 1999; Fenn et al. 2000). Another contributing factor may be that high precipitation during the growing season in the MCAB may transport a considerable amount of the atmospherically-deposited N from the stands (as leachate and surface runoff) before the trees are able to take up this N, thus reducing the impact of N deposition on the N status of the trees. However, this possibility has not been studied.

Individual *P. hartwegii* trees were fertilized with N at the four study sites at the beginning of the growing season in 1999, and foliage was sampled at the end of the growing season to determine foliar biomass and foliar nutrient responses to the fertilizer N as described by Kiefer and Fenn (1997). Foliar growth-response data from all four study sites except DL suggest a weak but positive growth response, but differences were not statistically significant between the control treatments and the fertilized trees. However at DL, fertilization caused a statistically significant ($p = 0.07$) negative foliar growth response (unpublished data). The only significant nutritional or N-status response of the foliage to the N fertilization occurred at PC, where foliar N concentrations decreased in the fertilized trees (15.0 versus 14.0 g/kg; $p = 0.01$). These results indicate that there was little response to N in three sites and a possible antagonistic growth effect of excess N at DL. Nitrogen may not be deficient in these forests, even at sites with low N deposition; or if N deficiency does occur to some degree in these sites, it appears to be much less pronounced than in most temperate forests. However, considering the extremely high phosphorus fixation capacity of these andisol soils (Brady and Weil 1999), it is probable that P deficiency limits tree growth responses to added N. Available P levels in soil have been reported at low levels in several studies in the Basin (Cruz-Cisneros, and Valdés 1990; López-López et al. 1998; Manske et al. 2000; Marin et al. 2002). It could also be that the weak response to N fertilization is partially because the dense understory of tussock grass (*zacatón*) out-competed the pine trees for the added N. However, fertilization with highly soluble ammonium nitrate, in two applications of 150 kg/ha each, should have provided abundant N for both understory and overstory uptake. Preliminary data from the year after the fertilization showed that N concentration in the grass was higher in the fertilized areas compared to grass growing under the unfertilized trees at ZOQ ($p = 0.01$), but not at DL. Further studies are needed to determine what effects N deposition has on the growth and physiology of forest vegetation, to determine the interactions of N enrichment and O_3 in plants, and to determine whether other nutrients are limiting plant response to N in these forests.

Atmospheric deposition measurements and studies of the nutrient responses of other vegetation types at DL, such as sacred fir (*Abies religiosa*), have not been investigated, except in studies addressing nutritional factors in sacred fir decline (see Chapter 10). Sacred fir stands are much more dense than pines or hardwood species, and even individual trees have a much greater leaf surface area (Fenn and Bytnerowicz 1997; Van Ek and Draaijers 1994). Thus the capacity for atmospheric deposition inputs is greater in sacred fir stands, as was demonstrated by large inputs of Pb to soil under sacred fir canopies (Watmough and Hutchinson 1999). Furthermore, fir trees at DL are frequently located in ravines, which serve as effective drainages and reservoirs of incoming air pollution masses (Alvarado-Rosales and Hernández-Tejeda 2002). Further studies are needed to determine the effects of N and S deposition on nutrient relations and decline of sacred fir stands in DL, in consideration of the possibility of greater effects of atmospheric deposition in fir forests.

Effects of N Deposition on Streamwater Chemistry

In a preliminary study of NO_3^- concentrations in streamwater, it was reported that nitrate levels were higher in streams from DL (1–45 μeq/L) compared to ZOQ (0–11 μeq/L; Fenn et al. 1999). Subsequent studies have confirmed that nitrate concentrations in streams and springs are much higher at DL than at low-deposition sites at PC and ZOQ. Of the two intermittent streams sampled at AJ, nitrate concentrations were relatively low in the AJ Experimento stream (3.6–13.4 μeq/L) and moderately high in the AJ Puente stream (6.2–27.5 μeq/L, most values >20); (Fig. 13.6). The nitrate levels reported for the AJ Puente stream indicate leaching of excess N from this watershed, presumably as a result of chronic N deposition. Concentrations rarely exceed 10 μeq/L in streams from undisturbed watersheds with low levels of N deposition (Fenn and Poth 1999; Stoddard 1994).

Nitrate levels in two of the streams at DL were several-fold higher than the AJ Puente stream and the other two streams at DL. Many nitrate values for these two streams were between 50 and 100 μeq/L (Fig. 13.6). Sixteen springs in the DL national park were sampled from June, 1999, to June, 2000. Nitrate concentrations in the springs ranged from 1.2 to 131.7 μeq/L with a median value of 24 μeq/L (Fig. 13.7). Nitrate concentrations were greater than 50 μeq/L in 32 of the samples.

Peak nitrate concentrations recorded in two streams in the SBM were greater than 350 μeq/L. Three-year average nitrate concentrations in four of the streams in the SBM ranged from 65 to 151 μeq/L, demonstrating that nitrate concentrations in forested streams and springs at DL are several-fold lower than in the high-

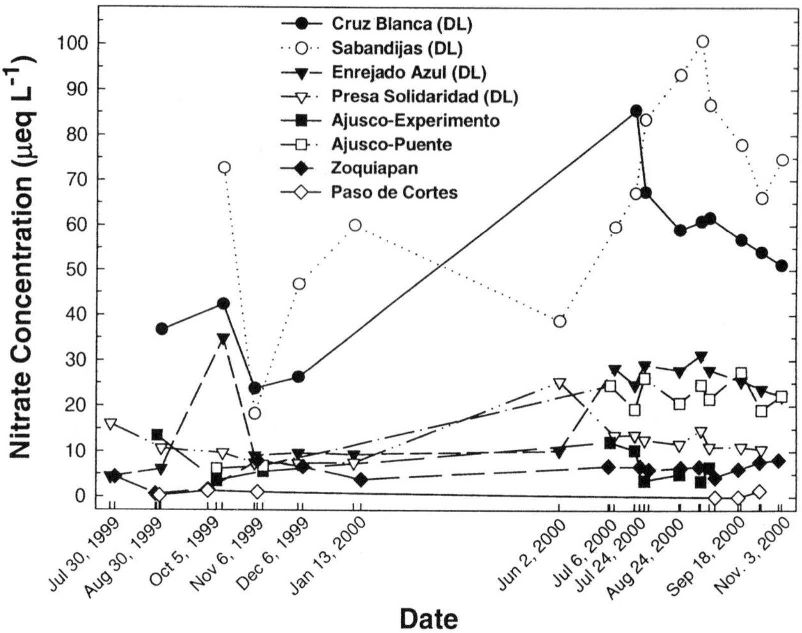

Figure 13.6. Nitrate concentrations in streamwater from sites across an air pollution gradient.

deposition watersheds of the SBM (Fenn and Poth 1999). This may indicate that N-saturation symptoms are more severe in the SBM than in forests downwind of Mexico City. However, in the absence of streamwater flux data at DL, the total N export into streamwater isn't known. The consistently high levels of nitrate in springs and streams at DL during the growing season is good evidence of a considerable pool of excess available N in these watersheds. Based on comparisons with the other study sites in the Basin and a lack of other major N inputs into these watersheds, it seems likely that the excess N in runoff at DL is a result of elevated atmospheric N deposition, as is reported to be true of N-saturated watersheds in North America and Europe (Dise and Wright 1995; Fenn et al. 1998).

Nitrate concentrations in streams and springs in DL are comparable to values reported from the Fernow Experimental Forest in West Virginia, an area experiencing severe symptoms of N excess or N saturation (Peterjohn et al. 1996). Thus at the watershed level, nitrate export from DL appears to be typical of mesic forests receiving chronic N deposition and considerable precipitation during the summer. In contrast, the abnormally high nitrate levels in streams in the SBM in southern California is believed to be largely a function of the fact that the major hydrologic runoff in this Mediterranean climate occurs during the winter off-

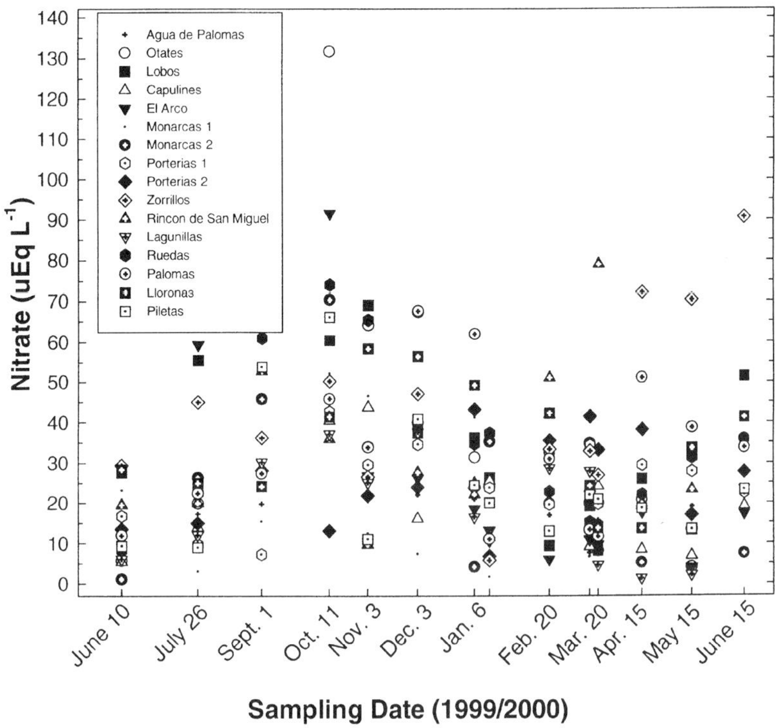

Figure 13.7. Nitrate concentrations in springs in Desierto de los Leones National Park.

season, when biological capacity for N retention is at its lowest (Fenn and Poth 1999).

Summary

Our understanding of atmospheric N and S deposition effects on forests in the MCAB and in the SBM near Los Angeles, California, is based largely on air pollution gradient studies. At DL, a forested site in the MCAB located 23 km directly downwind (southwest) of Mexico City, deposition of both N and S was highly elevated, with throughfall deposition of both compounds equaling approximately 20 kg/ha/yr in a relatively open *P. hartwegii* stand, based on a one-year study in the period from 1996 to 1997 (Fenn et al. 1999). Deposition to more dense stands (such as those of *Abies religiosa* (sacred fir), the other major tree species in the DL national park), is expected to be considerably higher because of the greater surface area for collecting atmospheric pollutants. Throughfall deposition of N at DL was similar to that reported for highly exposed stands in the SBM downwind of Los Angeles, but S deposition was seven-fold greater at the Mexican site.

The degree of N saturation and the degree of N enrichment of the soil, foliage, and drainage waters were more severe in the mixed conifer forests and watersheds in southern California than at DL and el Ajusco, two high-pollution sites in the MCAB. Total N levels in soil and C:N ratios suggest that these Mexican pine forests, even with low N deposition, are inherently much less N deficient than is the case with most northern temperate forests. Chronic N deposition at DL, is believed to be the most likely factor responsible for elevated nitrate levels in streams and springs, but peak and average nitrate concentrations in streamwater were 2.5-fold lower at DL than for the most polluted streams in the SBM in California.

In California, nitrate concentrations in streamwater are highest in the winter wet season when runoff volume is greatest, and the N demand of vegetation is limited by low temperatures and short days. In contrast, in the MCAB, precipitation occurs predominantly during the summer growing season when vegetative and microbial N-demand are expected to be greatest, resulting in lower nitrate losses in runoff. However, the fact that elevated nitrate concentrations are consistently measured in streams at DL in summer indicates that there is a significant degree of N excess in these watersheds as a result of chronic atmospheric inputs of N. Further research is needed to understand N cycling and retention mechanisms in the contrasting climates and forest types in California and the Basin of Mexico. With a greater understanding of ecosystem processes and responses to chronic ozone exposure and N deposition, it will be possible to implement effective ecosystem management practices for these severely human-impacted systems.

References

Aber, J.D., Nadelhoffer, K.J., Steudler, P., and Melillo, J.M. 1989. Nitrogen saturation in northern forest ecosystems. *Biosci.* 39:378–386.

Aguilar-A., S., Bravo-A., H., Saavedra-R., Ma.I., and Torres-J., R. 1985. Acid precipitation in the Mexico City Basin. In *Proceedings of the Technical Conference: Urban Climatology and its Applications with Special Regard to the Tropical Areas*, pp. 3, 4. November 26–30, 1984. (WMO/TD–No. 7; Abstracts) Mexico City: World Climate Programme/World Meteorological Organization.

Alvarado-Rosales, D., and Hernández-Tejeda, T. 2002. Decline of sacred fir in the Desierto de los Leones National Park. In *Urban Air Pollution and Forests: Resources at Risk in the Mexico City Air Basin*. Ecol. Stud., eds. M.E. Fenn, L.I. de Bauer, and T. Hernández-Tejeda. New York: Springer-Verlag.

Andersen, H.V., and Hovmand, M.F. 1994. Measurements of ammonia and ammonium by denuder and filter pack. *Atmos. Environ.* 28:3495–3512.

Anlauf, K.G., Fellin, P., Wiebe, H.A., Schiff, H.I., MacKay, G.I., Braman, R.S., and Gilbert, R. 1985. A comparison of three methods for measurement of atmospheric nitric acid and aerosol nitrate and ammonium. *Atmos. Environ.* 19:325–333.

Baez, A.P., Belmont, R.D., Azpra, E., Torres, M.B., and Padilla, H.G. 1997a. Cloudwater and precipitation chemistry at Popocatepetl volcano: A high elevation site. *J. Environ. Qual.* 26:253–258.

Brady, N.C., and Weil, R.R. 1999. *The Nature and Properties of Soils*, 12th edn. Upper Saddle River, New Jersey: Prentice Hall.

Bravo, A.H. 1987. *La Contaminación del Aire en México*. México, D.F.: Ed. Fundación Universo Veintiuno.

Butler, T.J., and Likens, G.E. 1995. A direct comparison of throughfall plus stemflow to estimates of dry and total deposition for sulfur and nitrogen. *Atmos. Environ.* 29:1253–1265.

Cadle, S.H., Marshall, J.D., and Mulawa, P.A. 1991. A laboratory investigation of the routes of HNO_3 dry deposition to coniferous seedlings. *Environ. Pollut.* 72:287–305.

Cruz-Cisneros, R., and Valdés, M. 1990. Ecological aspects of the actinorhizal plants growing in the Basin of Mexico. *Nitrogen Fixing Tree Res. Rep.* 8:42–47.

de Bauer, L.I. 1991. La influencia urbana en el área del Ajusco. In *Proceedings of the First National Symposium, Agricultura Sostenible: Una Opción para el Desarrollo sin Deterioro Ambiental*, pp. 174–183. Montecillo, Mexico: Comisión de Estudios Ambientales, Colegio de Postgraduados and M.O.A. International.

Del Pozzo, A.L.M., Espinasa, R., and Butron, M.A. 1996. Popocatepetl's 1994–1996 activity: eruptive products (abstr.). *Eos* 77:F809.

Dise, N.B., and Wright, R.F. 1995. Nitrogen leaching from European forests in relation to nitrogen deposition. *For. Ecol. Manage.* 71:153–161.

Durka, W., Schulze, E.D., Gebauer, G., and Voerkelius, S. 1994. Effects of forest decline on uptake and leaching of deposited nitrate determined from ^{15}N and ^{18}O measurements. *Nature* 372:765–767.

Edgerton, E.S., Lavery, T.F., and Boksleitner, R.P. 1992. Preliminary data from the USEPA dry deposition network: 1989. *Environ. Pollut.* 75:145–156.

Edgerton, S.A., Bian, X., Doran, J.C., Fast, J.D., Hubbe, J.M., Malone, E.L., et al. 1999. Particulate air pollution in Mexico City: A collaborative research project. *J Air Waste Manage. Assoc.* 49:1221–1229.

Emmett, B.A., Boxman, D., Bredemeier, M., Gundersen, P., Kjonaas, O.J., Moldan, F., et al. 1998. Predicting the effects of atmospheric nitrogen deposition in conifer stands: Evidence from the NITREX ecosystem-scale experiments. *Ecosystems* 1:352–360.

Febo, A., Perrino, C., and Allegrini, I. 1993. Field intercomparison exercise on nitric acid and nitrate measurement (Rome, 1988): A critical approach to the evaluation of the results. *Sci. Total Environ.* 133:39–71.

Fenn, M. 1991. Increased site fertility and litter decomposition rate in high-pollution sites in the San Bernardino Mountains. *For. Sci.* 37:1163–1181.

Fenn, M.E., and Bytnerowicz, A. 1993. Dry deposition of nitrogen and sulfur to ponderosa and Jeffrey pine in the San Bernardino National Forest in southern California. *Environ. Pollut.* 81:277–285.

Fenn, M.E., and Bytnerowicz, A. 1997. Summer throughfall and winter deposition in the San Bernardino Mountains in southern California. *Atmos. Environ.* 31:673–683.

Fenn, M.E., and Poth, M.A. 1999. Temporal and spatial trends in streamwater nitrate concentrations in the San Bernardino Mountains, southern California. *J. Environ. Qual.* 28:822–836.

Fenn, M.E., Poth, M.A., and Johnson, D.W. 1996. Evidence for nitrogen saturation in the San Bernardino Mountains in southern California. *For. Ecol. Manage.* 82:211–230.

Fenn, M.E., Poth, M.A., Dunn, P.H., and Barro, S.C. 1993. Microbial N and biomass, respiration and N-mineralization in soils beneath two chaparral species along a fire-induced age gradient. *Soil Biol. Biochem.* 25:457–466.

Fenn, M.E., Poth, M.A., Aber, J.D., Baron, J.S., Bormann, B.T., Johnson, D.W., et al. 1998. Nitrogen excess in North American ecosystems: predisposing factors, ecosystem responses, and management strategies. *Ecol. Appl.* 8:706–733.

Fenn, M.E., de Bauer, L.I., Quevedo-Nolasco, A., and Rodriguez-Frausto, C. 1999. Nitrogen and sulfur deposition and forest nutrient status in the Valley of Mexico. *Water Air Soil Pollut.* 113:155–174.

Fenn, M.E., Poth, M.A., Schilling, S.L., and Grainger, D.B. 2000. Throughfall and fog deposition of nitrogen and sulfur at an N-limited and N-saturated site in the San Bernardino Mountains, southern California. *Can. J. For. Res.* 30:1476–1488.

Friedland, A.J., Miller, E.K., Battles, J.J., and Thorne, J.F. 1991. Nitrogen deposition, distribution and cycling in a subalpine spruce-fir forest in the Adirondacks, New York, USA. *Biogeochem.* 14:31–55.

Garten, C.T., Schwab, A.B., and Shirshac, T.L. 1998. Foliar retention of ^{15}N tracers: Implications for net canopy exchange in low- and high-elevation forest ecosystems. *For. Ecol. Manage.* 103:211–216.

Gundersen, P., and Rasmussen, L. 1990. Nitrification in forest soils: effects from nitrogen deposition on soil acidification and aluminum release. *Rev. Environ. Contam. Toxicol.* 113:1–45.

Hanson, P.J., and Garten, C.T. 1992. Deposition of H $^{15}NO_3$ vapour to white oak, red maple and loblolly pine foliage—experimental observations and a generalized model. *New Phytol.* 122:329–337.

INEGI. 1999. *Estadísticas del Medio Ambiente del Distrito Federal y Zona Metropolitana.* Mexico City: Instituto Nacional de Estadística, Geografía e Informática.

Johnson, D.W. 1992. Nitrogen Retention in Forest Soils. *J. Environ. Qual.* 21:1–12.

Kiefer, J.W., and Fenn, M.E. 1997. Using vector analysis to assess nitrogen status of ponderosa and Jeffrey pine along deposition gradients in forests of southern California. *For. Ecol. Manage.* 94:47–59.

Kress, V. 1997. Magma mixing as a source for Pinatubo sulphur. *Nature* 389:591–593.

López-López, M.A., Velázquez-Mendoza, J., Velázquez-Martínez, A., González-Romero, V., and Cetina-Alcalá, V.M. 1998. Estado nutrimental de *Abies religiosa* en una área con problemas de contaminación ambiental. *Agrociencia* 32:53–59.

Lovett, G.M. 1992. Atmospheric deposition and canopy interactions of nitrogen. In *Atmospheric Deposition and Forest Nutrient Cycling*. Ecol. Stud. 91, eds. D.W. Johnson, and S.E. Lindberg, pp. 152–166. New York: Springer-Verlag.

Lovett, G.M. 1994. Atmospheric deposition of nutrients and pollutants in North America: An ecological perspective. *Ecol. Appl.* 4:629–650.

Lovett, G.M., and Lindberg, S.E. 1993. Atmospheric deposition and canopy interactions of nitrogen in forests. *Can. J. For. Res.* 23:1603–1616.

Manske, G.G.B., Ortiz-Monasterio, J.I., Van Ginkel, M., González, R.M., Rajaram, S., Molina, E., and Vlek, P.L.G. 2000. Traits associated with improved P-uptake efficiency in CIMMYT's semidwarf spring bread wheat grown on an acid Andisol in Mexico. *Plant Soil* 221:189–204.

Marín, L.E., Escolero-Fuentes, O., and Trinidad-Santos, A. 2002. Physical geography, geology, hydrogeology and forest soils of the Basin of Mexico. In *Urban Air Pollution and Forests: Resources at Risk in the Mexico City Air Basin*. Ecol. Stud., eds. M.E. Fenn, L.I. Bauer, and T. Hernández-Tejeda. New York: Springer-Verlag.

Marshall, J.D., and Cadle, S.H. 1989. Evidence for trans-cuticular uptake of HNO3 vapor by foliage of eastern white pine (*Pinus strobus* L.). *Environ. Pollut.* 60:15–28.

Miller, P.R., Bauer, M.L., Quevedo-Nolasco, A., and Hernádez-Tejeda, T. 1994. Comparison of ozone exposure characteristics in forested regions near Mexico City and Los Angeles. *Atmos. Environ.* 28:141–148.

Nohrstedt, H.O., Sikström, U., Ring, E., Nãsholm, T., Högberg, P., and Persson, T. 1996. Nitrate in soil water in three Norway spruce stands in southwest Sweden as related to N-deposition and soil, stand, and foliage properties. *Can. J. For. Res.* 26:836–848.

Parker, G.G. 1983. Throughfall and stemflow in the forest nutrient cycle. *Adv. in Ecol. Res.* 13:57–133.

Peterjohn, W.T., Adams, M.B., and Gilliam, F.S. 1996. Symptoms of nitrogen saturation in two central Appalachian hardwood forest ecosystems. *Biogeochem.* 35:507–522.

Ruijgrok, W., Davidson, C.I., and Nicholoson, K.W. 1995. Dry deposition of particles: Implications and recommendations for mapping of deposition over Europe. *Tellus* 47B:587–601.

Ruiz, Ma.E., and Gasca, J. 2002. Emissions trends and pollution control strategies in the Mexico City metropolitan area. In *Urban Air Pollution and Forests: Resources at Risk in the Mexico City Air Basin*. Ecol. Stud., eds. M.E. Fenn, L.I. de Bauer, and T. Hernández-Tejeda. New York: Springer-Verlag.

Rzedowski, J. 1978. *Vegetación de México*. México, D.F.: Editorial Limusa.

Schulze, E.-D. 1989. Air pollution and forest decline in a spruce (*Picea abies*) forest. *Science* 244:776–783.

Sickles, J.E., Hodson, L.L., and Vorburger, L.M. 1999. Evaluation of the filter pack for long-duration sampling of ambient air. *Atmos. Environ.* 33:2187–2202.

Stevenson, F.J., and Cole, M.A. 1999. *Cycles of Soil: Carbon, Nitrogen, Phosphorus, Sulfur, Micronutrients*, 2nd edn. New York: John Wiley and Sons, Inc.

Stoddard, J.L. 1994. Long-term changes in watershed retention of nitrogen: Its causes and aquatic consequences. In *Environmental Chemistry of Lakes and Reservoirs*, Advances in Chemistry Series 237, ed. L.A. Baker, pp. 223–284. Washington, D.C.: American Chemical Society.

Thimonier, A. 1998. Measurement of atmospheric deposition under forest canopies: Some recommendations for equipment and sampling design. *Environ. Monitor. Assess.* 52:353–387.

Van Ek, R., and Draaijers, G.P.J. 1994. Estimates of atmospheric deposition and canopy exchange for 3 common tree species in the Netherlands. *Water Air Soil Pollut.* 73:61–82.

Watmough, S.A., and Hutchinson, T.C. 1999. Change in the dendrochemistry of sacred fir close to Mexico City over the past 100 years. *Environ. Pollut.* 104:79–88.

Young, J.R., Ellis, E.C., and Hidy, G.M. 1988. Deposition of air-borne acidifiers in the western environment. *J. Environ. Qual.* 17:1–26.

Zeller, K., Donev, E., Bojinov, H., and Nikolov, N. 1997a. Air pollution status of the Bulgarian Govedartsi ecosystem. *Environ. Pollut.* 98:281–289.

Zeller, K., Cerny, M., Bytnerowicz, A., Smith, L., Sestak, M., Michalec, M., et al. 1997b. Air pollution status of a representative site in the Czech Republic Brdy Mountains. *Environ. Pollut.* 98:291–297.

Zöttl, H. 1960. Methodische Untersuchungen zur Bestimmung der Mineralstickstoffnachlieferung des Waldbodens. *Forstwissenschaftliches Centralblatt* 79:72–90.

14. Effects of Oxidants in Ambient Air on Annual Crops in the Basin of Mexico

Carlos Fredy Ortíz-García, Héctor D. Laguette-Rey, and L.I. de Bauer

The Importance of Annual Crops in the Basin of Mexico

Although urbanization has reduced the agricultural and livestock production zone within the Basin of Mexico, agriculture still remains an important activity and partially meets the demand for primary products for the megalopolis of Mexico City. Vegetables and legumes are among the principal agricultural crops of the region. These include, among others, green beans (*Phaseolus spp.*), squash (*Cucurbita pepo*), cucumber (*Cucumis sativus*), muskmelon (*Cucumis melo*), lettuce (*Lactuca sativa*), cabbage (*Brassica oleraceae*), radishes (*Raphanus sativus*), spinach (*Spinacea oleraceae*), potatoes (*Solanum tuberosum*), cereals such as corn (*Zea mays*), oats (*Avena sativa*), barley (*Hordeum vulgare*) and wheat (*Triticum aestivum*), and plant species used as forage, such as alfalfa (*Medicago sativa*). In addition to annual crops, a few ornamentals (e.g., roses, *Rosa* spp.) and fruit crops are also grown (e.g., plums, *Prunus americana;* pears, *Pyrus communis;* and apples, *Malus sylvestris*). Some of these plants, as well as others cultivated within the Basin of Mexico, have been reported to be sensitive to ozone (O_3) (Hernández-Tejeda and de Bauer 1989).

Both irrigated and rain-fed agriculture is practiced in the Basin of Mexico. Irrigated agriculture exists in small areas (e.g., in Xochimilco), and rain-fed zones are distributed throughout the basin. In the semiarid areas of the basin, located to the SSE, in an area called Milpa Alta, the early use of wild prickly-pear cactus by human populations has evolved into ever increasing areas of commercial plantations of *nopalitos* (the cladodes of *Opuntia ficus-indica*), which are consumed as vegetables. According to Flores-Valdez (1995), the plantations cover over 10,000 ha producing 80 to 90 t/ha/yr. The *nopalitos* produced in Milpa Alta and the nearby area of Tlanepantla, represent 70% of the total production of Mexico.

Foliar Symptoms of Ozone Injury in Plants

Ozone is taken up by plants through the stomata and can directly affect various physiological processes including those which are directly related to productivity, such as the photosynthetic activity of leaves (Manning and Feder, 1980). Ozone causes growth reduction and crop productivity losses and also affects the harvest quality of crops, as has been demonstrated with bean seeds (Laguette-Rey et al. 1986a). The symptoms caused by O_3 are diverse because they depend on the plant species and variety, the concentration of O_3, the exposure time, and the environmental conditions, as well as other factors (Hernández-Tejeda and de Bauer 1989; Pell 1979; Temple et al. 1983). The foliar symptoms caused by O_3 in dicotyledonous plants generally begin as whitish or dark lesions on the upper leaf surface. The lesions may coalesce, forming interveinal necrotic areas, often resulting in premature leaf abscission (Hill et al. 1961; Lacasse and Treshow 1976; Taylor 1968). These types of symptoms are frequently observed in sensitive cultivars of beans and soybeans near the Chapingo-Montecillo area in the State of Mexico (de Bauer et al. 1997; Laguette-Rey et al. 1985b; Ortiz-García 1988; Rodríguez-Frausto 1996). The same symptoms are observed on the pods of sensitive leguminous plants (Hofstra et al. 1978; Taylor 1968). At the microscopic level, damaged cells of different tissues suffer lysis; however, their walls remain, conserving the foliar structure (Hofstra et al. 1978; Weaver and Jackson 1968).

The sensitivity of plants to air pollutants is highly variable, even within a given species. Sensitivity to O_3 is also affected by environmental growth conditions, such as the photoperiod, light intensity (Taylor 1968), temperature, relative humidity, and plant water status (Heck 1968; Rodríguez-Frausto 1995). Another important factor which influences the sensitivity of plants is the physiological development stage during which they are exposed (Carnahan et al. 1978; Hass 1970; Heck 1968; Heck et al. 1983; Lacasse and Treshow 1976; Manning and Feder 1980).

The Use of Indicator Plants in the Basin of Mexico

Plant varieties which readily respond to O_3 exposure by developing characteristic visible injury symptoms are used as bioindicators of the potential for O_3-induced plant effects (Jacobson and Hill 1970). These indicator plants can also be used to compare air quality in different sites during predetermined time intervals (Manning and Feder 1980). Selected annual crop varieties are the most commonly used bioindicator plants, partly because they are easily cultured in the greenhouse. In particular, some varieties of tobacco (*Nicotiana tabacum*) and of beans (*Phaseolus vulgaris*) are highly O_3 sensitive and useful as bioindicators (Krupa, 1997; Krupa et al. 1982; Manning and Feder 1980).

Hernández-Tejeda and de Bauer (1989) published a literature review of the sensitivity of many plant species to atmospheric pollutants, and many of the crops grown in the Basin of Mexico are among those reported to be sensitive to O_3. Some genera and species of trees, shrubs, and ornamental plants found in the Basin of Mexico, are reported to be O_3 resistant, although much more research is needed to evaluate the O_3 sensitivity of the diverse vegetation occurring within the Basin (de Bauer et al. 2000).

The first study on the effect of air pollutants on plants in the Basin of Mexico was carried out by de Bauer (1972), who in a series of two-week periods from April through October of 1971, exposed several types of plants to ambient air in the metropolitan and surrounding zone of Mexico City. The plants were: young lettuce plants of the variety Parris Island, spinach var. Viroflay and H-7, bean var. Pinto 133, tobacco var. Bel W-3, and petunia var. Snowsturm. At five of the six exposure sites, the bean plants developed leaf bronzing, and the tobacco plants developed white stippling and foliar necrotic lesions. The observed symptoms are known to be induced by O_3 in these plant varieties (Manning 1998). Similarly, the lettuce and petunia plants exhibited glazing, bronzing, and collapse of the underside of the leaf tissues as described for peroxyacetyl nitrate (PAN) injury (Temple et al. 1998). These same exposures were continued for several years thereafter, and in all cases, a trend of increasing injury was observed in subsequent years.

In 1984, Laguette-Rey exposed several indicator plants (Bel W-3 tobacco, Clintland oats, petunia var. Snow Sturm, and the bean varieties Pinto 111, Canario 107, Bayomex and Cacahuate) to ambient oxidants in Mexico City. The purpose of the experiment was to confirm the nature of the symptoms previously observed in 1983 in Montecillo in the State of México (Fig. 14.1). All of the plants had been grown and kept in a greenhouse supplied with charcoal-filtered air at Chapingo, State of Mexico prior to exposure. The plants were exposed to ambient air from April 28 through July 8, 1984, in the backyard of a house in a residential zone in the southern (generally downwind) area of Mexico City. Periodically, their conditions were assessed. After 12 days of exposure to ambient air, the expanded leaves of var. Bel W-3 tobacco plants developed bifacial necrotic lesions of a light straw color that had the consistency of paper. This symptomatology, attributable to O_3, persisted in expanded leaves during the entire exposure period. The Clintland oat and Pinto 111 plants showed visible symptoms after 37 days of exposure, at the onset of flowering for the Pinto 111 bean plants. In the Canario 107 bean variety, O_3 injury was not observed until day 43, when the plants were in the pod formation stage. Foliar symptomatology observed in Clintland oat and in Pinto 111 and Canario 107 bean plants was similar to that observed in the field tests carried out at Montecillo in 1983. In addition to the

Figure 14.1. Map of the country and the Chapingo and Montecillo study sites within the Basin of Mexico.

bean plants, these field tests included some potted Clintland oat plants. Petunia var. Snowsturm and the bean varieties Bayomex and Cacahuate did not show symptoms commonly associated with O_3 (Laguette-Rey 1985; Laguette-Rey et al. 1986a).

In order to further confirm the presence of phytotoxic levels of O_3 at Montecillo, Ortiz-García (1988) exposed Tempo bean plants to ambient air on a weekly basis during the spring and summer of 1987. The seedlings developed necrotic lesions on the upper surfaces of the primary leaves. Ortiz-García confirmed that during the spring and summer, the O_3 concentration is high enough in Montecillo to induce symptoms ranging from stippling to bronzing of the primary leaves of Tempo bean plants. During this 1987 study, foliar bronzing symptoms, previously described for O_3 injury, were also observed in soybean and bean plants in the experimental field plots of the Postgraduate College in Montecillo (Fig. 14.1). These results demonstrate that some annual plants of the Basin of Mexico, native or introduced, could be used for biomonitoring the presence of oxidants in ambient air within the Basin.

Ozone Effects on Bean Plants

Young Sanilac bean seedlings exposed to 0.16 ppm O_3 for 0.5, 1.0, 1.5, and 2.0 hours developed progressively more severe symptoms with longer exposure time (Weaver and Jackson 1968). Symptoms consisted mainly of chlorosis and pigmentation followed by bronzing of the leaf, the development of necrotic lesions, or general necrosis and leaf abscission. The same authors reported that bean plants in the flowering stage, developed symptoms of stippling and necrosis when they were exposed to 0.40 ppm ozone for 0.5 hour. Similar injury occurs in plants exposed during the pod formation stage, indicating that premature senescence and defoliation can occur in bean plants when the ozone concentrations exceed 0.10 ppm (Blum and Heck 1980; Schwartz 1980). The symptoms described above are readily observed in sensitive bean varieties in the MCAB during the summer months, as pointed out by several authors (Laguette-Rey 1985; Ortiz-García 1988; Rodríguez-Frausto 1996).

Ozone Injury in Bean Varieties in the Basin of Mexico

Beans were chosen to be the subjects of a study on the effects of O_3 on annual crops within the Basin of Mexico (Laguette-Rey et al. 1986a; Ortiz-García 1988) based on several considerations: previously observed O_3 injury symptoms, the importance of beans in the diet of the Mexican people, and the extensive cultivation of beans in the Basin (CAEVAMEX 1981). The first field observations of symptoms of O_3 injury in beans in the Valley of Mexico involved the local bean varieties Amarillo 153 and Amarillo 154 in the Chapingo area (adjacent to Montecillo) in 1980 (de Bauer and Hernández-Tejeda 1986). In 1985, air pollution monitoring was conducted in Montecillo by the Center for Atmospheric Studies of the National Autonomous University of Mexico. On 70% of the days from August 7 to September 18, concentrations of O_3 fluctuated from 0.050 to 0.108 ppm during four hours of measurement. Daily monitoring in 1995 showed that in Montecillo, hourly-average O_3 concentrations from June to September varied from 0.031 to 0.127 ppm (Rodríguez-Frausto 1996), concentrations similar to those recorded 10 years earlier.

Well-developed primary leaves of certain bean varieties are highly sensitive to air pollutants (Carnahan et al. 1978; Hass 1970; Heck 1968; Manning and Feder 1980; Weidensaul 1980). In general, the bean plant is more sensitive in the flowering and pod formation developmental stages. Plant injury from air pollution affects yield by reducing the number of pods (Blum and Heck 1980; Davis and Kress 1974) or by decreasing the number of seeds (Weaver and Jackson 1968).

In light of the O_3 sensitivity, dietary importance, and extensive cultivation of beans, in the spring and summer of 1983, Laguette-Rey (1985) evaluated the sensitivity of 21 bean genotypes to ambient oxidants. The bean plants were directly seed-planted in an experimental field of the Postgraduate College in Montecillo. Potted Clintland oat and Cherry Belle radish plants were also evaluated in the field tests. The common bean (*Phaseolus vulgaris*) genotypes tested by Laguette-Rey (1985) included four "bush-type" or determined-growth varieties (Canario 107, Bayomex, Cacahuate, and Tempo), six "semirunner" or semi-undetermined growth varieties (Jamapa, Sataya 425, Bayo 107, Ojo de Negro 400, Negro Mecetral, and Michoacan 12-A-3), and seven "runner-type" or undetermined-growth varieties (Amarillo 153 and 154, Negro Puebla, Negro 150, Flor de Mayo X-16441, Cacahuate Bola, and Pinto 111). Bean plants of the species *Phaseolus accutifolius*, *P. lunatus*, *P. coccineus* (*ayocote*) and *P. dumosus* (*acalate*), all of undetermined-growth habit, were also evaluated.

The first clear symptom of foliar injury induced by ambient O_3 were seen 49 days after seeding in 13 of the species or varieties tested. The observed symptoms began as fine reddish-brown pigmented lesions that developed along the veins on the upper side of trifoliated leaves, producing a bronzed appearance. The plants were apparently more sensitive during the developmental flowering and pod formation stages. In some cases, symptomatic leaves rapidly turned senescent, and leaf abscission occurred. Based on the observed symptomatology, an injury index was developed to assess plant sensitivity to oxidant air pollution. The injury index for each bean material was calculated based on four individual evaluations of ambient oxidant symptoms, during the period from August through October of 1983. For the bush and semirunner varieties of *P. vulgaris*, the evaluation considered the mean percentages of symptomatic plants, symptomatic leaves per plant, and individual leaf area damaged. For the *P. vulgaris* runner varieties and for *P. coccineus*, *P. lunatus*, *P. dumosus* and *P. acutifolius*, the evaluation was based only on the mean percentages of symptomatic leaves and individual leaf area damaged. Due to the entanglement of the runner stems, individual plants could not be distinguished from each other in these bean varieties. For each bean variety, the mean percentages obtained from the evaluations described were expressed in decimal form and then progressively multiplied by each other to obtain a comprehensive injury index. The injury index and the sensitivity categories assigned to each of the bean varieties are presented in Table 14.1. The more sensitive varieties of *P. vulgaris* were Tempo and Pinto 111. Common bean varieties Amarillo 153 and 154, as well as Negro Puebla and Canario 107, exhibited intermediate sensitivity. The observed symptoms generally resembled those described by Jacobson and Hill (1970) and Lacasse and Treshow (1976) as chronic, that is, related to the exposure of plants to relatively low O_3 concentrations for long periods.

Table 14.1. Sensitivity of bean genotypes (*Phaseolus* spp.) exposed to atmospheric oxidants in the field in Montecillo, Mexico, in 1983

Species/varieties	Injury index	Oxidant sensitivity
P. vulgaris var. Tempo	0.6405	Very high
P. vulgaris var. Pinto 111	0.6223	Very high
P. vulgaris var. Amarillo 153	0.0681	Intermediate
P. vulgaris var. Negro Puebla	0.0293	Intermediate
P. vulgaris var. Canario 107	0.0210	Intermediate
P. vulgaris var. Amarillo 154	0.0133	Intermediate
P. vulgaris var. Bayo 107	0.0084	Low
P. coccineus	0.0079	Low
P. vulgaris var. Ojo de Cabra 400	0.0014	Low
P. vulgaris var. Jamapa	0.0008	Low
P. dumosus	0.0007	Low
P. vulgaris var. Negro Mecentral	0.0003	Low
P. vulgaris var. Sataya 425	0.0001	Low

Source: Laguette-Rey (1985).

Some varieties of the common bean (*Phaseolus vulgaris*) are sensitive to photochemical oxidants (Lacasse and Treshow 1976; Manning and Feder 1980; Taylor 1968; Weaver and Jackson 1968). Some show injury when exposed to 0.05 ppm of O_3 for several hours (Lacasse and Treshow 1976). Heck (1968) reported that when bean plants are exposed to 0.05 ppm O_3 for seven hours per day, yield is reduced by 3.1%. MacLean and Schneider (1976) reported a reduction of 24% and 27% in the number and fresh weight of marketable pods produced by the variety Tendergreen when plants were exposed to an average O_3 concentration of 0.0413 ppm for 43 days (the range was 0.017–0.090 ppm). Kohut and Laurence (1983) reported that when bean plants were exposed to 0.06 and 0.09 ppm O_3 during the pod filling stage, the bean yield (seeds/plant) of the exposed plants was 24% less and 27% less, respectively than the yield of unexposed plants. Hofstra et al. (1972) reported similar reductions in bean yield in response to O_3 in southeastern Ontario, Canada. Davis and Kress (1974) measured foliar injury in 10 bean varieties exposed to ozone and reported responses ranging from highly sensitive to tolerant. Krupa et al. (1982) pointed out that plants exposed to recurrent phytotoxic concentrations of O_3 could suffer a yield reduction without manifestatimg any visible symptoms.

In order to confirm the O_3 sensitivity determinations of bean genotypes made by Laguette-Rey during 1983 at Montecillo, seed samples were sent to Professor William J. Manning of the University of Massachusetts at Amherst in the United States. In a preliminary field trial, Professor Manning exposed plants from most of the previously mentioned bean varieties to ambient oxidants known to occur in Amherst, Massachusetts. Afterwards, under greenhouse controlled conditions, Professor Manning exposed *P. vulgaris* plants of the varieties Sataya 425, Negro 150, and Canario 107 to O_3 concentrations from 0.06 to 0.08 ppm for six weeks. Based on foliar symptomatology, Professor Manning considered the Sataya 425, Negro Puebla, and Amarillo 154 varieties of *P. vulgaris* to be somewhat sensitive to ambient O_3 levels at Amherst, Massachusetts. Moreover, among field-exposed and greenhouse-exposed plants of the Sataya 425 variety, Professor Manning detected symptoms of light metallic stippling and chlorosis in older leaves, corresponding to a mild response. The Negro 150 and Canario 107 varieties exhibited more pronounced metallic stippling and chlorosis (Laguette-Rey 1985).

Experimental Studies of Ozone Injury to Beans in the Basin of Mexico

Few experimental studies of the effects of air pollution on plants have been carried out in the Basin of Mexico. The earliest studies focused on identifying the causal agents of foliar injury observed in the basin (de Bauer 1972). Later studies investigated the possible impact of ambient oxidants on the yield and quality of beans. Laguette-Rey et al. (1986b) quantified the impact of photochemical oxidants on the growth and yield of two bean (*Phaseolus vulgaris*) varieties (Pinto 111 and Canario 107) using open-top chambers and N-[2-(2-oxo-1-imidazolidinyl)ethyl]-N′-phenylurea (ethylenediurea, EDU), a protectant compound. There were three treatments in this study: (1) plants were grown in open-top chambers with charcoal filters to remove ambient oxidants and particles; (2) plants were grown in open-top chambers without filtration; (3) the foliage of plants grown outside of chambers were sprayed with EDU at a concentration of 500 μg/ml as a chemical protectant against O_3 injury (Carnahan et al. 1978); and (4) control plants were directly exposed to ambient air. The trials were established during the summer of 1984. Laguette-Rey and collaborators (1986b) indicated that in the first month of experimentation, symptoms of O_3 injury were observed on the upper surface of trifoliate leaves of the control plants of the Pinto 111 variety. The symptoms were small reddish brown necrotic lesions, which gave the leaves a bronzed appearance. The Canario 107 variety was asymptomatic throughout the experiment. At harvest, the Canario 107 and Pinto 111 bean plants were analyzed for several parameters, including plant height, number of pods with normal seeds, number of nodes, and dry weights of pods, seeds, and stems (Laguette-Rey 1985). The plants treated with EDU foliar sprays had greater growth and yield than the control plants. Treatment with EDU resulted in a 40.7% greater yield for the Pinto 111 variety and a 4.5% greater yield for the Canario 107 variety. Based on foliar symptomatology and the nature of the protective effects of EDU, the smaller yields of the untreated plants were likely due to ambient O_3. Pollution effects could not be assessed with certainty for the plants in the open-top chambers due to apparent artifactual chamber effects.

Ortiz-García (1988) confirmed the presence of phytotoxic levels of O_3 at Montecillo by exposing seedlings of the bean variety Tempo during the stage of primary leaf development. These plants exhibited symptoms consisting of small reddish-brown necrotic lesions and necrotic spots on the upper leaf surface. The impact of atmospheric oxidants in the Montecillo area on the agronomic yield of beans was also investigated by exposing Tempo bean plants to ambient oxidants on three sequential planting dates during the spring and summer of 1987 after treating the plants with EDU. Injury assessments were based on the number of normal pods, average pod length, number of seeds per pod, and the dry weight of 100 seeds. The relationship between trifoliate leaf injury and agronomic yield was analyzed. Levels of foliar injury were evaluated with a visual injury scale based on the extent of the foliar area showing symptoms in terms of bronzing and/or reddish stippling and abscission. This scale resembles the one proposed by Manning and Feder (1980) for O_3-induced injury on primary leaves of bean and is described in Table 14.2 and Figure 14.2 (see color plate). Eight plants per treatment for each of three planting dates were visually evaluated for foliar injury. The evaluations were made at the beginning of the flowering period.

The mean injury index for each treatment was useful as a comparative statistic. The untreated plants exhibited less photosynthetic leaf area (due to reddish brown necrotic stippling, necrotic areas on upper surfaces, and leaf abscission), and this was related to their lesser yields compared with those of the EDU-treated plants. The severity of foliar injury in the untreated plants was statistically higher than in the treated plants. Also, the level of foliar injury was generally related to the physiological condition and stage of plant development:

- Flowering stage and onset of pod formation—injury values from light to moderate.
- End of pod formation period—foliar injury was light to severe.
- The physiologically mature stage—injury levels were moderate to severe. However, it appears that foliar injury at this stage did not further impact bean yield.

When the yield of the EDU-treated Tempo bean plants was compared with that of the untreated plants, the yield of untreated plants from the first sowing date was 22.6% less, and that of untreated plants from the second sowing date was 25.7% less. However, the yield of plants from the third sowing date was not

Table 14.2. Visual scale to assess severity of injury on bean trifoliate leaves at the flowering stage

Injury level	Injury index	Percent foliar area affected (%)
Absent	0	0
Light	1	1–25
Moderate	2	26–50
Moderate/severe	3	51–75
Severe	4	76–99
Total	5	100 or abscission

Source: Ortíz-García (1988).

Figure 14.2. Range of ozone-induced injury in Tempo trifoliate bean (*Phaseolus vulgaris* L.) leaves based on upper surface purple stipple/bronzing and bifacial necrosis. (See color plate.)

statistically different from that of the EDU-treated and the untreated plants. By this time of the year, the rainy season had begun, and cloudy days were quite frequent. As noted by several authors (e.g., Legassicke and Ormrod 1981) and confirmed by these results, EDU probably does not give total plant protection. Thus, the actual impact on bean yield in the Montecillo area is probably higher than that shown by the data.

It is important to point out that there might be considerable year-to-year differences in O_3 levels at Montecillo and at other sites in the basin. Meteorological and climatic conditions are expected to be the most important factors affecting O_3 levels from year to year. For example, the rainy period usually begins in May, but it can begin as late as July in some years, such as in 1998 when there was an extended drought. Extended dry and sunny conditions also favor high O_3 levels in the Basin and thus increase the potential for crop losses. This is especially true if favorable plant water status is maintained by irrigation, thus enhancing the potential for stomatal O_3 uptake.

Various factors (e.g., soil moisture content) affect the amount of O_3-induced injury in plants. Rodríguez-Frausto (1996) conducted an experiment in Montecillo to assess the yield of *Phaseolus vulgaris* L. cv. Tempo in the presence of O_3 under varying conditions of soil moisture availability. Three moisture levels were tested, and four replicates were evaluated for each one. EDU was sprayed in half of the plots. Yield fluctuations were found to be responsive both to the different levels of soil moisture and to the effects of ambient oxidants, particularly O_3. The highest yield was obtained at 80% soil moisture. Ozone exposure made a difference in the yield of plots with soil moisture levels of 60% and 80%, but it did not affect the yield of plots with 40% soil moisture, in which the yields of EDU-treated and untreated plants were also similar. The results supported the hypothesis that crop response was adversely affected by ambient O_3 concentrations only when soil moisture levels were adequate.

Summary

Ozone and related oxidants in ambient air have significant impacts on vegetation within the Basin of Mexico. Various studies demonstrate that although the Montecillo/Chapingo area in the State of Mexico is generally upwind from Mexico City, the area's ambient O_3 concentrations are phytotoxic and can significantly reduce the yield of some crops. Daily hourly-average O_3 concentrations registered in Montecillo in 1995 fluctuated from 0.031 to 0.127 ppm. These may represent common concentrations during the summer.

Additional conclusions are:

- All crops and vegetation within most if not all of the Basin of Mexico are exposed to recurrent ambient oxidant concentrations that can potentially cause growth and yield reduction. However, the severity of the impact is expected to vary widely based on geographic location within the Basin, and on genetic, physiological, and environmental interactions.
- Under chronic ambient concentrations of O_3 and other air pollutants, some plant species in the Basin of Mexico may show no visible symptoms, but yield may nonetheless be adversely affected.

It may be worthwhile to conduct a sensitivity screening program under local ambient conditions of the main annual crop varieties commonly used in the Basin of Mexico. Research on *Opuntia ficus-indica* would also be of interest, particularly in areas downwind of Mexico City (for instance, in Milpa Alta). Given its extensive cultivation within the Basin of Mexico, an evaluation of its response (as a xerophyte with characteristic physiology), to ambient O_3 concentrations would be of scientific and economic interest. This is an opportunity to study a plant species with "crassulacean acid metabolism" over a large area subjected to phytotoxic oxidant concentrations.

References

Blum, U., and Heck, W.W. 1980. Effects of acute ozone exposure on snap bean at various stages of its life cycle. *Env. Exp. Bot.* 20:73–85.

CAEVAMEX (Campo Agrícola Experimental del Valle de México). 1981. *Guía para la Asistencia Técnica Agrícola: Area del Campo Agrícola Experimental "Valle de Mexico"*, 1st Edn. Chapingo, Mex., México: Instituto Nacional de Investigaciones Agrícolas, Secretaría de Agricultura y Recursos Hidráulicos (SARH).

Carnahan, J.E., Jenner E.L., and Wat, E.K.W. 1978. Prevention of ozone injury to plants by a new protectant chemical. *Phytopathology* 68:1225–1229.

Davis, D.D., and Kress, L. 1974. The relative susceptibility of ten bean varieties to ozone. *Pl. Dis. Reptr.* 58:14–16.

de Bauer, L.I. 1972. Uso de plantas indicadoras de aeropolutos en la Ciudad de México. *Agrociencia. Serie* 9(D):139–141.

de Bauer, L.I., and Hernandez-Tejeda, T. 1986. *Contaminación: Una Amenaza para la Vegetación en México*. Chapingo, Edo. de México: Colegio de Postgraduados.

de Bauer, L.I., Hernández-Tejeda, T., and Skelly, J.M. 2000. Air pollution in the forested areas of Mexico and Central America. In *Air Pollution and the Forests of Developing and Rapidly Industrializing Countries*, Report 4 of the IUFRO Task Force on Environmental Change, eds. J.L. Innes and A.H. Haron, pp. 35–61. Wallingford, U.K.: CABI Publishing in association with the International Union of Forestry Research Organizations.

de Bauer, L.I., Rodríguez-F., C., Quevedo-N., A., Arteaga-R., R., and Hernandez-Tejeda, T. 1997. Behavior of edible beans (*Phaseolus vulgaris* L.) at three soil moisture levels under the influence of ozone. In *Proceedings of the Dahlia Greidinger International Symposium on Fertilization and the Environment*, ed. J.J. Mortvedt, pp. 88–99. 24–27 March, 1997, Haifa, Israel. Haifa, Israel: Technion—Israel Institute of Technology.

Flores-Valdez, C.A. 1995. "Nopalitos": Production, processing and marketing. In *Agroecology, Cultivation and Uses of Cactus Pear*, eds. G. Barbera, P. Inglese, and E. Pimienta-Barrios, pp. 92–99. Rome, Italy: Food and Agriculture Organization of the United Nations (FAO). 132.

Hass, J.H. 1970. Relationship to crop maturity and physiology of air pollution incited bronzing of *Phaseolus vulgaris* L. *Phytopathology* 60:407–410.

Heck, W.W. 1968. Factors influencing expression of oxidant damage to plants. *Ann. Rev. Phytopath.* 6:165–168

Heck, W.W., Adams, R.M., Cure, W.W., Heagle, A.S., Heggestad, H.E., Kohut, R.J., et al. 1983. A reassessment of crop loss from ozone. *Env. Sci. Tech.* 17:522–581.

Hernández-Tejeda, T., and de Bauer, L.I. 1989. *La Supervivencia Vegetal ante la Contaminación Atmosférica*. Montecillo, México: Colegio de Postgraduados.

Hill, A.C., Pack, M.R., Treshow, M., Downs, R.J., and Transtrum, L.G. 1961. Plant injury induced by ozone. *Phytopathology* 51:356–363.

Hofstra, G., Littlejohns, D.A., and Wukasch, R.T. 1978. The efficacy of the antioxidant ethylene-diurea (EDU) compared to carboxin and benomyl in reducing yield losses from ozone in navy bean. *Pl. Dis. Reptr.* 62:350–352.

Jacobson, J.S., and Hill, A.C. 1970. *Recognition of Air Pollution Injury to Vegetation.* Pittsburgh, Pennsylvania, U.S.A.: Air Pollution Control Association.

Kohut, R., and Laurence, J.A. 1983. Yield response of red kidney bean (*Phaseolus vulgaris* L.) to incremental ozone concentrations in the field. *Environ. Pollut.* 32: 233–240.

Krupa, S.V. 1997. *Air Pollution, People, and Plants: An Introduction.* St. Paul, Minnesota, U.S.A.: APS Press.

Krupa, S.V., Pratt, G.C., and Teng, P.S. 1982. Air pollution: An important issue in plant health. *Pl. Dis.* 66:429–434.

Lacasse, N.L., and Treshow, M. (eds.) 1976. *Diagnosing Vegetation Injury Caused by Air Pollution.* Research Triangle Park, North Carolina, U.S.A.: Air Pollution Training Institute, U.S. Environmental Protection Agency. Contract 68-02-1344.

Laguette-Rey, H.D. 1985. *Impacto de los Oxidantes Ambientales en el Cultivo de Frijol, en Montecillos Edo. de México.* M.Sc. Thesis. Chapingo, México: Colegio de Postgraduados.

Laguette-Rey, H.D., de Bauer, L.I., Kohashi-Shibata, J., and Marban, M.N. 1986a. Impacto de los oxidantes ambientales en el cultivo de frijol, en Montecillo, Edo. de México. *Agrociencia* 66:83–95.

Laguette-Rey, H.D., de Bauer, L.I., Kohashi, S.J., and Ortega, D.M.L. 1986b. Impact of ambient oxidants on the bean crop at a locality within the Valley of Mexico. *Ann. Rep. Bean Improv. Coop.* 29183–29184.

Legassicke, B.C., and Ormrod, D.P. 1981. Supression of ozone-injury on tomatoes by ethylene diurea in controlled environments and in the field. *Hort. Sci.* 16:561–568.

MacLean, D.C., and Schneider, R.E. 1976. Photochemical oxidants in Yonkers, New York: Effects on yield of bean and tomato. *J. Env. Qual.* 5:75–78.

Manning, W.J. 1998. The use of plants as bioindicators of ozone. In *Proceedings of the International Symposium on Air Pollution and Climate Effects on Forest Ecosystems*, Technical coordinators A. Bytnerowicz, M.J. Arbaugh, and S.L. Schilling, pp. 19–26. Gen. Tech. Rep. GTR-166. Feb 5–9, 1996, Riverside, CA. Albany, California: USDA Forest Service, Pacific Southwest Research Station. Electronic version is available at *http://www.rfl.psw.fs.fed.us/pubs/psw-gtr-164/index.html.*

Manning, W.J., and Feder, W.A. 1980. *Biomonitoring Air Pollutants with Plants.* London: Appl. Sci. Pub.

Ortíz-García, C.F. 1988. *Effecto de los Oxidantes Ambientales en el Rendimiento Agronómico del Frijol* (*Phaseolus vulgaris L.*). M.Sc. Thesis. Chapingo, Edo. de México, México: Colegio de Postgraduados.

Pell, E. 1979. How pollutants induce disease. In *Plant Disease an Advanced Treatise. How Pathogens Induce Disease*, Vol. IV, eds. J.G. Horsfall, and E.B. Cowling, pp. 273–291. New York: Academic Press, Inc.

Rodríguez-Frausto, C. 1996. *Influencia del Ozono y la Humedad del Suelo en el Rendimiento del Frijol* (*Phaseolus vulgaris L.*) *var. Tempo en Montecillo, México.* Tesis de Licenciatura. Chapingo, Edo. de México, México: Universidad Autónoma de Chapingo.

Schwartz, H.F. 1980. Problemas misceláneos. In *Problemas de Producción de Frijol*, eds. H.F. Schwart, and G.E. Gálvez, pp. 327–340. Cali, Colombia: Centro Int. de Agricultura Tropical.

Taylor, O.C. 1968. Effects of oxidant air pollutants. *J. Occ. Med.* 10:485–495.

Temple, P.J., Sun, J.E.-J., and Krause, G.H.M. 1998. Peroxacyl nitrates (PANs) and other minor pollutants. pp. 6-1–6-11 In Recognition of Air Pollution Injury to Vegetation: A Pictorial Atlas. 2nd ed. R.B. Flagler, A.H. Chappelka, W.J. Manning, P.M. McCool, and S.R. Shafer (eds.). Pittsburgh, PN, USA. Air & Waste Management Association.

Temple, P.J., Taylor, O.C. Reang, L.A., and Lennox, R.W. 1983. Smog damage to cotton in the San Joaquín Valley. University of California. Division of Agriculture and Natural Resources. *Rep. Prog. Res. Cal. Agric.* 37:4–5.
Weaver, G.W., and Jackson, H.O. 1968. Relationship between bronzing in white bean and phytotoxic level of atmospheric ozone in Ontario. *Can. J. Pl. Sci.* 48:561–568.
Weidensaul, T.C. 1980. N-[2-(2-oxo-1-imidaxolidinil)ethyl]-N-phenylurea as a protectant against ozone injury to laboratory fumigated pinto bean plants. *Phytopathology* 70:42–45.

4. Summary, Synthesis, and Perspectives

15. Summary of Air Pollution Impacts on Forests in the Mexico City Air Basin

Mark E. Fenn, L.I. de Bauer, and Tomás Hernández-Tejeda

Oxidant air pollution symptoms were first reported in bioindicator plants in the Mexico City Air Basin (MCAB) in 1971 (de Bauer 1972). Classic injury symptoms on well-known bioindicator plants strongly supported the presumption that symptoms were caused by photochemical oxidants, of which ozone (O_3) is the primary pollutant. Symptoms in indicator plants characteristic of injury caused by peroxyacetyl nitrate (PAN), ethylene, and sulfur dioxide (SO_2) were also reported (de Bauer 1972; de Bauer and Hernández-Tejeda 1986). These discoveries were followed in 1976 by the observation of O_3 injury symptoms in native pine species in forests in Ajusco (AJ), just south of Mexico City (de Bauer and Hernández-Tejeda 1986; Krupa and de Bauer 1976). Then in the 1980s, a dramatic and severe decline occurred in sacred fir (*Abies religiosa*) stands in the highly polluted area of the Desierto de los Leones (DL) National Park southwest of Mexico City (Alvarado-Rosales and Hernández-Tejeda 2002). Thousands of trees died, leaving dead zones called cemeteries. Air pollution was widely believed to be an important causal factor in the widespread mortality of sacred fir within the park (Ciesla and Macias-Samano 1987). Topographic conditions, including volcanic mountain ranges circumscribing much of the Basin, thermal atmospheric inversions, and prevailing winds which carry pollutants from the urban zone to forested areas to the south-southwest (SSW), create conditions favoring high pollution exposure for these forests (Bravo and Torres 2002; Jáuregui 2002).

Continuing increases in population and in the number of motor vehicle miles driven in the Basin and increasing traffic congestion (Enríquez 2000) indicate that chronic and severe air pollution exposures will be a problem in these forests for years to come, or at least until more stringent emission-control regulations are in place and newer less-polluting vehicles become more prevalent. The ultimate solution to air pollution in the MCAB would be greater use of less-polluting mass-transit systems and a technological breakthrough in developing an economically viable and readily available clean energy source for motor vehicles. Fuel efficient and low polluting hybrid (electric and gasoline) vehicles are now coming on to the market in a few countries. However, in order for this or other new technologies to have a major impact on fossil fuel emissions in the MCAB, it would be necessary for millions of people to purchase new vehicles to replace older high-polluting vehicles, a cost-prohibitive alternative for many inhabitants of the MCAB. In addition, a large fraction of the emissions are from heavy-duty vehicles such as semitrailers, trucks, buses, and vans from which emissions are high.

Although current pollution control measures in the MCAB are insufficient to reduce the concentrations of all pollutants to levels below the governmental standards, progress has been made. For example, levels of SO_2 have decreased significantly since 1993 and rarely exceed the 24-hour standard. However, annual average levels of SO_2 are still slightly above the annual standard (Bravo-Alvarez and Torres-Jardón 2002). With the introduction of low-lead gasoline in 1986 and unleaded gasoline in 1990, similar progress has been made in reducing emissions of lead. The reduced incidence of of emergency levels of O_3 (from 177 days in 1992 to 7 days in 1999) suggests that limited progress has been made in reducing O_3 levels. (SMA 1999; http://sma.df.gob.mx/publicaciones/aire/inf_cal_aire99/ indice.htm). However, O_3 levels continue to be a major air pollution problem. The O_3 air quality standard is exceeded by more than 4 to 5 hours per day for at least 300 days per year (Bravo-Alvarez and Torres-Jardón 2002). Nitrogen (N) and sulfur (S) deposition levels also remain highly elevated in forests downwind of Mexico City (Fenn et al. 1999), notwithstanding the greatly reduced S content of fuel.

Major Pollutants with Potential to Impact Forests in the MCAB

The major air pollution types in the MCAB that have the potential to impact sensitive plant species or some aspect of forest ecosystem function and the major characteristics of these types of air pollution are listed in Table 15.1. The available evidence indicates that under current air pollution exposures, the pollutants with the greatest ecological and environmental effects are O_3 and N pollutants. Sulfur dioxide probably was an important pollutant with observable biotic effects in sensitive organisms such as lichens and some plant species, particularly

Table 15.1. Characteristics and potential forest effects of the major pollutant types in the Mexico City Air Basin (MCAB)

Pollutant	Characteristics	Major effects on forests in the MCAB
Ozone	A secondary gaseous pollutant, a product of photochemical reactions of NO_x and hydrocarbons; highly phytotoxic	Severe visible foliar and crown injury in sensitive pine species; may be an important factor in sacred fir decline; symptoms in some understory species; crop loss also documented
Nitrogen oxides	Primary pollutants, mainly from fossil fuel emissions, ozone precursor	Main effects may be as an ozone precursor; source of N compounds deposited to forested watersheds which "fertilizes" the forest and increases nitrate levels in runoff
Nitrate and nitric acid vapor	A secondary pollutant from NOx	Increases N fertility of the forests and increases nitrate in runoff; direct injury from nitric acid hasn't been studied in the MCAB
Peroxyacetyl nitrate	A secondary gaseous photochemical oxidant, occurring at lower concentrations than ozone	Concentrations not usually high enough to cause visible injury to vegetation in rural areas; the MCAB may be an exception, but data is lacking in rural areas
Sulfur dioxide	Mainly from combustion of fossil fuels, petroleum and natural gas industries, smelting and refining processes; a gaseous pollutant	Likely effects on sensitive organisms such as lichens; more severe in the north of the Basin; concentrations are lower in forests in the south of the Basin, but there is little data from forested areas; levels have decreased as a result of low sulfur fuels
Sulfate	A secondary pollutant from SO_2 emissions	Causes S enrichment of the forests; increased S content of plants; possible role in soil acidification and cation leaching in more exposed forests
Heavy metals	Mainly from motor vehicles and industry. No data available on deposition to forests	Lead was the major heavy metal. Emissions have greatly decreased since the mid-1990s. Phytotoxic symptoms not reported, unknown effects on more sensitive organisms and soil processes; Potential water quality problem
Ammonium-ammonia	Mainly from intensive livestock operations; some industrial sources	Also increases N fertility of forests; may contribute to soil acidification due to increased nitrification rates
Volatile organic compounds (VOC)	From fossil fuels, industry and emissions from plants	VOC are precursors of ozone and other oxidants. Direct effects of VOC on vegetation in the MCAB has not been studied

in the northwestern industrial sector (Zambrano et al. 2002), but concentrations have decreased since the early 1990s. Sulfur dioxide concentrations are higher in the northern part of the Basin, while forests are found predominantly in the southern portion of the Basin. Much of the SO_2 emitted from the industrial areas in the north is presumably transformed to sulfate aerosols and transported to the southwest by the predominant winds (Bravo-Alvarez and Torres-Jardón 2002). Phytotoxic symptoms attributable to SO_2 have not been reported for forest species in the MCAB. However, pollutants can have significant physiological effects on plants without inducing visible symptoms (Bussotti and Ferretti 1998; Evans et al. 1996; Krupa et al. 1982). Furthermore, controlled air pollution exposures have not been carried out for the major overstory species in the MCAB with any of the major pollutants. For example, it has not yet been confirmed which, if any, pollutants may cause the foliar symptoms observed in declining sacred fir trees (Alvarado-Rosales and Hernández-Tejeda 2002). Thus, the lack of reported SO_2 injury symptoms in forests of the MCAB could possibly reflect the lack of study or knowledge of symptom development in the native species. Considering the historically high S emissions from the MCAB, and the, as yet, high atmospheric sulfate deposition inputs in throughfall, it seems clear that S deposition in high-pollution forests of the MCAB has increased the S concentration in foliage (Fenn et al. 1999, López-López et al. 1998) and is probably affecting nutrient cycling processes to some degree (see Fenn et al. 2002a).

Although we are not aware of direct measurements of atmospheric deposition rates for lead (Pb) and other heavy metals in forested areas of the MCAB, data on metal concentrations in forest soils and vegetation (Castro-Servín et al. 1997; Fenn et al. 2002a) and recent dendrochemical analyses (Watmough and Hutchinson 1999) demonstrate that heavy-metal deposition, particularly in the case of Pb, was high in the past. Lichen studies also demonstrate high deposition rates of heavy metals in some parts of the Basin (Zambrano et al. 2002). With the elimination of leaded gasoline, the ecological effects of heavy-metal deposition may now be less of an issue than it was a few years ago. However, because of the accumulation of metals in soil and soil organic matter, heavy metals can be biologically important even after deposition levels decrease (Bargagli 1998). Lead that has accumulated in soil and organic matter throughout the watersheds can continue to be exported in drainage waters for many years, and there is preliminary evidence that this is occurring in the Basin of Mexico (Fenn et al. 2002a).

As mentioned, O_3 and N compounds appear to be the pollutants with the greatest potential for ecological and environmental impacts in the MCAB. Nitrogen pollutants can include substances in several physical (gaseous, particulate, dissolved) and chemical (reduced, oxidized, ionic) forms. The main biotic effects of O_3 and N pollutants are usually quite different, inasmuch as O_3 is phytotoxic and decreases the growth and productivity of sensitive plant species. Nitrogen, on the other hand, is a growth stimulant and is also the plant nutrient most frequently limiting plant growth in temperate forests (Vitousek and Howarth 1991). On the other hand, phosphorus (P) is often limiting in sites with volcanic soils such as in the Basin of Mexico (Fenn et al. 2002b).

By the time nitrogen oxides from Mexico City have been transported to forested areas outside of the city, the N pollutants have been largely converted to particulate nitrate or nitric acid vapor (HNO_3). Nitrogen and S deposition occurs mainly as gaseous or particulate deposition in dry form or as wet deposition during precipitation events. Washoff from forest canopies of accumulated dry deposition occurring as throughfall, is a major mechanism for deposition of air pollutants to forest ecosystems. In high-pollution areas, a significant fraction of dry-deposited pollutants, especially nitrogenous compounds, are not washed off by precipitation but are retained within the canopy. Thus, throughfall measurements in areas of elevated N deposition commonly underestimate total N deposition by approximately 30% to 40% (Fenn et al. 2000; Lovett and Lindberg 1993).

Based on recent atmospheric measurements in Mexico City (Gaffney et al. 1999) and symptoms in bioindicator plants (de Bauer and Hernández-Tejeda 1986), phytotoxic levels of PAN occur in Mexico City, However, possible effects of PAN in forests in the MCAB have not been studied. Levels of PAN measured in north central Mexico City are the highest reported in the world since similar levels were reported near Los Angeles, California, in the late 1970s. Concentrations in the southwest of the Basin, in or near the major forested areas, are expected to be even higher (Bravo-Alvarez and Torres-Jardón 2002). PAN levels in Mexico City exhibit a strong diurnal pattern, indicating that this pollutant is transported out of the city almost completely during each diurnal period (Gaffney et al. 1999). This suggests that forests downwind of the city, such as the DL Park, may be exposed to high concentrations of PAN on a regular basis. Further research is needed to evaluate the impacts of PAN exposure on vegetation in forests of the MCAB.

Plants in the MCAB are exposed to many other organic pollutants besides PAN (Bravo-Alvarez and Torres-Jardón 2002). The effects of volatile organic compounds (VOC) on vegetation have not been investigated in the MCAB. Based on a review of the literature, it was concluded that the indirect effects of VOC, as precursors to O_3 formation, are more important than the direct effects on vegetation (Smidt 1994). However, this same author concluded that the direct effects of VOC on forest decline have not been well defined. The concentrations of formaldehyde (one example of an important atmospheric VOC) in the MCAB are among the highest reported in the literature. The peak concentration of formaldehyde (110 ppb) was reported in the spring of 1993 at a site to the southwest of the urban zone (Baez et al. 1995). However, common bean plants exposed to 400 ppb formaldehyde for 7 hours per day, 3 days per week for 4 weeks grew larger than the control plants (Mutters et al. 1993). Mutters et al. concluded that formaldehyde concentrations much greater those that occur in urban areas are required to injure bean foliage.

Air Pollution Impacts on Forest Health and Sustainability

Geographic Location of Air Pollution Studies in the Mexico City Air Basin

Much of the work on air pollution effects in the MCAB has focused on the DL National Park. However, the first reported observation of air pollution injury symptoms to forest species was that of O_3-induced injury to *Pinus leiophylla* and *P. hartwegii* in the AJ area (Fig. 15.1) (de Bauer and Hernández-Tejeda 1986; Krupa and de Bauer 1976). Air pollution gradient design studies have been useful in evaluating the effects of air pollutants on forests and forest species in the San Bernardino Mountains in southern California (Fenn and Bytnerowicz 1993; Fenn and Dunn 1989; Miller et al. 1986) and in the forests of the MCAB as well (de Bauer and Hernández-Tejeda 1986; Fenn et al. 1999; Fenn et al. 2002b; Hernández-Tejeda 1984). Hernández-Tejeda studied O_3 injury in pine species along a transect or air pollution gradient following a route parallel to the highway from Mexico City to Cuernavaca (de Bauer and Hernández-Tejeda 1986; Hernández-Tejeda 1984). The most exposed forest site in the MCAB is the DL National Park, which is adjacent to the urban zone and directly downwind (to the

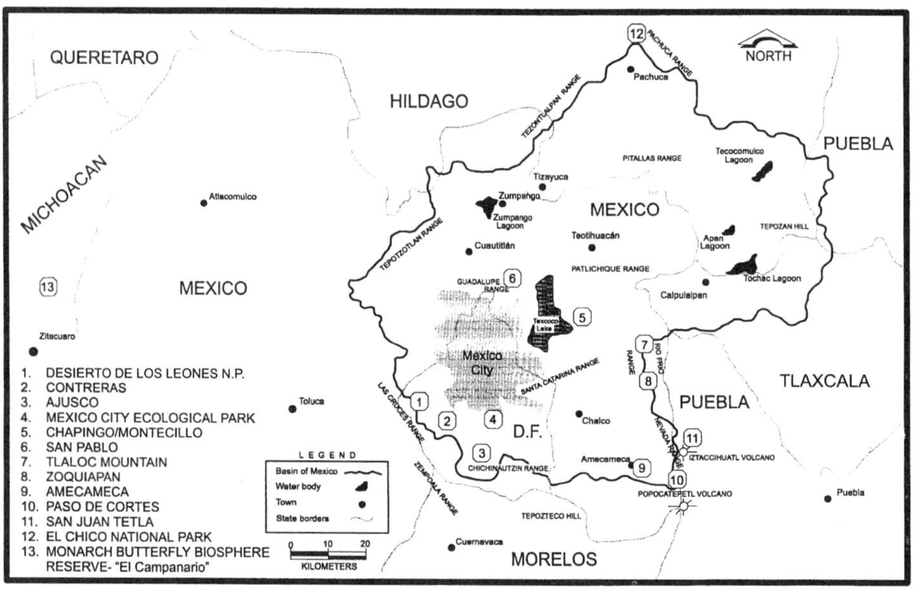

Figure 15.1. Location of the sites used by various researchers in studies of air pollution effects on forests in the Basin of Mexico. High-pollution sites are those to the SSW of Mexico City. Other sites indicated were mainly used as relatively low-pollution exposure sites. (Modified from Figure 3.1 in this volume.)

southwest) of Mexico City. The AJ area is SSW of the city (Fig. 15.1), and is also considered to be a high-pollution area, although it is apparently not as directly exposed to air pollution masses as DL. Lichen studies were carried out at high-pollution study sites including DL and the Mexico City Ecological Park located at the southwestern fringe of Mexico City (Fig. 15.1) (Zambrano et al. 2002). Forest sites which have been used as low-pollution sites (although their pollution levels are only low relative to the high-pollution sites) are found to the east or southeast of the city. Zoquiapan (ZOQ) National Park (Fenn et al. 1999), Paso de Cortes within the Iztaccíhuatl-Popocatépetl National Park (Fenn et al. 2002b), the municipality of Amecameca on the slopes of Popocatépetl at 13 km from Paso de Cortés (Alvarez et al. 1998), and San Juan Tetla in the state of Puebla (Fig. 15.1) Castro et al. 1997; Fenn et al. 2002a) are examples of such (relatively) low-pollution sites. In dry deposition and throughfall studies at DL and ZOQ, it was concluded that S deposition at ZOQ was higher than would have been expected based on the location of the site in relation to Mexico City because of volcanic outgassing from the Popocatépetl volcano during the sampling period (Fenn et al. 1999). Researchers using any forest sites within the zone of influence of Popocatépetl as low-pollution study sites should keep in mind that emissions from the volcano are likely to affect the atmospheric chemistry and deposition occurring in the area (Delmelle et al. 2001), thus possibly compromising the value of these sites as controls. Other control sites that have been used include the El Chico National Park just north of the boundary of the Basin of Mexico (Zambrano et al. 2002), the El Campanario study site within the Monarch Butterfly Biosphere Reserve in the state of Michoacán (Terrazas and Bernál-Salazar 2002), and San Pablo, located 45 km northeast of Mexico City (Watmough and Hutchinson 1999). Additional data is needed to better characterize air pollution exposures at these study sites. Recent developments in the use of passive samplers in conjunction with strategically located electronic monitors (Cox and Malcolm 1999; Koutrakis et al. 1993; Krupa and Legge 2000; Runeckles and Bowen 2000; Yamada et al. 1999) will facilitate collection of the data needed to better define pollution gradients in the MCAB.

Sacred Fir Decline

The most dramatic effect attributed to air pollution in the MCAB is the severe forest dieback of *Abies religiosa* in the DL National Park (Alvarado-Rosales et al. 1993; Cibrian-Tovar 1989; Ciesla and Macias-Samano 1987). Mortality of sacred fir as the end stage of this syndrome was first observed in 1981 (Cibrian-Tovar 1989) and increased throughout the 1980s. In recent years, mortality rates have declined, but symptoms of decline are still present in the stands. Sacred fir decline was so severe in the 1980s that entire stands were decimated, cleared, and replanted with alternative species. Replanting with sacred fir was not successful, possibly because of unfavorable conditions for seedling survival. The etiology of the decline appears to be complex, and the role of air pollution in the decline is not definitively known. Foliar symptoms begin as a whitish stippling. These lesions later coalesce and develop into a reddish-brown discoloration of the foliage. These symptoms are putatively caused by O_3, although controlled exposures to O_3 have not been performed with this species. However, the decline is particularly severe at the ends of ravines where air pollutants accumulate. Furthermore, foliar symptoms were prevented in branches of sacred fir enclosed in charcoal-filtered branch chambers or treated with an antitranspirant known to greatly reduce O_3 injury (Alvarado-Rosales et al. 1991, 1993). Based on the available evidence, oxidant air pollution is widely believed to play an important role in the decline of sacred fir in the park (Alvarado-Rosales and Hernández-Tejeda 2002; Ciesla and Macias-Samano 1987). Controlled exposures to O_3 will be carried out in the near future to determine whether O_3 causes the observed foliar symptoms. However, other phytotoxic pollutants, such as PAN, could also play a role in the observed foliar injury (Gaffney et al. 1999; Temple and Taylor 1983), although this has not been studied.

Stand densification and overmaturation of the trees are also believed to be contributing, and possibly critical, factors in the development of fir decline. The prohibition of silvicultural management of the forest in the park results in overcrowded and stressed trees. In neighboring forests, where thinning and harvesting for timber have occurred throughout the years, no decline has occurred. Drought, possibly exacerbated by the harvesting of water from springs throughout the park, may also contribute to the decline. Nutritional deficiencies, insect attacks, and pathogens have also been considered as contributing factors, but it is difficult to determine if these are casual factors in the decline or secondary factors occurring in declining trees (Alvarado-Rosales and Hernández-Tejeda 2002).

Ozone Injury of Pine

The primary forest species known to be severely impacted by air pollution is *Pinus hartwegii*, which is highly sensitive to O_3 injury (de Bauer and Hernández-Tejeda 1986; Miller et al. 2002). The development of injury symptoms in *P. hartwegii* is virtually identical to the classic symptoms described for *P. ponderosa* in the Los Angeles Air Basin in southern California (Miller et al. 2002), where high O_3 levels also occur in forested areas. Ozone-induced symptoms in both pine species include thinning of the crown, chlorotic mottling on the foliage, and premature senescence and abscission of the older foliage. In the San Bernardino Mountains in southern California, mature *P. ponderosa* trees exposed to elevated O_3 concentrations and N deposition have dramatically reduced fine root biomass, and greater foliar litterfall and litter accumulation (Grulke et al. 1998; Takemoto et al. 2001). These factors have not been studied in forests of the MCAB, but reduced root production as a result of O_3 exposure is a likely effect in the Mexican forests as well, considering that reduced root: shoot ratios are a general response of plants stressed by O_3 (Andersen et al. 1997;

Landolt et al. 1997; Scagel and Andersen 1997). Ozone-induced injury symptoms have also been reported on native understory species in the MCAB, including *Sambucus mexicana* (*Sauco*), *Sicyos* sp. (*Chayotillo*), *Eupatorium* sp., *Piqueria trinervia*, and *Solanum* (de Bauer and Hernández-Tejeda 1986).

In the MCAB, O_3 levels are high in both summer and winter, but in California, O_3 levels are very low in winter (Miller et al. 1994). Even more significant is the fact that for much of the summer, when O_3 levels are highest in California, soil moisture is low, thus severely limiting stomatal conductance and foliar uptake of O_3 (Temple and Miller 1998). The temporal asynchrony between the greatest available soil moisture and high O_3 levels limits the amount of O_3 injury incurred by ponderosa pine. In contrast, soil moisture levels are high in summer in the MCAB, and at the sametime, O_3 levels rise well above the threshold concentrations for inducing O_3 injury. Thus, because of concurrent high levels of O_3 and soil moisture, there is greater potential for O_3 injury in sensitive species in the MCAB, although O_3 injury scores were similar between high-pollution sites in the MCAB and the San Bernardino Mountains in the Los Angeles Air Basin (Miller et al. 1994; Miller et al. 2002).

Dendrochronology Studies

Chlorotic mottle on pine needles and other well-defined injury symptoms on indicator plant species provide strong evidence of areas where air pollution is exerting significant biological effects on sensitive forest species in the MCAB. Dendrochronological studies indicated that radial growth rates for *P. hartwegii* decreased significantly at DL and, to a lesser degree, at AJ from the 1970s onward (Alarcón et al. 1993). Similar growth trends have been demonstrated in several tree ring studies of *A. religiosa* at DL (Alvarado et al. 1993; Bernal-Salazar et al. 2002; Terrazas and Bernal-Salazar 2002). Xylem tracheid length and wall thickness also decreased after the 1970s, the period corresponding to the occurrence of severe air pollution problems in the MCAB. These and other tracheid modifications were similar to those previously reported for declining trees exposed to air pollutants (Terrazas and Bernal-Salazar 2002). These alterations in wood tissue structure in air pollution impacted trees of *A. religiosa* appear to be good indicators of physiological air pollution stress in trees.

Effects of Heavy Metals in Forests of the MCAB

Atmospheric deposition rates of Pb and other heavy metals have not been measured in forests in the MCAB, but data on metal concentrations in lichens, foliage, and soil from several studies demonstrate that inputs of heavy metals are greater in the forested areas to the SSW of Mexico City where exposure to O_3 and other pollutants is known to be higher. As a result of the introduction of unleaded fuels in 1990 and the phase out of leaded gasoline in 1997, Pb deposition to forested areas in the MCAB has undoubtedly decreased greatly in recent years. The available evidence suggests that levels of heavy metals in the plant species studied to date are not sufficient to cause noticeable phytotoxic effects. However, this is not to say that ecological effects on more sensitive species and processes have not occurred. Litter decomposition is known to be inhibited by elevated levels of heavy metals (Laskowski et al. 1994) and N transformation processes in soil can be affected at levels of heavy metals as low as three times background concentrations (Bääth 1989; Tyler et al. 1989).

Watmough and Hutchinson (1999) studied the trace metal chemistry of soil and tree rings of *A. religiosa* at DL and at San Pablo, a control site 45 km upwind (northeast) of Mexico City. Surface soils at DL contained elevated levels of trace metals, especially Pb, Cd and Zn. Lead and Cd levels were higher in tree rings formed since the 1960s. However, peaks in Cd and Pb were found in rings formed between 1920 and 1940, corresponding to the heartwood–sapwood boundary, especially in trees growing in contaminated soils with low pH. The authors suggested that Pb and Cd taken up by roots of sacred fir accumulate in the heartwood, whereas Pb and Cd entering through the bark are transported radially to a much lesser extent and more accurately reflect changes in trace-metal deposition. The accumulation of Cd and Pb in the heartwood–sapwood boundary limits the usefulness of this technique, at least with *A. religiosa*. However, it is possible that this technique may be more appropriate in other species as a methodology for determining temporal patterns in pollutant uptake in forests of the MCAB. Bargagli (1998) suggested that whenever possible, ring-porous trees such as oaks, elms, and ashes should be used for these types of studies because in these species, most of the water moves through only the current year's wood vessels.

There is a report of Pb concentrations in excess of the Mexican drinking water standard in spring water from the Ajusco area (Morales 1998). In subsequent streamwater studies at forest sites across the MCAB, including streams and springs in the high-pollution sites AJ and DL National Park, Pb concentrations tended to be higher at the two high-pollution sites (Fenn et al. 2002a), but concentrations were generally much lower than reported by Morales (1998). Further studies are needed to more fully assess the severity and extent of Pb export in drainage waters.

Air Pollution Effects on Lichen Communities

Epiphytic lichen diversity at DL is severely impoverished compared to El Chico National Park, a low-pollution site 100 km northeast of Mexico City. DL may have lost nearly 50% of its lichen species and lichen abundance is reduced by 60%, presumably as a result of the severe air pollution levels occurring at this site. Based on historical herbarium collections, the decline in lichen diversity appears to coincide with the period of accelerated industrial and population growth of Mexico City since the 1930s and 1940s (Zambrano et al. 2002). Concentrations of airborne metals such as Cu, Pb, and Zn in lichens from forests near the city indicate that metal deposition was more than twice as high in forests near the city as in forests far from the city. Short-term lichen transplant experiments in Sierra de las Cruces, a montane region just SSW and downwind of the urbanized zone, showed a 30% lower carbon fixation and 15% to 25% chlorophyll degradation compared to samples in El Chico National Park (Zambrano and Nash 2000; Zambrano et al. 1999). These results suggest that chronic air pollution is a major cause of lichen decline in forests surrounding the city, along with a variety of other anthropogenic disturbance factors.

Air Pollution Effects on Urban Forests

Because of the diversity of plant species grown in Mexico City and the surrounding urban zones, and considering the exposure of vegetation to elevated concentrations of several phytotoxic pollutants in the MCAB (Table 15.1), there is little doubt that sensitive species have been severely impacted. However, few studies have directly addressed pollution impacts on urban trees. One documented example is the discovery of O_3 injury in eucalyptus seedlings growing in Mexico City. That the observed injury symptoms in eucalyptus were in fact caused by O_3 was subsequently confirmed in controlled O_3 exposure studies (Hernández-Tejeda et al. 1981). Unfortunately, most reports of air pollution injury in urban vegetation are anecdotal or consist of observations without experimental confirmation of the cause of plant injury. Martínez and Chacalo (1994) discuss the major tree species grown in Mexico City and report on the air pollution sensitivity of the described species. However, the basis on which a particular species is reported to be sensitive to air pollution is not given, nor do they mention to which air pollutants the plants are sensitive or tolerant. This type of unsubstantiated general information is limited in its reliability, specificity, and utility.

More studies on the effects of air pollution on urban trees and vegetation are needed before clear guidelines can be established concerning the preferred species for use in urban plantings. In the absence of more information on air pollution sensitivity of the species used in the MCAB, possibly the best initial guide would be information from the literature on studies of the responses of similar species to air pollution. To this end, Hernández-Tejeda and de Bauer (1989) published a synopsis of air pollution effects on vegetation in the MCAB, along with lists of plant species, mainly from North America, Europe, and Mexico, and their relative sensitivity or tolerance to the major pollutants, O_3, PAN, nitrogen oxides, and SO_2.

Nitrogen Saturation: Effects of N Deposition on Nutrient Cycling and Water Quality

Nitrogen saturation can be defined as the long-term removal of N limitations on biotic activity in an ecosystem, accompanied by a decrease in N-retention capacity. Recent studies in forests of the northern hemisphere have shown that with chronic atmospheric N inputs, an excess of N accumulates in the system, and the system becomes N saturated. The classic symptom of an N-saturated watershed is the export of excess N as nitrate, which is leached through the soil profile and into the groundwater and surfacewater system. Thus, forests normally exhibiting a closed N cycle become open systems, and excess N is lost, mainly as leached nitrate, and to a lesser degree as nitrogenous trace gasses emitted from soil as the result of microbial processing of soil N. The latter process typically results in greater greenhouse gas emissions (e.g., N_2O, NO) from forested lands (Fenn et al. 1996; Fenn et al. 1998; Gasche and Papen 1999). The phenomenon of N-saturated watersheds demonstrates that air quality (atmospheric N pollutants) and water quality (nitrate levels) are intricately linked in forested watersheds receiving high atmospheric N deposition. Nitrogen deposition is also believed to stimulate forest growth, up to a point, although this is an open and complicated issue, especially because increased N availability can also impair mycorrhizal and root production and function and causes other disruptions of normal ecosystem processes (Fenn et al. 1998; Skeffington and Wilson 1988).

Nitrogen levels in vegetation and nitrate concentrations in runoff appear to be much more affected by N deposition in the San Bernardino Mountains east of Los Angeles, California, than in the DL National Park (Fenn 1991; Fenn and Poth 1999; Fenn et al. 1996; Fenn et al. 1999), based on comparisons with sites on the low end of N deposition gradients in both regions. Furthermore, there was greater foliar growth in pine trees fertilized with N in southern California, especially in N-limited sites with annual throughfall depositions of less than 12 kg/ha (Kiefer and Fenn 1997), but foliar growth of pine trees at sites with low or high N deposition in the MCAB did not respond to N fertilization. These results, and the inherently low C : N ratios and high total N content of the Mexican forest soils, suggest that the pine trees growing on these volcanic soils are not severely N limited. Some studies suggest that phosphorus may be limiting tree growth in forests in the MCAB (Fenn et al. 2002b), which would not be surprising considering the well-known high P fixation capacity of these andisols (Brady and Weil 1999; Fenn et al. 2002b). This may also explain why N concentrations in foliage often did not differ between sites across the pollution gradient in the MCAB. In a few instances, foliar N levels were significantly higher at the polluted sites, but differences were small compared to responses to N deposition in the San Bernardino Mountains in California (Fenn et al. 1996; Fenn et al. 1999).

In summary, it appears that the most significant effect of chronic N deposition to forests in the MCAB is to increase nitrate concentrations in springs and streams which provide drinking water for the adjacent urban population. However, preliminary evidence from several studies also suggests a relationship between greater soil acidity, Al mobilization, and loss of soil Ca and Mg at the more exposed high elevation pine sites in the DL (reviewed in Fenn et al. 2002a in this volume). More research is needed to evaluate these findings and to determine the possible role of atmospheric deposition of N and S in this phenomenon. Further study of the effects of N and S deposition on ecosystem processes is also warranted in vegetation types other than pine forests. For example, deposition of N, S, and other pollutants is expected to be greater under *A. religiosa* (e.g., Fenn et al. 2002a; Watmough and Hutchinson 1999) because of its greater surface area for collection of atmospheric deposition (Van Ek and Draaijers 1994), thus possibly increasing the magnitude of perturbations on nutrient cycling and plant function in fir stands.

Air Pollution Effects on Agronomic and Herbaceous Species

Oxidant air pollution injury symptoms have been observed in the MCAB on a number of agronomic species, on native understory species at forested sites, and on many well-known bioindicator species (de Bauer 1972; de Bauer and Hernández-Tejeda 1986; Laguette-Rey et al. 1986; Manning 1998; Ortíz-García et al. 2002). Oxidant air pollution injury has been reported for pinto beans, tobacco, lettuce, oats, soybeans, radish, petunia, eucalyptus seedlings, spinach, annual bluegrass (*Poa annua*), *Sicyos* sp., *Eupatorium* sp., *Piqueria trinervia*, and *Solanum verrucosum*, among others in the MCAB. In one study with pinto bean crops growing near Montecillo in the State of Mexico (a site usually upwind of Mexico City), bean production of an O_3-tolerant variety (Canario 107) was reduced by 4.5% when plants were exposed to ambient levels of oxidants. In contrast, bean production of an O_3-sensitive variety (Pinto 111) was reduced by 40.7% when plants were exposed to ambient levels of photochemical oxidants (Laguette-Rey et al. 1986; Ortíz-García et al. 2002). The observation of visible O_3 injury in so many different plant species in the MCAB demonstrates the widespread effects of oxidant air pollution on vegetation in the Basin. Furthermore, it is likely that physiological effects are occurring in other species before visible injury is apparent, or in some cases, when no visible foliar injury symptoms ever develop (Evans et al. 1996; Krupa et al. 1982).

Summary

The first evidence that air pollution was impacting vegetation in the MCAB was the observations of foliar injury symptoms in bioindicator plants (de Bauer 1972). These symptoms were attributed to O_3, PAN, SO_2, and possibly other pollutants. Subsequently, O_3-induced injuriy to foliage and crowns of pine trees were reported in forests to the south and southwest of Mexico City (de Bauer and Hernández-Tejeda 1986; Krupa and de Bauer 1976). Ozone is considered to be the pollutant that impacts most severely on vegetation within the urban zone and in forests downwind of the city. *P. hartwegii*, the most O_3-sensitive pine species, is severely impacted by the high O_3 exposures occurring SSW of the metropolitan area (Miller et al. 2002). The potential for O_3 injury is particularly high in the MCAB because O_3 levels are high during the summer rainy season, when soil moisture availability and stomatal conductance are greatest. These factors enhance O_3 uptake and injury. The decline of *A. religiosa* (*oyamel*) in the DL National Park is now a world-famous example of dramatic dieback and mortality of entire forest stands, presumably, at least in large part, as a result of air pollution stress (Alvarado-Rosales and Hernández-Tejeda 2002). However, other factors, such as the lack of stand thinning, are also believed to contribute, predisposing the trees to further decline.

Concentrations of PAN, and possibly concentrations of other oxidants, may also be damaging vegetation in some areas of the MCAB, but this has not been investigated. Aside from bioindicator evidence of PAN injury, the effects of organic pollutants in the MCAB or plants have not been studied, and little information is available to suggest whether organic pollutants are a significant threat to vegetation. Atmospheric deposition of Pb in the MCAB has undoubtedly decreased dramatically since the introduction of unleaded gasoline in September, 1990, and the phase out of leaded gasoline in August, 1997. The limited available evidence suggests that present foliar concentrations of heavy metals in forest species are not phytotoxic (Fenn et al. 2002a). Sulfur dioxide concentrations decreased in the early 1990s as a result of regulatory mandates limiting emissions. However, sensitive organisms in the northeast and northwest sectors of the urban zone, where concentrations are highest, may still be impacted by exposure to SO_2. Deposition of ionic forms of N and S in throughfall are equally high in forested areas southwest of the city. The effects of these chronic nutrient inputs to the forest are only beginning to be investigated and understood.

The most significant effect of elevated N deposition SW of Mexico City seems to be the increased levels of nitrate in runoff from forested watersheds downwind of Mexico City. This phenomenon has possible implications for the quality of water extracted from these forested watersheds that are expected to provide a relatively pristine water source for the adjacent urban populations. However, notwithstanding the increased streamwater nitrate concentrations, even the highest value reported to date from springs in DL (132 μeq/L) are still well below the Mexican and U.S. drinking water standard (714 μeq/L or 10 ppm as nitrate nitrogen). Further studies are needed to evaluate the extent of Pb export in drainage waters from forested watersheds in high-pollution sites to the SSW of Mexico City.

The ecological perturbations caused by severe exposures to air pollution in forests located downwind of Mexico City are expected to continue for the near future (the next 5–10 years), largely as a result of stubbornly high levels of O_3 and emissions of N oxides. The scenario over the longer term is more uncertain and depends largely on the effectiveness of regulatory emissions control strategies. Based on recent trends, it is hoped that pollutant levels will continue to fall. Forest responses to these downward trends will depend on how long it takes to reduce levels sufficiently to allow sensitive species to recover. Of course, some of the ecological damage, such as the loss of lichen diversity and of sensitive species, is probably irreversible (Zambrano et al. 2002). Preservation, and in some cases, prevention of further deterioration of the forests and green areas within the Basin of Mexico will certainly depend on the cooperation of government officials, regulatory agencies, and the general public. It is hoped that the increasing interest in ecological preservation and greater awareness of the importance of the ecological services provided by forests will make this goal achievable. Air pollution is but one of the anthropogenic factors threatening the preservation of the remaining forest ecosystems in the MCAB.

References

Alarcón, M.A., de Bauer, Ma. L.I., Jasso, J., Segura, G., and Zepeda, E.M. 1993. Patrón de crecimiento radial en árboles de *Pinus hartwegii* afectados por contaminación atmosférica en el suroeste del Valle de México. *Agrocienc. Serie Recursos Naturales Renovables* 3:67–80.

Alvarado-Rosales, D., and Hernández-Tejeda, T. 2002. Decline of sacred fir in the Desierto de los Leones National Park. In *Urban Air Pollution and Forests: Resources at Risk in the Mexico City Air Basin*, Ecol. Stud., eds. M.E. Fenn, L.I. de Bauer, and T. Hernández-Tejeda. New York: Springer-Verlag.

Alvarado-Rosales, D., de Bauer, L.I. de, and Galindo-Alonso, J. 1991. Declinación y muerte del bosque de oyamel (*Abies religiosa*) en el sur del Valle de México. *Agrocienc. Serie Recursos Naturales Renovables* 1:123–143.

Alvarado-Rosales, D., de Bauer, L.I., and Galindo, J.A. 1993. Decline of sacred fir (*Abies religiosa*) in a forest park south of Mexico City. *Environ. Pollut.* 80:115–121.

Alvarez, D., Laguna, G., and Rosas, I. 1998. Macroscopic and microscopic symptoms in *Abies religiosa* exposed to ozone in a forest near Mexico City. *Environ. Pollut.* 103:251–259.

Andersen, C.P., Wilson, R., Plocher, M., and Hogsett, W.E. 1997. Carry-over effects of ozone on root growth and carbohydrate concentrations of ponderosa pine seedlings. *Tree Physiol.* 17:805–811.

Bääth, E. 1989. Effects of heavy metals in soil on microbial processes and populations (A Review). *Water Air Soil Pollut.* 47:335–379.

Baez, A.P., Belmont, R., and Padilla, H. 1995. Measurements of formaldehyde and acetaldehyde in the atmosphere of Mexico City. *Environ. Pollut.* 89:163–167.

Bargagli, R. 1998. *Trace Elements in Terrestrial Plants: An Ecophysiological Approach to Biomonitoring and Biorecovery*. Heidelberg, Germany: Springer-Verlag and R.G. Landes Company.

Bernal-Salazar, S., Terrazas, T., and Alvarado, D. 2002. Wood anatomical characters and annual ring width of *Abies religiosa* (H.B.K.) Schl. et Cham. trees comparing two forests with contrasting health characteristics. *Environ. Pollut.* (in press).

Brady, N.C., and Weil, R.R. 1999. *The Nature and Properties of Soils*, 12th edn. Upper Saddle River, New Jersey: Prentice Hall.

Bravo-Alvarez, H., and Torres-Jardón, R. 2002. Air pollution levels and trends in the Mexico City metropolitan area. In *Urban Air Pollution and Forests: Resources at Risk in the Mexico City Air Basin*, Ecol. Stud., eds. M.E. Fenn, L.I. de Bauer, and T. Hernández-Tejeda. New York: Springer-Verlag.

Bussotti, F., and Ferretti, M. 1998. Air pollution, forest condition and forest decline in Southern Europe: an overview. *Environ. Pollut.* 101:49–65.

Castro-Servín, J.M., González-Kladiano, V., and Hernández-Tejeda, T. 1997. Contenido de metales pesados en algunos suelos y vegetación del Desierto de los Leones, Distrito Federal. *Cienc. For. México* 22:109–127.

Cibrian-Tovar, D. 1989. Air pollution and forest decline near Mexico City. *Environ. Monitor. Assess.* 12:49–58.

Ciesla, W.M., and Macias-Samano, J.E. 1987. Desierto de los Leones: A forest in crisis. *Am. For.* 93:29–31, 72–74.

Cox, R.M., and Malcolm, J.W. 1999. Passive ozone monitoring for forest health assessment. *Water Air Soil Pollut.* 116:339–344.

de Bauer, L.I. 1972. Uso de plantas indicadoras de aeropolutos en la Ciudad de México. *Agrociencia* 9 (D):139–141.

de Bauer, L.I., and Hernández-Tejeda, T. 1986. *Contaminación: Una Amenaza para la Vegetación en México*. Chapingo, México: Colegio de Postgraduados.

Delmelle, P., Stix, J., Bourque, C.P.A., Baxter, P.J., García-Alvarez, J., and Barquero, J. 2001. Dry deposition and heavy acid loading in the vicinity of Masaya Volcano, a major sulfur and chlorine source in Nicaragua. *Environ. Sci. Technol.* 35:1289–1293.

Enríquez, F.J. 2000. Challenges and perspectives of urban transportation for passengers in Mexico City. In *Urban Transportation and Environment, Proceedings of the International Conference*, eds. O.D. González-Palomas and C. Jamet, pp. 37–40. April 11–14, 2000, Mexico City. Rotterdam, Netherlands: A.A. Balkema.

Evans, L.S., Adamski, J.H., and Renfro, J.R. 1996. Relationships between cellular injury, visible injury of leaves, and ozone exposure levels for several dicotyledonous plant species at Great Smoky Mountains National Park. *Environ. Exp. Bot.* 36:229–237.

Fenn, M. 1991. Increased site fertility and litter decomposition rate in high-pollution sites in the San Bernardino Mountains. *For. Sci.* 37:1163–1181.

Fenn, M.E., and Bytnerowicz, A. 1993. Dry deposition of nitrogen and sulfur to ponderosa and Jeffrey pine in the San Bernardino National Forest in southern California. *Environ. Pollut.* 81:277–285.

Fenn, M.E., and Dunn, P.H. 1989. Litter decomposition across an air-pollution gradient in the San Bernardino Mountains. *Soil Sci. Soci. Am. J.* 53:1560–1567.

Fenn, M.E., and Poth, M.A. 1999. Temporal and spatial trends in streamwater nitrate concentrations in the San Bernardino Mountains, southern California. *J. Environ. Qual.* 28:822–836.

Fenn, M.E., Poth, M.A., and Johnson, D.W. 1996. Evidence for nitrogen saturation in the San Bernardino Mountains in southern California. *For. Ecol. Manage.* 82:211–230.

Fenn, M.E., Poth, M.A., Aber, J.D., Baron, J.S., Bormann, B.T., Johnson, D.W., Lemly, A.D., McNulty, S.G., Ryan, D.F., and Stottlemyer, R. 1998. Nitrogen excess in North American ecosystems: predisposing factors, ecosystem responses, and management strategies. *Ecol. Applic.* 8:706–733.

Fenn, M.E., de Bauer, L.I., Quevedo-Nolasco, A., and Rodriguez-Frausto, C. 1999. Nitrogen and sulfur deposition and forest nutrient status in the Valley of Mexico. *Water Air Soil Pollut.* 113:155–174.

Fenn, M.E., Poth, M.A., Schilling, S.L., and Grainger, D.B. 2000. Throughfall and fog deposition of nitrogen and sulfur at an N-limited and N-saturated site in the San Bernardino Mountains, southern California. *Can. J. For. Res.* 30:1476–1488.

Fenn, M.E., Castro-Servín, J.M., Hernández-Tejeda, T., Krage, N., Goodson, C., and Meixner, T. 2002a. Heavy metals in forest soils and vegetation in the Basin of Mexico. In *Urban Air Pollution and Forests: Resources at Risk in the Mexico City Air Basin*, Ecol. Stud., eds. M.E. Fenn, L.I. de Bauer, and T. Hernández-Tejeda, (this volume). New York: Springer-Verlag.

Fenn, M.E., de Bauer, L.I., Zeller, K., Quevedo-Nolasco, A., Perea, V., and Hernández-Tejeda, T. 2002b. Nitrogen and sulfur deposition in the Mexico City Air Basin: Impacts on forest nutrient status and nitrate levels in drainage waters. In *Urban Air Pollution and Forests: Resources at Risk in the Mexico City Air Basin*, Ecol. Stud., eds. M.E. Fenn, L.I. Bauer, and T. Hernández-Tejeda. New York: Springer-Verlag.

Gaffney, J.S., Marley, N.A., Cunningham, M.M., and Doskey, P.V. 1999. Measurements of peroxyacyl nitrates (PANS) in Mexico City: implications for megacity air quality impacts on regional scales. *Atmos. Environ.* 33:5003–5012.

Gasche, R., and Papen, H. 1999. A 3-year continuous record of nitrogen trace gas fluxes from untreated and limed soil of a N-saturated spruce and beech forest ecosystem in Germany. 2. NO and NO_2 fluxes. *J. Geophys. Res.* 104:18,505–18,520.

Grulke, N.E., Andersen, C.P., Fenn, M.E., and Miller, P.R. 1998. Ozone exposure and nitrogen deposition lowers root biomass of ponderosa pine in the San Bernardino Mountains, California. *Environ. Pollut.* 103:63–73.

Hernández-Tejeda, T., Krupa, S.V., Pratt, G.C., and de Bauer, L.I. 1981. Sensibilidad de plántulas de eucalipto (*Eucalyptus globulus* Labill.) al ozono. *Agrociencia* 43:89–95.

Hernández-Tejeda, T. 1984. *Efecto de los Gases Oxidantes Sobre Algunas Especies del Género Pinus Nativas del Valle de México*. Masters Thesis. Colegio de Postgraduados, Institución de Enseñanza e Investigación en Ciencias Agrícolas. Chapingo, México.

Hernández-Tejeda, T., and de Bauer, L.I. 1989. *La Supervivencia Vegetal ante la Contaminación Atmosférica*. Montecillo, Mexico: Colegio de Postgraduados.

Jáuregui, E. 2002. The climate of the Mexico City Air Basin: Its effects on the formation and transport of pollutants. In *Urban Air Pollution and Forests: Resources at Risk in the Mexico City Air Basin*, Ecol. Stud., eds. M.E. Fenn, L.I. de Bauer, and T. Hernández-Tejeda. New York: Springer-Verlag.

Kiefer, J.W., and Fenn, M.E. 1997. Using vector analysis to assess nitrogen status of ponderosa and Jeffrey pine along deposition gradients in forests of southern California. *For. Ecol. Manage.* 94:47–59.

Koutrakis, P., Sioutas, C., Ferguson, S.T., Wolfson, J.M., Mulik, J.D., and Burton, R.M. 1993. Development and evaluation of a glass honeycomb denuder filter pack system to collect atmospheric gases and particles. *Environ. Sci. Technol.* 27:2497–2501.

Krupa, S.V., and de Bauer, L.I. 1976. La ciudad daña los pinos del Ajusco. *Panagfa* 31:5–7.

Krupa, S.V., and Legge, A.H. 2000. Passive sampling of ambient, gaseous air pollutants: an assessment from an ecological perspective. *Environ. Pollut.* 107:31–45.

Krupa, S.V., Pratt, G.C., and Teng, P.S. 1982. Air pollution: An important issue in plant health. *Plant Dis.* 66:429–434.

Laguette-Rey, H.D., de Bauer, L.I., Kohashi-Shibata, J., and Marbán-Mendoza, N. 1986. Impacto de los oxidantes ambientales en el cultivo de frijol, en Montecillos, Estado de México. *Agrociencia* 66:83–95.

Landolt, W., Günthardt-Goerg, M.S., Pfenninger, I., Einig, W., Hampp, R., Maurer, S., and Matyssek, R. 1997. Effect of fertilization on ozone-induced changes in the metabolism of birch (*Betula pendula*) leaves. *New Phytol.* 137:389–397.

Laskowski, R., Niklińska, M., and Maryański, M. 1994. Heavy metals in forest litter: A chemical time bomb. In *Conference Papers 19: Climate and Atmospheric Deposition Studies in Forests*, eds. J. Solon, E. Roo-Zieliska, and A. Bytnerowicz, pp. 211–217. Warsaw, Poland: Institute of Geography and Spatial Organization, Polish Academy of Sciences.

López-López, M.A., Velázquez-Mendoza, J., Velázquez-Martínez, A., González-Romero, V., and Cetina-Alcalá, V.M. 1998. Estado nutrimental de *Abies religiosa* en una área con problemas de contaminación ambiental. *Agrociencia* 32:53–59.

Lovett, G.M., and Lindberg, S.E. 1993. Atmospheric deposition and canopy interactions of nitrogen in forests. *Can. J. For. Res.* 23:1603–1616.

Manning, W.J. 1998. The use of plants as bioindicators of ozone. In *Proceedings of the International Symposium, Air Pollution and Climate Change Effects on Forest Ecosystems*, technical coordinators A. Bytnerowicz, M.J. Arbaugh, and S.L. Schilling, pp. 19–26. Feb. 5–9, 1996, Riverside, California, U.S.A. Gen. Tech. Rep. PSW-GTR-166. Albany, California, USA: USDA Forest Service, Pacific Southwest Research Station. Electronic access to proceedings: *<http://www.rfl.psw.fs.gov/pubs/psw-gtr-164/index.html>*.

Martínez-González, L., and Chacalo-Hilu, A. 1994. *Los Arboles de la Ciudad de México*. Azcapotzalco, México, D.F.: Universidad Autónoma Metropolitana.

Miller, P.R., Taylor, O.C., and Poe, M.P. 1986. Spatial Variation of Summer Ozone Concentrations in the San Bernardino Mountains. In *Proceedings of the Air Pollution Control Association Annual Meeting*, pp. 3:86–39.2. June 22–27, Minneapolis, MN. Pittsburgh, Pennsylvania, U.S.A.: Air Pollution Control Association.

Miller, P.R., de Bauer, M.L., Quevedo-Nolasco, A., and Hernádez-Tejeda, T. 1994. Comparison of ozone exposure characteristics in forested regions near Mexico City and Los Angeles. *Atmos. Environ.* 28:141–148.

Miller, P.R., de Bauer, Ma. L.I., and Hernández-Tejeda, T. 2002. Oxidant exposure and effects on pine forests in the Mexico City and Los Angeles, California air basins. In *Urban Air Pollution and Forests: Resources at Risk in the Mexico City Air Basin*, Ecol. Stud., eds. M.E. Fenn, L.I. de Bauer, and T. Hernández-Tejeda, (this volume). New York: Springer-Verlag.

Mutters, R.G., Madore, M., and Bytnerowicz, A. 1993. Formaldehyde exposure affects growth and metabolism of common bean. Journal of the Air & Waste Management Association 43:113–116.

Ortíz-García, C.F., Laguette-Rey, H.D., and de Bauer, L.I. 2002. Effects of oxidants in ambient air on annual crops in the Basin of Mexico. In *Urban Air Pollution and Forests: Resources at Risk in the Mexico City Air Basin*, Ecol. Stud., eds. M.E. Fenn, L.I. de Bauer, and T. Hernández-Tejeda. New York: Springer-Verlag.

Pick, J.B., and Butler, E.W. 1997. *Mexico Megacity*. Boulder, Colorado: Westview Press.

Runeckles, V.C., and Bowen, P.A. 2000. The use of calibrated passive monitors to assess crop loss due to ozone in rural locations. In *Environmental Pollution and Plant Responses*, eds. S.B. Agrawal and M. Agrawal, pp. 307–316. Boca Raton, Florida, U.S.A.: Lewis Publishers, Inc.

Scagel, C.F., and Andersen, C.P. 1997. Seasonal changes in root and soil respiration of ozone-exposed ponderosa pine (*Pinus ponderosa*) grown in different substrates. *New Phytol.* 136:627–643.

Skeffington, R.A., and Wilson, E.J. 1988. Excess nitrogen deposition: Issues for consideration. *Environ. Pollut.* 54:159–184.

SMA (Secretaría del Medio Ambiente). 1999. *Informe Anual de la Calidad del Aire en el Valle de México*. México, D.F.: Secretaría del Medio Ambiente, Dirección General de Prevención y Control de la Contaminación. See also http://sma.df.gob.mx/publicaciones/aire/inf_cal_aire99/indice.htm.

Smidt, S. 1994. Endangering of forest trees by organic air pollutants. Zeitschrift Fur Pflanzenkrankheiten und Pflanzenschutz. *J. Plant Dis. Protect.* 101:423–445.

Takemoto, B.K., Bytnerowicz, A., and Fenn, M.E. 2001. Current and future effects of ozone and atmospheric nitrogen deposition on California's mixed conifer forests. *For. Ecol. Manage.* 144:159–173.

Temple, P.J., and Miller, P.R. 1998. Seasonal influences on ozone uptake and foliar injury to ponderosa and Jeffrey pines at a southern California site. In *Proceedings of the International Symposium, Air Pollution and Climate Change Effects on Forest Ecosystems*, technical coordinators A. Bytnerowicz, M.J. Arbaugh, and S.L. Schilling, pp. 221–228. Feb. 5–9, 1996, Riverside, California, U.S.A. Gen. Tech. Rep. PSW-GTR-166. Albany, California, U.S.A.: USDA Forest Service, Pacific Southwest Research Station. Electronic access to proceedings: <*http://www.rfl.psw.fs.gov/pubs/psw-gtr-164/index.html*>.

Temple, P.J., and Taylor, O.C. 1983. World-wide ambient measurements of peroxyacetyl nitrate (PAN) and implications for plant injury. *Atmos. Environ.* 17:1583–1587.

Terrazas, T., and Bernal-Salazar, S. 2002. Histological symptoms of air pollution injury in foliage, bark, and xylem of *Abies religiosa*, in the Basin of Mexico. In *Urban Air Pollution and Forests: Resources at Risk in the Mexico City Air Basin*, Ecol. Stud., eds. M.E. Fenn, L.I. de Bauer, and T. Hernández-Tejeda. New York: Springer-Verlag.

Tyler, G., Påhlsson, A.-M.B., Bengtsson, G., Bååth, E., and Tranvik, L. 1989. Heavy-metal ecology of terrestrial plants, microorganisms and invertebrates: A review. *Water Air Soil Pollut.* 47:189–215.

Van Ek, R., and Draaijers, G.P.J. 1994. Estimates of atmospheric deposition and canopy exchange for 3 common tree species in the Netherlands. *Water Air Soil Pollut.* 73:61–82.

Vitousek, P.M., and Howarth, R.W. 1991. Nitrogen limitation on land and in the sea–How can it occur? *Biogeochemistry* 13:87–115.

Watmough, S.A., and Hutchinson, T.C. 1999. Change in the dendrochemistry of sacred fir close to Mexico City over the past 100 years. *Environ. Pollut.* 104:79–88.

Yamada, E., Kimura, M., Tomozawa, K., and Fuse, Y. 1999. Simple analysis of atmospheric NO_2, SO_2, and O_3 in mountains by using passive samplers. *Environ. Sci. Technol.* 33:4141–4145.

Zambrano, A., and Nash III, T.H. 2000. Lichen responses to short-term transplantation in Desierto de los Leones, Mexico City. *Environ. Pollut.* 107:407–412.

Zambrano, A., Nash III, T.H., and Gries, C. 1999. Physiological effects of the Mexico city atmosphere on lichen transplants on oaks. *J. of Environ. Qual.* 28:1548–1555.

Zambrano, A., Nash III, T.H., and Herrera-Campos, M.A. 2002. Lichens and air pollution in the forests surrounding Mexico City. In *Urban Air Pollution and Forests: Resources at Risk in the Mexico City Air Basin*, Ecol. Stud., eds. M.E. Fenn, L.I. de Bauer, and T. Hernández-Tejeda. New York: Springer-Verlag.

16. Resources at Risk and Research Needs

L.I. de Bauer, Tomás Hernández-Tejeda, and Mark E. Fenn

Mexico City, the former Tenochtitlan and modern day megacity, will likely remain as the cultural, financial, communications, and population center of the country for the foreseeable future. Three intricately connected factors will continue to challenge efforts to improve the quality of life and the health of urban trees and of ecosystems within the Basin of Mexico: continued population growth, changing land use (including further urban expansion into the rural periphery), and atmospheric/environmental degradation. Overexploitation of water reservoirs is a notable example of the last. Researchers, resource managers and government agencies can contribute to the preservation of green areas within the urban zone and the sustainability of the surrounding forested zones by (a) continuing to implement more effective air pollution control measures and policies, (b) protecting wildlands and urban green areas by enforcing land-use zoning regulations, (c) developing strategies and programs to educate and inform the public of all ages and social levels about the important environmental issues affecting them, and (d) selecting and planting resistant species or varieties of plants (which may be the ones surviving under the polluted environment in Mexico City).

Mexico City has not functioned as a self-sustaining system for many years now (Ezcurra et al. 1999). The city serves as a large sink because much of the resources to sustain it are imported from outside the Basin. Since ancient times, the Basin has been hydrologically closed and functioned as a lacustrine system on which the Aztecs and their predecessors built a magnificent civilization. The degree of ecological alteration is probably best exemplified by the conversion from a lacustrine environment to one in which groundwater resources have been greatly over exploited causing severe land sinking. Water to sustain greater Mexico City must now be imported from outside the Basin, and sewage waste is exported in great quantities (Ezcurra and Mazari-Hiriart 1996). Land use change has also resulted in reduced aquifer recharge and greater soil erosion and runoff due to loss of forest cover and increased urbanization. While the focus of this book is on air pollution impacts on forests within the Mexico City Air Basin (MCAB), the effects of air pollution on forests and other natural resources must be considered within the context of other natural and anthropogenic stressors. Economic, sociological, and demographic conditions and trends are the ultimate drivers of ongoing human uses and alterations of the natural resources within the Basin of Mexico.

Detrimental effects of air pollution on forests constitute an additional impact of overpopulation on the natural resources of the Basin. Air pollution threatens the vigor and viability of desirable native species such as *Abies religiosa* and *Pinus hartwegii*. Recent studies also demonstrate that water quality in watersheds downwind of Mexico City is compromised as nitrogen oxides (NOx) from fossil fuel combustion eventually deposit from the atmosphere onto the forest, only to subsequently be flushed from the forest as nitrate in surface and subsurface runoff (Fenn et al. 1999). Lead (Pb), accumulated in soils and vegetation from years of atmospheric Pb deposition, may also be leaching from some watersheds, thus further threatening drinking water quality (Fenn et al. 2002a).

The effects of air pollution on forest health and water quality raise the question as to what can be done to mitigate these undesirable ecological and environmental effects. Two major approaches come to mind as possible strategies. Theoretically, the preferred strategy and the ultimate long-term solution for the protection of human health and forest sustainability is to greatly reduce atmospheric emissions of pollutants. Limited success in the MCAB has already been achieved in terms of greatly reduced emissions of Pb and to a lesser degree, of sulfur dioxide. However, concentrations of NOx, ozone (O_3), organic pollutants (including peroxyacetyl nitrate [PAN]), and fine particulates remain high in the MCAB (Bravo and Torres-Jardón 2002).

An alternative and complementary approach for managing air pollution impacts on forests and natural resources in the MCAB is to better understand what the ecological impacts are, to identify sensitive organisms and ecological processes, to delineate the geographic extent of the problem, and to develop silvicultural and management strategies to ameliorate air pollution effects. This strategy is also costly to implement, especially in the MCAB, where elevated concentrations of multiple pollutants occur, thus necessitating evaluation of combined and possibly interacting effects of various pollutants. A multifaceted approach is required including laboratory and controlled exposure studies, air pollution gradient studies, manipulative studies, long-term monitoring of forest condition, and air pollution monitoring. Unfortunately, these approaches have only been applied in a limited fashion in the MCAB because of financial constraints. To build upon the baseline data on air pollution effects in the forested regions of the MCAB, it would be of great benefit to future research efforts if key MCAB study sites were affiliated with coordinated research efforts such as the International Long-Term Ecological Research Network (Gosz et al. 2000; http://www.ilternet.edu). This would help ensure that long-term studies receive at least baseline financial support and provide a mechanism for maintaining a viable research program, while attracting more research partners to address a variety of important ecological effects research questions.

The current state of knowledge of air pollution effects on forest resources in the MCAB is summarized in the first 15 chapters of this work. In this final chapter, we attempt to highlight the most important gaps in our understanding of air pollution impacts on the forests surrounding the MCAB. It is our hope that this book, and in particular, this summary of research needs, will foster future research addressing important unanswered questions on the effects of atmospheric pollutants on forests within the MCAB.

Resources at Risk

The primary natural resources at risk due to air pollution in the MCAB include forest health and sustainability, biodiversity and preservation of species, agricultural crop production, water quantity and quality, health and vigor of urban forests and vegetation, and aesthetic values. The primary concern is not that severe air pollution exposures will cause forests to disappear completely, although mortality of entire sacred fir (*Abies religiosa*) stands has occurred in certain localities within the Desierto de los Leones (DL) National Park. In most cases, alternative and more pollution-tolerant vegetation will eventually replace the dying trees. However, the highly-valued tree species displaced and the understory species, lichens, invertebrates, fauna, and microorganisms associated with the native forest species displaced or lost may represent irreplaceable biotic components of these forests. Many of these associated species are probably as of yet unidentified. In addition, historical, cultural, sentimental, and aesthetic values associated with native tree species and other organisms in forests and parks provide incentive and additional rationale for seeking to preserve and protect these resources. The name of the conifer species most severely impacted in the DL (sacred fir) would seem to emphasize the high value placed on the native species which have grown in these forests for millennia. In the case of severe dieback, such as has occurred with sacred fir, it will take decades to replace the forest, and even that will occur in a sustainable fashion only if the replacement trees, of the same or alternate species, are successfully established over the long term. This is not a given if the replacement trees aren't well adapted to the climate, pathogens, insect pests, air pollution, and other stressors prevalent in these montane areas.

Loss of biodiversity in forests downwind of Mexico City is demonstrated by the lichen studies described in Chapter 12 of this volume (Zambrano et al. 2002).

Undoubtedly, other types of organisms have experienced similar losses and alterations in community composition as a result of many years of anthropogenic disturbances and exposure to high levels of a virtual potpourri of air pollutants. In addition, the composition of the air pollution exposures impacting these forests has changed and continues to change over time; and as a result, air pollution effects on the biota have also changed. Unfortunately, historical and current data on the concentrations and deposition rates of the major pollutants in forested areas in the MCAB is scanty and intermittent at best. Most of the data on atmospheric inputs and exposures in montane sites are from the DL site. Additional data is available from the Zoquiapan Experimental Forest on the eastern side of the Air Basin (Fenn et al. 1999) and from a few other study sites (Fenn et al. 2002c). Much of this data is based on field evaluations of ozone injury, bioindicator plants, or heavy-metal accumulation in vegetation and lichens (Fenn et al. 2002c; Miller et al. 1994; Zambrano et al. 2002).

Water is a vital and precious natural resource for the large population living within the MCAB. Recent studies in the MCAB demonstrate that water quality is directly impacted by the high levels of N deposition occurring in forests downwind of Mexico City—as evidenced by elevated nitrate levels in springwater and streamwater (Fenn et al. 1999, 2002b). In addition, preliminary studies indicate that export of Pb from montane watersheds SSW of Mexico City is also significantly higher than in unpolluted watersheds (Fenn et al. 2002a; Morales 1998). These findings highlight the importance of further research on the impacts of urban air quality on water quality from forested watersheds in the MCAB.

Major Research Needs

The research needs are many. For each of the major pollutants, elementary research questions concerning forest effects remain unanswered. On the one hand, this is not surprising considering the complexity of the problem, because even in Europe, the United States, and Canada, where air pollution studies have a long history, there are still many gaps in knowledge. However, for many pollutants and research areas in the MCAB, little investigation has been done on pollutant exposures or effects on forests or vegetation. The following research areas, among others, fit into this category: sulfur dioxide (SO_2) exposure responses of forest and crop species, physiological effects of air pollutants on native plant species, controlled pollution exposure studies, interactive effects of multiple pollutants on plants, pollutant effects on microbial processes including important symbionts such as mycorrhizae, effects of pollutants on other biotic and abiotic stresses, and pollutant transport and dynamics of pollutant exposures in montane areas. In the following pages, we will discuss a number of these priority research areas that need to be addressed if we are to advance our understanding of air pollution effects on forests within the zone of influence of greater Mexico City.

Air Pollution Monitoring

Monitoring of air pollution exposures and quantification of atmospheric deposition rates to forests is greatly needed. Except for an O_3 monitor at the DL National Park and limited studies on N and S deposition, there is little available data on air pollution levels in forested areas. There is no data on SO_2 concentrations outside of the urban area or of heavy-metal deposition rates in forests within the MCAB.

In 1990, the first O_3 and meteorological monitoring station in a forested area in the MCAB was established inside of a cell (room) of the "convent" in the DL National Park located immediately southwest of Mexico City. Ozone concentrations have been measured since then, and Miller and coworkers published a paper comparing the O_3 concentrations and meteorological conditions of this park with the San Bernardino National Forest in the Los Angeles California Air Basin (Miller et al. 1994). However the DL monitoring station remains the only forested location where routine monitoring has been done. Limited measurements of atmospheric concentrations of particulate and gaseous forms of N and S pollutants were also done intermittently at the DL site from 1992 to 1995 (Fenn et al. 2002b). Due to the inadequate performance of the first automatic atmospheric monitoring network installed in Mexico City in the 1970s, a second automatic network of air pollution monitors was established and began operation in 1986 (RAMA; Red Automática de Monitoreo Atmosférico). The following pollutants are currently being monitored: carbon monoxide (CO), Pb, NOx, O_3, SO_2, non-methane hydrocarbons, total suspended particles and suspended particles less than 10 μm in diameter (PM_{10}). Although data from the urban network indicates which pollutants are likely to be of importance in nearby forests, air pollution monitoring is needed in the forests surrounding Mexico City in order to determine exposure–ecological-response relationships and to better characterize the spatial extent and trends of air pollution exposures and impacts.

We have a general understanding of the gross-scale spatial patterns of where pollutant concentrations are highest, based on air pollution data from the urban monitoring network, dominant wind directions in the Basin, geographic patterns of O_3-induced injury symptoms, limited data with passive O_3 monitors, and data on throughfall inputs of N and S (at only two sites). Exposure and deposition data from forested regions of the MCAB are needed for the major pollutants that have the potential to have significant ecological impacts and for those that affect the chemical environment of soil and water. Primary among these would be O_3, N, and S pollutants in various forms, PAN, heavy metals, and organic pollutants.

Ultimately, it is hoped that the spatial extent of air pollution impacts in the MCAB can be better defined with Geographic Information System (GIS) analysis of the landscape patterns of air pollution exposure in relation to vegetation cover type. Passive monitors are the only cost-effective method of monitoring over a sufficiently dense network to obtain the data needed to characterize pollution exposures across forested areas in the MCAB. Such monitoring is also needed to determine whether pollution from the MCAB reaches important agricultural areas within the Basin and in tropical zones adjacent to the MCAB. This approach has recently been used to map O_3 exposure over a 73,000 km^2 area in the Sierra Nevada Mountains of California, by employing passive samplers at 90 sites (A. Bytnerowicz and M.A. Arbaugh, personal communication). Passive O_3 monitors have been used periodically in a few forested sites in the MCAB since 1991, but mainly for short periods. From July 1997 to December 1998, three different forested areas inside the MCAB were monitored with CanOxy plate passive O_3 monitors (Cox and Malcolm 1999) provided by the Canadian Forest Service. Further studies with passive monitors for O_3 and a variety of N and S pollutants are planned.

Bioindicator plants or lichens can also be useful in monitoring programs in forested areas. In fact, much of the original work establishing that air pollution is causing injury to forest species in the MCAB is based on observations of air pollution injury in native and well-known bioindicator plant species (de Bauer and Hernández 1986). Both sentinels and detector species can be effectively used. Sentinels are nonindigenous plant species that exhibit diagnostic reliable foliar symptoms when exposed to ambient O_3. Detector plants are indigenous to an area and may exhibit typical foliar injury symptoms to O_3 exposure *in situ*. They are useful in assessing the long-term or cumulative effects of O_3 (Manning 1998). A number of species native to the MCAB, such as Mexican black cherry *Prunus serotina* spp. *capuli* (Skelly et al. 1997), and *Sambucus mexicana* (see Chapter 9, Miller et al. 2002) show high potential as indicators of O_3 injury and could be developed for monitoring purposes. The presence of both genera in other regions of North America and their use in biomonitoring programs for O_3 injury in forests elsewhere (Temple 1999) indicate the value of these and other native species in biomonitoring.

Lichens have been widely used in many countries as bioindicators of air pollution. As described in Chapter 12 herein (Zambrano et al. 2002), patterns of lichen distribution, abundance, and diversity have been greatly impacted in high-pollution forested areas. Building upon the results of these earlier studies, lichen species could be selected and tested as bioindicators or bioaccumulators of particular pollutants for use in monitoring programs in forests of the MCAB. Foliage of native tree species can also serve as indicators of S accumulation in areas with elevated SO_2 exposure (Dmuchowski and Bytnerowicz 1995).

Dendrochronological and dendrochemical studies (Watmough and Hutchinson 1999) and histological studies of foliage and xylem have been used with promising results in the MCAB (Terrazas and Bernál-Salazar 2002). Dendrochronological studies (Alarcón et al. 1993) can be useful in evaluating historical growth trends in various pollution-tolerant and sensitive species at sites with varying pollution exposure. Such an approach can be used to determine long-term air pollution impacts on tree growth and forest productivity. Histological studies provides an additional tool for evaluating plant health and responses to air pollution, even in the absence of visual injury symptoms. Further application of these research approaches can help advance our understanding of air pollution impacts on forests in the MCAB.

Controlled Exposure Fumigation Experiments

Considering the extreme O_3 concentrations in the MCAB and the many reports of visual injury on native and agricultural species, further studies on the effects of oxidant gases on vegetation is a high-priority research area. Thus it is quite surprising, that no definitive controlled exposure experiments with O_3 or other pollutants have been done with the major trees species in the MCAB. This is due to a lack of fumigation facilities in Mexico and insufficient financial support. However, limited studies have been done with pollution exclusion branch chambers using charcoal filtration and with a protectant "clear-spray" applied to exposed branches (Alvarado-Rosales 1988; Alvarado et al. 1993; de Bauer and Hernández-Tejeda 1986; Hernández-Tejeda 1984). It is hoped that O_3 fumigation experiments with sacred fir, with a number of pine species, and with selected understory species will be realized in the near future. This is particularly important for determining the threshold concentrations at which visible injury and physiological effects begin to develop. Research is also needed on the growth effects of O_3 on tree species in the MCAB. Another critical need is to verify whether or not the foliar symptoms observed in declining sacred fir trees are due to the direct phytotoxicity of O_3. This is a major gap in the evaluation of the role of O_3 in the severe decline that has occurred in sacred fir stands in the high-pollution areas of the DL National Park located directly downwind of Mexico City. Studies should also address plant growth effects of O_3, as this has not been well studied.

Also needed are evaluations or rankings of plant species in terms of their sensitivity to O_3. There are 14 pine tree species, one fir species, one white cedar species, around nine oak species, and many native shrubs, herbs, and grasses growing in the MCAB. A ranking of the most important forest species with respect to their sensitivity to photochemical oxidants is needed in order to select the best species for reforesting the wildlands and urban areas of the MCAB. A ranking of pine species in order of sensitivity to O_3, based mainly on field evaluations (Hernández-Tejeda and Nieto de Pascual-Pola 1996) and observations, is given in Chapter 9 (Miller et al. 2002).

Data is also completely lacking for plant responses to SO_2 and PAN in forested areas. Before initiating fumigation experiments with SO_2 or PAN, however, it would be prudent to first evaluate what the current exposure levels for these pollutants are in forested areas of the MCAB. In the late 1970s and early 1980s, typical symptoms attributed to PAN were reported for several herbaceous species (de Bauer and Hernández-Tejeda 1986), particularly in the Xochimilco area of the MCAB. Gaffney et al. (1999) reported that levels of PAN in north-central Mexico City are the highest reported in the world. Levels in forested areas downwind of Mexico City may be even higher (Bravo and Torres-Jardón 2002). Biomonitoring with sensitive indicator species could be implemented at selected montane sites to evaluate the current potential for phytotoxic effects from PAN exposure. If symptoms are observed, it would be important to follow up with measurements with atmospheric concentrations of PAN in forested areas.

Similarly, measurements of atmospheric levels of SO_2 in forested areas with a combination of active monitors and passive monitors are needed. Controlled fumigations with SO_2 or PAN should be based on data or exposure profiles obtained from forested areas.

Nitrogen and Sulfur Deposition

The syndrome of ecosystem responses to chronic inputs of atmospheric N deposition to forests is known as "nitrogen (N) saturation". Briefly, N saturation can be defined as the long-term removal of N limitations on biotic activity, accompanied by a decrease in watershed N retention capacity (Fenn et al. 1998). The cardinal sign of N saturation at the watershed level is the export of abnormally high nitrate levels in surface or subsurface runoff (Fenn et al. 1998). Considering the number of NOx-emitting vehicles in the Mexico megacity and the atmospheric, climatic, and topographic factors leading to high pollution concentrations, it is not surprising that symptoms of N saturation occur in forests downwind of Mexico City (Fenn et al. 1999). Nitrogen and S deposition were equally high in a study at DL, with an annual deposition of ca. 20 kg/ha for N and S (Fenn et al. 1999).

Studies of N effects in the MCAB are yet in their infancy, with a number of basic questions yet to be addressed. Effects of chronic N deposition on tree growth and other vegetation responses in these forests dominated by volcanic soils aren't well understood. Recent fertilization studies with *Pinus hartwegii* at sites with varying N deposition levels suggest that these trees may not be N limited, even in low-deposition areas, possibly as a result of phosphorus limitations (Fenn et al. 2002b). Further research is needed to confirm this. If greater N availability does not increase forest productivity in the MCAB, this would certainly be a departure from studies in northern temperate forests (Fenn et al. 1998).

Considering the many years of high atmospheric inputs of N and S at DL, it is not implausible to expect that elevated leaching of plant nutrients from soil has occurred in the more exposed sites. In fact, some studies suggest that depletion of Ca and Mg pools in soil, soil acidification, and increased Al availability, possibly as a result of atmospheric deposition, are occurring in some areas of the DL park (Fenn et al. 2002a). Based on these reports, further studies of the effects of chronic atmospheric deposition on nutrient pools and plant/soil nutrient relations in pine and fir forests in the MCAB are in order. If this phenomenon is confirmed, it raises questions in regards to the effects of accelerated nutrient leaching on plant health and vigor and the long-term sustainability of forest stands growing under these conditions.

Chronic N deposition can also affect mycorrhizae and affect the diversity and composition of mycorrhizal symbionts (Cairney and Meharg 1999). This has not been studied in forests in the MCAB. Nitrogen deposition and O_3 stress can impact fine-root production (Grulke et al. 1998), but neither have been investigated in the MCAB. The effects of N deposition on the composition of plant, microbial, and lichen communities and on soil fluxes of important greenhouse gases such as nitrous oxide (N_2O), nitric oxide (NO), methane (CH_4) and carbon dioxide (CO_2) are additional neglected areas of study. In summary, chronic inputs of N can alter many key ecological processes and degrade water quality from N-saturated watersheds. We are only beginning to understand the responses of forests in the MCAB to high inputs of both N and S.

Air Pollution Effects on Water Quality

The quality of streamwater or groundwater coming from N-saturated watersheds is compromised because excess soil N is transported from the watershed in runoff as nitrate (Fenn et al. 1998). Although, nitrate levels in the springs and streams sampled to date in the MCAB are not above the Mexican and US drinking water standard (10 ppm N as nitrate), nitrate levels in streams from watersheds with high pollution exposure in the MCAB are orders of magnitude greater than in forested watersheds upwind of Mexico City (Fenn et al. 2002b). Nitrogen cycling in forests is normally highly conservative, and runoff from forested watersheds is expected to be relatively pristine. However, in watersheds receiving chronic N inputs from air pollution, nitrate levels are sometimes within the range expected from the runoff of fertilized agricultural fields (Fenn and Poth 1999).

Future research should determine the geographic extent of forested areas receiving sufficient N deposition to result in elevated N export in streamwater and should also determine what those threshold N deposition levels are. One issue which merits further investigation, considering that the most important aquifers in the MCAB are also located to the SSW of the Basin, where N deposition is highest, is the extent to which the N exported from forested watersheds impacts drinking water quality in the MCAB.

Preliminary data suggests that Pb contamination of drainage waters may also be an important environmental problem in some watersheds to the SSW of Mexico City (Fenn et al. 2002a; Morales 1998). More in-depth studies are needed to evaluate the severity of Pb export and its geographic extent. The health risks from Pb in runoff from watersheds exposed to high atmospheric pollutant loads must be ascertained. Morales (1998) reported that Pb concentrations in water from springs in the Ajusco area (0.168–0.259 ppm) were an order of magnitude higher than the Mexican drinking-water standard (0.025 pm). However, in a subsequent preliminary survey of montane springs and streams (Fenn et al. 2002a), Pb concentrations were found to be much lower. The highest concentration was 0.013 ppm, which is approximately 50% as high as the drinking-water standard. Lead emissions have decreased dramatically in recent years as a result of the use of unleaded fuels (Bravo and Torres-Jardón 2002). Thus, data is needed on current levels of atmospheric deposition of Pb (and of other heavy metals), on the extent of Pb accumulation in soil, litter, and vegetation, and most importantly, on the rate at which Pb is being released into the hydrologic system. Information is also lacking on the spatial and temporal trends of Pb concentrations in drainage waters within forested areas and on how long Pb is expected to continue leaching from these watersheds.

Sacred Fir Decline in the Desierto de los Leones National Park

Air pollution is believed to play a major role in the dramatic decline of sacred fir at DL (Alvarado-Rosales and Hernández-Tejeda 2002; López-López et al. 1998), but more work is needed to clarify the etiology of the decline and to develop management alternatives for areas impacted by dieback of sacred fir. Excessive stand density and, possibly, overmaturation of trees, seem to be important factors in the decline, suggesting the need for thinning trials in affected areas if park management policy can be adapted to allow for active management (Alvarado-Rosales and Hernández-Tejeda 2002). The decline is a complex problem, and research into its causal factors must include comprehensive studies of the possible roles of a variety of abiotic and biotic factors. A major gap in the investigation of the basic cause of the decline is the lack of controlled exposure fumigation studies to determine if the observed foliar symptoms are in fact caused by O_3, as is generally assumed. Information on the O_3 dose response of forest species in the MCAB is extremely scarce. The roles of other gaseous pollutants such as SO_2 and PAN and of elevated N and S deposition in the decline of sacred fir have also not been investigated sufficiently. In the case of SO_2, atmospheric levels have decreased considerably in recent years. Thus, even if SO_2 was a factor in sacred fir decline in previous years, and there is no available evidence to support this, current levels may not be sufficient to cause significant damage to plants. Similar reasoning also applies in the case of temporal trends in atmospheric Pb inputs to forests downwind of Mexico City.

In summary, basic questions still remain concerning the role of various stressors in the development of sacred fir decline. Complex interacting factors may be involved, making it difficult to unravel the causal mechanisms leading to decline and mortality. However, it seems clear that an important starting point is to carry out controlled pollutant exposure studies with O_3, and possibly PAN, to determine if one or both of these pollutants are the cause of the characteristic foliar symptoms observed in declining trees (Alvarado-Rosales and Hernández-Tejeda 2002).

Multiple Pollutant Exposure

Air pollution exposures and their effects on forests in the MCAB are sometimes compared to those of forests downwind of Los Angeles, California, because both of these large metropolitan centers experience severe smog episodes dominated by photochemical oxidants including O_3 (Miller et al. 1994). Forests in both areas are also exposed to elevated N deposition (Fenn et al. 1999, 2000). However a major difference between these two regions is that in the MCAB, other pollutants with potential for biological impacts, such as SO_2, heavy metals, and PAN, may also occur at high enough levels to affect important ecological processes (http://sma.df.gob.mx/publicaciones/aire/menu_aire.htm). Since emissions of SO_2 and Pb have decreased significantly in recent years, the biological impacts of these pollutants is presumably decreasing. However, little is known of how these pollutants have impacted forests in the MCAB when deposition rates were much higher, or of what their importance is now and in the future if downward trends continue.

A thorough investigation of air pollution impacts on forest ecosystems requires studies of the combined effects of all the biologically relevant pollutants. Interactions between pollutants and other biotic and abiotic stresses, such as nutrient imbalance, drought, climate change, increased CO_2, insects, and diseases, must also be considered. For practical reasons, most studies of the effects of air pollution on plants focus on a single pollutant, although the reality is that forests are exposed to a complex of pollutants. For example, in areas of high urban smog (e.g., Mexico City and Los Angeles, California), elevated O_3 concentrations are accompanied by high N deposition as well. This is not surprising considering that NOx are precursors of the secondary pollutants known as photochemical oxidants, one of which, O_3, is a major pollutant. However, only recently have the effects of N deposition been investigated in the MCAB (Fenn et al. 1999, 2002b). We suggest that studies of the effects of O_3 be considered not in isolation, but in conjunction with the effects of N pollutants. Nitrogen and O_3 have opposite overall effects on plant growth, inasmuch as O_3 is a phytotoxicant and N is a well-known growth enhancer, although both O_3 and high N fertility commonly reduce allocation of carbon to roots (Grulke et al. 1998). Nitrogen deposition and soil fertility can alter plant responses to ozone, although knowledge of N x ozone interactions in plants, and especially in forest species, is limited (Davison and Barnes 1998).

In the San Bernardino Mountains in the Los Angeles Air Basin, it has been reported that the combined effects of O_3 and N deposition result in unusual effects on carbon and nutrient cycling and in major alterations in physiological functioning of ponderosa pine, an O_3 sensitive species (Takemoto et al. 2001). Ponderosa pine trees exposed to high O_3 and elevated N deposition have greatly reduced fine-root biomass and reduced carbon allocation below ground (Grulke et al. 1998). Foliar leaf litter is increased, and litter accumulates as a thick litter layer on the forest floor. The phenological behavior of mature ponderosa pine trees is also dramatically altered in that 95% of the foliar biomass is current-year foliage. In essence, the impacted trees function as "deciduous conifers" (Grulke and Balduman 1999). These impacts are dramatic alterations in ecological function of pine trees in response to chronic exposure to O_3 and N pollutants. *P. hartwegii* in the MCAB responds to O_3 in a manner similar to that of *P. ponderosa*, but analogous studies of the combined effects of O_3 and N have not been done on *P. hartwegii* in the MCAB.

Air Pollution Effects on Below-Ground Processes

Air pollution impacts on below-ground processes are an aspect often neglected in air pollution research, mainly because of methodological difficulties and because visible air pollution symptoms are observed above ground. However, below-ground processes are key to forest health and sustainability, and the physiological effects of air pollution on plants include alterations in nutrient transport and carbon allocation between roots and shoots, and alterations in fine-root production and turnover. Nitrogen and O_3 can also affect mycorrhizal root development and susceptibility to insect pests and pathogens (Cairney and Meharg 1999; Skeffington and Wilson 1988). These effects are not readily discerned as air pollution injury in the same diagnostic sense as the characteristic chlorotic mottle on pine foliage caused by O_3. Nonetheless, the impacts on tree vigor and forest sustainability can be highly significant, especially when forest sustainability is viewed within the context of the multiple stresses to which forest trees are subjected during their lifespan. The effects of air pollution on root health, mycorrhizae, and tree diseases in the MCAB is an important but neglected area of study.

Long-Term Monitoring of Forest Condition

Air pollution impacts on forests or other ecosystems should not be studied in isolation from the natural environment or in isolation from multiple stresses acting upon plants. Air pollution must be considered as an additional stress factor. Therefore it is important to monitor forest health over the long term and to monitor the symptoms of other abiotic and biotic stressors. For example, pest damage may interact with air pollution or other abiotic stressors (Fenn et al. 1989). In the DL, *Lophodermium* disease was reported to be more severe in pines under air pollution stress (Alvarado-Rosales and de Bauer 1991; Alvarado-Rosales et al. 1993). There have been several studies on forest diseases and insects in the MCAB and in the rest of the country, but they cover only selected areas and are sporadic and discontinuous. During the last 30 years, occasional observations of urban trees and adjacent forested areas in the MCAB have shown crown deterioration in trees of both native and introduced species. At this point, it is difficult to determine the extent to which the various interactive factors may be affecting tree vigor. For example, the drastic reduction in the level of the water table has caused dieback symptoms in urban trees in some areas (Martínez-González and Chacalo-Hilu 1994), although air pollution is certainly another stress factor for urban trees.

Plant community changes in response to air pollution, site disturbances, management practices, and other factors also need to be included in monitoring activities. According to field observations in southern Mexico City in the last 25 years, there have been significant plant community changes, mainly with respect to the native shrubs and herbs. In Ajusco, on slopes facing the city, the tussock grass known as *zacatón* (*Mühlenbergia* sp.) now dominates the understory vegetation, and its abundance in the Basin is reportedly a result of disturbance (Domínguez 1975; Rzedowski 1978). It is also very clear that the incidence of dwarf mistletoe (*Arceuthobium* sp.) in the DL National Park is more severe at those places directly exposed to the city and at higher elevation where the pines are being severely affected (D. Alvarado, personal communication). This situation resembles one aspect of the overall status of declining forests in other locations such as in central Europe (Halwachs 1988). In summary, temporal and spatial trends of the effects of air pollution on forests in the MCAB are not well defined because of the lack of a long-term monitoring program. Implementation of forest health monitoring programs in the future should also include data collection for air pollution injury evaluations (see Miller et al. 2002 in this volume).

Air Pollution Impacts on Urban Trees

Measurements of high air pollution levels in the urban zone, results of a few studies and limited surveys, symptoms of bioindicator plants (de Bauer and Hernández-Tejeda 1986), and anecdotal evidence strongly indicate that many tree species and other ornamental plants are significantly impacted by air pollution in Mexico City (Martínez-González and Chacalo-Hilu 1994). However, systematic surveys of air pollution injury, and experiments to verify that air pollution is the actual cause of injury (Hernández-Tejeda and Nieto de Pascual-Pola 1996; Hernández-Tejeda et al. 1981) have rarely been done. Admittedly, resources for research on impacts of pollution on urban trees are limited, and human health effects will always receive higher priority. Nevertheless, botanists, environmental scientists, or other researchers, by investigating which plants are best adapted for the polluted environment of greater Mexico City, could possibly provide advice and help government employees responsible for urban planting, and private citizens as well, to enhance the environment of the city and reduce costs. In fact, air pollution data from the urban air pollution-monitoring network (RAMA) could be used to suggest which plants should not be cultivated in specific sectors of the city where there are high concentrations of pollutants affecting those species. In many instances, ornamental and forest plantings fail because the species planted are ill adapted for that particular climate, environment (including air pollution) or habitat.

Summary

Hopefully, implementation of stricter pollution control regulations and technological improvements will eventually lead to reduced air pollution concentrations in the city, thus reducing the negative effects of air pollution on vegetation and all living organisms present in the MCAB. Ultimately, because of the huge population within the MCAB and because of the topographic and meteorological conditions favoring the buildup of air pollutants, it will be necessary to significantly reduce emissions to alleviate the severe air pollution problems for which Mexico City is famous. In the meantime, further research on the impacts of air pollution on forests, crops, and urban vegetation are needed to protect and manage these important resources.

The total impacts of air pollution on forests and associated natural resources have not been well documented or investigated in depth. However, the available evidence suggests that tree growth has declined since the early 1970s in forests to the SSW (downwind) of Mexico City where air pollution levels are highest

(Alarcón et al. 1993; Terrazas and Bernál-Salazar 2002). In addition, water from these forests is enriched with nitrate, presumably as a result of chronic atmospheric deposition of N pollutants. Thus, a portion of the nitrogenous pollutants emitted from motor vehicles and industry in the MCAB returns to the city as dissolved nitrate leached into the aquifers supplying drinking water for Mexico City. This demonstrates that urban air quality and the quality of water from the adjacent forested watersheds are linked. The nitrate levels in drainage waters from the forested watersheds downwind of Mexico City do not pose as great of a health threat to human health as do other microbiological and toxic pollutants in Mexico City's drinking water (Mazari and Mackay 1993). However, water emanating from forests exposed to high levels of N deposition, which is expected to be in relatively pristine condition, instead contains nitrates. Preliminary reports of Pb in springs and streams from forested watersheds downwind of Mexico City further illustrate the linkage between air quality and water quality.

In this chapter, we have reviewed a number of priority areas for research into the effects of air pollution on forests. Suggested areas for future or continuing research include: controlled air pollution fumigation studies, long term monitoring of forest condition, monitoring of air pollution exposures in forested regions, studies of the contributing factors or causes of sacred fir decline and management options in impacted areas, studies of urban tree sensitivity to air pollution, studies of the impacts of atmospheric deposition on water quality and nutrient cycling, studies of the effects of O_3 and other pollutants on plant physiology, studies on the effects of air pollution on tree growth rates and forest productivity, the development of bioindicators of air pollution stress, studies on the effects of air pollution on the diversity and abundance of communities of lichens and understory plant species, and studies on the impacts of air pollution on greenhouse gas emissions and critical belowground processes. It comes as no surprise that the primary limiting factor for carrying out the suggested research is financial and institutional support. Researchers must be resourceful and creative in finding ways to obtain funds, in developing appropriate hypotheses, and in performing the research needed to better understand the effects of pollution on forests. It is hoped that future research will lead to solutions that will make it possible and practical to mitigate the undesirable effects of air pollution on forests in the MCAB. It is our hope that this volume will aid in this process and lead to greater interest in furthering our understanding of the effects of air pollution on forest ecosystem function and sustainability in the MCAB and in forests elsewhere.

References

Alarcón-C., M.A., de Bauer, Ma. L.I., Jasso-M., J., Segura-W., G., and Zepeda-B., E.M. 1993. Patrón de crecimiento radial en árboles de *Pinus hartwegii* afectados por contaminación atmosférica en el suroeste del Valle de México. *Agrociencia. Serie Recursos Naturales Renovables* 3:67–80.

Alvarado-Rosales, D. 1989. *Declinación y Muerte del Bosque de Oyamel (Abies religiosa) en el sur del Valle de México.* M.Sc. Thesis. Montecillo, Edo. de México, México: Centro de Fitopatología, Colegio de Postgraduados.

Alvarado-Rosales, D., and Hernández-Tejeda, T. 2002. Decline of sacred fir in the Desierto de los Leones National Park. In *Urban Air Pollution and Forests: Resources at Risk in the Mexico City Air Basin*. Ecological Studies series, eds. M.E. Fenn, L.I. de Bauer, and T. Hernández-Tejeda, (this volume). New York: Springer-Verlag.

Alvarado-Rosales, D., and de Bauer, L.I. 1991. Ataque de *Lophodermium* sp. en poblaciones naturales de *Pinus hartwegii* de el Ajusco, México, bajo el efecto de gases oxidantes. *Micol. Neotrop. Apl.* 4:99–109.

Alvarado-Rosales, D., de Bauer, L.I., and Galindo-A., J. 1993. Decline of sacred fir (*Abies religiosa*) in a forest park south of Mexico City. *Environ. Pollut.* 80:115–121.

Bravo, H., and Torres-Jardón, R. 2002. Air pollution levels and trends in the Mexico City metropolitan area. In *Urban Air Pollution and Forests: Resources at Risk in the Mexico City Air Basin*. Ecol. Stud. series, eds. M.E. Fenn, L.I. de Bauer, and T. Hernández-Tejeda, (this volume). New York: Springer-Verlag.

Cairney, J.W.G., and Meharg, A.A. 1999. Influences of anthropogenic pollution on mycorrhizal fungal communities. *Environ. Pollut.* 106:169–182.

Cox, R.M., and Malcolm, J.W. 1999. Passive ozone monitoring for forest health assessment. *Water Air Soil Pollut.* 116:339–344.

Davison, A.W., and Barnes, J.D. 1998. Effects of ozone on wild plants. *New Phytol.* 139:135–151.

de Bauer, L.I., and Hernández-Tejeda, T. 1986. *Contaminación: Una Amenaza para la Vegetación en México.* Chapingo, Edo. de México: Colegio de Postgraduados.

Domínguez, V. 1975. *Estudio Ecológico del Volcán Popocatépetl, Estado de México.* M. Sc. Thesis. México, D.F.: Facultad de Ciencias, Universidad Autónoma Metropolitana.

Dmuchowski, W., and Bytnerowicz, A. 1995. Monitoring environmental pollution in Poland by chemical analysis of Scots pine (*Pinus sylvestris* L.) needles. *Environ. Pollut.* 87:87–104.

Ezcurra, E., and Mazari-Hiriart, M. 1996. Are megacities viable? A cautionary tale from Mexico City. *Environment* 38:6–15; 26–35.

Ezcurra, E., Mazari-Hiriart, M., Pisanty, I., and Aguilar, A.G. 1999. *The Basin of Mexico: Critical Environmental Issues and Sustainability.* New York: United Nations University Press.

Fenn, M.E., and Poth, M.A. 1999. Temporal and spatial trends in streamwater nitrate concentrations in the San Bernardino Mountains, southern California. *J. Environ. Qual.* 28:822–836.

Fenn, M.E., Dunn, P.H., and Wilborn, R. 1989. Black stain root disease in ozone-stressed ponderosa pine. *Plant Dis.* 74:426–430.

Fenn, M.E., de Bauer, L.I., and Hernández-Tejeda, T. 2002c. Summary of air pollution impacts on forests in the Mexico City Air Basin. In *Urban Air Pollution and Forests: Resources at Risk in the Mexico City Air Basin*. Ecol. Stud. series, eds. M.E. Fenn, L.I. de Bauer, and T. Hernández-Tejeda, (this volume). New York: Springer-Verlag.

Fenn, M.E., Poth, M.A., Aber, J.D., Baron, J.S., Bormann, B.T., Johnson, D.W., et al. 1998. Nitrogen excess in North American ecosystems: Predisposing factors, ecosystem responses, and management strategies. *Ecol. Applic.* 8:706–733.

Fenn, M.E., de Bauer, L.I., Quevedo-Nolasco, A., Rodriguez-Frausto, C. 1999. Nitrogen and sulfur deposition and forest nutrient status in the Valley of Mexico. *Water Air Soil Pollut.* 113:155–174.

Fenn, M.E., Poth, M.A., Schilling, S.L., and Grainger, D.B. 2000. Throughfall and fog deposition of nitrogen and sulfur at an N-limited and N-saturated site in the San Bernardino Mountains, southern California. *Can. J. For. Res.* 30:1476–1488.

Fenn, M.E., Castro-Servín, J. Ma., Hernández-Tejeda, T., Krage, N., Goodson, C., and Meixner, T. 2002a. Heavy metals in forest soils, vegetation, and drainage waters in the Basin of Mexico. In *Urban Air Pollution and Forests: Resources at Risk in the Mexico City Air Basin*. Ecol. Stud. series, eds. M.E. Fenn, L.I. de Bauer, and T. Hernández-Tejeda, (this volume). New York: Springer-Verlag.

Fenn, M.E., de Bauer, L.I., Zeller, K., Quevedo-Nolasco, A., Rodríguez, C., and Hernández, T. 2002b. Nitrogen and sulfur deposition in the Mexico City Air Basin: Impacts on forest nutrient status and nitrate levels in drainage waters. In *Urban Air Pollution and Forests: Resources at Risk in the Mexico City Air Basin*. Ecol. Stud. series, eds. M.E. Fenn, L.I. de Bauer, and T. Hernández-Tejeda, (this volume). New York: Springer-Verlag.

Gaffney, J.S., Marley, N.A., Cunningham, M.M., and Doskey, P.V. 1999. Measurements of peroxyacyl nitrates (PANS) in Mexico City: Implications for megacity air quality impacts on regional scales. *Atmos. Environ.* 33:5003–5012.

Gosz, J.R., French, C., Sprott, P., and White, M. 2000. The International Long Term Ecological Research Network: Perspectives from Participating Networks. Albuquerque, New Mexico. U.S.A.: Long Term Ecological Research Network Office.

Grulke, N.E., and Balduman, L. 1999. Deciduous conifers: High N deposition and O_3 exposure effects on growth and biomass allocation in ponderosa pine. *Water Air Soil Pollut.* 116:235–248.

Grulke, N.E., Andersen, C.P., Fenn, M.E., and Miller, P.R. 1998. Ozone exposure and nitrogen deposition lowers root biomass of ponderosa pine in the San Bernardino Mountains, California. *Environ. Pollut.* 103:63–73.

Halwachs, G. 1988. Die Belastung der Waelder—ein Multifaktorielles Problem. *Oesterr. Wasserwitschaft* 40:114–118.

Hernández-Tejeda, T. 1984. *Efecto de los Gases Oxidantes Sobre Algunas Especies del Género Pinus Nativas del Valle de México.* M.Sc. Thesis. Chapingo, México: Colegio de Postgraduados.

Hernández-Tejeda, T., and Nieto de Pascual-Pola, C. 1996. Effects of oxidant air pollution on *Pinus maximartinezii* Rzedowski in the Mexico City region. *Environ. Pollut.* 92:79–83.

Hernández-Tejeda, T., Krupa, S.V., Platt, G.C., and de Bauer, L.I. 1981. Sensibilidad de plántulas de eucalipto (*Eucalyptus globulus* Labill.) al ozono. *Agrociencia* 43:89–95.

López-López, M.A., Velázquez-Mendoza, J., Velázquez-Martínez, A., González-Romero, V., and Cetina-Alcalá, V.M. 1998. Estado nutrimental de *Abies religiosa* en una área con problemas de contaminación ambiental. *Agrociencia* 32:53–59.

Manning, W.J. 1998. The use of plants as bioindicators of ozone. In *Proceedings of the International Symposium on Air Pollution and Climate Change Effects on Forest Ecosystems*, technical coordinators A. Bytnerowicz, M.J. Arbaugh, and S.L. Schilling, pp. 19–26. February 5–9, 1996, Riverside, California U.S.A. Gen. Tech. Rep. PSW-GTR-166. Albany, California: Pacific Southwest Research Station, Forest Service, U.S. Department of Agriculture. Electronic version is available at *http://www.rfl.psw.fs.fed.us/pubs/psw-gtr-164/index.html.*

Martínez-González, L., and Chacalo-Hilu, A. 1994. *Los Árboles de la Ciudad de México.* Azcapotzalco, México, D.F.: Universidad Autónoma Metropolitana.

Mazari, M., and Mackay, D.M. 1993. Potential for groundwater contamination in Mexico City. *Environ. Sci. Technol.* 27:794–802.

Miller, P.R., de Bauer, Ma. L.I., and Hernández-Tejeda, T. 2002. Oxidant exposure and effects on pine forests in the Mexico City and Los Angeles, California Air Basins. In *Urban Air Pollution and Forests: Resources at Risk in the Mexico City Air Basin*. Ecol. Stud. series, eds. M.E. Fenn, L.I. de Bauer, and T. Hernández-Tejeda, (this volume). New York: Springer-Verlag.

Miller, P.R., de Bauer, L.I., Quevedo-Nolasco, A., and Hernández-Tejeda, T. 1994. Comparison of ozone exposure characteristics in forested regions near Mexico City and Los Angeles. *Atmos. Environ.* 28:141–148.

Morales, R. 1998. *Estudio Geohidrológico del Valle de Monte Alegre, Ajusco, México.* M.Sc. Thesis. Montecillo, Edo. de México, México: Colegio de Postgraduados.

Rzedowski, J. 1978. *Vegetación de México.* México, D. F.: Ed. Limusa.

Skeffington, R.A., and Wilson, E.J. 1988. Excess nitrogen deposition: Issues for consideration. *Environ. Pollut.* 54:159–184.

Skelly, J.M., Savage, J.E., de Bauer, L.I., and Alvarado, D. 1997. Observations of ozone-induced foliar injury on black cherry (*Prunus serotina*, var. *capuli*) within the Desierto de los Leones National Park, Mexico City. *Environ. Pollut.* 95:155–158.

Takemoto, B.K., Bytnerowicz, A., and Fenn, M.E. 2001. Current and future effects of ozone and atmospheric nitrogen deposition on California's mixed conifer forests. *For. Ecol. Manage.* 144:159–173.

Temple, P.J. 1999. Effects of ozone on understory vegetation in the mixed conifer forest. In *Oxidant Air Pollution Impacts in the Montane Forests of Southern California: A Case Study of the San Bernardino Mountains*. Ecol. Stud. 134, eds. P.R. Miller, and J.R. McBride, pp. 208–222. New York: Springer-Verlag.

Terrazas, T., and Bernál-Salazar, S. 2002. Histological symptoms of air pollution injury in foliage, bark, and xylem of *Abies religiosa* in the Basin of Mexico. In *Urban Air Pollution and Forests: Resources at Risk in the Mexico City Air Basin*. Ecol. Stud. series, eds. M.E. Fenn, L.I. de Bauer, and T. Hernández-Tejeda, (this volume). New York: Springer-Verlag.

Watmough, S.A., and Hutchinson, T.C. 1999. Change in the dendrochemistry of sacred fir close to Mexico City over the past 100 years. *Environ. Pollut.* 104:79–88.

Zambrano, A., Nash III, T.H., and Herrera-Campos, M.A. 2002. Lichens and air pollution in the forests surrounding Mexico City. In *Urban Air Pollution and Forests: Resources at Risk in the Mexico City Air Basin*. Ecol. Stud. series, eds. M.E. Fenn, L.I. de Bauer, and T. Hernández-Tejeda, (this volume). New York: Springer-Verlag.

Index

Ecological Studies

Volumes published since 1995

Volume 111
Peatland Forestry: Ecology and Principles (1995)
E. Paavilainen and J. Päivänen

Volume 112
Tropical Forests: Management and Ecology (1995)
A.E. Lugo and C. Lowe (Eds.)

Volume 113
Arctic and Alpine Biodiversity: Patterns, Causes and Ecosystem Consequences (1995)
F.S. Chapin III and C. Körner (Eds.)

Volume 114
Crassulacean Acid Metabolism: Biochemistry, Ecophysiology and Evolution (1995)
K. Winter and J.A.C. Smith (Eds.)

Volume 115
Islands: Biological Diversity and Ecosystem Function (1995)
P.M. Vitousek, H. Andersen, and L. Loope (Eds.)

Volume 116
High-Latitude Rainforests and Associate Ecosystems of the West Coast of the Americas: Climate, Hydrology, Ecology and Conservation (1995)
R.G. Lawford, P.B. Alaback, and E.R. Fuentes (Eds.)

Volume 117
Anticipated Effects of a Changing Global Environment on Mediterranean-Type Ecosystems (1995)
J.M. Moreno and W.C. Oechel (Eds.)

Volume 118
Impact of Air Pollutants on Southern Pine Forests (1995)
S. Fox and R.A. Mickler (Eds.)

Volume 119
Freshwaters of Alaska: Ecological Synthesis (1997)
A.M. Milner and M.W. Oswood (Eds.)

Volume 120
Landscape Function and Disturbance in Arctic Tundra (1996)
J.F. Reynolds and J.D. Tenhunen (Eds.)

Volume 121
Biodiversity and Savanna Ecosystem Processes: A Global Perspective (1996)
O.T. Solbrig, E. Medina, and J.F. Silva (Eds.)

Volume 122
Biodiversity and Ecosystem Processes in Tropical Forests (1996)
G.H. Orians, R. Dirzo, and J.H. Cushman (Eds.)

Volume 123
Marine Benthic Vegetation: Recent Changes and the Effects of Eutrophication (1996)
W. Schramm and P.H. Nienhuis (Eds.)

Volume 124
Global Change and Arctic Terrestrial Ecosystems (1996)
W.C. Oechel (Ed.)

Volume 125
Ecology and Conservation of Great Plains Vertebrates (1997)
F.L. Knopf and F.B. Samson (Eds.)

Volume 126
The Central Amazon Floodplain: Ecology of a Pulsing System (1997)
W.J. Junk (Ed.)

Volume 127
Forest Design and Ozone: A Comparison of Controlled Chamber and Field Experiments (1997)
H. Sanderman, A.R. Wellburn, and R.L. Heath (Eds.)

Volume 128
The Productivity and Sustainability of Southern Forest Ecosystems in a Changing Environment (1998)
R.A. Mickler and S. Fox (Eds.)

Volume 129
Pelagic Nutrient Cycles (1997)
T. Andersen

Volume 130
Vertical Food Web Interactions (1997)
K. Dettner, G. Bauer, and W. Völkl (Eds.)

Volume 131
The Structuring Role of Submerged Macrophytes in Lakes (1998)
E. Jeppesen, M. Søndergaard, M. Søndergaard, and K. Christoffersen (Eds.)

Volume 132
Vegetation of the Tropical Pacific Islands (1998)
D. Mueller-Dombois and F.R. Fosberg

Volume 133
Aquatic Humic Substances (1998)
D.O. Hessen and L.J. Tranvik

Volume 134
Oxidant Air Pollution Impacts in the Montane Forests of Southern California: A Case Study of the San Bernardino Mountains (1999)
P.R. Miller and J.R. McBride (Eds.)

Volume 135
Predation in Vertebrate Communities: The Bialowieza Primeval Forest as a Case Study (1998)
B. Jedrzejewska and W. Jedrzejewski

Volume 136
Landscape Disturbance and Biodiversity in Mediterranean-Type Ecosystems (1998)
P.W. Rundel, G. Montenegro, and F.M. Jaksic (Eds.)

Volume 137
Ecology of Mediterranean Evergreen Oak Forests (1999)
F. Roda, J. Retana, C.A. Gracia, and J. Bellot (Eds.)

Volume 138
Fire, Climate Change, and Carbon Cycling in the Boreal Forest (2000)
E.S. Kasischke and B.J. Stocks (Eds.)

Volume 139
Responses of Northern U.S. Forests to Environmental Change (2000)
R.A. Mickler, R.A. Birdsey, and J. Hom (Eds.)

Volume 140
Rainforest Ecosystems of East Kalimantan: El Nino, Drought, Fire and Human Impacts (2000)
E. Guhardja, et al. (Eds.)

Volume 141
Activity Patterns in Small Mammals: An Ecological Approach (2000)
S. Halle and N.-C. Stenseth (Eds.)

Volume 142
Carbon and Nirtogen Cycling in European Forest Ecosystems (2000)
E.-D. Schulze (Ed.)

Volume 143
Global Climate Change and Human Impacts on Forest Ecosystems: Postglacial Development, Present Situation and Future Trends in Central Europe (2001)
P. Joachim and U. Bernhard

Volume 144
Coastal Marine Ecosystems of Latin America (2001)
U. Seeliger and B. Kjerfve (Eds.)

Volume 145
Ecology and Evolution of the Freshwater Mussels Unionoida (2001)
B. Gerhard and W. Klaus (Eds.)

Volume 146
Inselbergs: Biotic Diversity of Isolated Rock Outcrops in Tropical and Temperate Regions (2001)
S. Porembski and W. Barthlott (Eds.)

Volume 147
Ecosystem Approaches to Landscape Management in Central Europe (2001)
J.D. Tenhunen, R. Lenz, and R. Hantschel (Eds.)

Volume 148
A Systems Analysis of the Baltic Sea (2001)
F.V. Wulff, L.A. Rahm, and P. Larsson (Eds.)

Volume 149
Banded Vegetation Patterning in Arid and Semiarid Environments: Ecological Processes and Consequences for Management (2001)
D.J. Tongway, C. Valentin, and J. Seghieri (Eds.)

Volume 150
Biological Soil Crusts: Structure, Function and Management (2001)
J. Belnap and O.L. Lange (Eds.)

Volume 151
Ecological Comparisons of Sedimentary Shores (2001)
K. Reise (Ed.)

Volume 152
Global Biodiversity in a Changing Environment: Scenarios for the 21st Century (2001)
F.S. Chapin III, O.E. Sala, and E. Huber-Sannwald (Eds.)

Volume 153
UV Radiation and Arctic Ecosystems (2002)
D.O. Hessen (Ed.)

Volume 154
Geoecology of Antarctic Ice-Free Coastal Landscapes (2002)
L. Beyer and M. Bölter (Eds.)

Volume 155
Conserving Biodiversity in East African Forests: A Study of the Eastern Arc Mountains (2002)
W.D. Newmark (Ed.)

Volume 156
Urban Air Pollution and Forests: Resources at Risk in the Mexico City Air Basin (2002)
M.E. Fenn, L.I. de Bauer, and T. Hernández-Tejeda (Eds.)